AF334938

Cardiovascular Genetics and Genomics for the Cardiologist

EDITED BY

Victor J. Dzau, MD

James B. Duke Professor of Medicine
Director, Mandel Center for Hypertension
and Atherosclerosis Research
Chancellor for Health Affairs
Duke University
Durham, NC, USA

Choong-Chin Liew, PhD

Professor Emeritus, Department of Laboratory
Medicine and Pathobiology,
University of Toronto
Toronto
Ontario, Canada
and (formerly) Visiting Professor of Medicine
Brigham and Women's Hospital
Harvard Medical School
Boston, MA, USA

Blackwell Futura is an imprint of Blackwell Publishing
Blackwell Publishing, Inc., 350 Main Street, Malden, Massachusetts 02148-5020, USA
Blackwell Publishing Ltd, 9600 Garsington Road, Oxford OX4 2DQ, UK
Blackwell Science Asia Pty Ltd, 550 Swanston Street, Carlton, Victoria 3053, Australia

First published 2007

1   2007

ISBN: 978-1-4051-3394-4

Library of Congress Cataloging-in-Publication Data

Cardiovascular genetics and genomics for the cardiologist / edited by Victor J. Dzau,
Choong-Chin Liew.
    p.   ;   cm.
  Includes bibliographical references and index.
  ISBN-13: 978-1-4051-3394-4 (alk. paper)
  ISBN-10: 1-4051-3394-5 (alk. paper)
  1. Cardiovascular system–Diseases–Genetic aspects.   2. Cardiovascular
system–Molecular aspects.   3. Genomics.   I. Dzau, Victor J.   II. Liew, Choong-Chin.
  [DNLM: 1. Cardiovascular Diseases–genetics.   2. Cardiovascular Diseases–therapy.
3. Genomics.   WG 120 C26745 2007]

  RC669.C2854 2007
  616.1′042–dc22

                                                                          2007005634

A catalogue record for this title is available from the British Library

Commissioning Editors: Steve Korn and Gina Almond
Development Editors: Vicki Donald and Beckie Brand
Editorial Assistant: Victoria Pittman
Production Controller: Debbie Wyer

Set in 9.5/12pt Minion by Graphicraft Limited, Hong Kong
Printed and bound in Singapore by Fabulous Printers Pte Ltd

For further information on Blackwell Publishing, visit our website:
www.blackwellcardiology.com

Contents

Contributors

Victor J. Dzau, MD
Duke University
Medical Center
Durham, NC, USA

Mitra Esfandiarei, PhD
James Hogg iCAPTURE Centre
Providence Health Care Research
Institute UBC St. Paul's Hospital
Vancouver, BC, Canada

Frank J. Giordano, MD
Cardiovascular Gene Therapy Program
Yale University School of Medicine
New Haven, CT, USA

Julie A. Johnson, PharmD
Departments of Pharmacy Practice and Medicine
  (Cardiovascular Medicine)
Colleges of Pharmacy and Medicine, and Center for
  Pharmacogenomics
University of Florida
Gainesville, FL, USA

Ion S. Jovin, MD
Cardiovascular Gene Therapy Program
Yale University School of Medicine
New Haven, CT, USA

Kamilla Kelemen, MD
Departments of Anesthesiology, and Division of Clinical
  Pharmacology
Vanderbilt University School of Medicine
Nashville, TN, USA

Sabina Kupershmidt, PhD
Assistant Professor
Anesthesiology Research Divison
Vanderbilt University
Nashville, TN, USA

Choong-Chin Liew, PhD
GeneNewsCorporation
Toronto, ON, Canada

Ali J. Marian, MD
Center for Cardiovascular Genetic Research
The Brown Foundation Institute of Molecular Medicine
The University of Texas Health Science Center
Texas Heart Institute at St. Luke's Episcopal Hospital
Houston, TX, USA

Bruce M. McManus, MD, PhD, FRSC
The James Hogg iCAPTURE Centre for Cardiovascular and
  Pulmonary Research
St. Paul's Hospital/Providence Health Care
Department of Pathology and Laboratory Medicine
University of British Columbia
Vancouver, BC, Canada

Ruth McPherson, MD, PhD, FRCPC
Departments of Medicine and Biochemistry
University of Ottawa Heart Institute
Ottawa, ON, Canada

Markus Meyer, MD
Departments of Medicine and Molecular Physiology and
  Biophysics
University of Vermont
College of Medicine
Burlington, VT, USA

Steve Mohr, PhD
GeneNews Corporation
Toronto, ON, Canada

Tadashi Nakajima, MD, PhD
Department of Anesthesiology
Vanderbilt University School of Medicine
Nashville, TN, USA

Päivi Pajukanta, MD, PhD
Department of Human Genetics
David Geffen School of Medicine at UCLA
Los Angeles, CA, USA

Emerson C. Perin, MD, PhD
New Cardiovascular Interventional Technology
Texas Heart Institute
Baylor Medical School
Houston, TX, USA

Guilherme V. Silva, MD
Stem Cell Center
Texas Heart Institute
Baylor Medical School
Houston, TX, USA

Kiat Tsong Tan, MD, MRCP, FRCR
Department of Radiology
University of Bristol
Bristol, UK

Peter VanBuren, MD
Departments of Medicine and Molecular Physiology and
  Biophysics
University of Vermont
College of Medicine
Burlington, VT, USA

Robert Yanagawa, BSc, PhD
The James Hogg iCAPTURE Centre for Cardiovascular and
  Pulmonary Research
St. Paul's Hospital/Providence Health Care
Department of Pathology and Laboratory Medicine
University of British Columbia
Vancouver, BC, Canada

Issam Zineh, PharmD
Departments of Pharmacy Practice and Medicine
  (Cardiovascular Medicine)
Colleges of Pharmacy and Medicine, and Center for
  Pharmacogenomics
University of Florida
Gainesville, FL, USA

Foreword

In medicine, new developments seem to creep along until they add up to a momentous shift, such as those created by the development of X-ray technology, the discovery of penicillin, or the advent of open heart surgery. More recently, dramatic shifts in clinical practice have stemmed from the development of minimally invasive surgical techniques and the identification of lifestyle factors that significantly affect disease risk, particularly for heart disease and diabetes. But perhaps no development of the last decade will prove more revolutionary to medicine as we know it today than the completion of the Human Genome Project in 2003. This international effort provided not just the sequence of our genetic building blocks, but a raft of new technologies and computational abilities that made the project possible.

Research done "the old fashioned way" – without the technologies of the Human Genome Project – already has resulted in treatment advances that target the genetic problems of "single-gene" cardio-vascular conditions. With our genetic sequence known and these technologies available, however, researchers' hunt for genetic factors underlying common, genetically complex diseases will be significantly accelerated. It is a natural extension that this research will bring new treatment, prevention and diagnostic strategies to medicine. As a result, tomorrow's medical graduates will be well-versed in genetics, and today's practicing physicians will need to be as well.

This book will allow cardiologists and others to "catch up" with the genetic revolution and to prepare for the impact the Human Genome Project will have on the practice of cardiovascular medicine. Get ready. Change is coming, and in many cases it's already here.

Peter Agre, MD, Nobel Laureate 2003
Vice Chancellor for Science and Technology
Duke University Medical Center
Durham, NC, USA

Introduction

Until recently, a modest knowledge of genetics has been more than adequate for the day to day practice of clinical medicine and cardiology. However, this situation is rapidly changing. Advances in genetics and genomics over the past two decades, including the sequencing of the human genome, are shaping medicine to a greater extent than any other basic science endeavor. Indeed, the past 20 years of research has witnessed genetics becoming a foundational science. Information on genes, genetics, genetic testing, genomics, pharmacogenetics and related subjects has moved from highly specialized publications to the general medical literature. Genomics and genetic science is changing the practice of medicine in fundamental ways. In cardiovascular medicine, the genetic basis of several forms of dyslipidemia, hypertension, diabetes, cardiomyopathies and vascular diseases have been identified. Pharmacogenetic studies have demonstrated the influence of genetics on the effectiveness and safety of drugs in anticoagulants and congestive heart failure and other disorders.

Most cardiologists receive limited education in genetics and genomics during their training. Thus, there is an unmet need for education in genetics and genomics for the clinician. This textbook will serve to introduce the concepts of cardiovascular genetics and genomics to cardiologists and to prepare them for the new science that promises to reshape the way that cardiology is practiced. The book is organized into first, a mainly historic overview of genetics and genomics concepts, and a specific application of blood-based microarray technology; second, the single-gene cardiovascular disorders; third, polygenic cardiovascular disorders; and the last section deals with genetic and genomic-based cardiovascular therapeutics.

Chapter 1 aims to introduce some of the basic concepts of genetics and genomics in a historic context. What is a gene? How did ideas about the gene change over the twentieth century? What are the principles of the structure and function of the gene? We conclude the chapter with a discussion of genes of particular interest to the cardiologist.

The next part of the book is devoted to the single-gene cardiovascular disorders. Classically, monogenic disorders are those in which the disease phenotype is brought about by a defect in a single gene. The single-gene disorders are exemplified in Chapter 2 by monogenic hypercholesterolemia in which patients exhibit a severe phenotype of high plasma levels of low density lipoprotein, together with xanthoma tissue deposits, early onset atherosclerosis and in some cases premature death. Chapter 2 describes how our understanding of the genetic and molecular mechanisms underlying these rare disorders has been applied to developing therapeutic approaches for hypercholesterolemia. Familial hypercholesterolemia is the most common form of monogenic hypercholesterolemia. It was the study of this gene defect that led to the identification of low density lipoprotein receptor pathways and to the development of the statin drug class, an important therapeutic advance in cardiovascular disease. This chapter also discusses other less well-known examples of monogenic hypercholesterolemias that have shed light on various aspects of intracellular protein trafficking and cellular cholesterol handling.

Chapter 3 is a comprehensive description of the genetic and clinical aspects of hypertrophic cardiomyopathy. Identification of the genetic mutations underlying this disorder represents a historic landmark in cardiovascular disease genetics. The cause of hypertrophic cardiomyopathy was for decades a mystery following its first recognition as a clinical entity in the 1950s. It was not until the 1990s when the disorder was identified with a mutation in the beta myosin heavy chain gene that it was subsequently elucidated as a disorder of the sarcomeric proteins. Yet another longstanding puzzle of hypertrophic cardiomyopathy has been that the clinical manifestations of the disorder vary strikingly even within members of the same family with identical mutations. This chapter places the causal single-gene mutations within the context of modifier genes, gene–gene interactions, environmental and other factors as well as diseases such as hypertension. All have effects on the phenotypic manifestations making hypertrophic cardiomyopathy a truly complex – yet "single-gene" – disorder.

Chapter 4 provides an overview of the large number of single-gene cardiomyopathies other than hypertrophic cardiomyopathy. Dilated cardiomyopathy, restrictive cardiomyopathy and arrhythmogenic right ventricular cardiomyopathy represent a vast array of disturbances in numerous genes, having differing effects on protein function, all of which yield cardiomyopathy as their final common expression. Advances in molecular genetics have led to tools to identify single-gene defects underlying many of the cardiomyopathies. Understanding how mutations in genes affect cardiac function may further the search for potential targets for therapeutic intervention.

Long QT syndrome is discussed in Chapter 5. Long QT is a major cause of sudden cardiac death, including sudden infant death syndrome. Long QT is yet another puzzling disorder whose etiology was finally illuminated over the course of a series of genetic studies. These investigations, carried out during the 1990s, have led to an understanding of long QT as predominantly an ion channelopathy and to insights into cardiac electrophysiology and into mechanisms leading to arrhythmias. Knowledge of specific mutations has led to rational drug and other therapies for long QT. Acquired long QT is also a syndrome of great interest, and the pharmaceutical industry faces major challenges in identifying drug-induced life-threatening repolarization abnormalities.

The next section of the book deals with the cardiovascular polygenic disorders hypertension, atherosclerosis and heart failure. These are diseases and disorders that involve multiple genes interacting with epigenetic and environmental factors to produce the disease phenotype. Polygenic disorders are actually syndromes or complex disorders, described phenotypically rather than on a mechanistic basis. Polygenic multifactorial diseases are much more common in populations than are the single-gene disorders which are mostly very rare. Onset in polygenic disorders is usually delayed until later in life. Polygenic diseases/disorders are highly complex, involving multiple genes, gene pathways and interactions. Their complexity renders them much harder to understand at the molecular level.

Chapter 6 addresses atherosclerosis, one of the most complex of the polygenic cardiovascular disorders and the most common cause of mortality in the western world. Atherosclerosis involves the actions of perhaps 400 genes, which, together with multiple risk factors, determine the likelihood of atherosclerosis. The identification of atherosclerosis as an inflammatory disease, coupled with detailed molecular biology studies of the atherosclerotic process, have provided numerous clues to novel diagnostic, prognostic and therapeutic approaches. Work is also underway in determining further polymorphisms in genes that are specifically associated with atherosclerosis.

Chapter 7 is an introduction to the complexities of heart failure. Heart failure involves a multitude of genes: from the initial myocardial insult through the remodeling process that leads to heart failure. An understanding of gene regulation is key to understanding heart failure and this chapter describes the roles of neurohormonal factors, proteins in myocardial calcium handling, interstitial alterations, energy metabolism, apoptosis, reactive oxygen species and myriad other processes. Gene array studies are beginning to contribute to an understanding of heart failure, and to the identification of diagnostic or prognostic biomarkers. In the future, a patient's genetic profile may help to tailor individualized therapies in chronic heart failure.

An overview of the genes in hypertension is the subject of Chapter 8. Blood pressure is a polygenic trait with genetic and environmental factors as well as age contributing to the phenotype. Many studies have been undertaken to identify the genes of essential hypertension but with little success. However, studies of different forms of monogenic hypertension have identified the involvement of specific genes and have led to advances in understanding the pathophysiology of this condition. Most such disorders are caused by mutations of genes involved in renal sodium handling. A large number of patients have not achieved optimal blood pressure control or have side effects from medications. Pharmacogenomics and personalized medicine approaches hold promise in overcoming the challenges of hypertension management.

The next section deals with future therapeutic advances in cardiovascular disease promised by studies in genetics and genomics. Ultimately, the aim of gene science as applied to the practice of medicine is to produce clinically applicable results in the form of new medications, new vaccines, new technologies, new diagnostic and prognostic tests. It is predicted that significant practical changes will occur in medical practice in a short time period based on advances in pharmacogenetics and pharmacogenomics.

Chapter 9 examines the fundamental concepts, potentials and challenges of gene therapy. It also provides an updated review of cardiovascular gene therapy, specifically, in angiogenesis in peripheral vascular disease and ischemic coronary heart disease, strategies in heart failure, cardiac rhythm disturbances and atherosclerosis. Gene therapy research advanced dramatically in the past decade and now includes techniques to carry out targeted corrections of DNA mutations, engineered transcriptional factors to regulate endogenous gene expression as well as technologies to silence gene expression. Although still in its infancy, gene therapy holds promise as an effective approach to treat common diseases and potentially cure monogenic cardiovascular disorders.

Chapter 10 on stem cell therapy for cardiovascular disease explores exciting new evidence that the adult heart may harbour endogenous healing mechanisms. This concept has led to an intense interest in stem cell therapy to reverse the processes of heart disease by harnessing the heart's capacity to heal itself. This chapter reviews the basics of stem cells, provides an overview of animal studies, delivery mechanisms and early human clinical studies in cardiovascular stem cell therapy. The search continues for an ethically acceptable, easily accessible, high yield source of stem cells and for optimal means of delivering cells to therapeutic targets. Several strategies in humans have been tested clinically in the immediate post-myocardial infarction phase as well as in the chronic phase of ischemic heart disease but these studies are still in the very early stages of investigation. Many issues remain before the full potential of stem cell therapy can be realized.

Chapter 11 concerns the emerging new field of pharmacogenetics and pharmacogenomics. An important tool for the practice of personalized medicine, pharmacogenetics characterizes the effects of genetic variations on an individual's responses to specific drugs. Drug therapy in cardiovascular disease is frequently made on an empirical or on a protocol-driven basis. Thus, pharmacogenetic strategies should reduce the number of patients who fail specific courses of treatment or experience side effects. This chapter provides an up-to-date review of pharmacogenetic influences of genetic polymorphisms on statin and nonstatin therapy of hyperlipidemia, hypertension, heart failure, thrombosis and arrhythmia. Warfarin is likely to be the first of the cardiovascular disorders whose utilization might be improved by pharmacogenetic strategies; already there is a warfarin-dosing algorithm. In most cases though, it is likely to be 10–15 years before pharmacogenetic information can be used to guide the use of therapeutics in cardiovascular disease. Chapter 12 highlights the potential of blood-based microarray diagnostics, or bloodomics, a methodology that is transforming microarray and leading the way to a systems based biology.

In summary, this book presents cardiovascular disease in the context of the genetics and genomics. Such a book is long overdue. We are proud to have been involved in the editing of this book and we hope that it will be of service to cardiologists of the twenty-first century.

C.C. Liew and V.J. Dzau

CHAPTER 1

The gene in the twenty-first century

Choong-Chin Liew, PhD, *& Victor J. Dzau,* MD

Introduction

When the word was first used in 1909, "gene" was a hypothesis necessary to explain puzzling observations about heredity. As the century progressed, the hypothesis began to acquire reality as the structure and functions of the gene were gradually elucidated. Earlier and simpler concepts became superseded as evidence led to better understanding of the gene. Today the gene is recognized to be a highly complex entity. The genomics revolution is well underway but there is much that remains for twenty-first century science to learn before the potential of molecular biology and technology can be fully realized.

The search for the gene

Much of the science that laid the foundation for the genetics and genomics revolution took place in the very near past; 1900 is the date often considered to be the beginning of modern genetics. In that year, three botanists working on plant hybridization, independently, in three different countries, published their rediscovery of Gregor Mendel's (1822–1884) rules of inheritance, first presented in 1865 and then largely forgotten [1]. Carl Correns (1864–1933) in Germany, Hugo de Vries (1848–1935) in the Netherlands and Erich von Tschermak (1871–1962) in Austria each published their findings in the *Berichte der Deutsche Botanischen Gesellschaft* (Proceedings of the German Botanical Society) [2–4]. The botanists recognized that Mendel's concept of dominant and recessive traits could be used to explain how traits can skip generations, appearing and disappearing through the years. Hugo de

Vries named the transmitted substances "pangens"; he later coined the term "mutation" to signify the appearance of a new pangen [5].

Cambridge evolutionist William Bateson (1861–1926) translated Mendel into English and worked vigourously to promote Mendel's ideas in the English-speaking scientific world. Bateson himself coined the term "genetics" in 1906 [6]. The word "gene" was not introduced until 1909, when Wilhelm Johannsen (1857–1927), a Danish botanist, offered this term in preference to earlier terms [7].

A next major step towards an elucidation of the gene came with the discovery that genes have physical locations on chromosomes in studies on *Drosophila* carried out by Thomas Hunt Morgan (1866–1945) and his colleagues at the zoology department of Columbia University [8,9]. Morgan's student, Alfred Sturtevant (1891–1970) was able to show that the gene for a trait was localized in a fixed location or locus arranged "like beads on a string" in his often quoted metaphor [10]. Later, Calvin Bridges was able to visualize this arrangement using light microscopy to show in detail the parallel bands on the chromosomes of the salivary gland cells of larval fruit flies [11].

In 1927, another of Morgan's students, Herman J. Muller (1890–1967) proved in studies at the University of Texas, Austin, that ionizing radiation from X-rays and other mutagens could be used to create genetic mutations in fruit flies, and that some of these mutations were able to pass to offspring [12].

Muller believed as early as the 1920s that genes were "the basis of life" [13]. However, it was not until the 1940s that researchers began to work out the physical and material properties of genes. In 1944, Rockefeller University researchers, Oswald Avery (1877–1955), Colin MacLeod (1909–1972) and Maclyn McCarty (1911–2005) demonstrated that it was DNA that was the carrier of genetic information [14]. In 1952, Alfred Hershey (1908–1997) and his laboratory assistant Martha Chase (1928–2003) confirmed Avery's findings [15].

Against this background can be understood the importance mid-century of James Watson (b. 1928) and Francis Crick's (1916–2004) double helix. In their landmark paper, published in *Nature* in 1953, Watson and Crick presented for the first time a comprehensible model of a unit of heredity [16]. Briefly, their double helix is composed of two long polymers of alternating sugar-phosphate deoxyribose molecules, like the sides of a twisted spiral ladder. To these molecules Watson and Crick attached the ladder's rungs, four nucleotide bases: adenine and guanine (A and G) and cytosine and thymine (C and T). The property of each base is such that it attracts and bonds to its complementary base forming arrangements known as base pairs: "A" can only pair with "T" and "C" can only pair with "G." The DNA bases are loosely attached to each other by weak bonds; they are released from each other by disrupting the bonds. Thus, every time a cell divides it copies its DNA program, in the human cell, it copies its entire three billion base pair human genome.

Once the structure of the gene was described the molecule could take its place in the scientific ontology of the twentieth century. With the double helix, classic genetics began to shift to molecular genetics [8].

Gene function

Studies of gene function proceeded largely independently of investigations into gene structure. The first clue to the biologic behaviour of the gene in the organism came in 1902, with the work of London physician Archibald Garrod (1857–1936) [17]. In his famous paper published in the *Lancet* in 1902, Garrod hypothesized that alkaptonuria was a consequence of some flaw in body chemistry that disrupts one of the chemical steps in the metabolism of tyrosine [18]. He explained alkaptonuria as a recessive disorder, using the terms of the new

Mendelian genetics and conjectured that it is an absence of the enzyme involved that leads to alkaptonuria and other "inborn errors of metabolism" [19].

Garrod's hypothesis was given experimental support in an important series of studies on *Neurospora crassa* carried out at Stanford University by George Beadle (1903–1989) and Edward Tatum (1909–1975) between 1937 and 1941 [20]. Because biochemical processes are catalyzed by enzymes and because mutations affect genes, reasoned Beadle and Tatum, then genes must make enzymes: the "one-gene-one-enzyme" hypothesis, later made famous by Beadle.

The hypothesis was further developed in studies on sickle cell anemia. In 1949, the hereditary basis of the disorder was shown by James Neel (1915–2000) [21]. Also in 1949, Linus Pauling (1901–1994) and Harvey Itano showed that the disease was linked to a modification in hemoglobin, such that the hemoglobin in sickled cells carries a charge different to the charge of the molecule in normal cells [22].

Eight years later, Vernon Ingram (1924–2006) and Francis Crick demonstrated that this difference was caused by the replacement of a single amino acid, glutamic acid, by another, valine, at a specific position in the long hemoglobin protein [23]. Sickle cell anemia was the first disease explicitly identified as a disorder flowing from a derangement at the molecular level, or as Pauling himself put it, "a molecular disease" [22].

Understanding of gene function sped up once Watson and Crick had elucidated the structure of the double helix. As summarized in Crick's famous central dogma of 1958, information flows from DNA to RNA to protein [24]. The central dogma captured the imagination of biologists, the public and the media in the 1960s and 1970s [25]. "Stunning in its simplicity," Evelyn Fox Keller writes, the central dogma allows us to think of the cell's DNA as "the genetic program, the lingua prima, or perhaps, best of all, the book of life" [25].

The gene since 1960

By 1960, the definition of a gene was that implied by the central dogma: a gene is a segment of DNA that codes for a protein [26]. The first significant challenge to that definition arose with the work of microbiologists Francois Jacob (b. 1920) and Jacques Monod (1910–1976) of the Institut Pasteur in Paris [27].

Monod and Jacob's operon model explained gene function in terms of gene cluster. However, such a model adds levels of complexity to the gene and makes it more difficult to determine precisely what is a gene. What should be included in one gene? Its regulatory elements? Its coding elements? What are the boundaries of the gene? [25]. Furthermore, the Monod/Jacob gene loses some of its capacity for self-regulation: on the operon model the gene acts, not autonomously, but in response to proteins within the cell and between the cell and its environment [28].

Although Monod famously asserted that what was true for *Escherichia coli* would be true for the elephant, in fact the operon model of gene regulation characterizes prokaryotes (simple unicellular organisms without nuclei). In eukaryotes (animals and plants whose cells contain nuclei), gene regulation is far more complicated. Later research showed that in some cases, regulatory elements were scattered at sites far away from the coding regions of the gene; in other cases, regulator genes were found to be shared by several genes; gene regulation included further levels of control including positive control mechanisms, attenuation mechanisms, complex promoters, enhancers and multiple polyadenylation sites, making it even more difficult to clarify the boundaries of the gene (for discussion on difficulties in defining the gene see [25,29–31]).

Later, in 1970, Howard Temin (1934–1994) and David Baltimore (b. 1938) also posed challenges to the one way DNA-to-protein pathway implicit in the central dogma. In their work on viruses that can cause cancer they discovered an enzyme, reverse transcriptase, which uses RNA as the template to synthesize DNA [32,33].

Another unexpected finding to shake the central dogma occurred with the discovery of the split gene. In 1977, Phillip Sharp (b. 1944) at the Massachusetts Institute of Technology and Richard Rogers (b. 1943) at Cold Spring Harbor Laboratory showed that not all genes are made of one continuous series of nucleotides. Their electron microscopy comparisons of adenovirus DNA and mRNA showed that some genes are split, or fragmented into regions of coding pieces of DNA interrupted

by stretches of non-coding DNA [34,35]. Walter Gilbert (b. 1932) of Harvard University later coined the terms "exon" and "intron" to describe these regions [36].

Split genes can be spliced, or alternatively spliced, in different ways: exons can be excised out, some introns can be left in, or the primary transcript can be otherwise recombined (for review see [37]). The proteins thereby produced are similar although slightly different isoforms. Split genes play havoc with the straightforward one-gene-one-enzyme hypothesis. As Keller has pointed out "one gene – many proteins" is an expression common in the literature of molecular biology today [25].

Other nontraditional genes discovered (or become accepted by the research community) since the 1960s include transposons, moveable genes that travel from place to place in the genome of a cell where they affect the expression of other genes discovered by Barbara McClintock [38]; nested genes, whose exon sequences are contained within other genes; and pseudogenes which are "dead" or nonfunctional gene remnants, overlapping genes, repeated genes and other gene types (these and other "nonclassical" genes are reviewed in [30]).

Most recently, with the discovery of nonprotein coding RNAs, the idea that genes necessarily make proteins at all has been called into question. As far back as 1968, Roy Britten and David Kohne published a paper in *Science* reporting that long stretches of DNA do not seem to code for proteins at all [39]. Large areas of genome – hundreds to thousands of base pairs – seemed to consist of monotonous nucleotide sequence repetition of DNA. Such noncoding DNA, which includes introns within genes and areas between coding genes, represents a surprising fraction of the genomes, at least genomes of higher organisms. In humans, 98% of human DNA appears not to code for anything. Only a tiny percentage – about 2% – of the three billion base pairs of the human genome corresponds to the 20,000–25,000 protein coding genes tallied by the International Human Genome Sequencing Consortium [40]. Much of it consists of repeated DNA; some elements are repeated over 100,000 times in the genome with no apparent purpose. Such noncoding elements were long dismissed as parasitic or "junk" DNA: a chance by-product of evolution with no discernible function.

Table 1.1 Recently discovered noncoding RNA families and their functions.

Family		Processes affected
miRNAs	microRNAs	translation/regulation
siRNAs	small interfering RNAs	RNA interference/gene silencing
snRNAs	small nuclear RNAs	RNA processing/spliceosome components
st RNAs	small temporal RNAs	temporal regulation/translation
snoRNAs	small nucleolar RNAs	ribosomal RNA processing/modification
cis-antisense RNAs		transcription elongation/RNA processing/stability/mRNA translation

Adapted from Storz *et al.* [44].

Since the sequencing of the human and other genomes, however, and with the availability of transcriptomes and novel genomic technologies such as cDNA cloning approaches and genome tiling microarrays, researchers have begun to explore intronic and intragenic space. Increasingly since 2001 it appears that far from being junk, these stretches of DNA are rich in "gems" [41]: small genes that produce RNAs, called noncoding RNAs (reviewed in [42–44]) (Table 1.1).

Noncoding RNAs are not messenger RNAs, transfer RNAs or ribosomal RNAs, RNA species whose functions have long been known. They vary in size from tiny 20–30 nucleotide-long microRNAs to 100–200 nucleotide-long nonprotein coding RNAs (ncRNAs) in bacteria to more than 10,000 nucleotide-long RNAs involved in gene silencing. Many of these intriguing ncRNAs are highly conserved through evolution, and many seem to have important structural, catalytic and regulatory properties [45].

Noncoding RNAs were thought at first to be unusual; however, over the past 5 years increasing numbers of these intriguing elements have been emerging. The number of ncRNAs in mammalian transcriptomes is unknown, but there may be tens of thousands; it has been estimated that some 50% of the human genome transcriptome is made up of ncRNAs [46].

The function of these elements is only beginning to be explored; and their structural features are beginning to be modeled [47]. Nonprotein coding RNAs seem to be fundamental agents in primary molecular biologic processes, affecting complex regulatory networks, RNA signaling, transcription initiation, alternative splicing, developmental timing, gene silencing and epigenetic pathways [44].

One class of ncRNAs has been the focus of much research attention. MicroRNAs are hairpin-shaped RNAs first discovered in *Caenorhabditis elegans* [48,49]. These tiny, approximately 22 nucleotide elements seem to control aspects of gene expression in higher eukaryote plants and animals. Many microRNAs are highly conserved through evolution; others are later evolutionary elements. For example, of some 1500 microRNAs in the human genome [46], 53 are unique to primates [50]. Each microRNA may regulate as many as 200 target genes in a cell, or one-third of the genes in the human genome [51].

In animals, microRNAs appear to repress translation initiation or destabilize messenger RNA. In animals, microaRNAs so far characterized seem to be involved in developmental timing, neuronal cell fate, cell death, fat storage and hematopoietic cell fate [49]. The potential effects of these RNA elements on gene expression have led to the hypothesis that these elements may be involved in disease processes. For example, microRNAs have been suggested to be involved in cancer pathogenesis, acting as oncogenes and tumor suppressors [52]. Calin *et al.* [53] recently reported a unique microRNA microarray signature, predicting factors associated with the clinical course of human chronic lymphocytic leukemia.

In 2006 Andrew Fine of Stanford University and Craig Mello of the University of Massachusetts Medical School shared the Nobel Prize in Physiology or Medicine for their work in RNA interference gene silencing by double-stranded RNA.

The Human Genome Project

By this first decade of the twenty-first century the simple "bead on a string," "one-gene-one-enzyme" concept of the gene has given way to a far more detailed understanding of genes and gene function. In addition, there has been a fundamental shift in scientific emphasis since 2000 from gene to the genome: the whole complement of DNA of an organism and includes genes as well as intergenic and intronic space [25,30,54].

In February 2001, two independent drafts of the genome sequence were published simultaneously in the journals *Science* [55] and *Nature* [56]. The work highlighted in *Science* had been carried out by Celera, Rockville, Maryland, a company founded by Craig Venter; that in *Nature* was work by the International Human Genome Sequencing Consortium. The Human Genome Project, the culmination of decades of discussion, had officially begun in 1990 by the US National Institutes of Health and US Department of Energy. Completing the sequencing project and determining the location of the protein encoding genes opened a new "era of the genome" in the biologic sciences. Hopes have continued high that the project would provide the tools for a better and more fundamental level of understanding of human genetic diseases, of which there are some 4000 known, as well as providing new insights into complex multifactorial polygenic diseases.

Also in 2001, our team working at the University of Toronto was the first to describe the total number of genes expressed in a single organ system, the cardiovascular system [57]. This work had developed out of our 1990s research project using the expressed sequence tag (EST) strategy to identify genes in human heart and artery. We sequenced more than 57,000 ESTs from 13 different cardiovascular tissue cDNA libraries and in 1997 published a comprehensive analysis of cardiovascular gene expression, the largest existing database for a single human organ [58].

Even when the first draft of the genome was published in 2001 – still incomplete and with many gaps – researchers were surprised at the small number of genes in the human genome: approximately 30,000–40,000 genes and far fewer than the original (and often quoted) 100,000 genes that had been informally calculated by Walter Gilbert in the 1980s

Table 1.2 Genes in the genome.

Organism	Number of genes
Maize (*Zea mays*)	50,000
Mustard (*Arabidopsis thaliana*)	26,000
Human (*Homo sapiens*)	20,000–25,000
Nematode worm (*Caenorhabditis elegans*)	19,000
Fruit fly (*Drosophila melanogaster*)	14,000
Baker's yeast (*Saccharomyces cerevisiae*)	6000
Bacterium (*Escherichia coli*)	3000
Human immunodeficiency virus	9

Adapted from Functional and Comparative Genomics Factsheet. Human Genome Project. http://www.ornl.gov/sci/techresources/Human_Genome/faq/compgen.shtml#compgen and Human Genome Program, US Department of Energy, Genomics and Its Impact on Science and Society: A Primer, 2003. http://www.ornl.gov/sci/techresources/Human_Genome/publicat/primer2001/index.shtml

[59]. When in 2004 the almost completed final sequence of the genome appeared in *Nature*, our species' total gene count was further reduced to 20,000–25,000 [40]. Furthermore, when compared with the genomes of other organisms, humans seem to have surprisingly few genes: only about twice as many genes as fruit flies; and only half as many genes as the corn plant (Table 1.2).

Certainly a challenge for the Human Genome Project and a major challenge in the transition from structural to functional genomics was to identify the entire set of human genes in the genome. About 98% of the DNA in the genome does not code for any known functional gene product and only 2% encodes protein producing genes. In 1991, Mark Adams and J. Craig Venter and colleagues at the National Institutes of Health [60] had proposed the EST approach to gene identification. In this approach individual clones are randomly selected from cDNA libraries representing the genes expressed in a cell type, tissue or organ of interest. Selected clones are amplified and sequenced in a single pass from one or both ends, yielding partial gene sequences known as ESTs. These are then compared with gene sequences in existing nucleotide databases to determine whether they match known genes, or whether they represent uncharacterized genes.

Venter and his colleagues used automated fluorescent DNA sequencing technology to increase the efficiency and scale of EST generation; they were able to rapidly generate ESTs representing over 600 cDNA clones randomly selected from a human brain cDNA library [60]. More than half of these were human genes that had previously been unknown. Venter argued that this strategy could lead to the identification and tagging of 80–90% of human genes in a short period of time and at dramatically less cost than complete genome sequencing, a full decade before the proposed date of completion of the human genomic nucleotide sequence [61]. At about the same time at our laboratory at the University of Toronto we launched the first human heart EST project as we began our catalog of the complete set of genes expressed in the cardiovascular system [58,62,63].

The EST approach ultimately overcame skepticism [64,65] and became recognized as an important and powerful strategy complementing complete genome sequencing. It has been found that ESTs are an efficient vehicle for new gene discovery; ESTs provide information on gene expression levels in different cells/tissues and EST sequences can be used to design PCR primers for physical mapping of the genome. ESTs may also be useful in the search for new genes involved in genetic disease. Chromosomal localization of ESTs increases the ability to identify novel disease genes. Such positional candidate strategies were used, for example, to identify novel candidates for a familial Alzheimer's disease gene [66].

Early EST-based strategies for gene expression investigation were expensive and labor-intensive. Another important technology to emerge from the Human Genome Project, microarray technology enables data similar to EST data to be produced for thousands of genes, simultaneously, in a single experiment. Indeed, while ESTs have been useful for monitoring gene expression in different tissues or cells, their primary utility is now to provide materials for cDNA microarrays [67]. By tagging and identifying thousands of genes, EST repositories presently serve as the primary source of cDNA clones for microarrays.

The two types of microarray systems in widespread use are the photolithographic synthesis of oligodeoxynucleotides directly on to silicon chips and an X-Y-Z robotic system, which spots DNA onto coated standard glass microscope slides or nylon membranes [67–70]. Microarray will be discussed more fully in Chapter 2. DNA microarray technology can profile and compare thousands of genes between mRNA populations simultaneously.

The DNA microarray is also a novel tool to pinpoint differences in expression between single genes on a large scale. A series of transcript profiling experiments can be analyzed to determine relationships between genes or samples in multiple dimensions. A set of expression fingerprints, or profiles, similarities and differences in gene expression are used in order to group different mRNA populations or genes into discrete related sets or clusters. Clusters of co-regulated genes often belong to the same biologic pathways, or the same protein complex, whereas the clusters of mRNA populations are defined by their "expression fingerprint" providing a means to define differences between samples. Thus, the microarray is a powerful technique.

For example, a molecular profile of cancer has been a subject for cDNA microarray analysis. Perou *et al.* [71] compared transcript profiles between cultured human mammary epithelial cells subjected to a variety of growth factors or cytokines and primary breast tumors. Interestingly, a correlation between two subsets of genes with similar expression patterns *in vitro* and in the primary tumors was found, suggesting that these genes could be used for tumor classification. Other transcriptomal cancer studies have also yielded findings, such as new candidate genes that may now be further investigated in population based studies [72–74].

Microarrays are also increasingly being used to investigate gene expression in heart failure – a condition that has complex etiologies and secondary adaptations that make it difficult to study at the level of cellular and molecular mechanisms [75]. A few cardiovascular-based microarray studies have been published. For example, Friddle *et al.* [76] used microarray technology to identify gene expression patterns altered during induction and regression of cardiac hypertrophy induced by administration of angiotensin II and isoproterenol in a mouse model. The group identified 55 genes during induction or regression of cardiac hypertrophy. They confirmed 25 genes or pathways previously shown to be altered by hypertrophy and further identified 30 genes whose expression had not previously been associated with cardiac hypertrophy or regression. Among the 55 genes, 32 genes were altered only during induction, and eight were altered only during regression. This study used a genome-wide approach to show that a set of known and novel genes was involved in cardiac remodeling during regression and that these genes were distinct from those expressed during induction of hypertrophy.

In the first reported human microarray study in end stage heart failure, Yang *et al.* [77] used high density oligonucleotide arrays to investigate failing and nonfailing human hearts (end stage ischemic and dilated cardiomyopathy). Similar changes were identified in 12 genes in both types of heart failure, which, the authors maintain, indicate that these changes may be intrinsic to heart failure. They found altered expression in cytoskeletal and myofibrillar genes, in genes involved in degradation and disassembly of myocardial proteins, in metabolism, in protein synthesis and genes encoding stress proteins.

Our "CardioChip" microarray, an in-house 10,848-element human cardiovascular-based expressed sequence tag glass slide cDNA microarray, has also proved highly useful in helping elucidate molecular and genetic events leading to end stage heart failure. Our group used the CardioChip to explore expression analysis in heart failure [78,79]. We compared left ventricle heart transplant tissue with nonfailing heart controls and found some 100 transcripts that were consistently differentially expressed in dilated cardiomyopathy samples by more than one and a half times.

Microarrays have revolutionized our approach to studying the molecular aspects of disease. The whole genome scan opens a window through which we can monitor molecular pathways of interest and determine how gene expression changes in response to various stimuli (such as drug therapy). These comparisons offer the ability to study disease as it evolves over different time points and to compare patients with different epigenetic risk factor profiles and under different environmental influences. By examining tissue biopsies or cell samples, researchers can identify a whole-genome "portrait" of gene expression, extract candidate genes and conduct targeted follow-up studies that directly relate to specific cellular functions. Current micro-

array studies typically utilize tissue samples, and of necessity rely on tissue biopsy. In many cases, however, such as in the cardiovascular studies above, tissue samples can only be obtained in very late stage disease, at transplant or after death. The need for a simple noninvasive cost-effective method to replace tissue biopsy to identify early stage disease is clear. Hence, research interest has begun to turn to investigating the use of blood based gene expression profiling. Blood samples have a number of advantages over tissue samples, in particular that blood can be obtained early during disease development and causes little discomfort to patients.

There is a growing body of evidence that the blood contains substantial bioinformation and that biomarkers derived from blood RNA may provide an alternative to tissue biopsy for the diagnosis and prognosis of disease [80]. Recent studies have shown that blood cell gene expression profiles reflect individual characteristics [81,82], and alterations in blood cell transcriptomes have been found to characterize a wide range of diseases and disorders occurring in different tissues and organs, including juvenile arthritis [83], hypertension [84–86], colorectal cancer [87], chronic fatigue syndrome [88] and neuronal injuries [89,90]. Circulating blood cells also show distinctive expression patterns under various environmental pressures and stimuli, such as exercise [91], hexachlorobenzene exposure [92], arsenic exposure [93] and smoking [94].

Such research findings provide convincing support to the hypothesis that circulating blood cells act as a "sentinels" which detect and respond to microenvironmental changes in the body. Our laboratory, Gene News Corp., in Toronto has developed a methodology to establish the *Sentinel Principle*™. We have profiled gene expression from peripheral blood and we have identified mRNA biomarkers for different diseases. In an initial study, blood samples were drawn from patients with coronary artery disease and gene expression compared with healthy control samples [95]. Differentially expressed genes identified in the circulating blood successfully discriminated the coronary patients from healthy control subjects [95]. We have also used the principle to discriminate successfully between patients with schizophrenia and those with bipolar disorder and between patients and controls [96], which findings have been verified in later studies

[97]. Our group has also identified biomarkers in blood that have utility in identification of early osteoarthritis [98] and bladder cancer [99].

The new technologies of the Human Genome Project allow us to view the entire genome of an organism and permit better characterization of disease as a dynamic process. Although at an early stage as yet, the possibility of using blood samples as the basis for microarray studies of biology and disease opens up new vistas of research for the future.

Conclusions

The twentieth century opened with the start of the search for the gene. The concept grew in stature and importance with the double helix and the central dogma. However, research since 1960 has led to changes in traditional ideas about the gene. No longer is the gene the autonomous self-replicating unit of inheritance of 1953; rather it requires the assistance of a host of accessory regulatory proteins [25]. Indeed, when in 1986 Walter Gilbert proposed the "RNA world hypothesis": that RNA, which can self-replicate, must be the primary molecule in evolution, the traditional gene even lost its ascendancy over other molecules as "the basis of life" [100].

Since 2001, the date of the first draft of the Human Genome Project, and since the release of the genome sequencing projects of other organisms, floods of new genome data have been generated and novel technologies have been developed to attempt to make sense of that data. High throughput microarray technology has provided a "new kind of microscope" [101] for post genomic analysis. It is now possible to look at thousands of base pair sequences simultaneously. The one gene at a time paradigm has been replaced, or at least supplemented, with a more holistic model of the gene in its surrounding molecular landscape.

For example, the central dogma presupposes a correspondence between genes and complexity and one of the big surprises of the Human Genome Project has been the scarcity of genes in the genome. The human genome contains in fact very few protein coding genes and fewer than many "simpler" organisms, a mere one-quarter to one-fifth of the original estimates [40]. To begin to explain the paradoxical genome data, researchers have had to shift their emphasis away from genes and proteins and towards gene regulation. Why do humans have so few genes [102] has been replaced by the question: How do so few genes create such complexity?

Clearly, it is not genes themselves, per se, that confer complexity. Rather complexity occurs as a result of gene–gene interactions and programs – molecular pathways that modulate development. Alternative splicing is one possible mechanism that might allow the cell to produce numerous proteins from one basic gene, and the mechanisms, pathways and regulators governing alternative splicing and spliceosomes are the subject of intensive research investigation. In addition, the large amount of noncoding DNA in genomes suggests that noncoding DNA may have functional biological activity [103]. In particular, ncRNAs may prove to be the programmers controlling complexity [42].

Science in the post genome era recognizes that gene activity does not occur in isolation. Rather, a full understanding of the development, the disease and decay of organisms will be found when the "genes," including the protein gene, the RNA gene or any other genes that might be discovered, are considered together with gene regulatory factors, gene–gene interactions, gene–cell interactions, epigenetic factors and signaling pathways in gene expression. Understanding signaling pathways in gene expression is a major research focus.

Gene function is beginning to be understood in different ways, with different ways to pose the problems. For example, rarely today do we speak of a gene as causing a particular disease or giving rise to a specific trait; diseases, even the so-called single gene diseases, and traits are, rather, understood to be the results of hundreds and even thousands of genes operating in complex regulatory networks. This is especially true in cardiology, where such complex multifactorial diseases as coronary artery disease, heart failure, hypertension and atherosclerosis are caused by genetic factors together with a host of environmental and other factors. Even in the case of the "single" gene diseases, such as hypertrophic cardiomyopathy, dilated cardiomyopathy and other disorders considered to be the result of mutations of a single gene, it is becoming increasingly clear that such disorders are actually far more complex than previously thought [104–107].

Already with microarray and other novel technologies, holistic approaches to investigating the health and disease of organisms are becoming possible. As Evelyn Fox Keller put it, the twenty-first century will be "the century of the genome" [25].

A closer look at some genes of importance in cardiology

Cardiac myosin heavy chain genes

A family of genes of major importance in cardiology are the myosin heavy chain genes [108]. Myosin, the contractile protein of muscle, makes up the thick filaments of cardiac and skeletal muscle. Conventional myosin contains two heavy chains (220,000 kDa) which form the helical coiled rod region of the molecule and four light chains (26,000 and 18,000 kDA) which form the pair-shaped head regions. Striated muscle contraction is generated by interaction between myosin and thin filament actin. Upon fibre activation the myosin head binds to actin, which slides a short distance along the thick filament. Linkage is broken by adenosine triphosphate (ATP) hydrolysis whereupon actin and myosin dissociate. By this process the filaments are pulled along each other, rachet-like, in the classic sliding filament motion.

Myosin heavy chain genes are highly conserved and structurally similar [109–111]. Mammalian myocardial genes are large and complex, spanning approximately 24 kb and split into 40–41 exons and approximately the same number of introns, of various sizes [112]. Two isoforms of myosin heavy chain gene are expressed in myocardial cells, α-MYH and β-MYH, extending over 51 kb on chromosome 14 in humans; α-MYH and β-MYH are separated intragenically by about 4.5 kb; similar in overall structure, their sequences in the 5′ flanking regions are quite different, suggesting independent regulation of these genes [113].

The α and β cardiac heavy chain genes are tandemly linked, and are arranged in order of their expression during fetal development. The β-MHC is located 5′ upstream of the α-MHC sequence and is expressed first during heart development, followed by α-MHC gene expression. Despite the fact that there is almost 93% sequence identity between α-MYH and β-MYH, their ATPase activity differs by twofold suggesting functional differences.

Table 1.3 Response to stimuli of cardiac myosin heavy chain genes.

	α-MYH	β-MYH
+ Thyroid (T3)	Upregulated	Downregulated
– Thyroid (T3)	Downregulated	Upregulated
Exercise	Upregulated	Downregulated
Pressure	Downregulated	Upregulated
Aging	Downregulated	Upregulated

Adapted from Weiss & Leinwald [108].

α-MYH and β-MYH isoforms are tissue specific and differentially developmentally regulated (reviewed in [114]). Thus, α-MYH and β-MYH are both expressed at high levels throughout the cells of the developing fetal heart tube at about 7.5–8 days post coitum [115]. As ventricular and atrial chambers begin to form, isoform expression patterns begin to diverge: β-MYH begins to be restricted to ventricular myocytes in humans, and α-MHC levels diminish in ventricular cells, but continue to be expressed in adult human atrial cells [116]. Cardiac myosin heavy chain gene expression and proportion of α-MYH and β-MYH expressed is regulated by a number of factors, including thyroid hormone during development, pressure or volume overload, diabetes, catecholamine levels and aging (Table 1.3) [108,114]. Regulatory elements in cardiac myosin heavy chain genes have been studied extensively (reviewed in [108]).

Disease mutations associated with MYH genes include, most notably, hypertrophic cardiomyopathy. Hypertrophic cardiomyopathy, a primary disorder of the myocardium and an important cause of heart failure, was first associated with mutations in the β myosin heavy chain gene in 1990 when a missense mutation in R403Q was identified [117]. Subsequently, more than 80 mutations linked with hypertrophic cardiomyopathy have been identified in the β myosin heavy chain gene, and the list continues to grow [118].

In addition to mutations in the β myosin heavy chain gene, researchers have identified hundreds of mutations in at least 10 other genes, all encoding for proteins involved in the cardiac contractile apparatus including α-myosin heavy chain gene, cardiac myosin binding protein C, cardiac troponin T2, C and I, α-tropomyosin, myosin regulatory and essential light chains, actin and titin [119]. Because all of the genes identified as being causal in primary hypertrophic cardiomyopathy encode for the sarcomeric proteins, primary hypertrophic cardiomyopathy is now widely recognized as a disorder of the sarcomere [105].

Primer of genes and genomics

DNA

The deoxyribonucleic acid (DNA) of a living organism contains all of the genetic information necessary to construct a specific organism and to direct the activity of the organism's cells.

DNA is a very long, twisted, double stranded molecule made up of two chains of nucleotides. Each DNA nucleotide contains one of the four DNA bases: *guanine* (G), *adenine* (A), *thymine* (T) and *cytosine* (C). These bases are arranged side by side (for example, AAGTTAAG) and it is their sequence arrangement that will determine the protein constructed by the gene.

Gene

The basic unit of heredity, a gene is an ordered sequence of DNA nucleotides that can be decoded to produce a gene product. The overwhelming majority of genes of the human genome are protein-coding genes; noncoding genes produce RNA molecules, mainly involved in gene expression.

Gene expression

Gene expression is the complex process by which information in the gene is transcribed into RNA and translated into proteins. Gene expression is carried out in two stages: transcription and translation. During transcription genetic information is transcribed into an mRNA copy of a gene, which must then be translated into a protein.

Although each cell of the human body contains a complete genome and set of 20,000–25,000 genes, only a subset of these genes are expressed or turned on, depending on cell type. Such cell-specific gene expression determines whether a cell will be a brain cell, a heart cell or a liver cell, for example. Some genes that carry out basic cellular functions are expressed all the time in all the cells – they are called housekeeping genes. Others are expressed only under certain conditions, such as when activated by signals such as hormones. Researchers study changes in gene expression to gain understanding as to how cells behave in response to changes in stimuli.

Gene structure

The gene is a structured molecule comprising exons, introns and regulatory sequences. The region of the gene that codes for a gene product (usually a protein) is called the exon; between the exons are sequences of noncoding DNA, called introns. Introns must be edited out of the gene during transcription and before translation of the protein.

Stretches of DNA indicate the beginning and end of genes. Coding begins with the initiation codon or start codon "ATG" and ends with termination or stop codons: TAA, TAG or TGA.

Genome

Genome is a word compound of "gene" and "chromosome." A genome is the complete DNA required to build a living organism, and an organism's genome is contained in each of its cells. Some genomes are small, such as bacterial genomes which may contain less than a million base pairs and some are very large: the human genome comprises about three billion base pairs.

Human Genome Project

The Human Genome Project is an international consortium to sequence all of the three billion base pairs of the human genome. The Human Genome Project formally commenced in 1990, led by the US Department of Energy and the National Institutes of Health. The project was completed in April 2003 with the announcement that the human genome contains some 20,000–25,000 genes.

The benefits of the Human Genome Project are beginning to make themselves felt. As a result of the research project, powerful and novel technologies and resources have been developed which will lead eventually to an understanding of biology at the deepest levels. Major advances in diagnosis and treatment of many diseases, and disease prevention is expected as a result of Human Genome Project efforts.

How many genes in the human genome?

As of October 2004, the latest estimate from the Human Genome Project is that the human genome contains some 20,000–25,000 protein-coding genes.

Genomics

Genome is a word combining "gene" and "chromosome", and the genome includes the entire set of an organism's protein coding genes and all of the DNA sequences between the genes. Genomics uses the techniques of molecular biology and bioinformatics to study not just the individual genes of an organism but of the whole genome.

Metabolomics

By analogy with genomics and proteomics, metabolomics is the large-scale study of the all the metabolites of an organism. Understanding the metabolome offers an opportunity to understand genotype–phenotype and genotype–environment interactions.

Microarray

Microarray is an enabling technology that allows researchers to compare gene portraits of tissue samples at a snapshot in time. A microarray is a slide or membrane to which is attached an orderly array of DNA sequences of known genes. The researcher pipettes samples of mRNA onto the slide, containing unknown transcripts obtained from a tissue of interest. mRNA has the property that it is complementary to the DNA template of origin. Thus, mRNA binds or hybridizes to the slide DNA and can be calculated by computer to provide a portrait or snapshot of which genes are active in the sample.

By monitoring and comparing thousands of genes at a time – instead of one by one – a microarray gene chip data can be used to see which genes in a tissue are turned on or expressed and which are turned off.

Microarray gene expression profiling

Understanding gene function is crucially important to understanding health and disease. Most of the common and serious diseases afflicting humans are polygenic: that is, it takes hundreds if not thousands of genes interacting with each other and with the environment to cause such diseases as cancer and heart disease. By monitoring and comparing thousands of genes at a time – instead of one by one – microarray gene expression profiling can be used

to determine which genes in a tissue are turned on and which are turned off – and how actively the genes are producing proteins.

Such gene "portraits" can identify patients with early stage diseases as compared to no disease or late stage disease, to distinguishing patients with different diseases, or patients with different stages of disease for disease prognosis, drug effect monitoring and other clinical applications. As microarray technology advances researchers will be able to ask increasingly probing and important biologic questions.

Mutation

A mutation is a change in the DNA sequence of a gene. If the mutation is significant, then the protein produced by the gene will be defective in some way and unable to function properly. Not all mutations are harmful; some may be beneficial and some may have no discernible effect.

There are different types of mutations: base substitution, in which a single base is replaced by another: deletion, in which base(s) are left out; or insertion in which base(s) are added.

Mutations can be caused by radiation, chemicals or may occur during the process of DNA replication. Some mutations can be passed on through generations.

Protein

A protein is a large molecular chain of amino acids. Proteins are the cell's main structural building blocks and proteins are involved in all cellular functions. Information in the gene encodes for the protein and most of the genes of living organisms produce proteins. Humans are calculated to have about 400,000 proteins, far more than our 20,000 or so genes.

Proteomics

An understanding of cellular biology depends fundamentally on understanding protein structure and behaviour. Proteomics is the large-scale comprehensive study of the proteome, the complement of all of the proteins expressed in a cell, a tissue or an organism. Proteomics uses technology similar to genomics technologies, such as protein microarrays, to explore the structure and function of proteins and protein behaviour in response to changing environmental signals.

RNA

The relationship between a gene and its protein is not straightforward. DNA does not construct proteins directly; rather, genes set in motion intermediate processes that result in amino acid chains. The main molecule involved in this process is called ribonucleic acid (RNA). RNA nucleotides contain bases: adenine (A) uracil (U) guanine (G) and cytosine (C). Thus, RNA is chemically very similar to DNA, except that RNA has a uracil base rather than thymine.

The process of producing a protein from DNA template begins in the cell nucleus via the intermediary messenger (m) RNA. mRNA copies the relevant piece of DNA in a process called transcription. The short, single-stranded mRNA transcript is then transported out of the cell nucleus by transfer RNA and into the cytoplasm where it is translated into a protein by the ribosome. (Ribosomal RNA (rRNA), is involved in constructing the ribosomes.) Since the 1990s many new non-coding RNA genes have been discovered, such as microRNA.

Single nucleotide polymorphism

A single nucleotide polymorphism (SNP) is a base alteration in a single nucleotide in the genome. Unlike mutations, which are rare, single nucleotide polymorphisms are common alterations in populations, occurring in at least 1% of the population. SNPs make up 90% of all human genetic variation and occur every 100–300 human genome bases. In time researchers hope to be able to develop SNP patterns that can be used to test individuals for disease susceptibility or drug response.

Further information

National Center for Biotechnology Information. A Science Primer. http://www.ncbi.nlm.nih.gov/About/primer/genetics_molecular.html

National Institutes of Health. NIGM. Genetics Basics http://publications.nigms.nih.gov/genetics/science.html

Welcome Trust. Gene Structure. http://genome.wellcome.ac.uk/doc_WTD020755.html

Human Genome Project. http://www.google.ca/search?q=%22%22+what+is+a+gene%22+&hl=en&lr=&c2coff=1&start=30&sa=N

Microarrays http://www.ncbi.nlm.nih.gov/About/primer/microarrays.html

Introduction to proteomics: http://www.childrenshospital.org/cfapps/research/data_admin/Site602/mainpageS602P0.html

Bio-pro. Proteomics http://www.bio-pro.de/en/life/thema/01950/index.html

The human metabolome project http://www.metabolomics.ca/

References

1 Mendel G. Experiments in Plant Hybridization (1865) Read at the meetings of the Brünn Natural History Society, February 8 and March 8, 1865. (Available online at www.mendelweb.org)

2 Correns C. Mendel's law concerning the behavior of progeny of varietal hybrids. (Trans: Piernick LK) Electronic Scholarly Publications. 2000. http://www.esp.org/foundations/genetics/classical/holdings/c/cc-00.pdf (Originally: Mendel's Regel über das Verhalten der Nachkommenschaft der Rassenbastarde. *Ber Dtsch Botanisch Gesellschadt* 1900; 18: 158–168.)

3 De Vries H. Concerning the law of segregation of hybrids. (Trans: Hannah A.) Electronic Scholarly Publications. 2000 (http://www.esp.org/foundations/genetics/classical/holdings/v/hdv-00.pdf) (originally: Das Spaltungsgesetz der Bastarde. *Ber Dtsch Botanisch Gesellschaft* 1900; 18: 83–90.)

4 Tschermak E. Concerning artificial crossing in Pisum sativum. *Genetics* 1950: 35: 42–47. (Originally: Über Künstliche Kreuzung bei Pisum sativum. *Ber Dtsch Botanisch Gesellschaft* 1900; 18: 232–239.)

5 De Vries H. Intracellular pangenesis. Including a paper on Fertilization and Hybridization. (Trans: Gager CS.) Open Court Publishing, Chicago, 1910. http://www.esp.org/books/devries/pangenesis/facsimile/title3.html

6 Bateson W. The progress of genetic research. In: *Report of the Third International Conference on Genetics 1906.* Royal Horticultural Society. London. 1907; 90–97.

7 Johannsen W. *Elemente der Exakten Erblichkeitslehre.* Gustav Fischer, Jena, 1909.

8 Carlson EA. *Mendel's legacy: the origin of classical genetics.* Cold Spring Harbor Laboratory Press, Cold Spring Harbor, New York, 2004.

9 Sturtevant AH. *A History of Genetics.* Harper and Row, New York, 1965.

10 Sturtevant AH. The linear arrangement of six sex linked traits in drosophila, as shown by their mode of association. *J Exp Zool* 1913; 14: 43–59.

11 Bridges CE. Salivary chromosome maps: with a key to the banding of the chromosomes of *Drosophila melanogaster. J Hered* 1935; 26: 60–64.

12 Muller HJ. Artificial transmutation of the gene. *Science* 1927; 66: 84–87.

13 Muller HJ. The gene as the basis of life. *Proc Int Cong Plant Science* 1929; 1: 897–921.

14 Avery OT, MacLeod CM, McCarty M. Studies on the chemical nature of the substance inducing transformation of pneumoccocal types. *J Exp Med* 1944; 79: 137–158.

15 Hershey AD, Chase M. Independent functions of viral proteins and nucleic acid in growth of bacteriophage. *J Gen Physiol* 1952: 36: 39–56.

16 Watson JD, Crick FHC. Molecular structure of nucleic acids. *Nature* 1953; 171: 737–738.

17 Olby RC. *The Path to the Double Helix*. Macmillan, London, 1974.

18 Garrod AE. The incidence of alkaptonuria: A study in chemical individuality. *Lancet* 1902; ii: 1616–1620.

19 Garrod AE. *Inborn Errors of Metabolism*, 2nd edn. Henry Frowde and Hodder & Stoughton, London, 1923.

20 Beadle G, Tatum E. Genetic control of biochemical reactions in Neurospora. *Proc Natl Acad Sci USA* 1941; 27: 499–506.

21 Neel JV. The inheritance of sickle cell anemia. *Science* 1949; 110: 64–66.

22 Pauling L, Itano H, Singer SJ, Wells I. Sickle cell anemia, a molecular disease. *Science* 1949; 110: 543–548.

23 Ingram VM. Gene mutations in human haemoglobin: the chemical difference between normal and sickle-cell haemoglobin. *Nature* 1957; 180: 326–328.

24 Crick FHC. On protein synthesis. *Symp Soc Exp Biol* 1958; XII: 139–163.

25 Keller EF. *The Century of the Gene*. Harvard University Press, Cambridge, 2000.

26 Morange M. Century of the gene. Isuma. *Can J Policy Res* 2001; 2: 22–27.

27 Jacob F, Monod J. Genetic regulatory mechanisms in the synthesis of proteins. *J Mol Biol* 1961; 3: 318–356.

28 Morange M. What history tells us. The operon model and its legacy. *J Biosci* 2005; 30: 313–316.

29 Morange M. *The Misunderstood Gene*. Harvard University Press, Cambridge, 2001.

30 Portin P. The concept of the gene: Short history and present status. *Q Rev Biol* 1993; 68: 173–223.

31 Maas WK. *Gene Action*. Oxford University Press, Oxford, 2001.

32 Temin HM, Mizutani S. RNA-dependent DNA polymerase in virions of Rous sarcoma virus. *Nature* 1970: 226: 1211–1213.

33 Baltimore D. RNA dependent DNA polymerase in virions of RNA tumor viruses. *Nature* 1970; 226: 1209–1211.

34 Chow LT, Gelinas RE, Broker TR, Roberts RJ. An amazing sequence arrangement at the 5′ ends of adenovirus 2 messenger RNA. *Cell* 1977; 12: 1–8.

35 Berget SM, Moore C, Sharp PA. Spliced segments at the 5′ terminus of adenovirus 2 late mRNA. *Proc Natl Acad Sci USA* 1977; 74: 3171–3175.

36 Gilbert W. Why genes in pieces? *Nature* 1978; 271: 501.

37 Lopez AJ. Alternative splicing of pre-mRNA: Developmental consequences and mechanisms of regulation. *Annu Rev Genet* 1998; 32: 279–305.

38 McClintock B. The origin and behavior of mutable loci in maize. *Proc Natl Acad Sci USA* 1950; 36: 344–355.

39 Britten RJ, Kohne DE. Repeated sequences in DNA. *Science* 1968; 161: 529–540.

40 International Human Genome Sequencing Consortium. Finishing the euchromatic sequence of the human genome. *Nature* 2004; 431: 931–945.

41 Gibbs WW. The unseen genome: gems among the junk. *Sci Am* 2003; 289: 46–53.

42 Mattick JS. The hidden genetic program of complex organisms. *Sci Am* 2004; 4: 60–67.

43 Eddy SR. Non-coding RNA genes and the modern RNA world. *Nat Rev Genet* 2001; 2: 919–929.

44 Storz G, Altuvia S, Wassarman KM. An abundance of RNA regulators. *Annu Rev Biochem* 2005; 74: 199–217.

45 Eddy SR. Computational genomics of noncoding RNA genes. *Cell* 2002; 109: 137–40.

46 Mattick JS. The functional genomics of noncoding RNA. *Science* 2005; 309: 1527–1528.

47 Noller HF. RNA structure: reading the ribosome. *Science* 2005; 309: 1508–1514.

48 Zamore PD, Haley B. Ribo-gnome: the big world of small RNAs. *Science* 2005; 309: 1519–1524.

49 Ambros V. The functions of animal microRNAs. *Nature* 2004; 431: 350–355.

50 Bentwich I, Avniel A, Karov Y *et al.* Identification of hundreds of conserved and nonconserved human microRNAs. *Nat Genet* 2005; 37: 766–770.

51 Lewis BP, Burge CB, Bartel DP. Conserved seed pairing, often flanked by adenosines, indicates that thousands of human genes are microRNA targets. *Cell* 2005; 120: 15–20.

52 Chen CZ. MicroRNAs as oncogenes and tumor suppressors. *N Engl J Med* 2005; 353: 1768–1771.

53 Calin GA, Ferracin M, Cimmino A *et al.* A microRNA signature associated with prognosis and progression in chronic lymphocytic leukemia. *N Engl J Med* 2005; 353: 1793–1801.

54 Falk R. Long live the genome! So should the gene. *Hist Philos Life Sci* 2004; 26: 105–121.

55 Venter JC, Adams MD, Myers EW *et al.* The sequence of the human genome. *Science* 2001; 291: 1304–1351.

56 International Human Genome Sequencing Consortium. Initial sequencing and analysis of the human genome. *Nature* 2001; 409: 860–921.

57 Dempsey AA, Dzau VJ, Liew CC. Cardiovascular genomics: Estimating the total number of genes expressed in

the human cardiovascular system. *J Mol Cell Cardiol* 2001; 33: 1879–1886.

58 Hwang DM, Dempsey AA, Wang RX *et al.* A genome-based resource for molecular cardiovascular medicine: toward a compendium of cardiovascular genes. *Circulation* 1997; 96: 4146–4203.

59 Pennisi E. The human genome. *Science* 2001; 291: 1177–1180.

60 Adams MD, Kelley JM, Gocayne JD *et al.* Complementary DNA sequencing: expressed sequence tags and human genome project. *Science* 1991; 252: 1651–1656.

61 Roberts L. Gambling on a shortcut to genome sequencing. *Science* 1991; 252: 1618–1619.

62 Liew CC. A human heart cDNA library: the development of an efficient and simple method for automated DNA sequencing. *J Mol Cell Cardiol* 1993; 25: 891–894.

63 Liew CC, Hwang DM, Fung YW *et al.* A catalogue of genes in the cardiovascular system as identified by expressed sequence tags (ESTs). *Proc Natl Acad Sci USA* 1994: 91: 10645–10649.

64 Roberts L. Genome patent fight erupts. *Science* 1991: 254: 184–186.

65 Marshall E. The company that genome researchers love to hate. *Science* 1994; 266: 1800–1802.

66 Levy-Lahad E, Wasco W, Poorkaj P, *et al.* Candidate gene for the chromosome 1 familial Alzheimer's disease locus. *Science* 1995; 269: 973–977.

67 Schena M, Shalon D, Davis RW *et al.* Quantitative monitoring of gene expression patterns with a complementary DNA microarray. *Science* 1995; 270: 467–470.

68 Lockhart DJ, Dong H, Byrne MC *et al.* Expression monitoring by hybridization to high-density oligonucleotide arrays. *Nat Biotechnol* 1996; 14: 1675–1680.

69 Bowtell DDL. Options available – from start to finish – for obtaining expression data by microarray. *Nat Genet* 1999; Supplement 21: 25–32.

70 Lipshutz RJ, Fodor SPA, Gingeras TR *et al.* High density synthetic oligonucleotide arrays. *Nature Genetics.* 1999; Supplement 21: 20–24.

71 Perou CM, Jeffrey SS, van de Rijn M *et al.* Distinctive gene expression patterns in human mammary epithelial cells and breast cancers. *Proc Natl Acad Sci USA* 1999; 96: 9212–9217.

72 Rhodes DR, Chinnaiyan AM. Integrative analysis of the cancer transcriptome. *Nat Genet* 2005; 37 Supplement: S31–S37.

73 Segal E, Friedman N, Kaminski N *et al.* From signatures to models: understanding cancer using microarrays. *Nat Genet* 2005; 37 Supplement: S38–S45.

74 Mohr S, Leikauf GD, Keith G, Rihn BH. Microarrays as cancer keys: an array of possibilities. *J Clin Oncol* 2002; 20: 3165–3175.

75 Liew CC, Dzau VJ. Molecular genetics and genomics of heart failure. *Nat Rev Genet* 2004; 5: 811–825.

76 Friddle CL, Koga T, Rubin EM, Bristo J. Expression profiling reveals distinct sets of genes altered during induction and regression of cardiac hypertrophy *Proc Natl Acad Sci USA* 2000; 97: 6745–6750.

77 Yang J, Moravec CS, Sussman MA. Decreased SLIM1 expression and increased gelsolin expression in failing human hearts measured by high-density oligonucleotide arrays. *Circulation* 2000; 102: 3046–3052.

78 Barrans JD, Stamatiou D, Liew CC. Construction of a human cardiovascular cDNA microarray: portrait of a failing heart. *Biochem Biophys Res Commun* 2001; 280: 964–969.

79 Barrans JD, Allen PD, Stamatiou D *et al.* Global gene expression profiling of end stage dilated cardiomyopathy using a human cardiovascular based cDNA microarray. *Am J Pathol* 2002; 160: 2035–2043.

80 Liew CC. Expressed genome molecular signatures of heart failure. *Clin Chem Lab Med* 2005; 43: 462–469.

81 Whitney AR, Diehn M, Popper SJ *et al.* Individuality and variation in gene expression patterns in human blood. *Proc Natl Acad Sci USA* 2003; 100, 1896–1901.

82 Radich JP, Mao M, Stepaniants S *et al.* Individual-specific variation of gene expression in peripheral blood leukocytes. *Genomics* 2004; 83: 980–988.

83 Barnes MG, Aronow BJ, Luyrink LK *et al.* Gene expression in juvenile arthritis and spondyloarthropathy: pro-angiogenic ELR+ chemokine genes relate to course of arthritis. *Rheumatol (Oxf)* 2004; 43: 973–979.

84 Okuda T, Sumiya T, Mizutani K *et al.* Analyses of differential gene expression in genetic hypertensive rats by microarray. *Hypertens Res* 2002; 25: 249–255.

85 Chon H, Gaillard CA, van der Meijden BB *et al.* Broadly altered gene expression in blood leukocytes in essential hypertension is absent during treatment. *Hypertension* 2004; 43: 947–951.

86 Bull TM, Coldren CD, Moore M *et al.* Gene microarray analysis of peripheral blood cells in pulmonary arterial hypertension. *Am J Respir Crit Care Med* 2004; 170: 827–828.

87 DePrimo SE, Wong LM, Khatry DB *et al.* Expression profiling of blood samples from an SU5416 Phase III metastatic colorectal cancer clinical trial: a novel strategy for biomarker identification. *BMC Cancer* 2003; 3: 3.

88 Whistler T, Unger ER, Nisenbaum R, Vernon SD. Integration of gene expression, clinical, and epidemiologic data to characterize chronic fatigue syndrome. *J Transplant Med* 2003; 1: 10.

89 Tang Y, Lu A, Aronow BJ, Sharp FR. Blood genomic responses differ after stroke, seizures, hypoglycemia, and hypoxia: blood genomic fingerprints of disease. *Ann Neurol* 2001; 50: 699–707.

90 Tang Y, Nee AC, Lu A *et al.* Blood genomic expression profile for neuronal injury. *J Cereb Blood Flow Metab* 2003; 23: 310–319.

91 Connolly PH, Caiozzo VJ, Zaldivar F *et al.* Effects of exercise on gene expression in human peripheral blood mononuclear cells. *J Appl Physiol* 2004; 97: 1461–1469.

92 Ezendam J, Staedtler F, Pennings J *et al.* Toxicogenomics of subchronic hexachlorobenzene exposure in Brown Norway rats. *Environ Health Perspect* 2004; 112: 782–791.

93 Wu MM, Chiou HY, Ho IC *et al.* Gene expression of inflammatory molecules in circulating lymphocytes from arsenic-exposed human subjects. *Environ Health Perspect* 2003; 111: 1429–1438.

94 Ryder MI, Hyun W, Loomer P, Haqq C. Alteration of gene expression profiles of peripheral mononuclear blood cells by tobacco smoke: implications for periodontal diseases. *Oral Microbiol Immunol* 2004; 19: 39–49.

95 Ma J, Liew CC. Gene profiling identifies secreted protein transcripts from peripheral blood cells in coronary artery disease. *J Mol Cell Cardiol* 2003; 35: 993–998.

96 Tsuang MT, Nossova N, Yager T *et al.* Assessing the validity of blood-based gene expression profiles for the classification of schizophrenia and bipolar disorder: a preliminary report. *Am J Med Genet B Neuropsychiatr Genet* 2005; 133: 1–5.

97 Glatt SJ, Everall IP, Kremen WS *et al.* Comparative gene expression analysis of blood and brain provides concurrent validation of SELENBP1 upregulation in schizophrenia. *Proc Natl Acad Sci USA* 2005; 102: 15533–15538.

98 Marshall KW, Zhang H, Yager T *et al.* Blood-based biomarkers for detecting mild osteoarthritis in the human knee. *Osteoarthritis Cartilage* 2005; 13: 861–871.

99 Osman I, Bajorin D, Sun TT *et al.* Novel blood biomarkers of human urinary bladder cancer. *Clin Cancer Res* 2006; 12: 3374–3380.

100 Gilbert, W. Origin of life: The RNA world. *Nature* 1986; 319: 618.

101 Brown PO. Website (http://biochemistry.stanford.edu/research/brown.html)

102 Pennisi E. Why do humans have so few genes? *Science* 2005; 309: 80.

103 Ruddle F. Hundred-year search for the human genome. *Annu Rev Genomics Hum Genet* 2001; 2: 1–8.

104 Hughes SE. The pathology of hypertrophic cardiomyopathy. *Histopathology* 2004; 44: 412–427.

105 Seidman JG, Seidman C. The genetic basis for cardiomyopathy: from mutation identification to mechanistic paradigms. *Cell* 2001; 104: 557–567.

106 Towbin JA, Bowles NE. The failing heart. *Nature* 2002; 415: 227–233.

107 Bonne G, Carrier L, Richard P. Familial hypertrophic cardiomyopathy: from mutations to functional defects. *Circ Res* 1998; 83: 580–593.

108 Weiss A, Leinwald LA. The mammalian myosin heavy chain gene family. *Annu Rev Cell Dev Biol* 1996; 12: 417–439.

109 Liew CC, Jandreski MA. Construction and characterization of the α form of a cardiac myosin heavy chain cDNA clone and its developmental expression in the Syrian hamster. *Proc Natl Acad Sci USA* 1986; 83: 3175–3179.

110 Jandreski MA, Liew CC. Construction of a human ventricular cDNA library and characterization of a β-myosin heavy chain cDNA clone. *Hum Genet* 1987; 76: 47–53.

111 Jandreski MA, Sole MJ, Liew CC. Two different forms of β-myosin heavy chain are expressed in human striated muscle. *Hum Genet* 1987; 77: 127–131.

112 Strehler EE, Strehler-Page MA, Perriard JC *et al.* Complete nucleotide and encoded amino acid sequence of a mammalian myosin heavy chain gene. Evidence against intron dependent evolution of the rod. *J Mol Biol* 1986; 190: 291–317.

113 Yamauchi-Takihara K, Sole MJ, Liew J *et al.* Characterization of human cardiac myosin heavy chain genes. *Proc Natl Acad Sci USA* 1989: 86: 3504–3508.

114 Morkin E. Control of cardiac myosin heavy chain gene expression. *Microsc Res Tech* 2000; 50: 522–531.

115 Lyons GE, Ontell M, Cox R *et al.* The expression of myosin genes in developing skeletal muscle in the mouse embryo. *J Cell Biol* 1990; 111: 1465–1476.

116 Lompre AM, Nadal-Ginard B, Mahdavi V. Expression of the cardiac ventricular α- and β-myosin heavy chain genes is developmentally and hormonally regulated. *J Biol Chem* 1984; 259: 6437–6446.

117 Geisterfer-Lowrence AA, Kass S, Tanigawa G *et al.* A molecular basis for familial hypertrophic cardiomyopathy a β cardiac myosin heavy chain gene mis-sense mutation. *Cell* 1990; 62: 999–1006.

118 Seidman C. For an updated list go to: Sarcomere Protein Gene Mutation Database. (http://genetics.med.harvard.edu/~seidman/cg3/)

119 Ahmad F, Seidman JG, Seidman C. The genetic basis for cardiac remodeling. *Annu Rev Genomics Hum Genet* 2005; 6: 185–216.

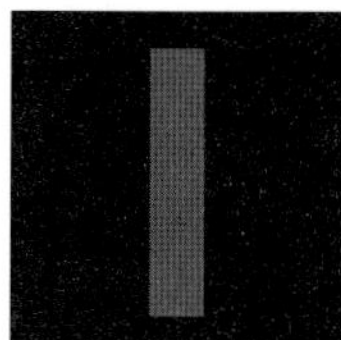

PART I
Cardiovascular single gene disorders

β-Myosin heavy chain (MYH7) chromosome localization

Chromosome 14

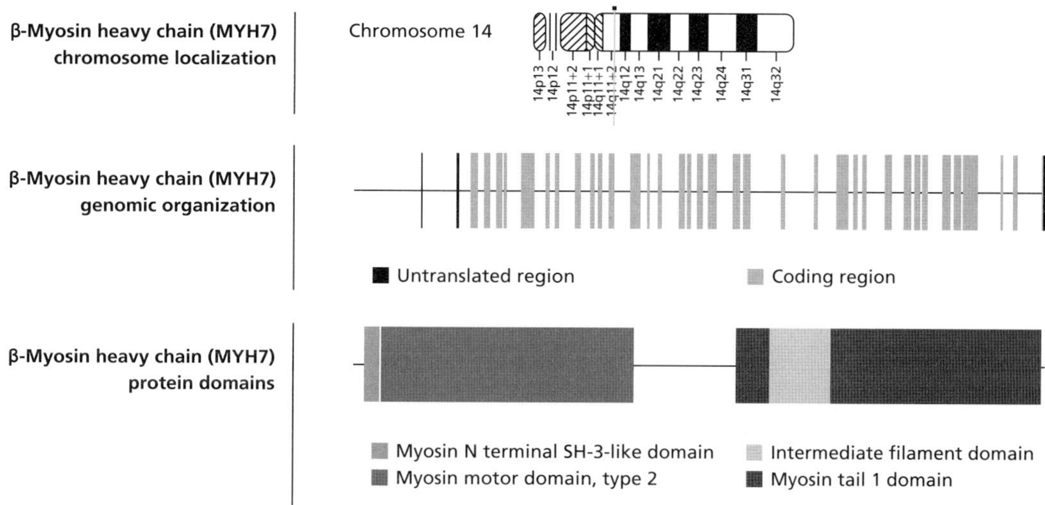

The chromosomal localization, genomic organization and protein domains of the human β-myosin heavy chain (MYH7). The mRNA sequence of the human β-myosin heavy chain was first sequenced and characterized in 1990 by Liew *et al.* [1]. The genomic sequence and organization of the human β-myosin heavy chain was first identified and characterized in 1989 by Yamauchi-Takihara *et al.* [2].

1. Liew CC, Sole MJ, Takihara KY *et al.* Complete sequence and organization of the human cardiac β-myosin heavy chain gene. *Nucleic Acids Res* 1990; 18: 3647–3651.
2. Yamauchi-Takihara K, Sole MJ, Liew J, Ing D, Liew CC. Characterization of human cardiac myosin heavy chain genes. *Proc Natl Acad Sci USA* 1989; 86: 3504–3508.

CHAPTER 2

Monogenic hypercholesterolemia

Ruth McPherson, MD, PhD, FRCPC

Introduction

Plasma concentrations of low-density lipoprotein (LDL) cholesterol are directly related to the incidence of coronary artery disease (CAD). Approximately half of interindividual variation in LDL-cholesterol is attributable to genetic factors [1,2]. The major part of this is believed to be oligogenic, the cumulative result of variations in several genes, or polygenic, resulting from a large number of genetic variants, each contributing a small effect. Plasma concentrations of LDL-cholesterol are also strongly influenced by environmental factors including diet and lifestyle as well as a number of endogenous and exogenous hormonal influences and various disease states. Very high plasma concentrations of LDL-cholesterol may be the consequence of rare monogenic disorders with severe clinical sequelae including tissue deposition of cholesterol, producing cutaneous xanthomas and premature atherosclerosis. Early diagnosis of these disorders is essential both to permit the prompt application of vigorous cholesterol lowering therapies required for CAD prevention and to alert the clinician to the need to screen first degree relatives.

Insights gained from the study of rare monogenic causes of hypercholesterolemia have also contributed significantly to our knowledge of intracellular protein trafficking and cellular cholesterol metabolism [3–5]. This review focuses on recent developments in our understanding of the genetic and molecular etiology of known Mendelian disorders of LDL-cholesterol metabolism (Table 2.1) and indicate how this knowledge been applied to develop effective therapies for both rare and common forms of hypercholesterolemia.

Overview of LDL metabolism

Cholesterol is a structural component of vertebrate plasma membranes and is a precursor for steroid hormone synthesis. Cholesterol is transported in plasma in the form of lipoproteins with unesterified cholesterol as a surface component and cholesteryl esters packaged in the core of spherical lipoproteins. The liver synthesizes and secretes very low-density lipoprotein (VLDL), which are triglyceride-rich lipoproteins, containing one molecule of apoB. The triglycerides and phospholipids of circulating VLDL are hydrolyzed by lipases at vascular endothelial surfaces, leaving a cholesterol-enriched VLDL remnant, which may be removed directly by the liver or converted to LDL, a process that involves remodeling by hepatic lipase and cholesteryl ester transfer protein (CETP). LDL particles are largely cleared from the circulation by the liver after binding to LDL receptors by receptor mediated endocytosis (Plate 2.1) [5]. When LDL receptors are absent or dysfunctional, LDL accumulates in the plasma and eventually crosses into the subendothelial space, where following oxidation or other types of enzymatic modification, it may be taken up by macrophage scavenger receptors, leading to foam cell formation [6].

Domain organization of the LDL receptor

The protein domain structure of the LDL receptor (LDLR) is illustrated in Plate 2.2. This receptor is a glycoprotein of 839 amino acids with a single transmembrane domain. Seven LDL receptor type A (LA) molecules at the amino terminal end are responsible for lipoprotein binding via apoB or apoE [7]. Mutations in apoB (esp Arg3500Gln)

Table 2.1 Monogenic disorders of LDL-cholesterol (LDL-C) metabolism.

	Gene defect	Causative sequence variants	Molecular etiology	Prevalence (heterozygotes)	LDL-C concentration
Autosomal dominant					
FH	LDLR 19p13.1–13.3	>1000	Impaired synthesis or secretion or function of LDLR	~1/500	7–10 mmol/L in heterozygous FH 15–30 mmol/L in homozygous FH
FDB	APOB 2p23–24	Most due to single missense mutation Arg3500Gln	Impaired interaction of apoB with LDLR	~1/1000	6–8 mmol/L in heterozygous FBD
FH3	PCSK9 1p32	S127R F216L	Gain of function mutations leading to decreased cell surface expression of LDLR	~1/2500 8–25 mmol/L	Variable
Autosomal recessive					
ARH	ARH 1p35–36.1	>10	Impaired LDLR mediated endocytosis	~1/5 × 10^6	10–14 mmol/L
Sitosterolemia	ABCG5 or ABCG8	>25 require 2 mutations in either ABCG5 or ABCG8	Increased absorption of plant sterols	~1/5 × 10^6	Variable

ARH, autosomal recessive hypercholesterolemia; FDB, familial defective apoB; FH, familial hypercholesterolemia; LDLR, LDL receptor.

impair the interaction of LDL with the LDLR leading to familial dysbetalipoproteinemia (FDB). Adjacent to this is a region with homology to the epidermal growth factor precursor (EGFP) consisting of two EGF like repeats, a YWTD domain and a third EGF repeat. This region of the LDLR is implicated in the release of internalized lipoproteins in acidic endosomes at low pH [8]. Interspersed between the EGFR and the plasma membrane is a region rich in serine and threonine which undergoes N-linked glycosylation. This O-linked sugar domain is followed by the transmembrane domain and a 50 AA cytoplasmic tail required for required for receptor localization in clathrin coated pits and a NPxY motif required for receptor internalization [9] (reviewed in [10]).

Cellular itinerary of the LDLR

Following synthesis, the LDLR undergoes folding in the endoplasmic reticulum, a process facilitated by the molecular chaperone, receptor associated protein (RAP) [11]. PCKS9 is a serine protease which appears to function in the post translational regulation of LDLR processing [12–14]. Gain of function mutations in PCKS9 result in decreased LDLr cell surface expression and have been identified as a rare cause of autosomal dominant hypercholesterolemia [15]. The LDLR is secreted from the endoplasmic reticulum as a 120-kDa protein which undergoes extensive O-linked glycosylation in the Golgi to form the mature 160 kDa receptor, which is transported to the cell surface. The LDLR binds with high affinity to apoB and apoE containing lipoproteins at the cell surface. Receptor lipoprotein complexes enter the cell via clathrin coated pit mediated endocytosis, a process that requires the NPXY sequence in the cytoplasmic tail. ARH-1 encodes a protein with a PTB domain capable of binding the NPXY sequence in the LDLR, a canonical clathrin binding sequence, LLDLE and a sequence recognized by the β2 adaptin subunit of AP-2, a major structural component of clathrin coated pits [16]. Mutations in ARH-1 have been identified as a cause of autosomal recessive hypercholesterolemia (ARH) [17]. Following clathrin coated pit mediated endocytosis, LDLR and cargo

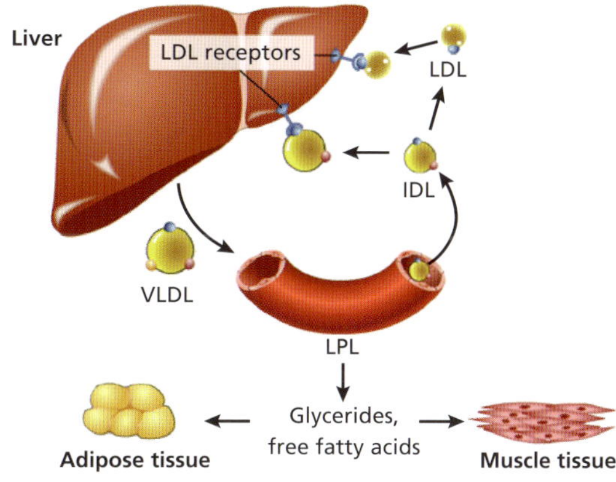

Plate 2.1 Overview of low density lipoprotein (LDL) metabolism. The liver synthesizes and secretes very low density lipoprotein (VLDL), which are triglyceride-rich lipoproteins, containing one molecule of apoB. The triglycerides and phospholipids of circulating VLDL are hydrolyzed by lipases at vascular endothelial surfaces. Free fatty acids may be taken up by adipose tissue and stored in lipid droplets or oxidized in skeletal muscle or other tissues. The remaining cholesterol-enriched intermediate density lipoprotein (IDL) remnant may be removed directly by the liver or converted to LDL, a process which involves remodeling by hepatic lipase and cholesteryl ester transfer protein (CETP). LDL are largely cleared from the circulation by the liver after binding to LDL receptors by receptor-mediated endocytosis [5].

Plate 2.2 Domain organization of the low density lipoprotein (LDL) receptor. This receptor is a glycoprotein of 839 amino acids with a single transmembrane domain. Seven LDL receptor type A (LA) molecules at the amino terminal end are responsible for lipoprotein binding via apoB or apoE [7]. The ligand binding domain is the most frequent site of mutations leading to familial hypercholesterolemia (FH). Mutations in apoB (esp Arg3500Gln) impair the interaction of LDL with the LDLr leading to familial dysbetalipoproteinemia (FDB). Adjacent to this is a region with homology to the epidermal growth factor precursor (EGFP) consisting of two EGF-like repeats, a YWTD domain and a third EGF repeat. This region of the LDLr is implicated in the release of internalized lipoproteins in acidic endosomes at low pH [8]. Interspersed between the epidermal growth factor receptor (EGFR) and the plasma membrane is a region rich in serine and threonine which undergoes N-linked glycosylation. This O-linked sugar domain is followed by the transmembrane domain and a 50 AA cytoplasmic tail required for receptor localization in clathrin coated pits and a NPxY motif required for receptor internalization [9] (reviewed in [10]).

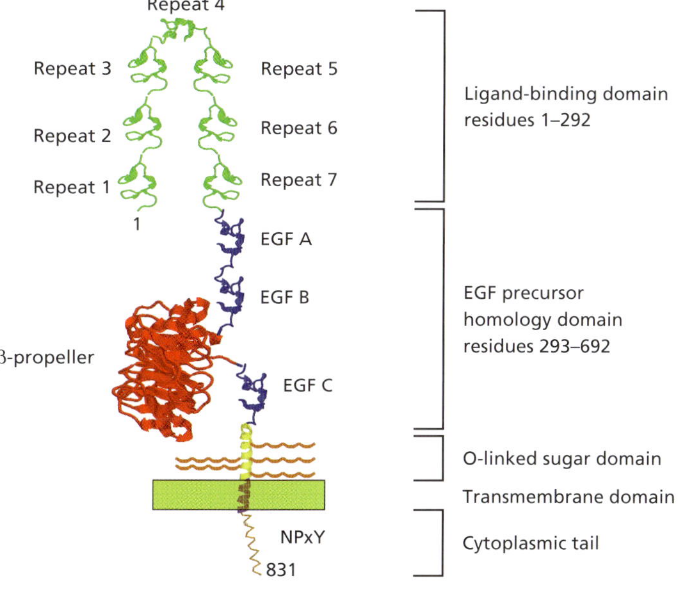

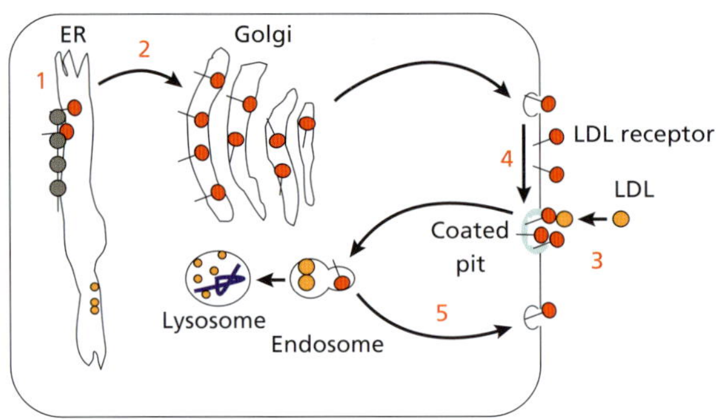

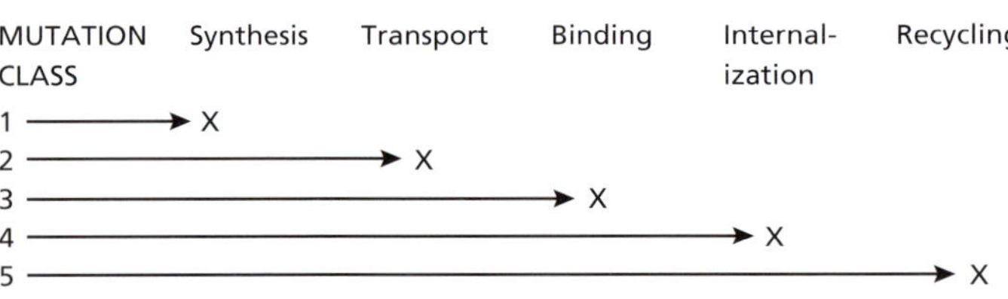

Adapted from Hobbs *et al.* Ann. Rev. Genet. 1990; 24: 133–170.

Plate 2.3 Classification of low density lipoprotein (LDL) receptor mutations that cause familial hypercholesterolemia (FH). Mutations of the LDL receptor that result in FH have been classified based on how they perturb LDL receptor intracellular trafficking or function. Class 1: The LDLr is not synthesized (e.g., deletions in the promoter region or splice defects). Class 2: the receptor is synthesized but not transported to the cell surface (predominately mutations in the EGF precursor homology domain). Class 3: the receptor is presented on the cell surface but cannot bind ligands (mutations in the ligand-binding domain). Class 4: the receptor cannot localize in coated pits and, as a consequence, cannot mediate endocytosis (mutations or deletions in the cytoplasmic tail). Class 5: the LDL receptor–ligand complex fails to undergo pH-dependent dissociation and the LDL receptor does not recycle (deletion of the EGF precursor homology domain).

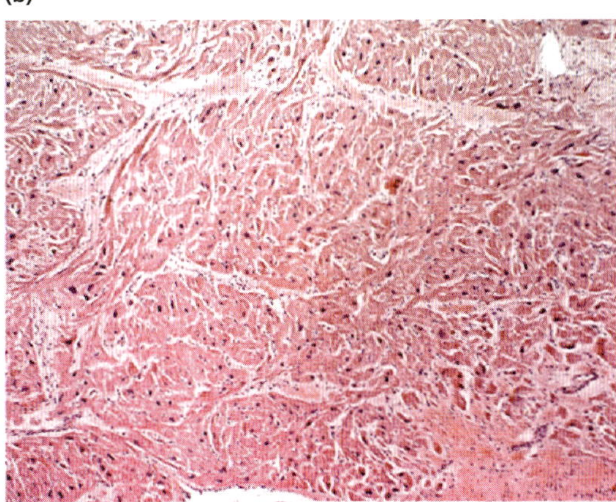

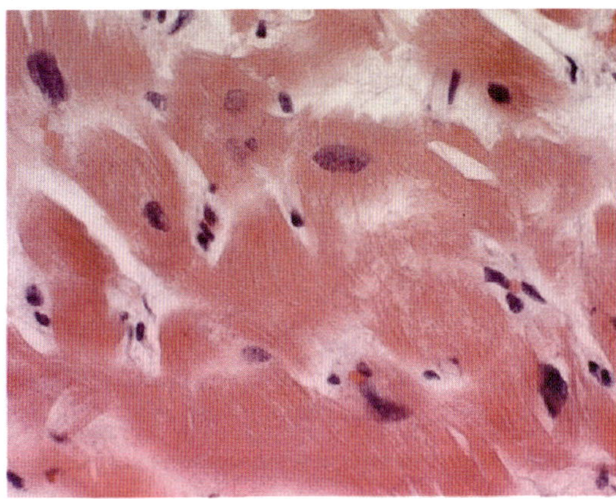

Plate 3.1 Gross anatomic and histologic phenotype in HCM. (a) Coronal section of the myocardium showing hypertrophic walls and small left ventricular cavity (courtesy of Sidney S. Murphree, MD, University of Louisville). (b) Low magnification view (× 6) of H&E stained myocardial section, showing disorganized architecture. (c) High magnification view of myocytes (× 60) on a H&E stained section.

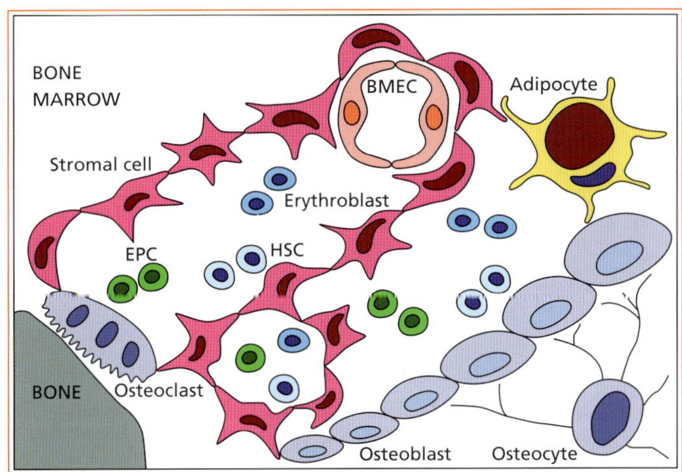

Plate 10.1 Bone marrow derived cells. Adapted from Bianco P, Cossu G. Uno, nessuno e centomila: searching for the identity of mesodermal progenitors. *Exp Cell Res*. 1999 Sept 15; (2): 257–63.

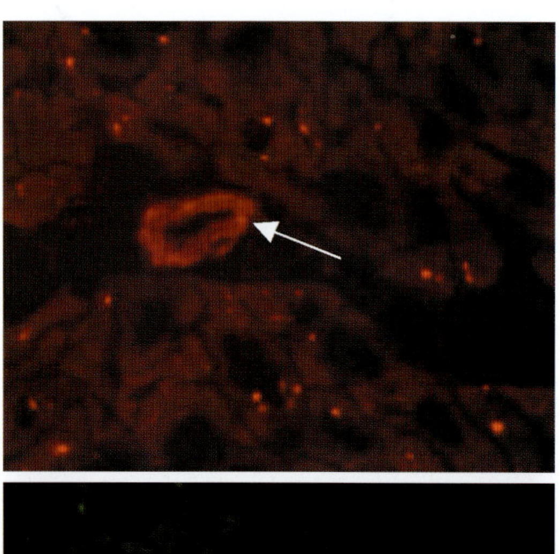

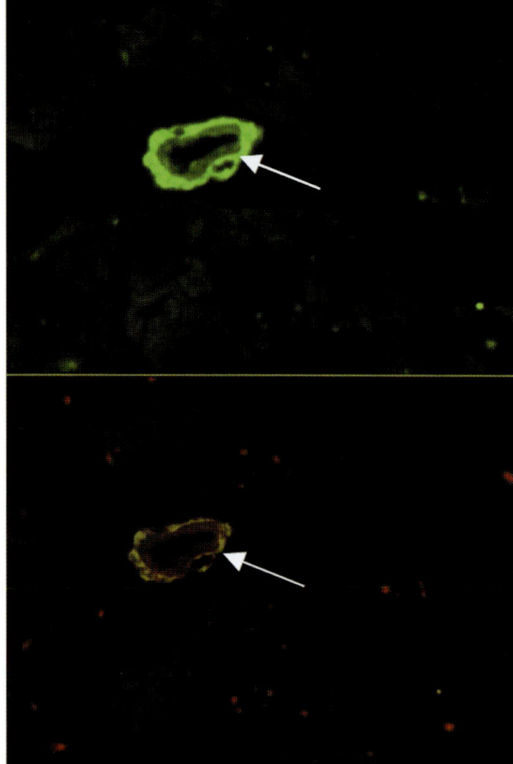

Plate 10.2 (Top) DiI-positive stem cells (red) in the mid-myocardium of the anterolateral wall. (Middle) x-Smooth muscle actin staining with Fitc (green) showing cross-section of vessel wall. (Bottom) Stained areas showing colocalization (yellow) of stem cells and smooth muscle cells, suggesting transformation of stem cells into smooth muscle cells. The vessel shown is in the myocardial interstitium. Arrows point to vessel media. Reprinted from *Circulation* 2005; 111: 150–156 with permission.

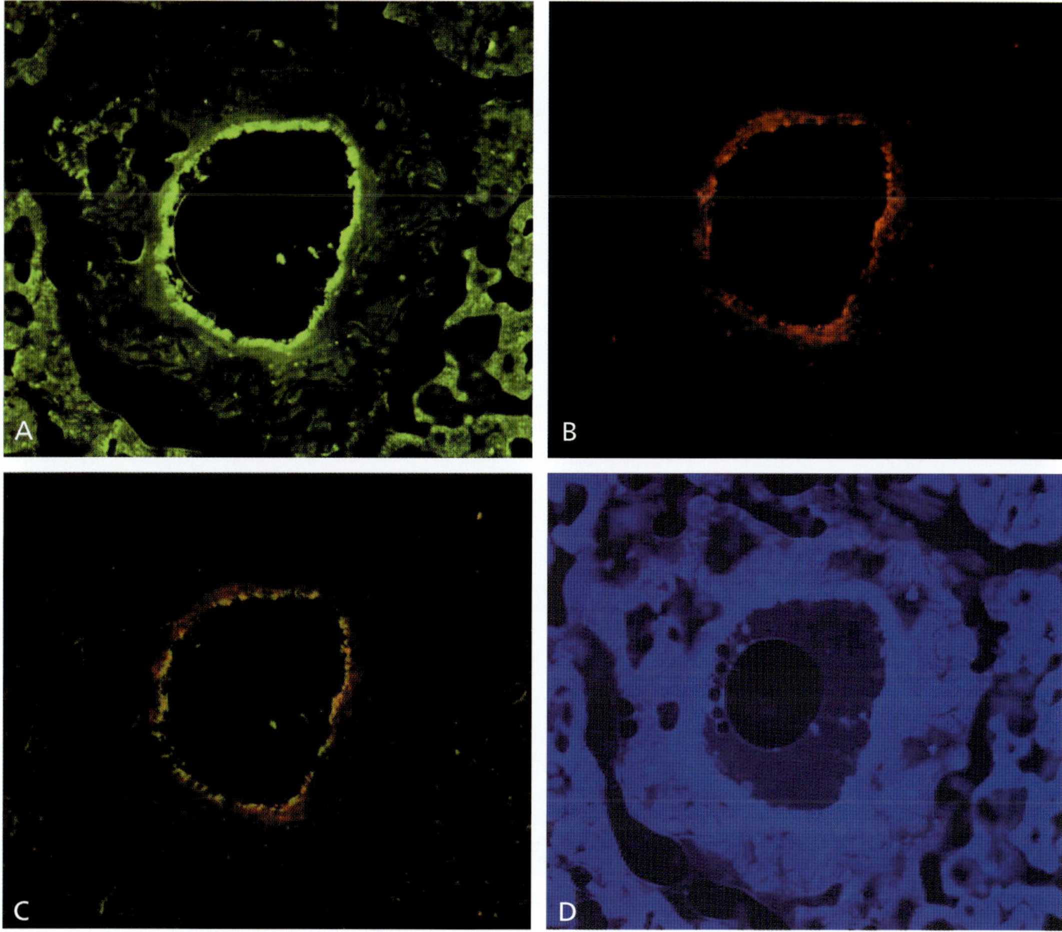

Plate 10.3 (A) Factor VIII staining with Fitc (green) showing a thin vessel wall. (B) Dil-positive mesenchymal stem cells (red) in a vessel of the anterolateral wall. (c) Colocalization (yellow) of MSCs and endothelial cells, indicating transformation of MSCs into endothelial cells. (D) DAPI stain showing labeled endothelial nuclei. Reprinted from *Circulation* 2005; 111: 150–156 with permission.

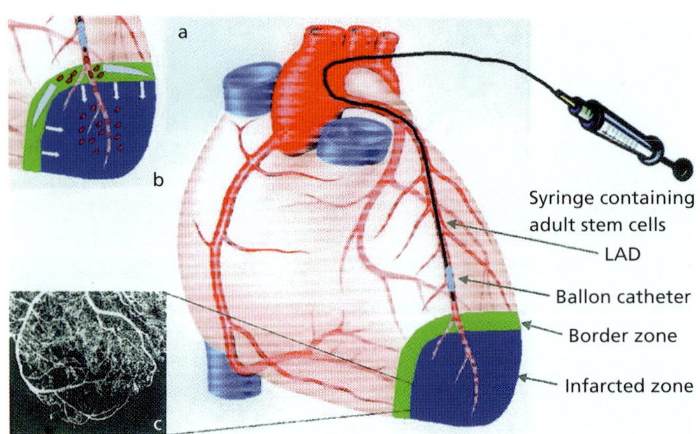

Plate 10.4 Technique for cardiac stem cell transplantation as treatment for myocardial infarction. (a) Balloon catheterization of infarct-related artery (LAD) above the infarct border zone followed by high-pressure infusion of stem cells into the artery. (b) Migration of stem cells (red dots) into infarcted zone via infarct-related blood vessels along suggest the possible route of migration. (c) Migration of cells to both infarct and border zone via existing blood flow within infarcted zone. From Straver BE, Brehm M, Zeus T, et al. Repair of infarcted myocardium by autologous intracoronary mononuclear bone marrow cell transplantation in humans. Reprinted from *Circulation* 2002; 106: 1913–8, with permission.

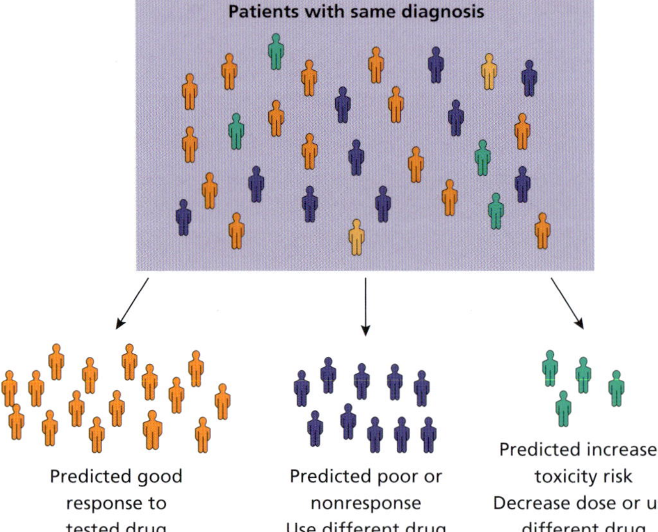

Plate 11.1 Clinical potential of pharmacogenetics. Patients with the same diagnosis (e.g., hypertension, dyslipidemia, heart failure) are typically treated either empirically (trial and error) or by a protocol-driven approach. However, their responses to drug therapy will not be the same, with some having an efficacious response, some having little to no response, and others having an adverse response. Pharmacogenetics has the potential to provide a tool for predicting those patients who are likely to have the desired response to the drug, those who are likely to have little or no benefit, and those at risk for toxicity. This would allow tailored therapy that should reduce adverse reactions to drugs, and increase efficacy rates. Reprinted from [191] with permission from Elsevier.

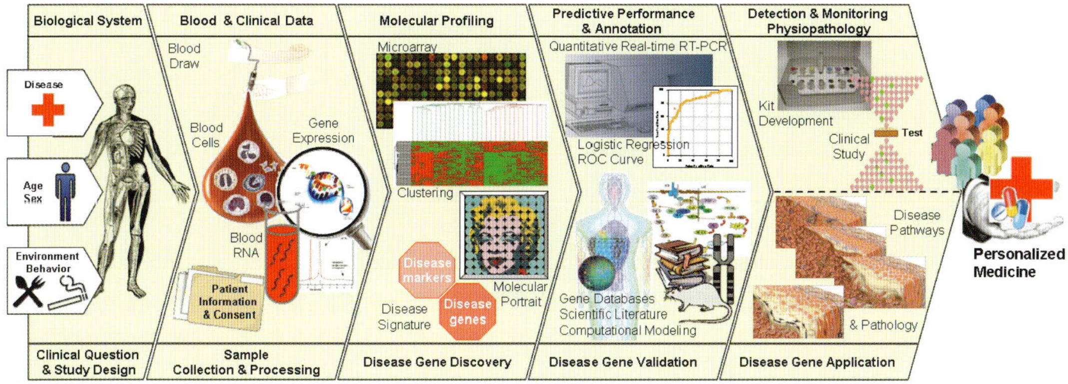

Plate 12.1 Work flow for systematic blood-based RNA biomarker discovery, validation and application.

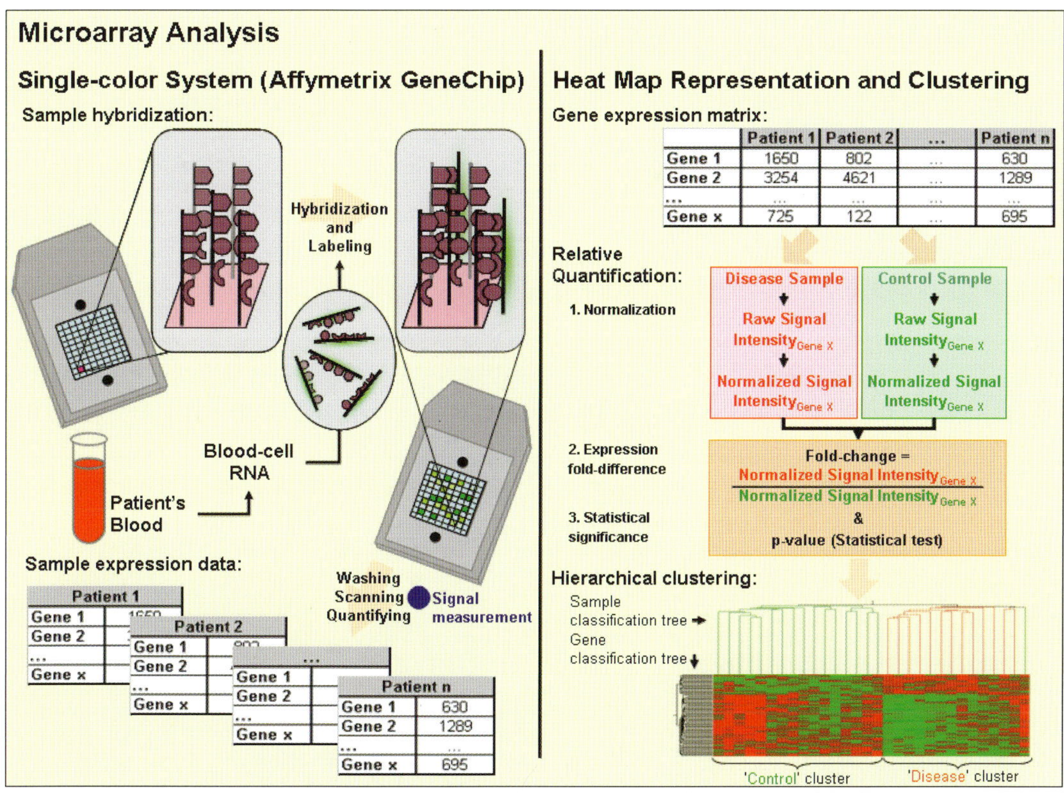

Sample expression data:

Patient 1	
Gene 1	1650
Gene 2	
...	
Gene x	

Patient 2	
Gene 1	802
Gene 2	
...	
Gene x	

Patient n	
Gene 1	630
Gene 2	1289
...	...
Gene x	695

Gene expression matrix:

	Patient 1	Patient 2	...	Patient n
Gene 1	1650	802	...	630
Gene 2	3254	4621	...	1289
...	...	...	...	...
Gene x	725	122	...	695

Fold-change = $\dfrac{\text{Normalized Signal Intensity}_{\text{Gene X}}}{\text{Normalized Signal Intensity}_{\text{Gene X}}}$ & p-value (Statistical test)

Plate 12.2 Disease gene discovery: Microarray gene expression profiling.

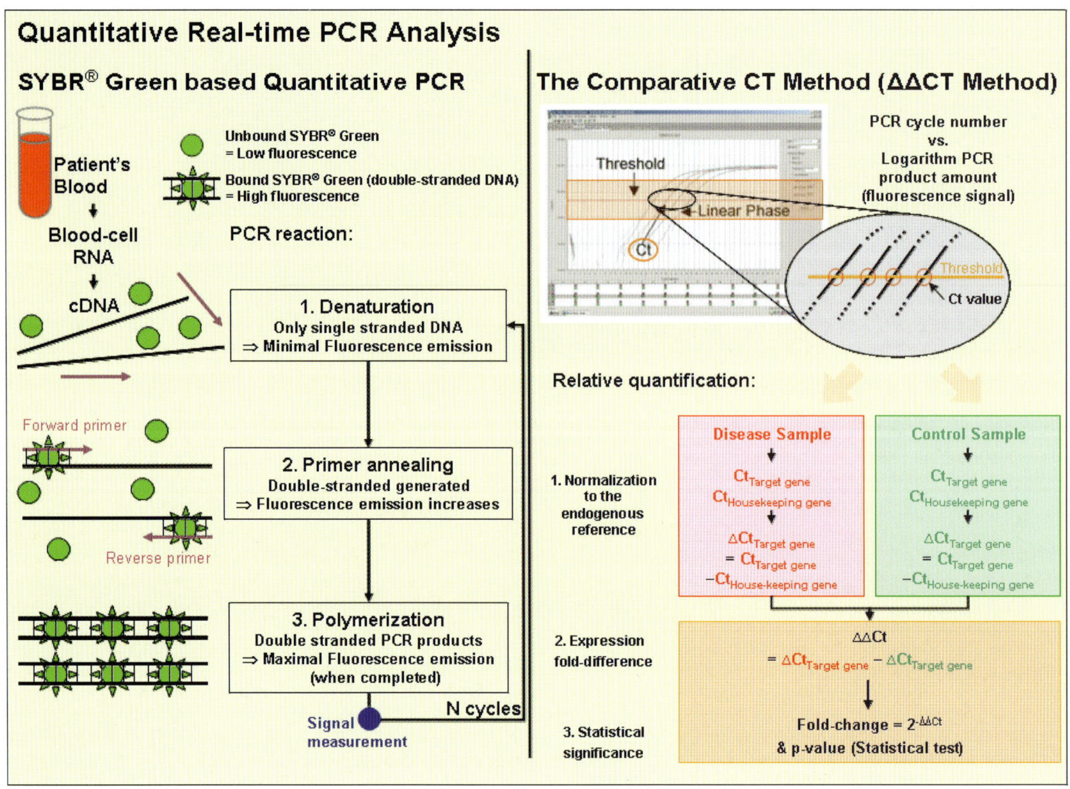

Plate 12.3 Disease gene validation: Real-time reverse transcriptase polymerase chain reaction (RT-PCR) relative quantification.

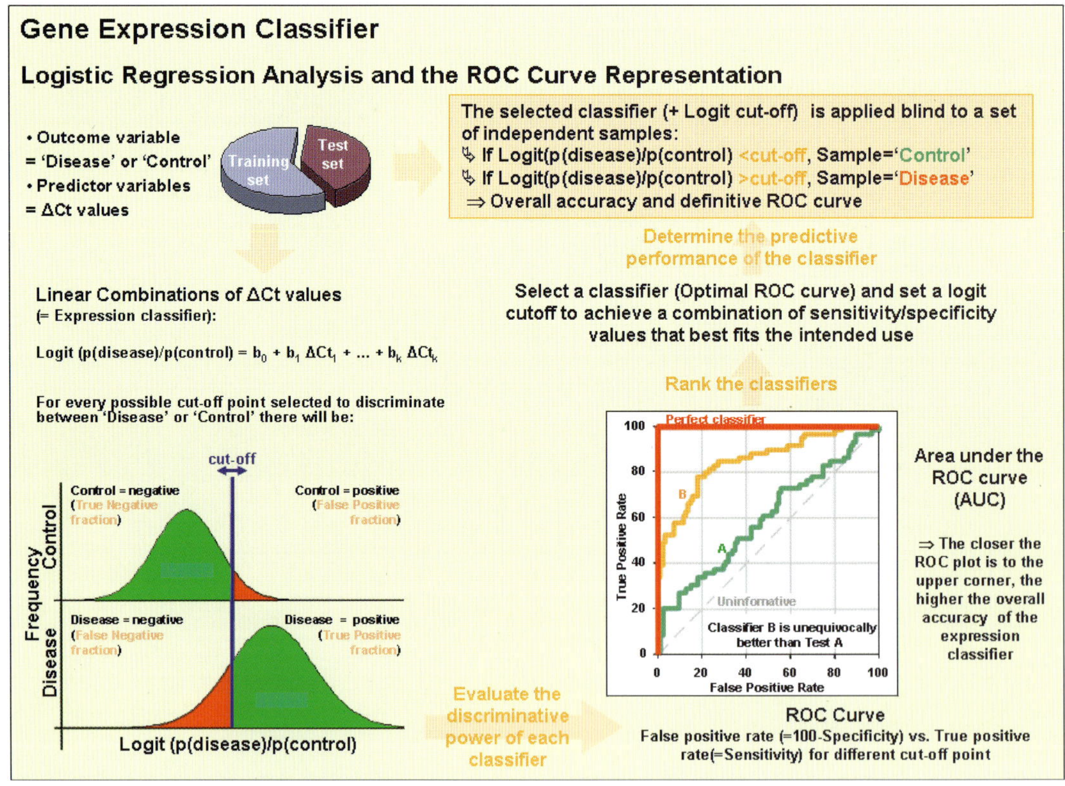

Plate 12.4 Disease gene application: Multigene expression classifier.

are delivered to acidic endosomes, where the low pH leads to release of the LDLR which recycles to the plasma membrane. The released lipoproteins traffic to lysosomes where the cholesteryl ester is hydrolyzed to free cholesterol and the protein moiety degraded. There is continual uptake and recycling of each LDLR with or without bound lipoproteins every 10–30 minutes with about 100 passages before degradation [10].

Familial hypercholesterolemia

Familial hypercholesterolemia (FH) was first brought to clinical attention almost 70 years ago when Müller described the familial clustering of a syndrome of cutaneous xanthomas, elevated cholesterol and premature CAD and proposed that this might be caused by a single gene defect. In the 1960s, Frederickson *et al.* [18] demonstrated impaired LDL metabolism in patients with FH and other investigators indicated a genetic link to a locus on chromosome 19. These studies eventually culminated in the discovery by Brown & Goldstein [5] that FH is the result of mutations in the LDL receptor gene.

Clinical diagnosis

Various paradigms for the diagnosis of heterozygous FH have been developed by different groups including the US MedPed Program, the UK Simon Broome Register Group and the Dutch lipid Clinic Network (reviewed in [19,20]). Because total cholesterol levels increase with age, the cutpoints for diagnosis of FH in individuals with a family history of FH in a first degree relative range from >220 mg/dL (5.7 mmol/L) for a family member <20 years of age to >290 mg/dL (7.5 mmol/L) for those >40 years of age [21]. DNA based evidence of a mutation in the LDLR or APOB gene, the presence of clinical stigmata of FH including tendon xanthomas, inferior corneal arcus at an early age and a family history of CAD in a first degree relative before the age of 50 years will facilitate the diagnosis.

In adults with heterozygous FH, plasma cholesterol levels are typically 9–11 mmol/L (350–430 mg/dL). Tendon xanthomas are rarely present until after 20 years of age. In untreated FH, CAD typically presents in the fifth decade in males and in the sixth decade in females. CAD onset can be much earlier in patients with other risk factors such as cigarette smoking or high plasma levels of lipoprotein (a), justifying early screening for both conventional and emerging risk factors [20].

Homozygous FH presents in childhood with cutaneous planar or tuberous xanthomas, tendon xanthomas and dense corneal arcus. Cholesterol concentrations are typically >600 mg/dL (15 mmol/L) and can be as high as 1000–1200 mg/dL (25–30 mmol/L). Atheroma of the aortic root and aortic valve develops by puberty with evidence for an aortic valvular gradient, angiographic narrowing of the aortic root and coronary osteal stenosis. Without treatment, sudden death or acute myocardial infarction typically occurs before the age of 30 years. The age of onset of CAD is dependent in part on the contributing mutations in the LDLR and degree of residual LDLR function (see below).

Prevalence

Heterozygous frequency of FH was typically estimated from the observed homozygous frequency assuming Hardy–Weinberg equilibrium. This approach may be flawed because genetic counseling has made couples who are both affected with heterozygous FH aware of the risk for their offspring. The prevalence of functional mutations in the LDLR gene causative of FH is estimated at approximately 1 in 500 in most North American and European populations but higher in certain groups such as South African Ashkenazi Jews (1 in 72), Lebanese Christians (1 in 85), Africaners (1 in 100), Tunisians (1 in 165) and French Canadians (1 in 270) as a result of founder effects (reviewed in [20]). A founder effect occurs when a subpopulation is formed through the immigration of a small number of individuals followed by population expansion. If certain of the founders had FH, these same, and limited number of mutations would be enriched in their descendents. For example, amongst French Canadians, 11 LDLR mutations account for 90% of FH, the most prevalent being a receptor negative mutation resulting from a 15-kb deletion [22,23].

In other populations, the prevalence of FH is apparently lower; for example, 1 in 950 in Denmark and 1 in 900 in Japan. Homozygous FH is proportionally rare, with a reported incidence of about 1 in 10^6 in North America.

Table 2.2 LDL receptor (LDLR) mutations in familial hypercholesterolemia.

Mutation	General location	Functional effects
Class 1	Disruptions of promoter sequence nonsense, frameshift or splicing mutations	No protein synthesis (null alleles)
Class 2	Primarily in the ligand-binding domain and EGF precursor regions	Disrupt transport of the LDLR from the ER to Golgi
Class 3	Ligand binding and EGF precursor regions	Interfere with cell surface binding of the LDLR to LDL
Class 4	Cytoplasmic domain or cytoplasmic and membrane spanning domains	Inhibit the clustering of the LDLR on the cell surface
Class 5	EGF precursor region	Prevent the release of LDL in endosome and thus impair recycling of the LDLR to the cell surface
(Class 6)	C-terminal end of the cytoplasmic tail towards the NPXY sequence (G823D)	Interferes with the proper sorting of the LDLR towards the basolateral membrane in polarized cells

EGF, epidermal growth factor; ER, endoplasmic reticulum.

Genetic variants

The LDLR gene maps to the short arm of chromosome 19 at 19p13.1–p13.3, spans 45 kb and has 18 exons. As of September 1, 2005 over 1000 LDLR variants have been identified in subjects with FH although not all have been proven to be functional. Gene variants are compiled online at two websites: http://www.ucl.ac.uk/fh/ [24] and www.umd.necker.fr/LDLR/research.html [25].

Functional mutations in the LDLR gene have been characterized according to their functional effects in human fibroblasts (Table 2.2; Plate 2.3) [26,27]. Null alleles and mutations in the ligand binding region (exons 2–6) demonstrate high penetrance and are prevalent in patients referred for DNA diagnosis. Exon 4 codes for repeat 5, which is required for both LDL binding via apoB and VLDL uptake via apoE and mutations in this region produce a particularly severe phenotype [28]. Approximately 5% of patients with FH have been identified to have various deletions or duplications in the LDLR gene, primarily in introns 1–8 and intron 12, associated with a high frequency of *Alu* sequences in these areas [29].

Treatment of heterozygous FH

Heterozygous FH responds well to strategies designed to upregulate the normal LDLR allele. The development of HMG-CoA reductase inhibitors (statins) was a direct result of the discovery that FH resulted from mutations in the LDLR gene and that expression of LDLR is regulated by cellular sterol via a sterol regulatory element (SRE) in its 5′ flanking sequence [4,30,31]. Statins decrease hepatic cholesterol synthesis, which leads to an increase in cell surface LDL receptors as well as decrease in VLDL secretion and hence reduced LDL production. Current recommendations advocate statin therapy in children with FH as young as 10 years and are supported by evidence of premature CAD in untreated patients [32]. Statin monotherapy can lower LDL-cholesterol by as much as 50% in FH. Combination therapy is required for many patients. Ezetimibe inhibits endogenous and dietary cholesterol absorption via an effect of the intestinal cholesterol transporter, NPCL1, and can reduce LDL-cholesterol by a further 20–25% [33,34]. Niacin is also a useful second or third agent, particularly in FH patients with low plasma levels of HDL-cholesterol or high plasma concentrations of lipoprotein (a).

Treatment of homozygous FH

Homozygous FH responds relatively poorly to statin and/or ezetimibe therapy although responses vary dependent upon the causative mutations and residual LDLR activity. Atorvastatin 80 mg/day reduced LDL-cholesterol by 18% in receptor negative and 41% in receptor defective patients [35]. Currently, the treatment of choice for homozygous FH is LDL apheresis, a process by which LDL particles and lipoprotein (a) are selectively removed from the body by extracorporeal binding to heparin (heparin extracorporeal LDL precipitation system, HELP), or dextran sulfate (dextran-sulfate cellulose absorption, DSA) or direct absorption of lipoproteins using hemoperfusion (DALI). LDL-cholesterol lowering efficacy ranges from 77% to 84%. Additional benefits include reduction of lipoprotein (a), various adhesion molecules, C-reactive protein and improved endothelial function [36,37]. LDL apheresis clearly elicits regression of xanthomas and attenuates atherosclerosis progression. Typically, LDL apheresis is performed at biweekly intervals with patients being maintained on maximally tolerated doses of lipid lowering agents. Where LDL apheresis is not available, plasmapheresis can be a useful alternative [38].

In children with homozygous FH, statin therapy should be introduced by the age of 1 year and titrated up to 1–2 mg/kg/day atorvastatin [35]. LDL apheresis is typically initiated at the age of 6 years or earlier [36,37]. To maintain adequate blood flow rates, creation of an arteriovenous fistula is normally required (reviewed in [39]).

Gene therapy would appear to be a logical approach in a single gene disorder. However, success has been hampered by the inability to achieve high level and long-term expression of the LDLR gene in liver and by the safety of viral vectors [40–42]. Liver transplantation to provide functional LDL receptors is the most definitive therapy currently available for homozygous FH but the surgical risks and need for lifelong immunosuppression have limited its popularity. However, improved surgical techniques and evidence of favorable outcomes of liver/cardiac transplants in adults [43] and living-donor liver transplants in children [44] suggest that this may become the intervention of choice in the future.

Familial defective apoB-100

A second relatively common cause of severe autosomal dominant hypercholesterolemia is familial defective apoB (FDB). This disorder was first identified in patients who presented similarly to FH, had reduced LDL apoB turnover rates by kinetic analyses but normal LDLR expression and function in isolated fibroblasts. FDB usually results from a missense mutation (Arg3500Gln) in the LDLR binding domain of apoB [45,46]. Other, less frequent mutations in apoB have been reported to cause FDB. The prevalence of FDB is approximately 1 in 1000 in individuals of Northern European descent and somewhat less in other populations. Patients with heterozygous FDB present with plasma cholesterol levels in the 275–350 mg/dL (7–9 mmol/L) range and hence somewhat lower than those typical of patients with FH. FDB homozygotes have cholesterol concentrations comparable to that of FH heterozygotes and a more benign course as compared to homozygous FH, with onset of clinical CAD commonly not until the fifth decade of life [47].

LDL from subjects with FDB have a 90% decrease in affinity for the LDLR and LDL clearance is markedly impaired. Kinetic studies indicate that LDL production is also decreased likely because of decreased VLDL secretion and increased clearance of apoE-rich VLDL remnants, mediated by both the LDLR and LDL receptor related protein (LRP) [48]. The genetic diagnosis of the Arg3500Gln variant is straightforward but other causative mutations have been described. Treatment is similar to that of FH with reliance on statin therapy, which both decreases VLDL production and enhances clearance of VLDL remnants. A second agent such as ezetimibe is frequently required for optimal lipid control [33].

PCKS9 nonsense mutations (FH3)

More recently, an additional cause of autosomal dominant FH, FH3 was discovered. Abifadel *et al.* [15] further mapped a region on chromosome 1 that had previously been linked to FH in a large Utah kindred, in 23 French families in whom LDLR or APOB sequence variants were excluded. In a region containing 41 genes, they identified PCKS9 (NARC-1) as a candidate gene and identified two missense mutations, S127R and F216L resulting in gain of function. In alter studies S127R and D374Y were shown to result in decreased cell surface expression of the LDLR [2]. In a later study, a third causative mutation, D374Y in the PCKS9 gene was

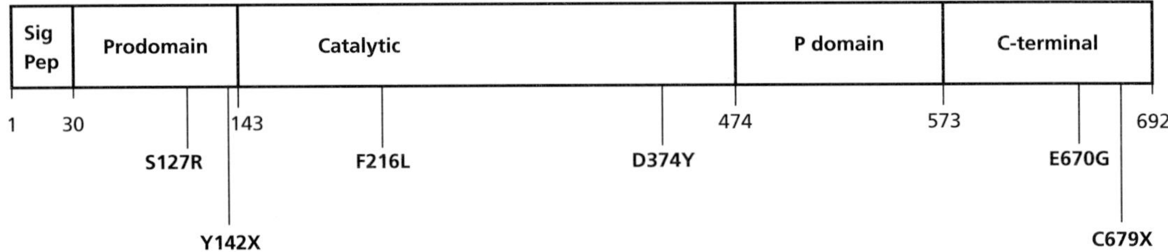

Figure 2.1 Schematic representation of PCKS9. The domain structure of PCKS9 includes a signal peptide of 30 amino acids (aa), a pro-peptide domain (aa 30–143), a catalytic domain (aa 143–474), containing the catalytic triad of aspartate (D), histidine (H) and serine (S) as well as a highly conserved arginine (N). This is followed by a P domain (aa 474–573) and the C-terminal domain (aa 573–692), which is cysteine rich. PCKS9 is synthesized as a zymogen and undergoes autocatalytic processing in the endoplasmic reticulum (ER). The cysteine-rich C-terminal domain is required for autoprocessing and a C-terminal deletion has been shown to impair the exit of PCKS9 from the ER [52]. The missense mutations (S127R, F216L, D374Y) are causative of severe hypercholesterolemia whereas the nonsense mutations (Y142X, C679X) have been associated with hypocholesterolemia. E670G has been associated with moderate increases in LDL-cholesterol. Adapted from [26] with permission from the *Annual Review of Genetics*, volume 24 © 1990 by Annual Reviews www.annualreviews.org.

identified in a Norwegian family [49] and in Utah kindreds [50]. Subjects with the S127R and F216L mutations displayed LDL-cholesterol levels which were two- to fivefold higher than age-matched controls, thus similar to FH [15]. Response to statin treatment was reportedly often inadequate [51].

PCKS9 (NARC-1) encodes neural apoptosis-regulated convertase 1 [52], a 691 amino acid protein belonging to the proteinase K family of subtilases, which is predominantly expressed in liver and intestine. The domain structure of PCKS9 (Fig. 2.1) includes a signal peptide of 30 amino acids (aa), a pro-peptide domain (aa 30–143), a catalytic domain (aa 143–474), containing the catalytic triad of aspartate (D), histidine (H) and serine (S) as well as a highly conserved arginine (N). This is followed by a P domain (aa 474–573) and the C-terminal domain (aa 573–692), which is cysteine rich. PCKS9 is synthesized as a zymogen and undergoes autocatalytic processing in the endoplasmic reticulum. The cysteine-rich C-terminal domain is required for autoprocessing and a C-terminal deletion has been shown to impair the exit of PCKS9 from the ER [52].

PCKS9 functions in the post translational regulation of LDLR processing, reducing LDLR number [3,14,53]. Adenoviral mediated overexpression of Pcks9 in mice [14,53,54] results in near complete depletion of LDL receptors whereas inactivation of the catalytic unit is without effect. Secreted PCSK9 decreases the number of LDL receptors in hepatocytes [4]. Other effects of autosomal dominant

PCKS9 mutations include decreased zymogen processing of PCKS9 and reduced LDLR density [54]. In kinetic studies, subjects with the S127R mutation were shown to have a 30% decrease in LDL apoB fractional catabolic rate, not dissimilar to that observed in subjects heterozygous for mutations in the LDLR gene. However, the most marked effect of S127R mutation was to increase VLDL apoB production rate by threefold and LDL apoB production by twofold [55], with evidence suggestive of direct synthesis of LDL by the liver. Secreted lipoproteins also had an abnormal composition with an elevated cholesteryl ester : triglyceride ratio, which would be expected to decrease their affinity for lipoprotein lipase. It has been speculated that PCKS9 may be involved in the post translational degradation of nascent VLDL [56]. Taken together, these studies suggest that PCKS9 normally functions to both increase VLDL and LDL production and attenuate cellular uptake of LDL. Mutations causative of an autosomal dominant form of hypercholesterolemia appear to result in "gain of function" [12,57,58,59].

Other sequence variants in PCKS9 have been shown to act in a non-Mendelian fashion. PCKS9 genetic variants have been linked to LDL-cholesterol levels in Japan [58] and in the TexGen study E670G in the C-terminal region was linked to plasma cholesterol concentrations [59]. Homozygotes for the rare allele (GG) had 20% higher plasma LDL-cholesterol compared with EE homozygotes. Conversely, other PCKS9 sequence variants

(Y142X and C679X) resulting in "loss of function" have recently been associated with low plasma concentrations of LDL-cholesterol in African-Americans, apparently accounting for this phenotype in 11% of subjects with LDL-cholesterol concentrations <58 mg/dL (1.5 mmol/L) [60]. Haplotype analyses suggested that each of these nonsense mutations arose from a single ancestor, indicating a founder effect. The mutation coding for Y142X, (426C to A) in exon 3 introduces a stop codon, which is predicted to delete the last 80% of the protein. A second nonsense mutation, (2037 C to A) introduces a stop codon at residue 679 (C679X) and is predicted to truncate the protein by 14 amino acids. Studies on the effect of inactivation of PckS9 on cellular processing of the LDLR and on LDL metabolism *in vivo* are required to clarify the relationship between these nonsense mutations and LDL cholesterol concentrations. Cohen *et al.* recently reported that three mutations in PCSK9 were associated with significant reductions in LDL-cholesterol concentrations and incident coronary heart disease (CHD) in the Atherosclerosis Risk in Communities (ARIC) study. Among African-American subjects, two nonsense mutations (Y142X and C679X) were associated with a 28% reduction in mean LDL-cholesterol concentrations and an 88% reduction in coronary heart disease (CHD). Among white subjects in the same study, 3.2% had a missense variant in PCSK9 (R46L); R46L carriers had a 15% reduction in LDL-cholesterol levels and a 47% reduction in the risk of CHD[5].

PCKS9 is regulated by sterols and expression is decreased by dietary cholesterol (Maxwell KN JLR 2003). Conversely, statins increase PCKS9 expression [61], an effect which may attenuate the LDL lowering effects of these agents [62]. Although a number of questions regarding the function of PCKS9 in the regulation of LDLR processing and in VLDL and LDL secretion remain unanswered, the identification of sequence variants resulting in both gain and loss of function have highlighted its unique and important role in LDL metabolism.

Autosomal recessive hypercholesterolemia

Khachadurian and Uthman [63] described a Lebanese family in 1973 including four members with severe hypercholesterolemia and tendon xanthomas typical of homozygous FH but with near-normal LDLR function in their cultured fibroblasts. A similar phenotype was later identified in a group of Sardinian patients, who were also found to have decreased LDL apoB fractional catabolic rate [64]. The disorder, termed autosomal recessive hypercholesterolemia (ARH), did not segregate with either the LDLR or ApoB genes and exhibited an autosomal recessive inheritance, with the parents of affected subjects having normal or slightly elevated LDL-cholesterol values.

ARH is brought about by mutations in ARH, a novel adaptor protein that functions in the internalization of LDLR and cargo. Ten mutations have been identified and all are predicted to introduce premature stop codons, either as a result of a point mutation or frameshift yielding no detectable ARH protein [65,66]. Interestingly, LDLR function is normal in ARH fibroblasts but is impaired in lymphocytes, macrophages and hepatocytes [67]. LDLR protein is present in these cells in normal amounts but is abnormally distributed with most of the receptor residing on the plasma membrane, where it binds LDL avidly but fails to internalize and degrade its cargo [68,69].

ARH contains a phosphotyrosine binding (PTB) domain similar to that found in several adaptor proteins including numb, Disabled-1 (Dab-1), Disabled-2 (Dab-2) and GULP. All of these bind directly to the acidic phospholipids phosphatidylinositol 4,5-biphosphate $\{PtdIns(4,5)P_2\}$ as well as to NPXY motifs in the cytoplasmic domain of signaling receptors and have a role in endocytosis or cell signaling. Of interest, Dab-2 colocalizes with the LDLR on the plasma membrane and binds to the FDNPVY internalization sequence of the LDLR [70], raising the possibility that Dab-2 may compensate for ARH deficiency and enable normal LDL endocytosis in ARH fibroblasts. The PTB domain of ARH-1 binds the NPXY sequence in the LDLR [70,71]. The distal segment of ARH also contains a canonical clathrin binding sequence, LLDLE, which permits direct interaction with the terminal domain of clathrin. Finally, ARH contains a sequence recognized by the β2 adaptin subunit of AP-2, a major structural component of clathrin coated pits. Thus, ARH-1 functions as an adaptor protein to link the LDLR to endocytic machinery by simultaneously binding to the NPxY sequence of the receptor cytoplasmic tail, clathrin and the AP-2 adaptor [72]. Retroviral expression of normal ARH in the immortalized lymphocytes of affected subjects was shown to restore LDLR internalization [67].

Clinical phenotype of ARH

ARH is apparently uncommon, with the largest series of affected subjects identified in Sardinia and Italy. There is notably marked heterogeneity in disease phenotype with total cholesterol levels varying between 350 and 1050 mg/dL (9–27 mmol/L) with near-normal triglyceride concentrations (reviewed in [65,66]). Large bulky tuberous, tendinous and/or planar xanthomas are typically present from early childhood. The pathophysiology of accelerated xanthoma formation in ARH is not clear but may represent accelerated uptake of modified LDL by scavenger receptors. Atherosclerosis and CAD appear in early adulthood but aortic stenosis is a somewhat less prominent feature compared with patients with homozygous FH.

Patients with ARH generally respond well to lipid-lowering medication in contrast to patients with homozygous FH [17]. This may be related to preserved LDLR activity in ARH fibroblasts. Combination therapy is most effective and decreases in total cholesterol of 70–75% have been reported with 80 mg atorvastatin in combination with 16 mg cholestyramine [65]. LDL apheresis remains an option for patients with an inadequate response to pharmacologic therapy.

The discovery of ARH has provided important insight into the LDLR endocytic pathway. This remains an intriguing disorder in part because of the cell specificity of its effects on LDLR function. ARH is also unusual in that there is significant heterogeneity in terms of severity of the clinical phenotype. This rare genetic disorder demonstrates that our current understanding of the clathrin mediated LDLR endocytic pathway is as yet incomplete.

Sitosterolemia

Noncholesterol sterols such as sitosterol are normally present in trace amounts in normal individuals. Sitosterol, like cholesterol, is taken up into enterocytes in the proximal intestine. Although 20–80% of dietary cholesterol is incorporated into chlylomicrons, <5% sitosterol is normally absorbed and that which is absorbed is rapidly resecreted into bile. Patients with sitosterolemia have markedly increased intestinal absorption of sitosterol and impaired resecretion of this sterol into bile [73]. Sitosterolemia is caused by over 25 mutations in either ABCG5 or ABCG8. These are ABC half-transporters and are encoded by two adjacent genes. Affected subjects have at least two mutations in either of ABCG5 or ABCG8 [74–77].

Clinical presentation

Patients with sitosterolemia typically have 50-fold elevations in plasma and tissue concentrations of plant and shellfish sterols [73,78,79]. Although variable, LDL-cholesterol concentrations are significantly elevated especially in childhood and can be similar to those of FH (~300 mg/dL or 8 mmol/L) [80]. Planar xanthomas are frequent and untreated patients develop aortic stenosis and premature CAD. The plasma cholesterol level in sitosterolemia is very responsive to dietary cholesterol intake and to bile acid sequestrant therapy, which increases the resecretion of cholesterol into bile in the form of bile acids. Ezetimibe, which inhibits the intestinal cholesterol transporter NPCL1, is particularly effective [33,81].

Conclusions

Although rare, the various types of monogenic hypercholesterolemia discussed above are important and treatable causes of premature cardiovascular disease. Their study has also contributed significantly to our knowledge of lipoprotein metabolism and atherosclerosis and have provided important insights into intracellular protein trafficking and cellular cholesterol metabolism [3–5]. The early studies of FH patients led to identification of the LDLR pathway and ultimately the development of statins, one of the most important clinical advances for the primary and secondary prevention of cardiovascular disease. More recent studies of the etiology of hypercholesterolemia in patients with mutations in PCKS9 and ARH have highlighted the complexity of the LDLR secretory and endocytic pathways.

References

1 Dubuc G, Chamberland A, Wassef H. Davignon J, Seidah NG, Bernier L, Prat A. Statins upregulate PCSK 9, the gene encoding the proprotein convertase neural apoptosis-regulated convertase-1 implicated in familial hypercholesterolemia. *Arterioscler Thromb Vasc Biol* 2004; 24: 1454–1459.

2 Cameron J, Holla OL, Ranheim T, Kulseth MA, Berge KE, Leren TP. Effect of mutations in the PCSK 9 gene on the cell surface LDL receptors. *Hum Mol Genet* 2006; 15: 1551–1558.

3 Maxwell KN, Breslow JL. Adenoviral-mediated expression of PCSK 9 in mice results in a low-density lipprotein receptor knockout phenotype. *Proc Natl Acad Sci USA* 2004; 101: 7100–7105.

4 Lagace TA, Curtis De, Garuti R, McNutt MC, Park SW, Prather HB, Anderson NN, Ho YK, Hammer RE, Horton JD. Secreted PCSK 9 decreases the number of LDL receptors in hepatocytes and in livers of parabiotic mice. J Clin Invest 2006; 116: 2995–3005.

5 Cohen JC, Boerwinkle E, Mosely TH, Hobbs HH. Sequence variations in PCSK 9, low LDL, and protection against coronary heart disease. *N Engl J Med* 2006; 354: 1265–1272.

6 Greaves DR, Gordon S. Thematic review series: the immune system and atherogenesis. Recent insights into the biology of macrophage scavenger receptors. *J Lipid Res* 2005; 46: 11–20.

7 Russell DW. Protein domains of the low density lipoprotein receptor. *Acta Med Scand Suppl* 1987; 715: 39–44.

8 Davis CG, van Driel IR, Russell DW, Brown MS, Goldstein JL. The low density lipoprotein receptor. Identification of amino acids in cytoplasmic domain required for rapid endocytosis. *J Biol Chem* 1987; 262: 4075–4082.

9 van Driel IR, Davis CG, Goldstein JL, Brown MS. Self-association of the low density lipoprotein receptor mediated by the cytoplasmic domain. *J Biol Chem* 1987; 262: 16127–16134.

10 Beglova N, Blacklow SC. The LDL receptor: how acid pulls the trigger. *Trends Biochem Sci* 2005; 30: 309–317.

11 Gent J, Braakman I. Low-density lipoprotein receptor structure and folding. *Cell Mol Life Sci* 2004; 61: 2461–2470.

12 Attie AD, Seidah NG. Dual regulation of the LDL receptor; some clarity and new questions. *Cell Metab* 2005; 1: 290–292.

13 Maxwell KN, Fisher EA, Breslow JL. Overexpression of PCSK9 accelerates the degradation of the LDLR in a post-endoplasmic reticulum compartment. *Proc Natl Acad Sci USA* 2005; 102: 2069–2074.

14 Park SW, Moon YA, Horton JD. Post-transcriptional regulation of low density lipoprotein receptor protein by proprotein convertase subtilisin/kexin type 9a in mouse liver. *J Biol Chem* 2004; 279: 50630–50638.

15 Abifadel M, Varret M, Rabes JP *et al.* Mutations in PCSK9 cause autosomal dominant hypercholesterolemia. *Nat Genet* 2003; 34: 154–156.

16 He G, Gupta S, Yi M, Michaely P, Hobbs HH, Cohen JC. ARH is a modular adaptor protein that interacts with the LDL receptor, clathrin, and AP-2. *J Biol Chem* 2002; 277: 44044–44049.

17 Arca M, Zuliani G, Wilund K *et al.* Autosomal recessive hypercholesterolaemia in Sardinia, Italy, and mutations in ARH: a clinical and molecular genetic analysis. *Lancet* 2002; 59: 41–47.

18 Fredrickson DS, Levy RI, Lees RS. Fat transport in lipoproteins: an integrated approach to mechanisms and disorders. *N Engl J Med* 1967; 276: 273–281.

19 Austin MA, Hutter CM, Zimmern RL, Humphries SE. Familial hypercholesterolemia and coronary heart disease: a HuGE association review. *Am J Epidemiol* 2004; 160: 421–429.

20 Austin MA, Hutter CM, Zimmern RL, Humphries SE. Genetic causes of monogenic heterozygous familial hypercholesterolemia: a HuGE prevalence review. *Am J Epidemiol* 2004; 160: 407–420.

21 Williams RR, Hunt SC, Schumacher MC *et al.* 1993. Diagnosing heterozygous familial hypercholesterolemia using new practical criteria validated by molecular genetics. *Am J Cardiol* 72: 171–176.

22 Couture P, Morissette J, Gaudet D *et al.* Fine mapping of low-density lipoprotein receptor gene by genetic linkage on chromosome 19p13.1–p13.3 and study of the founder effect of four French Canadian low-density lipoprotein receptor gene mutations. *Atherosclerosis* 1999; 143: 145–151.

23 Leitersdorf E, Tobin EJ, Davignon J, Hobbs HH. Common low-density lipoprotein receptor mutations in the French Canadian population. *J Clin Invest* 1990; 85: 1014–1023.

24 Heath KE, Gahan M, Whittall RA, Humphries SE. Low-density lipoprotein receptor gene (LDLR) world-wide website in familial hypercholesterolaemia: update, new features and mutation analysis. *Atherosclerosis* 2001; 154: 243–246.

25 Villeger L, Abifadel M, Allard D *et al.* The UMD-LDLR database: additions to the software and 490 new entries to the database. *Hum Mutat* 2002; 20: 81–87.

26 Hobbs HH. Russell DW, Brown MS, Goldstein JL. The LDL receptor locus in familial hypercholesterolemia: Mutational analysis of a membrane protein. *Annu Rev Genet* 1990; 24: 133–170.

27 Hobbs HH, Brown MS, Goldstein JL. Molecular genetics of the LDL receptor gene in familial hypercholesterolemia. *Hum Mutat* 1992; 1: 445–466.

28 Gudnason V, Day INM, Humphries SE. Effect on plasma lipid levels of different classes of mutations in the low-density lipoprotein receptor gene in patients with familial hypercholesterolemia. *Arterioscler Thromb* 1994; 14: 1717–1722.

29 Yamamoto T, Davis CG, Brown MS *et al.* The human LDL receptor: a cysteine-rich protein with multiple Alu sequences in its mRNA. *Cell* 1984; 39: 27–38.

30 Goldstein JL, Brown MS. Regulation of the mevalonate pathway. *Nature* 1990; 343: 425–430.

31 Goldstein JL, Hobbs HH, Brown MS. Familial hypercholesterolemia. In: Scriver CR, Beaudet AL, Sly WS, Valle D, eds. *The Metabolic and Molecular Basis of Inherited Disease.* McGraw-Hill, Inc., New York, 1995: 1981–2030.

32 Gotto AM. Efficacy and safety of statin therapy in children with familial hypercholesterolemia: a randomized controlled trial. *J Pediatr* 2005; 146: 144–145.

33 Patel SB. Ezetimibe: a novel cholesterol-lowering agent that highlights novel physiologic pathways. *Curr Cardiol Rep* 2004; 6: 439–442.

34 Yamamoto A, Harada-Shiba M, Endo M *et al.* The effect of ezetimibe on serum lipids and lipoproteins in patients with homozygous familial hypercholesterolemia undergoing LDL-apheresis therapy. *Atherosclerosis* 2006; 186: 126–131.

35 Marais AD, Firth JC, Blom DJ. Homozygous familial hypercholesterolemia and its management. *Semin Vasc Med* 2004; 4: 43–50.

36 Pokrovsky S, Straube R, Afanasieva O, Kukharchuk V, Konovalov G. Lp(a) apheresis for the treatment of severe CHD patients with Lp(a) hyperlipidemia. *Ther Apher Dial* 2005; 9: A40.

37 Hershcovici T, Schechner V, Orlin J, Harell D, Beigel Y. Effect of different LDL-apheresis methods on parameters involved in atherosclerosis. *J Clin Apher* 2004; 19: 90–97.

38 Thompson GR, Miller JP, Breslow JL. Improved survival of patients with homozygous familial hypercholesterolaemia treated with plasma exchange. *Br Med J (Clin Res Ed)* 1985; 291: 1671–1673.

39 Thompson GR. LDL apheresis. *Atherosclerosis* 2003; 167: 1–13.

40 Grossman M, Wilson JM. Frontiers in gene therapy: LDL receptor replacement for hypercholesterolemia. *J Lab Clin Med* 1992; 119: 457–460.

41 Rader DJ, Cohen J, Hobbs HH. Monogenic hypercholesterolemia: new insights in pathogenesis and treatment. *J Clin Invest* 2003; 111: 1795–1803.

42 Grossman M, Rader DJ, Muller DW *et al.* A pilot study of *ex vivo* gene therapy for homozygous familial hypercholesterolaemia. *Nat Med* 1995; 1: 1148–1154.

43 Alkofer BJ, Chiche L, Khayat A *et al.* Liver transplant combined with heart transplant in severe heterozygous hypercholesterolemia: report of the first case and review of the literature. *Transplant Proc* 2005; 37: 2250–2252.

44 Shirahata Y, Ohkohchi N, Kawagishi N *et al.* Living-donor liver transplantation for homozygous familial hypercholesterolemia from a donor with heterozygous hypercholesterolemia. *Transpl Int* 2003; 16: 276–279.

45 Boren J, Ekstrom U, Agren B, Nilsson-Ehle P, Innerarity TL. The molecular mechanism for the genetic disorder familial defective apolipoprotein B100. *J Biol Chem* 2001; 276: 9214–9218.

46 Boren J, Lee I, Zhu W, Arnold K, Taylor S, Innerarity TL. Identification of the low density lipoprotein receptor-binding site in apolipoprotein B100 and the modulation of its binding activity by the carboxyl terminus in familial defective apo-B100. *J Clin Invest* 1998; 101: 1084–1093.

47 Myant NB. Familial defective apolipoprotein B-100: A review, including some comparisons with familial hypercholesterolaemia. *Atherosclerosis* 1993; 104: 1–18.

48 Schaefer JR, Scharnagl H, Baumstark MW *et al.* Homozygous familial defective apolipoprotein B-100: Enhanced removal of apolipoprotein E-containing VLDLs and decreased production of LDLs. *Arterioscler Thromb Vasc Biol* 1997; 17: 348–353.

49 Leren TP. Mutations in the PCSK9 gene in Norwegian subjects with autosomal dominant hypercholesterolemia. *Clin Genet* 2004; 65: 419–422.

50 Timms KM, Wagner S, Samuels ME *et al.* A mutation in PCSK9 causing autosomal-dominant hypercholesterolemia in a Utah pedigree. *Hum Genet* 2004; 114: 349–353.

51 Naoumova RP, Tosi I, Patel D *et al.* Severe hypercholesterolemia in four British families with the D374Y mutation in the PCSK9 gene. Long-term follow-up and treatment response. *Arterioscler Thromb Vasc Biol* 2005; 25: 2654–2660.

52 Seidah NG, Benjannet S, Wickham L *et al.* The secretory proprotein convertase neural apoptosis-regulated convertase 1 (NARC-1): liver regeneration and neuronal differentiation. *Proc Natl Acad Sci USA* 2003; 100: 928–933.

53 Maxwell KN, Breslow JL. Adenoviral-mediated expression of Pcsk9 in mice results in a low-density lipoprotein receptor knockout phenotype. *Proc Natl Acad Sci USA* 2004; 101: 7100–7105.

54 Benjannet S, Rhainds D, Essalmani R *et al.* NARC-1/PCSK9 and its natural mutants: zymogen cleavage and effects on the low density lipoprotein (LDL) receptor and LDL cholesterol. *J Biol Chem* 2004; 279: 48865–48875.

55 Ouguerram K, Chetiveaux M, Zair Y *et al.* Apolipoprotein B100 metabolism in autosomal-dominant hypercholesterolemia related to mutations in PCSK9. *Arterioscler Thromb Vasc Biol* 2004; 24: 1448–1453.

56 Jirholt P, Adiels M, Boren J. How does mutant proprotein convertase neural apoptosis-regulated convertase 1 induce autosomal dominant hypercholesterolemia? *Arterioscler Thromb Vasc Biol* 2004; 24: 1334–1336.

57 Attie AD. The mystery of PCSK9. *Arterioscler Thromb Vasc Biol* 2004; 24: 1337–1339.

58 Shioji K, Mannami T, Kokubo Y *et al.* Genetic variants in PCSK9 affect the cholesterol level in Japanese. *J Hum Genet* 2004; 49: 109–114.

59 Chen SN, Ballantyne CM, Gotto AM Jr, Tan Y, Willerson JT, Marian AJ. A common PCSK9 haplotype, encompassing the E670G coding single nucleotide polymorphism, is a novel genetic marker for plasma low-density lipoprotein cholesterol levels and severity of coronary atherosclerosis. *J Am Coll Cardiol* 2005; 45: 1611–1619.

60 Cohen J, Pertsemlidis A, Kotowski IK, Graham R, Garcia CK, Hobbs HH. Low LDL cholesterol in individuals of African descent resulting from frequent nonsense mutations in PCSK9. *Nat Genet* 2005; 37: 161–165.

61 Dubuc G, Chamberland A, Wassef H *et al.* Statins upregulate PCSK9, the gene encoding the proprotein convertase neural apoptosis-regulated convertase-1 implicated in familial hypercholesterolemia. *Arterioscler Thromb Vasc Biol* 2004; 24: 1454–1459.

62 Rashid S, Curtis DE, Garuti R *et al.* Decreased plasma cholesterol and hypersensitivity to statins in mice lacking Pcsk9. *Proc Natl Acad Sci USA* 2005; 102: 5374–5379.

63 Khachadurian AK, Uthman SM. Experiences with the homozygous cases of familial hypercholesterolemia. A report of 52 patients. *Nutr Metab* 1973; 15: 132–140.

64 Zuliani G, Vigna GB, Corsini A, Maioli M, Romagnoni F, Fellin R. Severe hypercholesterolaemia: unusual inheritance in an Italian pedigree. *Eur J Clin Invest* 1995; 25: 322–331.

65 Soutar AK, Naoumova RP, Traub LM. Genetics, clinical phenotype, and molecular cell biology of autosomal recessive hypercholesterolemia. *Arterioscler Thromb Vasc Biol* 2003; 23: 1963–1970.

66 Soutar AK, Naoumova RP. Autosomal recessive hypercholesterolemia. *Semin Vasc Med* 2004; 4: 241–248.

67 Eden ER, Patel DD, Sun XM *et al.* Restoration of LDL receptor function in cells from patients with autosomal recessive hypercholesterolemia by retroviral expression of ARH1. *J Clin Invest* 2002; 110: 1695–1702.

68 Garcia CK, Wilund K, Arca M *et al.* Autosomal recessive hypercholesterolemia caused by mutations in a putative LDL receptor adaptor protein. *Science* 2001; 292: 1394–1398.

69 Jones C, Hammer RE, Li WP, Cohen JC, Hobbs HH, Herz J. Normal sorting but defective endocytosis of the low density lipoprotein receptor in mice with autosomal recessive hypercholesterolemia. *J Biol Chem* 2003; 278: 29024–29030.

70 Mishra SK, Watkins SC, Traub LM. The autosomal recessive hypercholesterolemia (ARH) protein interfaces directly with the clathrin-coat machinery. *Proc Natl Acad Sci USA* 2002; 99: 16099–16104.

71 Michaely P, Li WP, Anderson RG, Cohen JC, Hobbs HH. The modular adaptor protein ARH is required for low density lipoprotein (LDL) binding and internalization but not for LDL receptor clustering in coated pits. *J Biol Chem* 2004; 279: 34023–34031.

72 Wilund KR, Yi M, Campagna F *et al.* Molecular mechanisms of autosomal recessive hypercholesterolemia. *Hum Mol Genet* 2002; 11: 3019–3030.

73 Salen G, Shefer S, Nguyen L, Ness GC, Tint GS, Shore V. Sitosterolemia. *J Lipid Res* 1992; 33: 945–955.

74 Graf GA, Cohen JC, Hobbs HH. Missense mutations in ABCG5 and ABCG8 disrupt heterodimerization and trafficking. *J Biol Chem* 2004; 279: 24881–24888.

75 Heimer S, Langmann T, Moehle C *et al.* Mutations in the human ATP-binding cassette transporters ABCG5 and ABCG8 in sitosterolemia. *Hum Mutat* 2002; 20: 151.

76 Lee MH, Lu K, Hazard S *et al.* Identification of a gene, ABCG5, important in the regulation of dietary cholesterol absorption. *Nat Genet* 2001; 27: 79–83.

77 Patel SB, Salen G, Hidaka H *et al.* Mapping a gene involved in regulating dietary cholesterol absorption. The sitosterolemia locus is found at chromosome 2p21. *J Clin Invest* 1998; 102: 1041–1044.

78 Hidaka H, Nakamura T, Aoki T *et al.* Increased plasma plant sterol levels in heterozygotes with sitosterolemia and xanthomatosis. *J Lipid Res* 1990; 31: 881–888.

79 Bhattacharyya AK, Connor WE. Beta-sitosterolemia and xanthomatosis. A newly described lipid storage disease in two sisters. *J Clin Invest* 1974; 53: 1033–1043.

80 Wang J, Joy T, Mymin D, Frohlich J, Hegele RA. Phenotypic heterogeneity of sitosterolemia. *J Lipid Res* 2004; 45: 2361–2367.

81 Salen G, Von Bergmann K, Lutjohann D *et al.* Ezetimibe effectively reduces plasma plant sterols in patients with sitosterolemia. *Circulation* 2004; 109: 966–971.

CHAPTER 3

Hypertrophic cardiomyopathy

Ali J. Marian, MD

Introduction

Hypertrophic cardiomyopathy (HCM) is a primary disease of the myocardium with unique and fascinating clinical and pathologic manifestations [1]. HCM was first described by French pathologist Liouville [2] in 1869 as cardiac contraction below the aortic valve. In 1907, German pathologist Schmincke described the condition in two adult women as diffuse muscular "hyperplasia" at the left ventricular outflow tract [3]. Schmincke made the astute suggestion that outflow tract hypertrophy obstructs left ventricular ejection, which leads to further cardiac hypertrophy and hence, a vicious cycle. The disease as a clinical entity, however, remained largely unrecognized until the second half of the twentieth century, when detailed pathologic phenotypes, familial inheritance and the phenotype of sudden cardiac death (SCD) were described [4–6]. In the last 50 years, advances in modern diagnostic tools led to characterization of specific cardiac phenotypes and emphasized the heterogeneous nature of the disease. In the 1960s, HCM was described primarily as a hemodynamic entity characterized by outflow tract obstruction. Braunwald coined the term "idiopathic hypertrophic subaortic stenosis" (IHSS) to describe the phenotype [7]. Subsequently, in 1964 Braunwald *et al.* [8,9] published the most comprehensive report up to then on the clinical, hemodynamic and angiographic aspects of HCM in 64 patients. In the same year, Morrow *et al.* [10] described septal myectomy through a transaortic approach as an effective method for reduction of left ventricular outflow tract gradient in 10 of the above 64 patients. Collectively, these reports established HCM as a hemodynamic entity and surgery as the treatment of choice.

The initial reports on the utility of echocardiography in the diagnosis of HCM were published in late 1960s and were soon followed by echocardiographic findings of asymmetric septal hypertrophy [11,12]. The term asymmetric septal hypertrophy (ASH) was frequently used to emphasize the predominant involvement of the interventricular septum in patients with HCM [13,14]. Advances in Doppler echocardiography in the late 1970s and early 1980s led to a better understanding of the physiology of the left ventricular flow dynamics [15,16]. Accordingly, Doppler echocardiography became the desirable tool for assessment and quantification of left ventricular outflow tract obstruction, supplanting the routine use of cardiac catheterization [17,18]. Doppler echocardiography also led to a better appreciation of diastolic dysfunction in HCM [18].

The discovery of the first causal gene and mutation for HCM by Geisterfer-Lowrance *et al.* [19] in 1990 heralded the dawning of the molecular era. Since then, over a dozen genes and several hundred mutations have been identified and the genetic basis of HCM is all but delineated [20]. In 1995, Sigwart [21] described a new catheter-based intervention for the treatment of outflow tract obstruction in HCM. The procedure, accomplished through injection of ethanol into the main septal branches of left anterior descending coronary artery, has emerged as an effective method for reduction of the outflow tract gradient and has raised considerable debate on the choice of method for alleviation of outflow tract obstruction in HCM [22,23]. In addition, implantable defibrillators have emerged as effective tools in preventing SCD in high-risk patients [24]. Despite these advances, the pharmacologic treatment of HCM has remained largely unchanged during the past three decades. Conventional pharmacologic therapies are limited to the use of beta-blockers, calcium-channel blockers, disopyramide and amiodarone, the latter in a subset of patients with serious cardiac arrhythmias. Recent experimental studies have raised the potential utility of statins and inhibitors of renin–angiotensin–aldosterone system (RAAS) in prevention and regression of cardiac phenotype HCM [25–27]. However, the potential utility of these novel therapeutic and preventive interventions in humans has to await large-scale clinical studies.

The current research emphasis is to decipher the molecular pathogenesis of the HCM phenotype, to develop methods for genetic screening and diagnosis, to develop accurate risk stratification based on genetic and clinical tools, and to identify novel drug targets in order to develop pharmacologic intervention targeted to molecules involved in the pathogenesis of HCM. It is hoped that within the next few years, clinicians will be able to utilize and apply the recent advances to the care of patients with HCM. The ultimate goal of correcting the genetic defect has to await the development of specific strategies whereby a single base pair could be changed.

Definition

HCM is a primary disease of the myocardium characterized by cardiac hypertrophy in the absence of an increased external load and a hyperdynamic left ventricle with a small chamber. The above definition, based on the clinical detection of "primary" cardiac hypertrophy, is neither specific nor sufficiently sensitive for the accurate diagnosis of HCM. For example, by strict definition, the presence of systemic hypertension excludes the diagnosis of HCM. However, systemic hypertension is a common disease and hence could be present concomitantly in patients with HCM and even contribute to the phenotypic expression of HCM. Thus, the presence of hypertension, or similarly other overload conditions alone, is not sufficient to exclude HCM. In such situations, the extent, severity and the type of cardiac hypertrophy as well as other features, such as a hyperdynamic left ventricle, a small left ventricular cavity and the presence of outflow tract obstruction should be considered.

Similarly, the term "unexplained cardiac hypertrophy" is often used to define HCM. However, "unexplained cardiac hypertrophy" could also occur in phenocopy conditions, defined as a clinical phenotype grossly similar to HCM. Examples of HCM phenocopy include storage diseases, mitochondrial diseases, triplet repeat syndromes and many others [28]. It is often difficult to distinguish clinically between the phenocopy and true HCM, as gross phenotype is quite similar [29]. Features such as hyperdynamic left ventricle and small left ventricular chamber size would suggest true HCM. In contrast, the presence of a depressed global cardiac systolic function, conduction defects or involvement of other organs, such as deafness, neurologic abnormalities and skeletal myopathy would suggest a phenocopy. Endomyocardial biopsy and histologic examination is likely to provide a diagnosis that is more robust. Myocyte disarray is considered the pathologic hallmark of true HCM and is expected to be absent in HCM phenocopy. In contrast, special staining of the myocardial sections would help in the diagnosis of storage diseases. Finally, the pathogenesis of true HCM and HCM phenocopy are likely to differ considerably, as shown in HCM phenocopy caused by mutations in the γ2 subunit of adenosine monophosphate kinase [30–32]. The latter is characterized by cardiac hypertrophy because of deposition of glycogen in the heart, conduction defect and pre-excitation pattern on electrocardiogram (ECG) [33]. The distinction between true HCM and phenocopy is likely to gain clinical significance with the development of specific genetic and molecular treatment for HCM and its phenocopy.

Thus, clinical diagnosis, which is primarily based on the echocardiographic finding of cardiac hypertrophy, has considerable shortcomings. It is expected that genetic-based diagnosis will supplant clinical diagnosis of HCM upon further advances in molecular genetic techniques. An important value of genetic-based diagnosis is in the preclinical diagnosis of mutation careers, which could have considerable implications, not only for an early diagnosis and risk stratification, but also for the implementation of preventive measures. Nonetheless, clinicians should bear in mind that genetic-based diagnosis, while robust in identification of those with causal mutations, does not accurately predict the severity of clinical phenotypes. The management of patients should take into consideration not only the causal mutations and modifier genetic factors, but also the phenotypic expression of disease.

Prevalence

HCM is a relatively common disease with an estimated prevalence of approximately 1 in 500 in the general population [34,35]. The estimate is based on the presence of left ventricular wall thickness of 15 mm or greater on an echocardiogram in 23–35-year-old individuals [34]. The estimate is considered relatively conservative as many cases with HCM have a milder degree of cardiac hypertrophy and many mutations are nonpenetrant in the above age group. Accordingly, one would expect a higher prevalence if those with milder degrees of cardiac hypertrophy and an older population are included in the estimate. In addition, those with hypertension are commonly excluded in estimating the prevalence of HCM, which could lead to underestimation of the true prevalence. In contrast, inclusion of those with HCM phenocopy could lead to overestimation. Thus, determination of the true prevalence of HCM will require large-scale molecular epidemiologic studies.

Phenotypic manifestations

Patients with HCM exhibit a diverse array of clinical, morphologic and pathologic phenotypes. A significant number of patients with the clinical diagnosis of HCM are asymptomatic or minimally symptomatic. The most common symptoms are dyspnea, chest pain, palpitations, dizziness and lightheadedness. Syncope is an infrequent but a serious symptom that merits full investigation. It is often associated with serious cardiac arrhythmias and heralds SCD [36–38]. Atrial fibrillation and nonsustained ventricular tachycardia are the most common cardiac arrhythmias and are associated with adverse clinical outcome [39,40]. Electrocardiographic findings of Wolff–Parkinson–White (WPW) syndrome are present in a small percentage of patients with HCM and their presence raises the possibility of a phenocopy [30–33].

SCD is often the first manifestation of HCM in apparently young healthy individuals [41,42].

HCM is the most common cause of SCD in young competitive athletes, accounting for almost half of all cases of SCD in athletes younger than 35 years of age in the USA [41,42]. The risk of SCD is greater during or immediately after exercise. While there is no reliable predictor, several clinical, pathologic and genetic factors have been identified as potential predictors of the risk of SCD in patients with HCM (Table 3.1). Overall, the positive predictive value of each marker is relatively low. Therefore, the global risk, derived from the combination of all known risk factors, should be considered in counseling, risk stratification and management of patients with HCM [38,43]. In the absence of two or more risk factors for SCD, HCM is considered a relatively benign disease. The estimated annual mortality of patients with HCM is about 1% in the adult population [44–46].

Cardiac hypertrophy is the quintessential phenotype of HCM and the basis for the clinical diagnosis. Cardiac hypertrophy is concentric but commonly asymmetric, predominantly involving the interventricular septum. In approximately one-third of cases, cardiac hypertrophy is symmetric or shows an atypical feature, such as the predominant involvement of the apex, lateral wall or the posterior wall. HCM has been classified into four categories according to the site and the extent of involvement of the left ventricular walls [47]. In apical HCM, hypertrophy is localized to the apex of the heart, which is an uncommon form, with the unique feature of giant T-wave inversion in the precordial leads on the ECG [48]. It has a relatively benign prognosis, with a 15-year survival rate of approximately 95% in the North American population [49]. Because of concentric nature of hypertrophy, the left ventricular cavity size is small and left ventricular ejection fraction, a global index of systolic function, is increased or at least preserved. Cardiac diastolic function is impaired and left ventricular end-diastolic pressure is elevated, which is the primary reason for symptoms of heart failure. In a small fraction of patients with HCM, the phenotype evolves into that of a dilated heart with gradual decline in the left ventricular ejection fraction (i.e., evolving into dilated cardiomyopathy, DCM).

A characteristic phenotype of HCM is increased left ventricular ejection fraction (LVEF), which is often interpreted as evidence of enhanced myocardial

Table 3.1 Potential risk factors for sudden cardiac death (SCD) in patients with hypertrophic cardiomyopathy (HCM).

Predictor	Comments
Prior episode of aborted SCD	A major risk factor and mandates ICD implantation
Family history of SCD	ICD implantation is recommended if more than 1 SCD in the family. Additional risk factors should be considered If only 1 SCD in the family
History of syncope	A major risk factor in the presence of family history of SCD [38] or ventricular arrhythmias on Holter monitoring [181] requiring ICD implantation
Sustained and repetitive non-sustained ventricular tachycardia	A major risk factor requiring ICD implantation. Negative predictive value of EP studies is greater than the positive predictive value [181]
Cardiac hypertrophy	Severe hypertrophy increases the risk of SCD [38,65]. It is also associated with the risk of syncope [37]
Early onset of clinical manifestations (young age)	Early penetrance is associated with a higher incidence of SCD [42]
Causal mutations, including double mutations	Variable according to individual mutations and genetic background [74,75,94,126–131]
Modifier genes	The DD genotype of the angiotensin-converting enzyme 1 gene is associated with the risk of SCD [101]
Outflow tract gradient	Severe outflow tract gradient is associated with a higher mortality [160]
Severe interstitial fibrosis and myocyte disarray	Fibrosis could predisposes to ventricular arrhythmias [67]
Abnormal blood pressure response to exercise	Could cause syncope because of exercise induced hypotension [42]
Presence of myocardial ischemia	Significance unclear in adult population

EP, Electrophysiology; ICD, internal cardioverter-defibrillator.

contractility. LVEF is a load-independent index and alone is not robust evidence of myocardial contractility. It may be increased in patients with HCM simply because of a smaller left ventricular end-diastolic volume resulting from concentric hypertrophy and the resulting smaller afterload [50,51]. Assessment of myocardial contractile function by load-independent indices suggests reduced myocardial contractility and relaxation [52–54]. Given the diversity of HCM mutations and their diverse functional effects, the impact of the causal mutations on myocardial contractile performance are expected to vary, some leading to enhanced and others to reduced myocardial contractile performance. For example, mutations that enhance Ca^{2+} sensitivity of the myofibrils for ATPase activity or force generation could increase myocardial contractile performance, and vice versa [55–57]. Similarly, the impact of mutations on actin–myosin interactions could determine the effects on global systolic function [58–61]. Mutations that reduce actin–myosin interactions are expected to impair global cardiac systolic function. In contrast, mutations that reduce the inhibitory effects of cardiac troponin I on actin–myosin interactions would be expected to increase the ejection fraction.

Cardiac myocyte disarray, defined as malaligned, distorted and often short and hypertrophic myocytes oriented in different directions, is the pathologic hallmark of HCM (Plate 3.1). Other pathologic features of HCM include myocyte hypertrophy, interstitial fibrosis, thickening of media of intramural coronary arteries and, sometimes, malpositioned mitral valve with elongated leaflets. Myocyte disarray often comprises >20% of the ventricle, as opposed to <5% of the myocardium in normal hearts [62,63]. It is more prominent in the interventricular septum, but commonly found throughout the myocardium [63]. Cardiac hypertrophy, interstitial fibrosis and myocyte disarray are associated with the risk of SCD, mortality and morbidity in patients with HCM [64–67].

Molecular genetics

HCM is a genetic disease with an autosomal dominant mode of inheritance. An autosomal recessive form has also been described but is uncommon [68]. It is familial in approximately half to two-thirds of cases and sporadic in the remainder [69,70]. Familial aggregation of HCM was first described more than 50 years ago [4]. However, the molecular genetic basis of HCM was unknown until the seminal report of the R403Q mutation in the β-myosin heavy chain (MyHC) in a family with HCM by Geisterfer-Lowrance *et al.* in 1999 [19]. This seminal discovery was a watershed and led to subsequent identification of over 400 causal mutations in more than a dozen different genes, all encoding the sarcomeric proteins (Table 3.2). Accordingly, HCM is considered a disease of mutant sarcomeric proteins [71]. It is important to note that identification of a genetic variant in a patient with HCM alone does not establish causality. To fulfill the Koch postulates for causality and exclude the possibility of polymorphism, it is necessary to show co-segregation of the mutation with the phenotype, its absence in the general population and its functional and biologic effects [72]. The true mutation usually changes a highly conserved amino acid sequence, leading to a premature truncation or reducing the expression level of the protein.

Causal genes

Over a dozen genes encoding for sarcomeric proteins have been identified in patients and families with HCM (Table 3.2). The prevalence of causal genes and mutations varies among different populations. Overall, the collective results of genetic epidemiologic studies suggest that approximately two-thirds of the causal genes for HCM have been identified [73–78]. The most common causal genes are *MYH7*, *MYBPC3*, *TNNT2* and *TNNI3*, which encode the MyHC, myosin binding protein-C (MyBP-C), cardiac troponin T and I, respectively [73–76,79,80]. Collectively, mutations in the above four genes account for approximately two-thirds of all cases of HCM [73–75]. Mutations in *MYH7* and *MYBPC3* are the most common, each accounting for approximately 30% of cases of HCM [73–75]. Over 100 different mutations in the β-MyHC have been identified, the vast majority of which are missense mutations located in the globular head of β-MyHC. Mutations in the hinge and rod domain of β-MyHC have also been described but, in general, are less common [81–83]. Similarly, over 100

Table 3.2 Causal genes for hypertrophic cardiomyopathy (HCM).

Gene	Gene symbol	Locus	Frequency
β-Myosin heavy chain [19]	*MYH7*	14q12	~30%
Myosin binding protein-C [182,183]	*MYBPC3*	11p11.2	~30%
Cardiac troponin T [71]	*TNNT2*	1q32	~5%
Cardiac troponin I [184]	*TNNI*	19p13.2	~5%
α-tropomyosin [71]	*TPM1*	15q22.1	<5%
Cardiac α-actin [85]	*ACTA*	11q	<5%
Essential myosin light chain [185]	*MYL3*	3p21.3–p21.2	<5%
Regulatory myosin light chain [185]	*MYL2*	12q23.q24.3	<5%
Titin [84]	*TTN*	2q24.3	<5%
Tcap (Telethonin or titin-cap) [91]	*TCAP*	17q12	Rare
Cardiac myosin light peptide kinase [90]	*MYLK2*	20q13.3	Rare
α-Myosin heavy chain (association) [134]	*MYH6*	14q12	Rare
Cardiac troponin C (association) [88]	*TNNCI*	3p21.3–3p14.3	Rare
Caveolin 3 (association) [93]	*CAV3*	3p25	Rare [93]
Phospholamban (association) [92]	*PLN*	6p22.1	Rare [92]

Citations refer to the first report of the mutations.

different mutations in *MYBPC3* have been identified which are scattered through the genes [74,75]. Mutations in *TNNT2* and *TNNI3* are less common and together account for another 10–15% of the cases of HCM [73,76,78]. Mutations in *TPM1*, encoding α-tropomyosin; *TTN*, encoding titin; *ACTC*, encoding cardiac α-actin; *MYL3* and *MYL2*, encoding essential and regulatory light chains, respectively, have also been identified in patients and families with HCM [71,78,84–87]. Rare mutations in *TNNC1*, encoding cardiac troponin C [88]; *MYH6*, encoding α-MyHC [89]; *MYLK2*, encoding myosin light chain kinase [90]; *TCAP*, encoding telethonin [91]; *PLN*, encoding phospholamban [92]; and *CAV3*, encoding caveolin 3 [93] have been reported in patients with HCM.

The vast majority of HCM mutations are point or missense mutations that affect a highly conserved amino acid. A small fraction of the mutations are insertion/deletion or splice junction mutations, which are commonly found in *MYBPC3* [74,94]. Insertion/deletion mutations induce frame-shift and lead to premature truncation of the MyBP-C protein. Thus, such mutations impart significant functional impairment of the expressed protein. In addition, deletion mutations in troponin T involving the splice donor sites have been reported that could lead to truncation of the encoded protein [71]. The frequency of each specific HCM mutation is low and hence no single mutation predominates. Accordingly, the mutations are referred to as "private mutations." A few suggestive hot spots for mutations, such as codons 403 and 719 in *MYH7*, have been noted [95,96]. Finally, double mutations have been reported in a small fraction of patients with HCM [97,98].

Modifier genes

Patients with HCM exhibit variable clinical manifestations [1]. The variability in the phenotype is observed among individuals with different causal mutations as well as among individuals with identical causal mutations [99]. Even members of a single family, who share the same causal mutation as well as a significant portion of the genome, exhibit considerable variability in the phenotypic expression of HCM. The molecular basis for the interindividual variability in the phenotypic expression of HCM is largely unknown. Causal mutations, while important contributors to the phenotype, do not fully account for the variability in the phenotypic expression of HCM. The presence of multiple mutations could account for a small part of the variability [97,98]. However, multiple mutations are uncommon and alone are insufficient to explain the variability in the phenotype [97,98]. It is likely that environmental factors, such as heavy physical exercise, particularly isometric exercises, contribute to the phenotypic expression of HCM. However, their contribution remains largely unknown.

Single nucleotide polymorphisms (SNPs) are important determinants of the interindividual variability in the phenotypic expression of single gene disorders, as they are in determining the susceptibility to disease, response to therapy and clinical outcome. Accordingly, the phenotypic expression of HCM, a classic single-gene disorder with a Mendelian inheritance, is determined not only by the causal mutations, but also by the genetic background in which the mutations occur. SNPs in genes implicated in cardiac growth and hypertrophy are the prime candidates to affect phenotypic expression of HCM and hence are referred to modifier mutations or genes [100]. Unlike the causal mutations, modifier genes are neither necessary nor sufficient to cause HCM. Instead, functional variants of the modifier genes affect the severity of phenotype, such as the magnitude of cardiac hypertrophy and/or the risk of SCD. In view of complexity of the molecular biology of cardiac hypertrophic response, a large number of SNPs are expected to affect the phenotypic expression of HCM, each imparting a relatively small effect. The identities of the modifier genes have remained largely unknown. The gene encoding the angiotensin-converting enzyme 1 (ACE-1) is a potential modifier gene as its variants have been associated with the severity of cardiac hypertrophy and the risk of SCD [101–103]. Several other candidate modifiers have also been identified [104]; however, the results have been largely provisional pending replication and confirmation through experimentation.

Genetic basis of HCM phenocopy

The phenotype of "unexplained cardiac hypertrophy" is not unique to HCM caused by mutant

Table 3.3 Genes known to cause hypertrophic cardiomyopathy (HCM) phenocopy.

Gene	Gene symbol	Chromosome	Frequency
Protein kinase A, γ subunit [30,32]	PRKAG2	7q22–q31.1	1–2% [186]
α-Galactosidase A [187]	GLA	Xq22	3% [108]
Unconventional myosin 6 [188]	MOY6	6q12	Rare
Lysosome-associated membrane protein 2 [118]	LAMP2	Xq24	1–2% [118,189]
Mitochondrial genes [190]	MTTG, MTTI	MtDNA	Rare
Frataxin (Friedreich ataxia) [191]	FRDA	9q13	Rare
Myotonin protein kinase (myotonic dystrophy) [192]	DMPK, DMWD	19q13	Uncommon
Protein tyrosine phosphatase, nonreceptor type 11 [113]	PTPN11	12q24	Uncommon, higher in children [46]

sarcomeric proteins but also occurs in a variety of other conditions, such as storage diseases, mitochondrial disorders and triplet repeat syndromes. The prevalence of phenocopy in patients with the diagnosis of HCM is unclear but expected to be higher in children because of the early age of manifestations of many diseases that mimic true HCM. In one report, 28.8% of patients with the diagnosis of HCM had Noonan syndrome [46]. A list of conditions causing HCM phenotype and the causal genes and mutations are listed in Table 3.3.

A prototypic example of HCM phenocopy is Fabry disease, an autosomal recessive lysosomal storage disease characterized by the deficient activity of α-galactosidase A (α-Gal A), also known as ceramide trihexosidase [105,106]. The phenotype results from the deposits of glycosphingolipids in multiple organs, including the heart [107]. Fabry disease is caused by mutations in *GLA* gene on chromosome Xq22, which encodes lysosomal hydrolase α-Gal A protein [107]. Phenotypically, it is characterized by angiokeratoma, renal insufficiency, proteinuria, neuropathy, transient ischemic attack, stroke, anemia, corneal deposits and HCM phenocopy [107]. Cardiac manifestations of Fabry disease include hypertrophy, which is often indistinguishable from true HCM, high QRS voltage, conduction defects, cardiac arrhythmias, valvular regurgitation, coronary artery disease, myocardial infarction and aortic annular dilatation [29,108]. The disease predominantly affects males; female carriers could exhibit a milder form [29]. The diagnosis is established by measuring α-Gal A levels and activity in leukocytes. The estimated prevalence of Fabry disease in adult population with the clinical

diagnosis of HCM is approximately 3% [108]. The distinction between true HCM and Fabry disease is important because the latter can be treated with enzyme replacement therapy using human α-Gal A (agalsidase α) or recombinant human α-Gal A (agalsidase β) [105,106,109,110].

Another phenocopy is a glycogen storage disease caused by mutations in the *PRKAG2* gene [30–33]. The gene encodes the γ2 regulatory subunit of AMP-activated protein kinase (AMPK), which is considered the energy biosensor of the cell. Mutations in *PRKAG2* lead to cardiac hypertrophy, conduction defects and WPW [30–32]. Cardiac hypertrophy results predominantly from the storage of glycogen in myocytes. Hence, it differs considerably from true HCM.

HCM phenocopy also occurs in trinucleotide repeat syndromes, a group of genetic disorders caused by expansion of naturally occurring GC-rich triplet repeats in genes [111]. Friedreich ataxia is an example of trinucleotide repeat syndrome that causes HCM phenocopy. It is an autosomal recessive neurodegenerative disease caused by expansion of GAA repeat sequences in the intron of *FRDA* [112]. Cardiac involvement could manifest as either a hypertrophic or a dilated heart.

HCM phenocopy also occurs in patients with Noonan syndrome, an uncommon autosomal dominant disorder characterized by dysmorphic facial features, pulmonic stenosis, mental retardation, bleeding disorders and cardiac hypertrophy. The causal gene in approximately half of cases is protein-tyrosine phosphatase, nonreceptor type 11 (*PTPN11*) gene [113,114]. LEOPARD syndrome (lentigines, electrocardiographic conduction abnormalities, ocular hypertelorism, pulmonic stenosis, abnormal genitalia, retardation of growth, and deafness) is an allelic variant of Noonan syndrome.

Defective mitochondrial oxidative phosphorylation pathways could lead to a phenotype mimicking HCM. Kearns–Sayre syndrome (KSS) is an example of a mitochondrial disease that causes HCM phenocopy [115]. It is characterized by a triad of progressive external ophthalmoplegia, pigmentary retinopathy and cardiac conduction defects, and, less frequently, cardiac hypertrophy. Cardiac hypertrophy could occur in a variety of metabolic diseases, such as Refsum disease, glycogen storage disease type II (Pompe disease), Danon disease, Niemann–Pick disease, Gaucher disease, hereditary hemochromotasis and CD36 deficiency [116–119].

Genetic and non-genetic determinants of phenotype

The clinical phenotype of HCM is the consequence of the cardiac response to mutant sarcomeric proteins. Hence, a variety of genetic and non-genetic factors are expected to influence expression of the clinical phenotype in HCM (Fig. 3.1). Thus, the ensuing phenotype, whether it is hypertrophy, interstitial fibrosis or SCD, is determined not only by causal mutations, but also by modifier SNPs, interactions between genes (epistasis), transcrip-

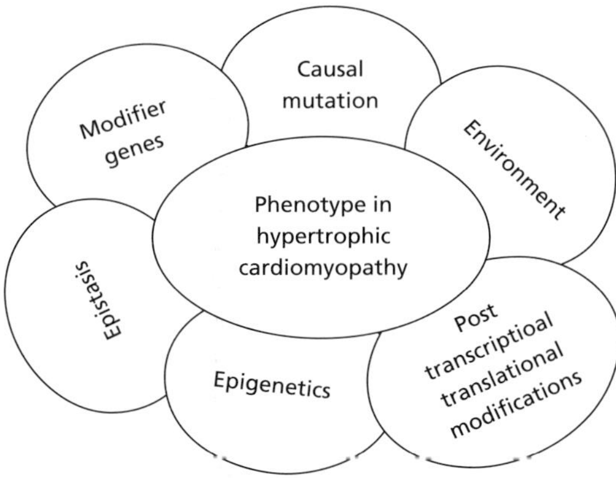

Figure 3.1 Determinants of cardiac phenotype in hypertrophic cardiomyopathy (HCM). While causal mutations are important determinants of the severity of the phenotypes, several others factors shown in the figure contribute to phenotypic expression and severity of the morphologic and clinical phenotypes in HCM.

tional and post transcriptional regulation of gene expression, post translational modification of proteins and environmental factors. Thus, the clinical phenotype of HCM, a single gene disorder, is in fact a complex phenotype and hence variable and diverse among affected individuals. The presence of a remarkable interindividual variability in the phenotypic expression of HCM restricts generalization of the findings in a particular subset of patients or extension of the results across subgroups. Accordingly, as in any other risk assessment or prediction of the phenotype, all components that contribute to the phenotypic expression of HCM should be considered as well as the limitations of applying the group data to an individual patient.

The topography of the causal mutations, and hence their impact on the structure and function of the respective proteins, is expected to be an important determinant of the ensuing phenotype. This point is illustrated in the case of mutations in β-MyHC, cTnT and cTnI, which could cause either HCM or DCM [120,121]. The basis of the divergence of the phenotypes resulting from mutations in a single gene remains unknown but likely pertains to the effect of the mutation on protein structure and function [57,122,123]. One could speculate that mutations that enhance Ca^{2+} sensitivity of the myofibrillar force generation or diminish the inhibitory effect of cTnI on actin–myosin interactions to cause HCM [57,122,123]. In contrast, mutations that reduce Ca^{2+} sensitivity of the myofibrillar force generation or increase the inhibitory effect of cTnI on actin–myosin interaction to cause DCM [57,122–125]. Thus far, no systematic genotype–phenotype correlation study, based on topographic classification of the mutations, has been performed. It is also noteworthy that no phenotype is specific to a genotype and there is considerable overlap in the phenotypic expression of HCM caused by different mutations. Similarly, because the ensuing clinical phenotype in HCM is a complex trait, "benign" or "malignant" phenotype could be observed for any HCM mutation.

Despite the presence of considerable variability in the phenotype expression of HCM and the limitations of the genotype–phenotype correlation studies, the data suggest that the causal mutations as well as the modifier genetic factors exert a significant impact on expression of cardiac hypertrophy and the risk of SCD [74,75,101,102,126–131]. In general, mutations in β-MyHC are associated with a high penetrance, an early onset of clinical phenotype, extensive hypertrophy and a relatively higher incidence of SCD than those in MyBP-C [94,132]. The phenotype in the latter group is generally characterized by a low penetrance, late onset of clinical phenotype, mild cardiac hypertrophy and a lower risk of SCD [74,94,132]. The relatively low penetrance of a mutation indicates that a phenotypically normal individual could express the clinical phenotype later in life [133]. Thus, unless genetic testing is performed to exclude the mutation, those at risk should be evaluated periodically. The phenotype imparted by mutations in cTnT, cTnI and α-tropomyosin, while variable, is generally characterized by a mild degree of cardiac hypertrophy, but a higher incidence of SCD and extensive myocyte disarray [67,130]. Concerning the impact of the modifier genes, the existing data suggest ACE-1 DD genotypes are associated with more extensive hypertrophy and a higher risk of SCD [101–103].

Another determinant of cardiac phenotype in HCM is the presence of double mutations [98] and concomitant diseases, such as hypertension. Hypertension provides extra stimulus for cardiac hypertrophic growth and so could increase the penetrance and accelerate the phenotypic expression of HCM. "Hypertensive hypertrophic cardiomyopathy of the elderly" is considered a form of HCM caused by low penetrant mutations, often in MyBP-C, whose phenotypic expression is enhanced because of concomitant hypertension [94,134]. It is also unclear whether heavy physical exercise, particularly isometric exercises, could enhance phenotypic expression of cardiac hypertrophy in HCM. Because hypertrophy is the response of the heart to the genetic defect, one could speculate that factors that promote cardiac growth enhance phenotypic expression of HCM. Despite the lack of conclusive data, because HCM is the most common cause of SCD in young competitive athletes [41], patients with HCM are advised not to participate in competitive or contact sports.

Genetic screening

There is considerable interest in genetic screening of those at risk for HCM mutations. The interest, expressed by patients and physicians alike, is primarily based on the potential utility of genetic testing in preclinical diagnosis of mutation careers. Family members who do not carry the causal mutation and thus are not at the risk of HCM could be identified. In addition, members who carry the mutation and hence are at risk of developing HCM could be diagnosed early, prior to and independent of the clinical phenotype [135]. Despite the need and in spite of identification of the majority of the causal genes for HCM, routine genetic screening has not been feasible. In addition, the clinical utility of genetic testing in risk stratification and treatment, despite its plausibility, remain to be established.

A number of issues have impeded the development and application of routine genetic screening in HCM. The diversity of the causal mutations and the low frequency of each specific causal mutation are the major limiting factors. In addition, routine genetic testing will require systematic screening of all known causal genes, which at best could lead to identification of the causal mutation in approximately two-thirds of cases. Furthermore, the genetic screening technique would need to be highly sensitive and specific and not very expensive. The current gold standard technique is direct sequencing of the genomic DNA, which has an excellent sensitivity and specificity. In the case of HCM, it would be necessary to sequence at least all coding exons and the exon–intron boundaries of a dozen sarcomeric genes, assuming no phenocopy. The current state of sequencing technology is not conducive to routine genetic screening because of the labor intensive and expensive nature of the task. Hence, implementation of genetic screening on a routine basis would require the development of new screening methods. An alternative approach is to screen the coding exons and exon–intron boundaries of the most common causal genes for HCM. The approach is costly, labor intensive with the current technology, but feasible. It could lead to detection of the causal mutations in about 60% of cases. Screening technology is rapidly evolving. One could anticipate that within the next few years, advanced mutation-screening techniques will become available and applied on a routine basis to screen individuals at risk of developing HCM.

The primary utility of genetic testing would be in the accurate diagnosis of those at risk of the disease (mutation carriers) and normal individuals (non-carriers). The utility of genetic testing in prognostication and identification of those at risk of SCD remains to be established and has to await large-scale clinical studies. The significance of identifying the causal mutations as the determinants of the clinical phenotypes cannot be overemphasized. In addition, information on the modifier genes, which remain largely unrecognized, as well as environmental factors and others would be necessary for accurate clinical management and counseling of patients with HCM. Thus, an integrated approach that utilizes genetic and nongenetic predictors should be used in counseling and treating those with HCM mutations.

Pathogenesis

Elucidation of the molecular genetic basis of HCM afforded the opportunity to perform a series of mechanistic experiments to delineate the molecular pathogenesis. The effects of the HCM mutations on protein, cell and organ structure and function have been studied through a large number of *in vitro* and *in vivo* studies. Over a dozen genetically modified animal models of HCM have been generated, which in part or fully recapitulate the phenotype of human HCM (Table 3.4). Collectively, the results have provided considerable insight into the pathogenesis of HCM and have led to partial understanding of the links between the causal genetic defect and the ensuing phenotypic expression of hypertrophy, fibrosis and myocyte disarray.

The sequence of events from the genetic defect to the clinical phenotype could be simplified into three major stages: the initial functional phenotypes; intermediary molecular phenotypes; and the ensuing morphologic phenotypes [136]. A change in the protein sequence (i.e., causal mutation) by changing the secondary structure or charge of the encoded protein affects interactions of the mutant protein with other protein components of the sarcomeres. The defective interaction affects the function of the entire multiprotein sarcomeric units. The resulting functional defects are diverse, which is reflective of the diversity of the causal genes and mutations. They include reduced ATPase activity of the myofibrils [137], impaired actin–myosin cross-bridging [138], enhanced Ca^{2+} sensitivity of myofibrils in the generation of contractile force [55]. A partial list of the initial functional phenotype is given in Table 3.5. Altered functional abnormalities of sarcomeres and myofibrils lead to expression and activation of intermediary molecular phenotypes, such as intracellular signaling molecules. The changes in gene expression and signaling activation instigates cardiac morphologic and histologic response, such as hypertrophy and fibrosis (Fig. 3.2). In addition, certain sarcomeric proteins, such as titin and sarcomere-associated proteins are directly involved in regulating muscle gene expression [139]. Mutations could also interfere with regulation of gene expression by the sarcomeric proteins. At organ level, impaired myocardial contraction and relaxation as well as impaired bioenergetics have been implicated in mediating cardiac hypertrophic response in HCM [140–143]. Tissue Doppler velocities of myocardial contraction and relaxation are reduced in human subjects with HCM mutations prior to the development of cardiac hypertrophy [52]. Similarly, the ratio of cardiac phosphocreatine (PCr) to adenosine triphosphate (ATP) in the heart is reduced in patients with HCM mutations [144]. Thus, a number of initial defects alone or collectively at protein, myofibrillar or organ level, contribute to the development of evolving cardiac hypertrophy in HCM. Accordingly, cardiac hypertrophy, the clinical hallmark of HCM, and interstitial fibrosis are considered secondary phenotypes because of activation of intermediary molecules and hence potentially reversible.

The pathogenesis of myocyte disarray, the pathologic hallmark of HCM, is unknown. Myocyte disarray is an early phenotype, independent of hypertrophy and fibrosis [137]. Observations in a mouse model expressing human cTnT-Q92 mutation implicate impaired myocyte–myocyte attachment at adherens junctions, as a potential mechanism for myocyte disarray [145]. The extracellular domains of cadherins form the intercellular bonds between the adjacent myocytes. Meanwhile, the cytoplasmic domains of cadherins attach to cytoskeletal actin through β-catenin and other effector proteins. Excess phosphorylation of β-catenin reduces complexing with N-cadherin at the adherens junctions, impairing proper alignment of the myocytes and hence myocyte disarray [137]. Several other alternative pathways could be responsible for myocyte

Table 3.4 Genetically engineered animal models of hypertrophic cardiomyopathy (HCM).

Knock out/in models	Phenotype
α-MyHC-Q403 mice	Myocyte disarray, interstitial fibrosis, hypertrophy mild and late, enlarged left atrium, premature death, neonatal dilated cardiomyopathy in homozygous mice [193], systolic and diastolic dysfunction, increased contractile performance in very early life, heterogeneous ventricular conduction, inducible ventricular tachycardia, reduced crossbridge kinetics, increased force generation of single myosin molecules and $[Ca^{2+}]_i$ sensitivity, reduced [PCr], and increased [Pi], increased actin-activated ATPase activity [138,180,193–202]
α-MyHC knock out mice	Embryonic lethality in –/–, +/– show fibrosis, sarcomere disarray, impaired contractility and relaxation [148]
α-Tropomyosin knock-out	Embryonic lethality in homozygotes, no phenotype in heteozygotes, normal cardiac function [150,203]
Transgenic mice	
α-MyHC-Q403/ΔAA468-527	Myocyte disarray, interstitial fibrosis, ventricular hypertrophy in female and dilatation in male mice, Increased ANF expression [204,205]
α-MyHC-ΔLCBD	Myocyte disarray, hypertrophy (only in homozygote), valvular thickening, decreased Ca^{2+} sensitivity and diastolic dysfunction [206]
cTnT-ΔC-terminus	Myocyte disarray, interstitial fibrosis, myocyte atrophy and drop out, cardiac atrophy, premature death, systolic and diastolic dysfunction [207]
cTnT-Q92	Myocyte disarray, interstitial fibrosis, myocyte drop out, cardiac atrophy, systolic and diastolic dysfunction, enhanced myofibrillar Ca^{2+} sensitivity [55,208,209]
cTnT-N179	Normal, normal survival, no hypertrophy, increased Ca^{2+} sensitivity of ATPase activity and force generation, increased rate of contraction and relaxation, lower maximum force/cross-section area and ATPase [210]
MyBP-C-ΔC-terminus	Myocyte disarray, sarcomere dysgenesis, interstitial fibrosis, no cardiac hypertrophy [211]
Truncated MyBP-C	Neonatal dilated cardiomyopathy in homozygous mice expressing <10% of the truncated protein, disarray, minimal or mild hypertrophy, decreased contractility and diastolic dysfunction [212,213]
ELC-V149 (human gene)	Papillary muscle hypertrophy, altered stretch-activation response [214]
ELC-V149 (mouse cDNA)	Normal, no hypertrophy, increased Ca^{2+} sensitivity and impaired relaxation [215]
α-Tropomyosin-N175	Myocyte disarray and hypertrophy, impaired contractility and relaxation. Increased Ca^{2+} sensitivity and decreased relaxation [216,217]
cTnI-G146	Myocyte disarray, interstitial fibrosis, premature death. Increased Ca^{2+} sensitivity, hypercontractility and diastolic dysfunction [218]
Transgenic rat	
cTnT-ΔExon 16	Normal, no cardiac hypertrophy, systolic and diastolic dysfunction, After 6 months of exercise hypertrophy, myofibrillar disarray [219]
Transgenic rabbit	
β-MyHC-Q403	Cardiac hypertrophy, myocyte disarray, interstitial fibrosis, increased mortality and SCD, systolic and diastolic dysfunction, preserved global systolic function, reduced myocardial contraction and relaxation velocities [220,221]
cTnI-G146	Subtle defects including mycoyte disarray, fibrosis, altered connexin43 organization and leftward shift in the force-pCa^{2+} curves [222]

$[Ca^{2+}]_i$, intracellular Ca^{2+} concentration; cTnT, cardiac troponin T; cTnI, cardiac troponin I; ELC, essential light chain; LCBD, light chain binding domain; MyBP-C, myosin binding protein C; MyHC, myosin heavy chain; [PCr], phosphocreatinine; ↑ [Pi], inorganic phosphate; –/–, null (homozygous for the deletion); +/–, heterozygous.

Table 3.5 Partial list of initial functional and biologic defects caused by hypertrophic cardiomyopathy (HCM) mutations.

Altered Ca⁺ and pH sensitivity of myofibrillar force generation [56,124,222–225]

Altered Ca⁺ and pH sensitivity of myofibrillar ATPase activity [56,57,226]

Impaired bioenergetics, higher energy cost (ATPase/force) and reduced free energy of ATPase activity [56,144,227,228]

Altered kinetics of actin–myosin cross-bridge cycling [224,229,230]

Altered phosphorylation, protein folding and proteolytic susceptibility [223,231–234]

Impaired interaction with other sarcomeric proteins [190,225,235]

Sarcomere dysgenesis and poor assembly [101,211,233,236]

---

disarray in HCM including activation of the integrin/wnt signaling pathway by increased extracellular matrix proteins [146].

A number of the causal mutations in HCM are splice-junction or frame-shift mutations that lead to null alleles. Thus, the mutant sarcomeric proteins are not expressed or if expressed are not expected to be incorporated into myofibrils. The net effect is "haplo-insufficiency," which is particularly relevant to frame-shift or insertion/deletion mutations in the MyBP-C protein [73,74,147]. Whether null alleles could cause HCM because of haploinsufficiency remains unclear. Gene-targeting experiments in mice suggest haplo-insufficiency could change sarcomere structure and function as well as myocardial dysfunction but may be gene specific [148–150]. Ablation of α-tropomyosin does not lead to discernible morphologic or functional abnormalities in heterozygous mouse [149,150], whereas heterozygosity of α-MyHC ablation leads to a cardiomyopathic phenotype [148]. Thus, the null mutations may lead to HCM only when compensatory mechanisms fail to overcome the haplo-insufficiency.

Treatment

The focus in the management of patients with HCM is on determining the risk of SCD, intervening to reduce the risk and the alleviation of symptoms. Fortunately, the vast majority of patients with HCM are at a relatively low risk of SCD and are asymptomatic or minimally symptomatic. In such patients, only periodic follow-up and assessment by ECG and two-dimensional and Doppler echocardiography is recommended. In an asymptomatic patient who is at high risk for SCD, typically defined by the presence of two or more major risk factors (Table 3.1), an internal cardioverter-defibrillator (ICD) should be implanted and has been shown to be effective [24]. Otherwise, there is no pharmacologic or nonpharmacologic intervention to prevent or slow the evolution of HCM phenotype.

Treatment options of symptomatic patients comprise pharmacologic therapy, ICD implantation in

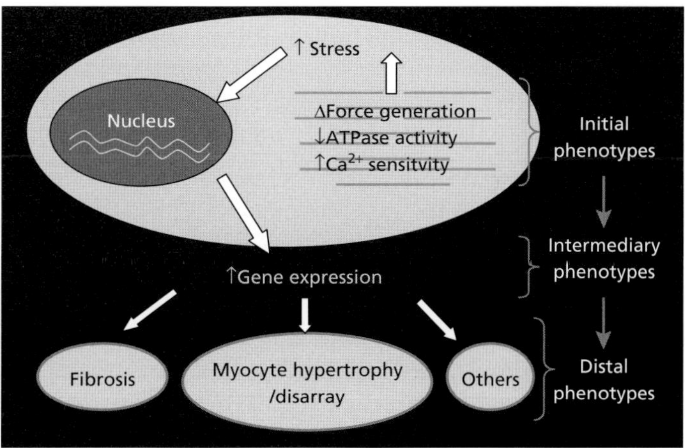

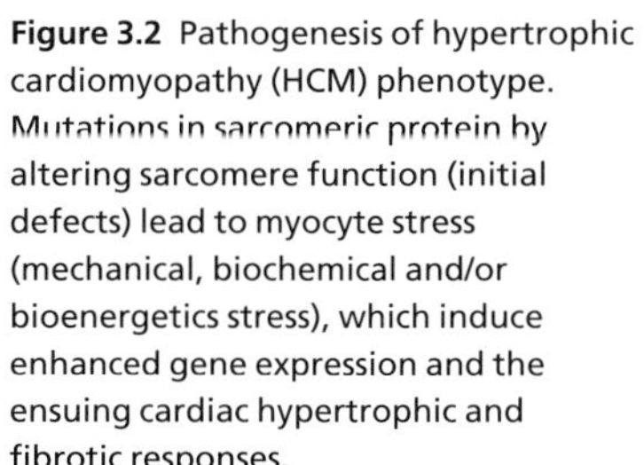

Figure 3.2 Pathogenesis of hypertrophic cardiomyopathy (HCM) phenotype. Mutations in sarcomeric protein by altering sarcomere function (initial defects) lead to myocyte stress (mechanical, biochemical and/or bioenergetics stress), which induce enhanced gene expression and the ensuing cardiac hypertrophic and fibrotic responses.

patients at high-risk of SCD and, in those with significant outflow tract obstruction, surgical myectomy and transcatheter septal ablation. Pharmacologic treatment includes the use of beta-blockers, verapamil, disopyramide, low-dose diuretics and amiodarone, the latter primarily for treatment of atrial and ventricular arrhythmias. Treatment with beta-blockers is preferred in patients with outflow tract gradient at rest. Verapamil is generally not recommended in such patients because of the risk of hypotension resulting from vasodilatation. Patients with symptoms of palpitations should undergo Holter monitoring and, if needed, electrophysiologic studies to determine the etiology and provide appropriate therapy. New onset atrial fibrillation is not well tolerated in those with severe hypertrophy or outflow tract obstruction. It often requires electrical cardioversion. Patients with chronic or intermittent atrial fibrillation should be anticoagulated in order to reduce to risk of systemic embolization and stroke. Beta-blockers, verapamil or amiodarone are used for treatment of patients with chronic or paroxysmal atrial fibrillation.

Syncope is a major risk factor for SCD and requires extensive investigation and proper treatment. It often indicates serious ventricular arrhythmias and, less frequently, exercise-induced hypotension. Patients with syncope should undergo Holter monitoring, electrophysiologic studies and, if needed, tilt table testing. Patients with repetitive bursts of nonsustained ventricular tachycardia on Holter monitoring and those with sustained ventricular tachycardia are candidates for automatic ICD implantation, regardless of the presence or absence of symptoms. Ventricular arrhythmias are the main cause of SCD in patients with HCM [24]. Implantation of an ICD as a preventive measure in patients at high risk of SCD reduces the risk of SCD [24]. An algorithm showing our current approach to management of HCM patients is shown in Fig. 3.3.

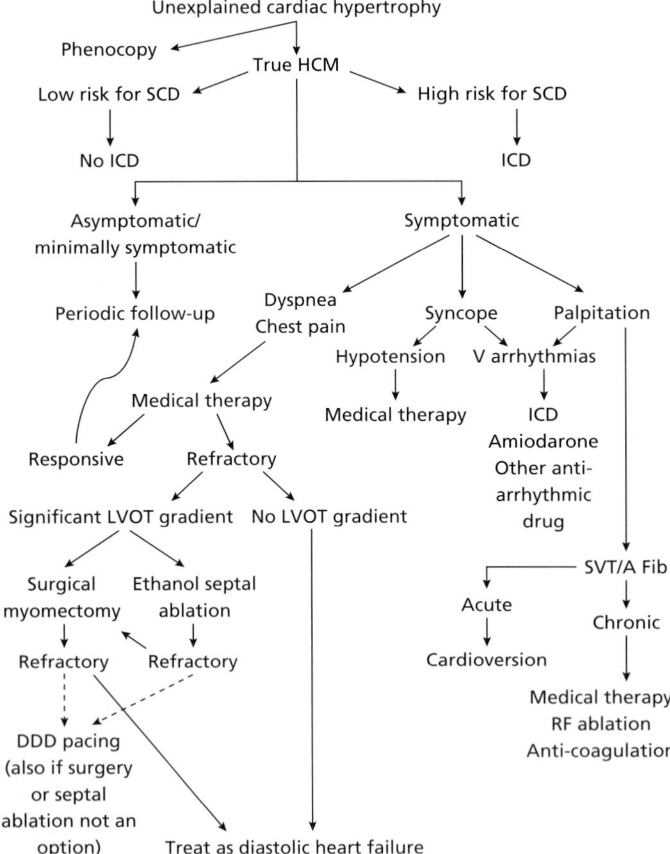

Figure 3.3 An algorithm for management of patients with hypertrophic cardiomyopathy (HCM). A Fib, atrial fibrillation; DDD, dual chamber pacing; ICD, internal cardioverter-defibrillator; LVOT, left ventricular outflow tract; RF, radio-frequency ablation; SCD, sudden cardiac death; SVT, supra-ventricular tachycardia; V, ventricular.

Table 3.6 Comparison of surgical myectomy and ethanol septal ablation.

	Surgical myectomy	*Ethanol septal ablation*
Approach	Cardiopulmonary bypass	Cardiac catheterization
Hospital stay	5–7 days	1–2 days
Perioperative mortality	1–5%	1–5%
Procedural success	>95%	>85%
Short-term symptomatic relief	Excellent	Excellent
Long-term symptomatic relief	Excellent	Excellent
Long-term safety	Established	Risk of ventricular arrhythmias
Impact on survival	Unknown	Unknown
Septal infarction/fibrosis	None	Present
Recurrence of LVOT gradient	Rare	Uncommon
Repeat procedure	Rare	Uncommon
Atrioventricular block requiring permanent pacemaker	~2%	~20%
Late ventricular arrhythmias	Rare	Uncommon
Postoperative atrial fibrillation	Uncommon	Rare
Significant aortic regurgitation	Rare	None
Ventricular septal defect	Rare	None
Correction of concomitant problems	Amenable	NA

LVOT, left ventricular outflow tract; NA, not applicable.

A small group of patients with HCM do not respond to medical therapy and remain significantly symptomatic (New York Heart Association Functional class III and IV for heart failure). These patients are candidates for invasive therapeutic interventions if they exhibit interventricular septal thickness of 15 mm or greater and significant outflow tract gradient obstruction (>50 mmHg at rest). The most commonly used invasive interventions are surgical myectomy and transcoronary septal ablation. While there are no prospective randomized studies to compare the clinical outcomes after surgical myectomy and percutaneous transcoronary septal ablation, several observational studies have shown efficacy of both techniques in reducing left ventricular outflow tract gradient and reduction of symptoms [151–157]. Nonetheless, there is considerable debate regarding the superiority of each intervention [158,159]. Overall, there is a general agreement on the suitability of transcoronary septal ablation as an alternative and effective option in symptomatic patients who are not candidates for surgical myectomy. Neither surgical myectomy nor transcoronary septal ablation is considered a cure for HCM. Close monitoring, follow-up and treatment of these patients are warranted. The advantages and disadvantages of these two techniques are summarized in Table 3.6.

Surgical myectomy (myomectomy)

The focus of surgical myectomy is to reduce outflow tract obstruction, a determinant of morbidity and mortality of patients with HCM [160]. Surgical myectomy is indicated in a small group of patients who have significant outflow tract obstruction at rest (typically >50 mmHg gradient) and are refractory to pharmacologic therapy. Surgical myectomy is the procedure of choice in HCM patients who have concomitant coronary artery disease or valvular disorders. The procedure involves resection of a small portion of the interventricular septum, commonly restricted to the base of the septum, through a transaortic approach (Morrow procedure) [161]. Mitral valve repair, placation and replacement are performed during surgery in those with significant mitral regurgitation. The overall surgical mortality is 1–5% [151,153,162,163]. It is somewhat higher in elderly patients and in those with concomitant surgeries, such as coronary bypass or valve surgery. Surgical myectomy is very effective in reducing or abolishing the outflow tract gradient and improving symptoms. The long-term results have been remarkable for sustained symptomatic relief, a low recurrence rate requiring a second intervention and excellent survival, almost matching that of the natural history of HCM [151,153,156,158,162–164]. However, the impact of surgical myectomy on reducing cardiovascular mortality and the risk of SCD remains to be established.

Transcoronary septal ablation

Transcoronary septal ablation is performed by injecting a small amount of pure ethanol (1–3 mL) into the main septal perforators of the left anterior descending artery [21]. It is an effective treatment option for partial ablation of the hypertophic interventricular septum and reduction of the left ventricular outflow tract gradient [154,165,166]. Injection of ethanol or microspheres into the septal branches induces local myocardial necrosis, emulating surgical myectomy. The procedure is best reserved for symptomatic patients who are refractory to medical therapy, have an interventricular septal thickness of 15 mm or greater and a left ventricular outflow tract gradient of 50 mmHg or more. In patients who have significant exertional dyspnea and a thick interventricular septum but no significant gradient at rest, exercise echocardiography could be used to provoke gradient. The merit of dobutamine-induced outflow tract gradient as an indication for septal ablation is uncertain.

Progressive left ventricular remodeling occurs in conjunction with enlargement of the left ventricular outflow tract following transcatheter septal ablation [159,167]. The remodeling could lead to continued improvement of symptoms. The procedure is well tolerated, with a relatively low perioperative morbidity and mortality [166]. A relatively high early and late mortality has been reported [153,168]. A major complication is the development of advanced conduction defect requiring permanent pacemaker implantation in approximately 20% of patients [168–170]. There is also a concern regarding the possibility of ventricular arrhythmias resulting from localized myocardial necrosis. This potentially serious complication appears to be uncommon, albeit documented, and could account for the higher late mortality reported [168,171–173]. Overall, the intermediary follow-up data for

transcoronary septal ablation are remarkable for an acceptable recurrence rate, but long-term follow-up data are not yet available [166].

Dual chamber pacing

Dual chamber pacing is a treatment modality reserved for rare situations where medical therapy fails and surgical myectomy or transcatheter septal ablation are not considered options. It reduces left ventricular outflow tract gradient by modifying left ventricular excitation pattern causing dyssynchronous depolarization of left ventricular contraction. Optimal timing of the AV interval is considered crucial for the effectiveness of the pacing strategy. Initial observational studies showed a significant improvement in symptoms along with a major reduction in outflow tract obstruction [174]. However, subsequent randomized multicenter studies showed a significant placebo effect without discernible improvement in exercise tolerance [175,176]. The poor results of two randomized clinical studies have reduced the overall enthusiasm in the clinical utility of dual chamber pacing in the treatment of patients with HCM [175,176]. Thus, dual chamber pacing is no longer recommended, except in severely symptomatic patients who are not candidates for either surgical or transcatheter septal ablation.

Potential new therapeutic approaches

Current pharmacologic interventions in human patients, largely restricted to the use of beta-blockers and verapamil, are empiric and none has been shown to reduce mortality or induce regression of cardiac hypertrophy, fibrosis or disarray in patients with HCM. Advances in the molecular genetics and biology of HCM have provided new therapeutic targets, which have been tested in animal models of HCM with encouraging results.

Experimental studies in a transgenic rabbit model, which recapitulate the phenotype of human HCM, have shown the potential utility of 3-hydroxy-3-methyglutaryl-coenzyme A (HMG-CoA) reductase inhibitors (statins) in prevention, attenuation and reversal of evolving phenotypes in HCM [26,27]. Statins exert considerable antihypertrophic effects by blocking geranyl geranylation of RhoA and Rac1, essential mediators of cardiac hypertrophic response [177]. In a randomized study in transgenic rabbits with established HCM phenotype, treatment with simvastatin reduced left ventricular mass, wall thickness, myocyte size and normalized interstitial fibrosis [26]. In addition, indices of cardiac diastolic function and filling pressure were improved significantly. The potential utility of statins in prevention of HCM phenotype has also been assessed in a prospective randomized study [27]. Treatment of transgenic rabbit early and prior to the development of cardiac phenotype prevented evolution of cardiac hypertrophy by reducing activation of hypertrophic signaling molecule [27]. Studies in other animal models of cardiac hypertrophy and heart failure have corroborated the potential beneficial effects of statins in prevention and attenuation of hypertrophy and fibrosis [177,178]. Clinical studies in humans are ongoing to test the potential utility of statins in the attenuation of cardiac phenotypes and symptoms in HCM.

Another potential therapeutic target for treatment for human HCM is inhibition of RAAS. Studies in genetic mouse models of HCM have supported the potential utility of angiotensin II receptor blockade by losartan and mineralocorticoid receptor blockade by aldosterone in complete resolution of interstitial fibrosis [25,145]. The findings are noteworthy and merit further investigation in human patients because conventionally these agents are not recommended in patients with obstructive HCM.

There has been significant controversy regarding the utility of calcineurin inhibitors in the treatment and prevention of cardiac hypertrophy in a variety of conditions [179]. A recent study in α-MyHC-Q403$^{+/-}$ mice showed that treatment with FK506 or cyclosporine, inhibitors of calcineurin, worsened cardiac and myocyte hypertrophy, myocyte disarray and interstitial fibrosis and increased mortality [180]. Pretreatment with diltiazem, an L-type Ca^{2+} channel blocker, prevented the exaggerated cardiac hypertrophic response to inhibitors of calcineurin [180]. These findings have reduced the overall enthusiasm for the potential utility of calcineurin inhibitors in HCM.

Acknowledgments

Supported by grants from the National Heart, Lung, and Blood Institute, Specialized Centers of Research P50-HL54313, RO1 HL68884, and a TexGen grant from Greater Houston Community Foundation.

References

1 Maron BJ. Hypertrophic cardiomyopathy: a systematic review. *JAMA* 2002; 287: 1308–1320.

2 Liouville H. Retrecissement cardiaque sous aortique. *Gazette Med Paris* 1869; 24: 161–165.

3 Schmincke A. Ueber linkseitige muskulose conus-stenosen. *Dtsche Med Wochnschr* 1907; 33: 2082.

4 Davies LG. A familial heart disease. *Br Heart J* 1952; 14: 206–212.

5 Brock R, Fleming PR. Aortic subvalvar stenosis; a report of 5 cases diagnosed during life. *Guys Hosp Rep* 1956; 105: 391–408.

6 Teare D. Asymmetrical hypertrophy of the heart in young adults. *Br Heart J* 1958; 20: 1–8.

7 Braunwald E, Ebert PA. Hemogynamic alterations in idiopathic hypertrophic subaortic stenosis induced by sympathomimetic drugs. *Am J Cardiol* 1962; 10: 489–495.

8 Braunwald E, Lambrew CT, Rockoff SD, Ross J Jr, Morrow AG. Idiopathic hypertrophic subaortic stenosis. 1. A description of the disease based upon an analysis of 64 pateints. *Circulation* 1964; 30 (Supplement 4): 3–119.

9 Pierce GE, Morrow AG, Braunwald E. Idiopathic hypertrophic subaortic stenosis. 3. Intraoperative studies of the mechanism of obstruction and its hemodynamic consequences. *Circulation* 1964; 30 (Supplement 4): 152.

10 Morrow AG, Lambrew CT, Braunwald E. Idiopathic hypertrophic subaortic stenosis. 2. Operative treatment and the results of pre- and postoperative hemodynamic evaluations. *Circulation* 1964; 30 (Supplement 4): 120–151.

11 Moreyra E, Segal BL. [Echocardiographic study in sub-aortic muscular stenosis patients]. *Prensa Med Argent* 1968; 55: 767–773.

12 Moreyra E, Elein JJ, Shimada H, Segal BL. Idiopathic hypertrophic subaortic stenosis diagnosed by reflected ultrasound. *Am J Cardiol* 1969; 23: 32–37.

13 Henry WL, Clark CE, Epstein SE. Asymmetric septal hypertrophy. Echocardiographic identification of the pathognomonic anatomic abnormality of IHSS. *Circulation* 1973; 47: 225–233.

14 Shah PM, Gramiak R, Kramer DH. Ultrasound localization of left ventricular outflow obstruction in hypertrophic obstructive cardiomyopathy. *Circulation* 1969; 40: 3–11.

15 Joyner CR, Harrison FS Jr, Gruber JW. Diagnosis of hypertrophic subaortic stenosis with a Doppler velocity flow detector. *Ann Intern Med* 1971; **74**: 692–696.

16 Boughner DR, Schuld RL, Persaud JA. Hypertrophic obstructive cardiomyopathy. Assessment by echocardiographic and Doppler ultrasound techniques. *Br Heart J* 1975; 37: 917–923.

17 Maron BJ, Gottdiener JS, Arce J, Rosing DR, Wesley YE, Epstein SE. Dynamic subaortic obstruction in hypertrophic cardiomyopathy: analysis by pulsed Doppler echocardiography. *J Am Coll Cardiol* 1985; 6: 1–18.

18 Takenaka K, Dabestani A, Gardin JM *et al.* Left ventricular filling in hypertrophic cardiomyopathy: a pulsed Doppler echocardiographic study. *J Am Coll Cardiol* 1986; 7: 1263–1271.

19 Geisterfer-Lowrance AA, Kass S, Tanigawa G *et al.* A molecular basis for familial hypertrophic cardiomyopathy: a beta cardiac myosin heavy chain gene missense mutation. *Cell* 1990; 62: 999–1006.

20 Marian AJ. Recent advances in genetics and treatment of hypertrophic cardiomyopathy. *Future Cardiol* 2005; 1: 341–353.

21 Sigwart U. Non-surgical myocardial reduction for hypertrophic obstructive cardiomyopathy. *Lancet* 1995; 346: 211–214.

22 Hess OM, Sigwart U. New treatment strategies for hypertrophic obstructive cardiomyopathy: alcohol ablation of the septum: the new gold standard? *J Am Coll Cardiol* 2004; 44: 2054–2055.

23 Maron BJ, Dearani JA, Ommen SR, *et al.* The case for surgery in obstructive hypertrophic cardiomyopathy. *J Am Coll Cardiol* 2004; 44: 2044–2053.

24 Maron BJ, Shen WK, Link MS *et al.* Efficacy of implantable cardioverter-defibrillators for the prevention of sudden death in patients with hypertrophic cardiomyopathy. *N Engl J Med* 2000; 342: 365–373.

25 Lim DS, Lutucuta S, Bachireddy P *et al.* Angiotensin II blockade reverses myocardial fibrosis in a transgenic mouse model of human hypertrophic cardiomyopathy. *Circulation* 2001; 103: 789–791.

26 Patel R, Nagueh SF, Tsybouleva N *et al.* Simvastatin induces regression of cardiac hypertrophy and fibrosis and improves cardiac function in a transgenic rabbit model of human hypertrophic cardiomyopathy. *Circulation* 2001; 104: 317–324.

27 Senthil V, Chen SN, Tsybouleva N *et al.* Prevention of cardiac hypertrophy by atorvastatin in a transgenic rabbit model of human hypertrophic cardiomyopathy. *Circ Res* 2005; 97: 285–292.

28 Elliott P, McKenna WJ. Hypertrophic cardiomyopathy. *Lancet* 2004; 363: 1881–1891.

29 Chimenti C, Pieroni M, Morgante E *et al.* Prevalence of Fabry disease in female patients with late-onset hypertrophic cardiomyopathy. *Circulation* 2004; 110: 1047–1053.

30 Gollob MH, Green MS, Tang AS *et al.* Identification of a gene responsible for familial Wolff–Parkinson–White syndrome. *N Engl J Med* 2001; 344: 1823–1831.

31 Gollob MH, Seger JJ, Gollob TN *et al.* Novel PRKAG2 mutation responsible for the genetic syndrome of ventricular preexcitation and conduction system disease with childhood onset and absence of cardiac hypertrophy. *Circulation* 2001; 104: 3030–3033.

32 Blair E, Redwood C, Ashrafian H *et al.* Mutations in the gamma(2) subunit of AMP-activated protein kinase cause familial hypertrophic cardiomyopathy: evidence for the central role of energy compromise in disease pathogenesis. *Hum Mol Genet* 2001; 10: 1215–1220.

33 Arad M, Benson DW, Perez-Atayde AR *et al.* Constitutively active AMP kinase mutations cause glycogen storage disease mimicking hypertrophic cardiomyopathy. *J Clin Invest* 2002; 109: 357–362.

34 Maron BJ, Gardin JM, Flack JM, Gidding SS, Kurosaki TT, Bild DE. Prevalence of hypertrophic cardiomyopathy in a general population of young adults. Echocardiographic analysis of 4111 subjects in the CARDIA Study. Coronary Artery Risk Development in (Young) Adults. *Circulation* 1995; 92: 785–789.

35 Miura K, Nakagawa H, Morikawa Y *et al.* Epidemiology of idiopathic cardiomyopathy in Japan: results from a nationwide survey. *Heart* 2002; 87: 126–130.

36 Kofflard MJ, ten Cate FJ, van der Lee C, van Domburg RT. Hypertrophic cardiomyopathy in a large community-based population: Clinical outcome and identification of risk factors for sudden cardiac death and clinical deterioration. *J Am Coll Cardiol* 2003; 41: 987–993.

37 Nienaber CA, Hiller S, Spielmann RP, Geiger M, Kuck KH. Syncope in hypertrophic cardiomyopathy: multivariate analysis of prognostic determinants. *J Am Coll Cardiol* 1990; 15: 948–955.

38 Elliott PM, Poloniecki J, Dickie S *et al.* Sudden death in hypertrophic cardiomyopathy: identification of high risk patients. *J Am Coll Cardiol* 2000; 36: 2212–2218.

39 Olivotto I, Cecchi F, Casey SA, Dolara A, Traverse JH, Maron BJ. Impact of atrial fibrillation on the clinical course of hypertrophic cardiomyopathy. *Circulation* 2001; 104: 2517–2524.

40 Monserrat L, Elliott PM, Gimeno JR, Sharma S, Penas-Lado M, McKenna WJ. Non-sustained ventricular tachycardia in hypertrophic cardiomyopathy: an independent marker of sudden death risk in young patients. *J Am Coll Cardiol* 2003; 42: 873–879.

41 Maron BJ, Shirani J, Poliac LC, Mathenge R, Roberts WC, Mueller FO. Sudden death in young competitive athletes. Clinical, demographic, and pathological profiles. *JAMA* 1996; 276: 199–204.

42 McKenna W, Deanfield J, Faruqui A, England D, Oakley C, Goodwin J. Prognosis in hypertrophic cardiomyopathy: role of age and clinical, electrocardiographic and hemodynamic features. *Am J Cardiol* 1981; 47: 532–538.

43 Marian AJ. On predictors of sudden cardiac death in hypertrophic cardiomyopathy. *J Am Coll Cardiol* 2003; 41: 994–996.

44 Cannan CR, Reeder GS, Bailey KR, Melton LJ III, Gersh BJ. Natural history of hypertrophic cardiomyopathy. A population-based study, 1976 through 1990. *Circulation* 1995; 92: 2488–2495.

45 Maron BJ, Olivotto I, Spirito P *et al.* Epidemiology of hypertrophic cardiomyopathy-related death: revisited in a large non-referral-based patient population. *Circulation* 2000; 102: 858–864.

46 Nugent AW, Daubeney PEF, Chondros P *et al.* Clinical features and outcomes of childhood hypertrophic cardiomyopathy: Results from a national population-based study. *Circulation* 2005; 112: 1332–1338.

47 Maron BJ, Wolfson JK, Ciro E, Spirito P. Relation of electrocardiographic abnormalities and patterns of left ventricular hypertrophy identified by 2-dimensional echocardiography in patients with hypertrophic cardiomyopathy. *Am J Cardiol* 1983; 51: 189–194.

48 Sakamoto T, Tei C, Murayama M, Ichiyasu H, Hada Y. Giant T wave inversion as a manifestation of asymmetrical apical hypertrophy (AAH) of the left ventricle. Echocardiographic and ultrasono-cardiotomographic study. *Jpn Heart J* 1976; 17: 611–629.

49 Eriksson MJ, Sonnenberg B, Woo A *et al.* Long-term outcome in patients with apical hypertrophic cardiomyopathy. *J Am Coll Cardiol* 2002; 39: 638–645.

50 Furubayashi K. Hemodynamic characterics of hypertrophic and congestive cardiomyopathies. *Jpn Circ J* 1981; 45: 1014–1024.

51 Hirota Y, Furubayashi K, Kaku K *et al.* Hypertrophic nonobstructive cardiomyopathy: a precise assessment of hemodynamic characteristics and clinical implications. *Am J Cardiol* 1982; 50: 990–997.

52 Nagueh SF, Bachinski L, Meyer D *et al.* Tissue Doppler imaging consistently detects myocardial abnormalities in patients with familial hypertrophic cardiomyopathy and provides a novel means for an early diagnosis prior to an independent of hypertrophy. *Circulation* 2001; 104: 128–130.

53 Beyar R. Hypertrophic cardiomyopathy: functional aspects by tagged magnetic resonance imaging. *Adv Exp Med Biol* 1995; 382: 293–301.

54 Nagueh SF, McFalls J, Meyer D *et al.* Tissue Doppler imaging predicts the development of hypertrophic cardiomyopathy in subjects with subclinical disease. *Circulation* 2003; 108: 395–398.

55 Solaro RJ, Varghese J, Marian AJ, Chandra M. Molecular mechanisms of cardiac myofilament activation: modulation by pH and a troponin T mutant R92Q. *Basic Res Cardiol* 2002; 97 (Supplement I): 102–110.

56 Hernandez O, Szczesna-Cordary D, Knollmann BC *et al.* F110I and R278C troponin T mutations that cause familial hypertrophic cardiomyopathy affect muscle contraction in transgenic mice and reconstituted human cardiac fibers. *J Biol Chem* 2005; 280: 37183–37194.

57 Chang AN, Harada K, Ackerman MJ, Potter JD. Functional consequences of hypertrophic and dilated cardiomyopathy causing mutations in alpha-tropomyosin. *J Biol Chem* 2005; 280: 34343–34349.

58 Cuda G, Fananapazir L, Epstein ND, Sellers JR. The *in vitro* motility activity of beta-cardiac myosin depends on the nature of the beta-myosin heavy chain gene mutation in hypertrophic cardiomyopathy. *J Muscle Res Cell Motil* 1997; 18: 275–283.

59 Deng Y, Schmidtmann A, Kruse S *et al.* Phosphorylation of human cardiac troponin I G203S and K206Q linked to familial hypertrophic cardiomyopathy affects actomyosin interaction in different ways. *J Mol Cell Cardiol* 2003; 35: 1365–1374.

60 Palmer BM, Fishbaugher DE, Schmitt JP *et al.* Differential cross-bridge kinetics of FHC myosin mutations R403Q and R453C in heterozygous mouse myocardium. *Am J Physiol Heart Circ Physiol* 2004; 287: H91–H99.

61 Fujita H, Sugiura S, Momomura S, Omata M, Sugi H, Sutoh K. Characterization of mutant myosins of *Dictyostelium discoideum* equivalent to human familial hypertrophic cardiomyopathy mutants. Molecular force level of mutant myosins may have a prognostic implication. *J Clin Invest* 1997; 99: 1010–1015.

62 Maron BJ, Roberts WC. Quantitative analysis of cardiac muscle cell disorganization in the ventricular septum of patients with hypertrophic cardiomyopathy. *Circulation* 1979; 59: 689–706.

63 Maron BJ, Anan TJ, Roberts WC. Quantitative analysis of the distribution of cardiac muscle cell disorganization in the left ventricular wall of patients with hypertrophic cardiomyopathy. *Circulation* 1981; 63: 882–894.

64 Shirani J, Pick R, Roberts WC, Maron BJ. Morphology and significance of the left ventricular collagen network in young patients with hypertrophic cardiomyopathy and sudden cardiac death. *J Am Coll Cardiol* 2000; 35: 36–44.

65 Spirito P, Bellone P, Harris KM, Bernabo P, Bruzzi P, Maron BJ. Magnitude of left ventricular hypertrophy

and risk of sudden death in hypertrophic cardiomyopathy. *N Engl J Med* 2000; 342: 1778–1785.

66 Varnava AM, Elliott PM, Baboonian C, Davison F, Davies MJ, McKenna WJ. Hypertrophic cardiomyopathy: histopathological features of sudden death in cardiac troponin T disease. *Circulation* 2001; 104: 1380–1384.

67 Varnava AM, Elliott PM, Mahon N, Davies MJ, McKenna WJ. Relation between myocyte disarray and outcome in hypertrophic cardiomyopathy. *Am J Cardiol* 2001; 88: 275–279.

68 Olson TM, Karst ML, Whitby FG, Driscoll DJ. Myosin light chain mutation causes autosomal recessive cardiomyopathy with mid-cavitary hypertrophy and restrictive physiology. *Circulation* 2002; 105: 2337–2340.

69 Greaves SC, Roche AH, Neutze JM, Whitlock RM, Veale AM. Inheritance of hypertrophic cardiomyopathy: a cross sectional and M mode echocardiographic study of 50 families. *Br Heart J* 1987; 58: 259–266.

70 Maron BJ, Nichols PF III, Pickle LW, Wesley YE, Mulvihill JJ. Patterns of inheritance in hypertrophic cardiomyopathy: assessment by M-mode and two-dimensional echocardiography. *Am J Cardiol* 1984; 53: 1087–1094.

71 Thierfelder L, Watkins H, MacRae C *et al.* Alpha-tropomyosin and cardiac troponin T mutations cause familial hypertrophic cardiomyopathy: a disease of the sarcomere. *Cell* 1994; 77: 701–712.

72 Marian AJ, Roberts R. On Koch's postulates, causality and genetics of cardiomyopathies. *J Mol Cell Cardiol* 2002; 34: 971–974.

73 Richard P, Charron P, Carrier L *et al.* Hypertrophic cardiomyopathy: distribution of disease genes, spectrum of mutations, and implications for a molecular diagnosis strategy. *Circulation* 2003; 107: 2227–2232.

74 Erdmann J, Raible J, Maki-Abadi J *et al.* Spectrum of clinical phenotypes and gene variants in cardiac myosin-binding protein C mutation carriers with hypertrophic cardiomyopathy. *J Am Coll Cardiol* 2001; 38: 322–330.

75 Charron P, Dubourg O, Desnos M *et al.* Clinical features and prognostic implications of familial hypertrophic cardiomyopathy related to the cardiac myosin-binding protein C gene. *Circulation* 1998; 97: 2230–2236.

76 Mogensen J, Murphy RT, Kubo T *et al.* Frequency and clinical expression of cardiac troponin I mutations in 748 consecutive families with hypertrophic cardiomyopathy. *J Am Coll Cardiol* 2004; 44: 2315–2325.

77 Van Driest SL, Jaeger MA, Ommen SR *et al.* Comprehensive analysis of the beta-myosin heavy chain gene in 389 unrelated patients with hypertrophic cardiomyopathy. *J Am Coll Cardiol* 2004; 44: 602–610.

78 Van Driest SL, Ellsworth EG, Ommen SR, Tajik AJ, Gersh BJ, Ackerman MJ. Prevalence and spectrum of thin filament mutations in an outpatient referral population with hypertrophic cardiomyopathy. *Circulation* 2003; 108: 445–451.

79 Andersen PS, Havndrup O, Bundgaard H *et al.* Genetic and phenotypic characterization of mutations in myosin-binding protein C (MYBPC3) in 81 families with familial hypertrophic cardiomyopathy: total or partial haploinsufficiency. *Eur J Hum Genet* 2004; 12: 673–677.

80 Torricelli F, Girolami F, Olivotto I *et al.* Prevalence and clinical profile of troponin T mutations among patients with hypertrophic cardiomyopathy in tuscany. *Am J Cardiol* 2003; 92: 1358–1362.

81 Marian AJ, Yu QT, Mares A Jr, Hill R, Roberts R, Perryman MB. Detection of a new mutation in the beta-myosin heavy chain gene in an individual with hypertrophic cardiomyopathy. *J Clin Invest* 1992; 90: 2156–2165.

82 Blair E, Redwood C, de Jesus OM *et al.* Mutations of the light meromyosin domain of the beta-myosin heavy chain rod in hypertrophic cardiomyopathy. *Circ Res* 2002; 90: 263–269.

83 Hougs L, Havndrup O, Bundgaard H *et al.* One third of Danish hypertrophic cardiomyopathy patients have mutations in MYH7 rod region. *Eur J Hum Genet* 2004; 13: 161–165.

84 Satoh M, Takahashi M, Sakamoto T, Hiroe M, Marumo F, Kimura A. Structural analysis of the titin gene in hypertrophic cardiomyopathy: identification of a novel disease gene. *Biochem Biophys Res Commun* 1999; 262: 411–417.

85 Mogensen J, Klausen IC, Pedersen AK *et al.* Alpha-cardiac actin is a novel disease gene in familial hypertrophic cardiomyopathy. *J Clin Invest* 1999; 103: R39–R43.

86 Flavigny J, Richard P, Isnard R *et al.* Identification of two novel mutations in the ventricular regulatory myosin light chain gene (MYL2) associated with familial and classical forms of hypertrophic cardiomyopathy. *J Mol Med* 1998; 76: 208–214.

87 Andersen PS, Havndrup O, Bundgaard H *et al.* Myosin light chain mutations in familial hypertrophic cardiomyopathy: phenotypic presentation and frequency in Danish and South African populations. *J Med Genet* 2001; 38: E43.

88 Hoffmann B, Schmidt-Traub H, Perrot A, Osterziel KJ, Gessner R. First mutation in cardiac troponin C, L29Q, in a patient with hypertrophic cardiomyopathy. *Hum Mutat* 2001; 17: 524.

89 Carniel E, Taylor MRG, Sinagra G *et al.* α-Myosin heavy chain: A sarcomeric gene associated with dilated and hypertrophic phenotypes of cardiomyopathy. *Circulation* 2005; 112: 54–59.

90 Davis JS, Hassanzadeh S, Winitsky S *et al.* The overall pattern of cardiac contraction depends on a spatial gradient of myosin regulatory light chain phosphorylation. *Cell* 2001; 107: 631–641.

91 Hayashi T, Arimura T, Itoh-Satoh M *et al.* Tcap gene mutations in hypertrophic cardiomyopathy and dilated cardiomyopathy. *J Am Coll Cardiol* 2004; 44: 2192–2201.

92 Minamisawa S, Sato Y, Tatsuguchi Y *et al.* Mutation of the phospholamban promoter associated with hypertrophic cardiomyopathy. *Biochem Biophys Res Commun* 2003; 304: 1–4.

93 Hayashi T, Arimura T, Ueda K *et al.* Identification and functional analysis of a caveolin-3 mutation associated with familial hypertrophic cardiomyopathy. *Biochem Biophys Res Commun* 2004; 313: 178–184.

94 Niimura H, Bachinski LL, Sangwatanaroj S *et al.* Mutations in the gene for cardiac myosin-binding protein C and late-onset familial hypertrophic cardiomyopathy. *N Engl J Med* 1998; 338: 1248–1257.

95 Anan R, Greve G, Thierfelder L *et al.* Prognostic implications of novel beta cardiac myosin heavy chain gene mutations that cause familial hypertrophic cardiomyopathy. *J Clin Invest* 1994; 93: 280–285.

96 Dausse E, Komajda M, Fetler L *et al.* Familial hypertrophic cardiomyopathy. Microsatellite haplotyping and identification of a hot spot for mutations in the beta-myosin heavy chain gene. *J Clin Invest* 1993; 92: 2807–2813.

97 Van Driest SL, Vasile VC, Ommen SR *et al.* Myosin binding protein C mutations and compound heterozygosity in hypertrophic cardiomyopathy. *J Am Coll Cardiol* 2004; 44: 1903–1910.

98 Blair E, Price SJ, Baty CJ, Ostman-Smith I, Watkins H. Mutations in *cis* can confound genotype–phenotype correlations in hypertrophic cardiomyopathy. *J Med Genet* 2001; 38: 385–388.

99 Marian AJ. On genetic and phenotypic variability of hypertrophic cardiomyopathy: nature versus nurture. *J Am Coll Cardiol* 2001; 38: 331–334.

100 Marian AJ. Modifier genes for hypertrophic cardiomyopathy. *Curr Opin Cardiol* 2002; 17: 242–252.

101 Marian AJ, Yu QT, Workman R, Greve G, Roberts R. Angiotensin-converting enzyme polymorphism in hypertrophic cardiomyopathy and sudden cardiac death. *Lancet* 1993; 342: 1085–1086.

102 Lechin M, Quinones MA, Omran A *et al.* Angiotensin-I converting enzyme genotypes and left ventricular hypertrophy in patients with hypertrophic cardiomyopathy. *Circulation* 1995; 92: 1808–1812.

103 Tesson F, Dufour C, Moolman JC *et al.* The influence of the angiotensin I converting enzyme genotype in familial hypertrophic cardiomyopathy varies with the disease gene mutation. *J Mol Cell Cardiol* 1997; 29: 831–838.

104 Brugada R, Kelsey W, Lechin M *et al.* Role of candidate modifier genes on the phenotypic expression of hypertrophy in patients with hypertrophic cardiomyopathy. *J Investig Med* 1997; 45: 542–551.

105 Wilcox WR, Banikazemi M, Guffon N *et al.* Long-term safety and efficacy of enzyme replacement therapy for Fabry disease. *Am J Hum Genet* 2004; 75: 65–74.

106 Eng CM, Guffon N, Wilcox WR *et al.* Safety and efficacy of recombinant human alpha-galactosidase A: replacement therapy in Fabry's disease. *N Engl J Med* 2001; 345: 9–16.

107 Desnick RJ, Brady R, Barranger J *et al.* Fabry disease, an under-recognized multisystemic disorder: Expert recommendations for diagnosis, management, and enzyme replacement therapy. *Ann Intern Med* 2003; 138: 338–346.

108 Sachdev B, Takenaka T, Teraguchi H *et al.* Prevalence of Anderson–Fabry disease in male patients with late onset hypertrophic cardiomyopathy. *Circulation* 2002; 105: 1407–1411.

109 Schiffmann R, Murray GJ, Treco D *et al.* Infusion of alpha-galactosidase A reduces tissue globotriaosylceramide storage in patients with Fabry disease. *Proc Natl Acad Sci USA* 2000; 97: 365–370.

110 Frustaci A, Chimenti C, Ricci R *et al.* Improvement in cardiac function in the cardiac variant of Fabry's disease with galactose-infusion therapy. *N Engl J Med* 2001; 345: 25–32.

111 Cummings CJ, Zoghbi HY. Trinucleotide repeats: mechanisms and pathophysiology. *Annu Rev Genomics Hum Genet* 2000; 1: 281–328.

112 Palau F. Friedreich's ataxia and frataxin: molecular genetics, evolution and pathogenesis. *Int J Mol Med* 2001; 7: 581–589.

113 Tartaglia M, Mehler EL, Goldberg R *et al.* Mutations in PTPN11, encoding the protein tyrosine phosphatase SHP-2, cause Noonan syndrome. *Nat Genet* 2001; 29: 465–468.

114 Tartaglia M, Kalidas K, Shaw A *et al.* PTPN11 mutations in Noonan syndrome: molecular spectrum, genotype-phenotype correlation, and phenotypic heterogeneity. *Am J Hum Genet* 2002; 70: 1555–1563.

115 Ashizawa T, Subramony SH. What is Kearns–Sayre syndrome after all? *Arch Neurol* 2001; 58: 1053–1054.

116 Mihalik SJ, Morrell JC, Kim D, Sacksteder KA, Watkins PA, Gould SJ. Identification of PAHX, a Refsum disease gene. *Nat Genet* 1997; 17: 185–189.

117 Raben N, Plotz P, Byrne BJ. Acid alpha-glucosidase deficiency (glycogenosis type II, Pompe disease). *Curr Mol Med* 2002; 2: 145–166.

118 Charron P, Villard E, Sebillon P *et al.* Danon's disease as a cause of hypertrophic cardiomyopathy: a systematic survey. *Heart* 2004; 90: 842–846.

119 Guertl B, Noehammer C, Hoefler G. Metabolic cardiomyopathies. *Int J Exp Pathol* 2000; 81: 349–372.

120 Watkins H, McKenna WJ, Thierfelder L *et al.* Mutations in the genes for cardiac troponin T and alpha-tropomyosin in hypertrophic cardiomyopathy. *N Engl J Med* 1995; 332: 1058–1064.

121 Li D, Czernuszewicz GZ, Gonzalez O *et al.* Novel cardiac troponin T mutation as a cause of familial dilated cardiomyopathy. *Circulation* 2001; 104: 2188–2193.

122 Morimoto S, Yanaga F, Minakami R, Ohtsuki I. Ca^{2+}-sensitizing effects of the mutations at Ile-79 and Arg-92 of troponin T in hypertrophic cardiomyopathy. *Am J Physiol* 1998; 275: C200–C207.

123 Morimoto S, Lu QW, Harada K *et al.* Ca^{2+}-desensitizing effect of a deletion mutation Delta K210 in cardiac troponin T that causes familial dilated cardiomyopathy. *Proc Natl Acad Sci USA* 2002; 99: 913–918.

124 Harada K, Potter JD. Familial hypertrophic cardiomyopathy mutations from different functional regions of troponin T result in different effects on the pH and Ca^{2+} sensitivity of cardiac muscle contraction. *J Biol Chem* 2004; 279: 14488–14495.

125 Lu QW, Morimoto S, Harada K *et al.* Cardiac troponin T mutation R141W found in dilated cardiomyopathy stabilizes the troponin T-tropomyosin interaction and causes a Ca^{2+} desensitization. *J Mol Cell Cardiol* 2003; 35: 1421–1427.

126 Watkins H, Rosenzweig A, Hwang DS *et al.* Characteristics and prognostic implications of myosin missense mutations in familial hypertrophic cardiomyopathy. *N Engl J Med* 1992; 326: 1108–1114.

127 Fananapazir L, Epstein ND. Genotype–phenotype correlations in hypertrophic cardiomyopathy. Insights provided by comparisons of kindreds with distinct and identical beta-myosin heavy chain gene mutations. *Circulation* 1994; 89: 22–32.

128 Epstein ND, Cohn GM, Cyran F, Fananapazir L. Differences in clinical expression of hypertrophic cardiomyopathy associated with two distinct mutations in the beta-myosin heavy chain gene. A 908Leu–Val mutation and a 403Arg–Gln mutation [see comments]. *Circulation* 1992; 86: 345–352.

129 Marian AJ, Mares A Jr, Kelly DP *et al.* Sudden cardiac death in hypertrophic cardiomyopathy. Variability in phenotypic expression of beta-myosin heavy chain mutations. *Eur Heart J* 1995; 16: 368–376.

130 Watkins H, McKenna WJ, Thierfelder L *et al.* Mutations in the genes for cardiac troponin T and alpha-tropomyosin in hypertrophic cardiomyopathy. *N Engl J Med* 1995; 332: 1058–1064.

131 Tesson F, Richard P, Charron P *et al.* Genotype–phenotype analysis in four families with mutations in beta-myosin heavy chain gene responsible for familial hypertrophic cardiomyopathy. *Hum Mutat* 1998; 12: 385–392.

132 Charron P, Dubourg O, Desnos M *et al.* Genotype–phenotype correlations in familial hypertrophic cardiomyopathy. A comparison between mutations in the cardiac protein-C and the beta-myosin heavy chain genes. *Eur Heart J* 1998; 19: 139–145.

133 Maron BJ, Niimura H, Casey SA *et al.* Development of left ventricular hypertrophy in adults in hypertrophic cardiomyopathy caused by cardiac myosin-binding protein C gene mutations. *J Am Coll Cardiol* 2001; 38: 315–321.

134 Niimura H, Patton KK, McKenna WJ *et al.* Sarcomere protein gene mutations in hypertrophic cardiomyopathy of the elderly. *Circulation* 2002; 105: 446–451.

135 Marian AJ, Roberts R. To screen or not is not the question: it is when and how to screen. *Circulation* 2003; 107: 2171–2174.

136 Marian AJ. Pathogenesis of diverse clinical and pathological phenotypes in hypertrophic cardiomyopathy. *Lancet* 2000; 355: 58–60.

137 Nagueh SF, Chen S, Patel R *et al.* Evolution of expression of cardiac phenotypes over a 4-year period in the β-myosin heavy chain-Q403 transgenic rabbit model of human hypertrophic cardiomyopathy. *J Mol Cell Cardiol* 2004; 36: 663–673.

138 Blanchard E, Seidman C, Seidman JG, LeWinter M, Maughan D. Altered crossbridge kinetics in the alphaMHC403/+ mouse model of familial hypertrophic cardiomyopathy. *Circ Res* 1999; 84: 475–483.

139 Lange S, Xiang F, Yakovenko A *et al.* The kinase domain of titin controls muscle gene expression and protein turnover. *Science* 2005; 308: 1599–1603.

140 Marian AJ, Zhao G, Seta Y, Roberts R, Yu QT. Expression of a mutant (Arg92Gln) human cardiac troponin T, known to cause hypertrophic cardiomyopathy, impairs adult cardiac myocyte contractility. *Circ Res* 1997; 81: 76–85.

141 Nagueh SF, Lakkis NM, Middleton KJ, Spencer WH III, Zoghbi WA, Quinones MA. Doppler estimation of left ventricular filling pressures in patients with hypertrophic cardiomyopathy. *Circulation* 1999; 99: 254–261.

142 Rust EM, Albayya FP, Metzger JM. Identification of a contractile deficit in adult cardiac myocytes expressing hypertrophic cardiomyopathy-associated mutant troponin T proteins. *J Clin Invest* 1999; 103: 1459–1467.

143 Watkins H, Seidman CE, Seidman JG, Feng HS, Sweeney HL. Expression and functional assessment of a truncated cardiac troponin T that causes hypertrophic cardiomyopathy. Evidence for a dominant negative action [see comments]. *J Clin Invest* 1996; 98: 2456–2461.

144 Crilley JG, Boehm EA, Blair E *et al.* Hypertrophic cardiomyopathy due to sarcomeric gene mutations is

characterized by impaired energy metabolism irrespective of the degree of hypertrophy. *J Am Coll Cardiol* 2003; 41: 1776–1782.

145 Tsybouleva N, Zhang L, Chen SN *et al.* Aldosterone, through novel signaling proteins, is a fundamental molecular bridge between the genetic defect and the cardiac phenotype of hypertrophic cardiomyopathy. *Circulation* 2004; 109: 1284–1291.

146 Frame S, Cohen P. GSK3 takes centre stage more than 20 years after its discovery. *Biochem J* 2001; 359: 1–16.

147 Rottbauer W, Gautel M, Zehelein J *et al.* Novel splice donor site mutation in the cardiac myosin-binding protein-C gene in familial hypertrophic cardiomyopathy. Characterization of cardiac transcript and protein. *J Clin Invest* 1997; 100: 475–482.

148 Jones WK, Grupp IL, Doetschman T *et al.* Ablation of the murine alpha myosin heavy chain gene leads to dosage effects and functional deficits in the heart. *J Clin Invest* 1996; 98: 1906–1917.

149 Blanchard EM, Iizuka K, Christe M *et al.* Targeted ablation of the murine alpha-tropomyosin gene. *Circ Res* 1997; 81: 1005–1010.

150 Rethinasamy P, Muthuchamy M, Hewett T *et al.* Molecular and physiological effects of alpha-tropomyosin ablation in the mouse. *Circ Res* 1998; 82: 116–123.

151 van der Lee C, ten Cate FJ, Geleijnse ML *et al.* Percutaneous versus surgical treatment for patients with hypertrophic obstructive cardiomyopathy and enlarged anterior mitral valve leaflets. *Circulation* 2005; 112: 482–488.

152 Nagueh SF, Ommen SR, Lakkis NM *et al.* Comparison of ethanol septal reduction therapy with surgical myectomy for the treatment of hypertrophic obstructive cardiomyopathy. *J Am Coll Cardiol* 2001; 38: 1701–1706.

153 Ralph-Edwards A, Woo A, McCrindle BW *et al.* Hypertrophic obstructive cardiomyopathy: Comparison of outcomes after myectomy or alcohol ablation adjusted by propensity score. *J Thorac Cardiovasc Surg* 2005; 129: 351–358.

154 Firoozi S, Elliott PM, Sharma S *et al.* Septal myotomy–myectomy and transcoronary septal alcohol ablation in hypertrophic obstructive cardiomyopathy. A comparison of clinical, haemodynamic and exercise outcomes. *Eur Heart J* 2002; 23: 1617–1624.

155 Qin JX, Shiota T, Lever HM *et al.* Outcome of patients with hypertrophic obstructive cardiomyopathy after percutaneous transluminal septal myocardial ablation and septal myectomy surgery. *J Am Coll Cardiol* 2001; 38: 1994–2000.

156 Heric B, Lytle BW, Miller DP, Rosenkranz ER, Lever HM, Cosgrove DM. Surgical management of hypertrophic obstructive cardiomyopathy. Early and late results. *J Thorac Cardiovasc Surg* 1995; 110: 195–206.

157 Kimmelstiel CD, Maron BJ. Role of percutaneous septal ablation in hypertrophic obstructive cardiomyopathy. *Circulation* 2004; 109: 452–456.

158 Maron BJ, Dearani JA, Ommen SR *et al.* The case for surgery in obstructive hypertrophic cardiomyopathy. *J Am Coll Cardiol* 2004; 44: 2044–2053.

159 Hess OM, Sigwart U. New treatment strategies for hypertrophic obstructive cardiomyopathy: Alcohol ablation of the septum: the new gold standard? *J Am Coll Cardiol* 2004; 44: 2054–2055.

160 Maron MS, Olivotto I, Betocchi S *et al.* Effect of left ventricular outflow tract obstruction on clinical outcome in hypertrophic cardiomyopathy. *N Engl J Med* 2003; 348: 295–303.

161 Morrow AG, Reitz BA, Epstein SE *et al.* Operative treatment in hypertrophic subaortic stenosis. Techniques, and the results of pre and postoperative assessments in 83 patients. *Circulation* 1975; 52: 88–102.

162 Merrill WH, Friesinger GC, Graham TP Jr *et al.* Long-lasting improvement after septal myectomy for hypertrophic obstructive cardiomyopathy. *Ann Thorac Surg* 2000; 69: 1732–1735.

163 Schonbeck MH, Brunner-La Rocca H, Vogt PR *et al.* Long-term follow-up in hypertrophic obstructive cardiomyopathy after septal myectomy. *Ann Thorac Surg* 1998; 65: 1207–1214.

164 Schulte HD, Bircks WH, Loesse B, Godehardt EA, Schwartzkopff B. Prognosis of patients with hypertrophic obstructive cardiomyopathy after transaortic myectomy. Late results up to twenty-five years. *J Thorac Cardiovasc Surg* 1993; 106: 709–717.

165 Seggewiss H. Current status of alcohol septal ablation for patients with hypertrophic cardiomyopathy. *Curr Cardiol Rep* 2001; 3: 160–166.

166 Faber L, Meissner A, Ziemssen P, Seggewiss H. Percutaneous transluminal septal myocardial ablation for hypertrophic obstructive cardiomyopathy: long term follow up of the first series of 25 patients. *Heart* 2000; 83: 326–331.

167 Mazur W, Nagueh SF, Lakkis NM *et al.* Regression of left ventricular hypertrophy after nonsurgical septal reduction therapy for hypertrophic obstructive cardiomyopathy. *Circulation* 2001; 103: 1492–1496.

168 Minamino T, Gaussin V, DeMayo FJ, Schneider MD. Inducible gene targeting in postnatal myocardium by cardiac-specific expression of a hormone-activated Cre fusion protein. *Circ Res* 2001; 88: 587–592.

169 Qin JX, Shiota T, Lever HM *et al.* Conduction system abnormalities in patients with obstructive hypertrophic cardiomyopathy following septal reduction interventions. *Am J Cardiol* 2004; 93: 171–175.

170 Chang SM, Lakkis NM, Franklin J, Spencer WH III, Nagueh SF. Predictors of outcome after alcohol septal

ablation therapy in patients with hypertrophic obstructive cardiomyopathy. *Circulation* 2004; 109: 824–827.

171 McGregor JB, Rahman A, Rosanio S, Ware D, Birnbaum Y, Saeed M. Monomorphic ventricular tachycardia: a late complication of percutaneous alcohol septal ablation for hypertrophic cardiomyopathy. *Am J Med Sci* 2004; 328: 185–188.

172 Boltwood CM Jr, Chien W, Ports T. Ventricular tachycardia complicating alcohol septal ablation. *N Engl J Med* 2004; 351: 1914–1915.

173 Kaplan SR, Gard JJ, Carvajal-Huerta L, Ruiz-Cabezas JC, Thiene G, Saffitz JE. Structural and molecular pathology of the heart in Carvajal syndrome. *Cardiovasc Pathol* 2004; 13: 26–32.

174 Fananapazir L, Epstein ND, Curiel RV, Panza JA, Tripodi D, McAreavey D. Long-term results of dual-chamber (DDD) pacing in obstructive hypertrophic cardiomyopathy. Evidence for progressive symptomatic and hemodynamic improvement and reduction of left ventricular hypertrophy. *Circulation* 1994; 90: 2731–2742.

175 Kappenberger L, Linde C, Daubert C *et al.* Pacing in hypertrophic obstructive cardiomyopathy. A randomized crossover study. PIC Study Group. *Eur Heart J* 1997; 18: 1249–1256.

176 Maron BJ, Nishimura RA, McKenna WJ, Rakowski H, Josephson ME, Kieval RS. Assessment of permanent dual-chamber pacing as a treatment for drug-refractory symptomatic patients with obstructive hypertrophic cardiomyopathy. A randomized, double-blind, crossover study (M-PATHY). *Circulation* 1999; 99: 2927–2933.

177 Delbosc S, Cristol JP, Descomps B, Mimran A, Jover B. Simvastatin prevents angiotensin II-induced cardiac alteration and oxidative stress. *Hypertension* 2002; 40: 142–147.

178 Takemoto M, Node K, Nakagami H *et al.* Statins as antioxidant therapy for preventing cardiac myocyte hypertrophy. *J Clin Invest* 2001; 108: 1429–1437.

179 Dorn GW. Calcineurin inhibition in hypertrophy: back from the dead! *Circulation* 2001; 104: 9–11.

180 Fatkin D, McConnell BK, Mudd JO *et al.* An abnormal Ca²⁺ response in mutant sarcomere protein-mediated familial hypertrophic cardiomyopathy. *J Clin Invest* 2000; 106: 1351–1359.

181 Fananapazir L, Chang AC, Epstein SE, McAreavey D. Prognostic determinants in hypertrophic cardiomyopathy. Prospective evaluation of a therapeutic strategy based on clinical, Holter, hemodynamic, and electrophysiological findings. *Circulation* 1992; 86: 730–740.

182 Bonne G, Carrier L, Bercovici J *et al.* Cardiac myosin binding protein-C gene splice acceptor site mutation is associated with familial hypertrophic cardiomyopathy. *Nat Genet* 1995; 11: 438–440.

183 Watkins H, Conner D, Thierfelder L *et al.* Mutations in the cardiac myosin binding protein-C gene on chromosome 11 cause familial hypertrophic cardiomyopathy. *Nat Genet* 1995; 11: 434–437.

184 Kimura A, Harada H, Park JE *et al.* Mutations in the cardiac troponin I gene associated with hypertrophic cardiomyopathy. *Nat Genet* 1997; 16: 379–382.

185 Poetter K, Jiang H, Hassanzadeh S *et al.* Mutations in either the essential or regulatory light chains of myosin are associated with a rare myopathy in human heart and skeletal muscle. *Nat Genet* 1996; 13: 63–69.

186 Murphy RT, Mogensen J, McGarry K *et al.* Adenosine monophosphate-activated protein kinase disease mimicks hypertrophic cardiomyopathy and Wolff–Parkinson–White syndrome: Natural history. *J Am Coll Cardiol* 2005; 45: 922–930.

187 Sawada K, Mizoguchi K, Hishida A *et al.* Point mutation in the alpha-galactosidase A gene of atypical Fabry disease with only nephropathy. *Clin Nephrol* 1996; 45: 289–294.

188 Mohiddin SA, Ahmed ZM, Griffith AJ *et al.* Novel association of hypertrophic cardiomyopathy, sensorineural deafness, and a mutation in unconventional myosin VI (MYO6). *J Med Genet* 2004; 41: 309–314.

189 Yang Z, McMahon CJ, Smith LR *et al.* Danon disease as an underrecognized cause of hypertrophic cardiomyopathy in children. *Circulation* 2005; 112: 1612–1617.

190 Lu QW, Morimoto S, Harada K *et al.* Cardiac troponin T mutation R141W found in dilated cardiomyopathy stabilizes the troponin T-tropomyosin interaction and causes a Ca²⁺ desensitization. *J Mol Cell Cardiol* 2003; 35: 1421–1427.

191 Rotig A, de Lonlay P, Chretien D *et al.* Aconitase and mitochondrial iron-sulphur protein deficiency in Friedreich ataxia. *Nat Genet* 1997; 17: 215–217.

192 Igarashi H, Momoi MY, Yamagata T, Shiraishi H, Eguchi I. Hypertrophic cardiomyopathy in congenital myotonic dystrophy. *Pediatr Neurol* 1998; 18: 366–369.

193 Fatkin D, Christe ME, Aristizabal O *et al.* Neonatal cardiomyopathy in mice homozygous for the Arg403Gln mutation in the alpha cardiac myosin heavy chain gene. *J Clin Invest* 1999; 103: 147–153.

194 Geisterfer-Lowrance AA, Christe M, Conner DA *et al.* A mouse model of familial hypertrophic cardiomyopathy. *Science* 1996; 272: 731–734.

195 Kim SJ, Iizuka K, Kelly RA *et al.* An alpha-cardiac myosin heavy chain gene mutation impairs contraction and relaxation function of cardiac myocytes. *Am J Physiol* 1999; 276: H1780–H1787.

196 Georgakopoulos D, Christe ME, Giewat M, Seidman CM, Seidman JG, Kass DA. The pathogenesis of familial hypertrophic cardiomyopathy: early and evolving effects from an alpha-cardiac myosin heavy chain mis-

sense mutation [see comments]. *Nat Med* 1999; 5: 327–330.

197 Berul CI, Christe ME, Aronovitz MJ *et al.* Familial hypertrophic cardiomyopathy mice display gender differences in electrophysiological abnormalities. *J Interv Card Electrophysiol* 1998; 2: 7–14.

198 Berul CI, Christe ME, Aronovitz MJ, Seidman CE, Seidman JG, Mendelsohn ME. Electrophysiological abnormalities and arrhythmias in alpha MHC mutant familial hypertrophic cardiomyopathy mice. *J Clin Invest* 1997; 99: 570–576.

199 Bevilacqua LM, Maguire CT, Seidman JG, Seidman CE, Berul CI. QT dispersion in alpha-myosin heavy-chain familial hypertrophic cardiomyopathy mice. *Pediatr Res* 1999; 45: 643–647.

200 Gao WD, Perez NG, Seidman CE, Seidman JG, Marban E. Altered cardiac excitation-contraction coupling in mutant mice with familial hypertrophic cardiomyopathy. *J Clin Invest* 1999; 103: 661–666.

201 Tyska MJ, Hayes E, Giewat M, Seidman CE, Seidman JG, Warshaw DM. Single-molecule mechanics of R403Q cardiac myosin isolated from the mouse model of familial hypertrophic cardiomyopathy. *Circ Res* 2000; 86: 737–744.

202 Spindler M, Saupe KW, Christe ME *et al.* Diastolic dysfunction and altered energetics in the alphaMHC403/+ mouse model of familial hypertrophic cardiomyopathy. *J Clin Invest* 1998; 101: 1775–1783.

203 Jin JP, Wang J, Ogut O. Developmentally regulated muscle type-specific alternative splicing of the COOH-terminal variable region of fast skeletal muscle troponin T and an aberrant splicing pathway to encode a mutant COOH-terminus. *Biochem Biophys Res Commun* 1998; 242: 540–544.

204 Vikstrom KL, Factor SM, Leinwand LA. Mice expressing mutant myosin heavy chains are a model for familial hypertrophic cardiomyopathy. *Mol Med* 1996; 2: 556–567.

205 Vikstrom KL, Bohlmeyer T, Factor SM, Leinwand LA. Hypertrophy, pathology, and molecular markers of cardiac pathogenesis. *Circ Res* 1998; 82: 773–778.

206 Welikson RE, Buck SH, Patel JR *et al.* Cardiac myosin heavy chains lacking the light chain binding domain cause hypertrophic cardiomyopathy in mice. *Am J Physiol* 1999; 276: H2148–H2158.

207 Tardiff JC, Factor SM, Tompkins BD *et al.* A truncated cardiac troponin T molecule in transgenic mice suggests multiple cellular mechanisms for familial hypertrophic cardiomyopathy. *J Clin Invest* 1998; 101: 2800–2811.

208 Tardiff JC, Hewett TE, Palmer BM *et al.* Cardiac troponin T mutations result in allele-specific phenotypes in a mouse model for hypertrophic cardiomyopathy. *J Clin Invest* 1999; 104: 469–481.

209 Oberst L, Zhao G, Park JT *et al.* Dominant-negative effect of a mutant cardiac troponin T on cardiac structure and function in transgenic mice. *J Clin Invest* 1998; 102: 1498–1505.

210 Miller T, Szczesna D, Housmans PR *et al.* Abnormal contractile function in transgenic mice expressing a familial hypertrophic cardiomyopathy-linked troponin T (I79N) mutation. *J Biol Chem* 2001; 276: 3743–3755.

211 Yang Q, Sanbe A, Osinska H, Hewett TE, Klevitsky R, Robbins J. A mouse model of myosin binding protein C human familial hypertrophic cardiomyopathy. *J Clin Invest* 1998; 102: 1292–1300.

212 McConnell BK, Jones KA, Fatkin D *et al.* Dilated cardiomyopathy in homozygous myosin-binding protein-C mutant mice. *J Clin Invest* 1999; 104: 1235–1244.

213 McConnell BK, Fatkin D, Semsarian C *et al.* Comparison of two murine models of familial hypertrophic cardiomyopathy. *Circ Res* 2001; 88: 383–389.

214 Vemuri R, Lankford EB, Poetter K *et al.* The stretch-activation response may be critical to the proper functioning of the mammalian heart. *Proc Natl Acad Sci USA* 1999; 96: 1048–1053.

215 Sanbe A, Nelson D, Gulick J *et al. In vivo* analysis of an essential myosin light chain mutation linked to familial hypertrophic cardiomyopathy. *Circ Res* 2000; 87: 296–302.

216 Muthuchamy M, Pieples K, Rethinasamy P *et al.* Mouse model of a familial hypertrophic cardiomyopathy mutation in alpha-tropomyosin manifests cardiac dysfunction. *Circ Res* 1999; 85: 47–56.

217 Evans CC, Pena JR, Phillips RM *et al.* Altered hemodynamics in transgenic mice harboring mutant tropomyosin linked to hypertrophic cardiomyopathy. *Am J Physiol Heart Circ Physiol* 2000; 279: H2414–H2423.

218 James J, Zhang Y, Osinska H *et al.* Transgenic modeling of a cardiac troponin I mutation linked to familial hypertrophic cardiomyopathy. *Circ Res* 2000; 87: 805–811.

219 Frey N, Franz WM, Gloeckner K *et al.* Transgenic rat hearts expressing a human cardiac troponin T deletion reveal diastolic dysfunction and ventricular arrhythmias. *Cardiovasc Res* 2000; 47: 254–264.

220 Marian AJ, Wu Y, Lim DS *et al.* A transgenic rabbit model for human hypertrophic cardiomyopathy. *J Clin Invest* 1999; 104: 1683–1692.

221 Nagueh SF, Kopelen HA, Lim DS *et al.* Tissue Doppler imaging consistently detects myocardial contraction and relaxation abnormalities, irrespective of cardiac hypertrophy, in a transgenic rabbit model of human hypertrophic cardiomyopathy. *Circulation* 2000; 102: 1346–1350.

222 Sanbe A, James J, Tuzcu V *et al.* Transgenic rabbit model for human troponin I-based hypertrophic cardiomyopathy. *Circulation* 2005; 111: 2330–2338.

223 Gomes AV, Harada K, Potter JD. A mutation in the N-terminus of troponin I that is associated with hypertrophic cardiomyopathy affects the Ca^{2+}-sensitivity, phosphorylation kinetics and proteolytic susceptibility of troponin. *J Mol Cell Cardiol* 2005; 39: 754–765.

224 Kruger M, Zittrich S, Redwood C *et al.* Effects of the mutation R145G in human cardiac troponin I on the kinetics of the contraction-relaxation cycle in isolated cardiac myofibrils. *J Physiol (Lond)* 2005; 564: 347–357.

225 Doolan A, Tebo M, Ingles J *et al.* Cardiac troponin I mutations in Australian families with hypertrophic cardiomyopathy: clinical, genetic and functional consequences. *J Mol Cell Cardiol* 2005; 38: 387–393.

226 Szczesna-Cordary D, Guzman G, Zhao J, Hernandez O, Wei J, Diaz-Perez Z. The E22K mutation of myosin RLC that causes familial hypertrophic cardiomyopathy increases calcium sensitivity of force and ATPase in transgenic mice. *J Cell Sci* 2005; 118: 3675–3683.

227 Chandra M, Tschirgi ML, Tardiff JC. Increase in tension-dependent ATP consumption induced by cardiac troponin T mutation. *Am J Physiol Heart Circ Physiol* 2005; 289: H2112–H2119.

228 Javadpour MM, Tardiff JC, Pinz I, Ingwall JS. Decreased energetics in murine hearts bearing the R92Q mutation in cardiac troponin T. *J Clin Invest* 2003; 112: 768–775.

229 Palmiter KA, Tyska MJ, Haeberle JR, Alpert NR, Fananapazir L, Warshaw DM. R403Q and L908V mutant beta-cardiac myosin from patients with familial hypertrophic cardiomyopathy exhibit enhanced mechanical performance at the single molecule level. *J Muscle Res Cell Motil* 2000; 21: 609–620.

230 Palmer BM, Fishbaugher DE, Schmitt JP *et al.* Differential cross-bridge kinetics of FHC myosin mutations R403Q and R453C in heterozygous mouse myocardium. *Am J Physiol Heart Circ Physiol* 2004; 287: H91–H99.

231 Kremneva E, Boussouf S, Nikolaeva O, Maytum R, Geeves MA, Levitsky DI. Effects of two familial hypertrophic cardiomyopathy mutations in α-tropomyosin, Asp175Asn and Glu180Gly, on the thermal unfolding of actin-bound tropomyosin. *Biophys J* 2004; 87: 3922–3933.

232 Kobayashi T, Dong WJ, Burkart EM, Cheung HC, Solaro RJ. Effects of protein kinase C dependent phosphorylation and a familial hypertrophic cardiomyopathy-related mutation of cardiac troponin I on structural transition of troponin C and myofilament activation. *Biochemistry* 2004; 43: 5996–6004.

233 Vang S, Corydon TJ, Borglum AD *et al.* Actin mutations in hypertrophic and dilated cardiomyopathy cause inefficient protein folding and perturbed filament formation. *FEBS J* 2005; 272: 2037–2049.

234 Barta J, Toth A, Jaquet K, Redlich A, Edes I, Papp Z. Calpain-1-dependent degradation of troponin I mutants found in familial hypertrophic cardiomyopathy. *Mol Cell Biochem* 2003; 251: 83–88.

235 Li MX, Wang X, Lindhout DA, Buscemi N, VanEyk JE, Sykes BD. Phosphorylation and mutation of human cardiac troponin I deferentially destabilize the interaction of the functional regions of troponin I with troponin C. *Biochemistry* 2003; 42: 14460–14468.

236 Wang Q, Moncman CL, Winkelmann DA. Mutations in the motor domain modulate myosin activity and myofibril organization. *J Cell Sci* 2003; 116: 4227–4238.

CHAPTER 4

Dilated cardiomyopathy and other cardiomyopathies

Mitra Esfandiarei, PhD, *Bobby Yanagawa,* PhD, *&*
Bruce M. McManus, MD, PhD, FRSC

Introduction

Cardiomyopathy or "sickness" of the heart muscle is a condition characterized by diastolic or systolic cardiac dysfunction in which the main abnormality lies in the myocardium itself. This group of disorders is responsible for acute and chronic heart failure and arrhythmias and, secondarily, disability and death [1]. The origins and pathogenesis of heart muscle disease are diverse, often reflecting a collision of genetic and environmental factors, or ecogenetics. Certain cardiomyopathies remain completely unexplained from an etiologic or mechanistic standpoint.

Classification

Cardiomyopathies have been traditionally divided into two main categories: primary (idiopathic) diseases of unknown causes and secondary diseases of known causes or associated with disorders of other systems. In defining the cardiomyopathies clinically, it is useful to recognize the varied pathophysiology that is expressed. Both the primary and secondary categories have three possible functional states:

1 *Hypertrophic:* hyperdynamic, characterized by massive left ventricular hypertrophy, predominantly in the septal region, variable dynamic outflow tract obstruction, diastolic and systolic dysfunction, and a familial propensity.

2 *Dilated:* congestive, ventricular dilatation, systolic dysfunction, often resulting in systolic heart failure, and less common familial tendency.

3 *Restrictive:* constrictive, stiff muscle and/or endocardial scarring and associated diastolic dysfunction.

To designate a cardiomyopathy as primary, acquired diseases of heart valves, coronary arteries, pericardium and aorta, and congenital cardiac defects must be excluded. Myocardial storage diseases and secondary endocardial diseases also must be sought and excluded as major causes of cardiac dysfunction.

In secondary cardiomyopathies, the cause of myocardial abnormalities is known and may be a manifestation of a systemic disease process [1]. Although therapy at times may be similar for primary and secondary myocardial diseases, treatment is often distinctly different and the specific diagnosis may carry a very different prognosis. In 1995, the World Health Organization/International Society and Federation of Cardiology (WHO/ISFC) Task Force recommended that the cardiomyopathies be classified into specific cardiomyopathies and primary cardiomyopathies [2]. This reorganization was necessitated by the discovery of new entities, not included in the 1980 WHO classification.

The specific cardiomyopathies include heart muscle diseases associated with myocarditis, termed inflammatory cardiomyopathy, as well as ischemic, valvular, hypertensive diseases and those diseases associated with cardiac or systemic disorders, such as amyloidosis and hemochromatosis. The primary cardiomyopathies are intrinsic to myocardium and include dilated cardiomyopathy (DCM), arrhythmogenic right ventricular cardiomyopathy (ARVC) and unclassified cardiomyopathies. Restrictive cardiomyopathies can be classified as both primary and secondary. Unclassified cardiomyopathies include left ventricular noncompaction, mitochondrial cardiomyopathies and endocardial fibroelastosis.

Characterization of cardiomyopathies

A normal healthy adult has a heart weight that correlates with his/her body size. The adult male heart weight is in the range 250–350 g, while the adult female heart is in the range 200–300 g. The normal diastolic thickness of the right ventricular muscle is <0.5 cm and that of the left <1.5 cm [3]. Ventricular thicknesses that are above these normal levels indicate hypertrophy and increased mass, while measurements below this index but with enlarged chambers imply dilatation. However, dilated hearts wherein the ventricular walls are of normal thickness still may have a marked increase in myocardial mass.

The myocardium has a spiral layered organization formed by a syncytial-like arrangement of cardiac myocytes [4]. Myocytes are elongated and joined to one another by intercellular junctions (mediating cellular adhesion and transmission of electrical impulses), accompanied by a rich supply of larger blood vessels and capillaries, and embedded in a dilated yet strong matrix of connective tissue [5]. The array of contractile elements between two adjacent Z-bands is known as a sarcomere and constitutes the contractile unit of cardiac muscle [3]. Z-bands are responsible for prominent cross-striations seen at the junction of sarcomeres and are usually well aligned within a cell. The intercalated discs at the cell junctions are typically undulant in conformation.

Contractile material occupies about 50% of the cytoplasm of myocytes and forms a continuous mass, which is separated into myofibrils of varying size by the interfibrillar matrix [3]. This cellular matrix contains mitochondria, sarcoplasmic reticulum, T tubules and glycogen particles, as well as other structures. Myofibrils are highly ordered arrays of contractile elements.

Age-related increases in heart weight relate to the presence of cardiac pathology. In the atria, a progressive increase in the amount of endocardial elastic tissue and collagen is observed with age, in addition to a proportionate decrease in muscle mass and an increase in fat [6]. Thus, the endocardium in the atria and over the atrial surface of the atrioventricular (AV) valves becomes thicker and more opaque. Not only is there proliferation of collagen, but also progressive fragmentation, disorganization and irregularity as shown with various stains [7]. The ventricles show a small but significant increase in collagen, and the amount of myocyte lipofuscin pigment increases substantially with age [8]. Basophilic myofiber degeneration is also strongly age-associated, appearing as amorphous, at times bubbly, basophilic masses in myocyte cytoplasm upon hematoxylin and eosin staining. Thickening and nodularity along the lines of apposition of the valve cusps and leaflets is prominent, as well as lipid accumulation appearing in the basal anterior mitral leaflet. The collagen of the aortic valve and mitral ring undergoes changes with maturity. Lipid begins to accumulate and calcification is often seen with maturity and senescence.

In the postnatal period, cardiac myocytes are typically believed to be terminally differentiated, nondividing cells. While recent discussions have suggested that myocytes may under certain conditions undergo cell division, in general the responses of myocytes to physiologic or pathologic stimuli include altered synthesis of DNA, RNA and protein, the evolution of increased myocyte heterogeneity [9], and modified enzymic and metabolic profile.

The question clearly remains as to when a myocyte in the heart becomes "sick." Thus, an increase in size of myocytes, commonly termed hypertrophy, may be an appropriate or inappropriate response to injury, leading to compensated or decompensated function. The relationship between hypertrophic and atrophic changes and altered ventricular mechanical geometry remains to be understood.

When myocardial injury leads to frank cell drop-out, the process involves one of the cell death pathways. By convention, the types of necrosis are usually designated coagulation (coagulative necrosis), contraction band (coagulative myocytolysis) and vacuolar degeneration (colliquative myocytolysis) [10]. The most common form of myocyte necrosis, coagulation type, is typically associated with myocardial ischemia and represented by the loss of striations, hypereosinophilia, drop-out of stainable nuclei and a generally amorphous appearance of dead tissues and cells. The loss of contractile capability leads to alterations in neighboring viable myocytes as well as in the geometry of the pumping chambers. The extent of loss resulting from coagulative necrosis will determine the ultimate degree of heart failure and the eventuation of circulatory instability. Contraction bands are more often associated with the margins of myocardial infarcts, as well as with ischemia-reperfusion injury [11].

The evolution of hypertrophy and atrophy of cardiac myocytes in a myopathic heart muscle represents the divergent pathways that a myocyte may take in response to injury. Thus, a myocyte may be lost through a process of diminished synthesis of cell constituents or may become enlarged through enhanced synthesis. The variability in myocyte size and configuration, as well as that of the nuclei, should be a signal, in part, to the nature of an injury or insult. Efforts to distinguish these responses have generally not been successful at the light microscopic level in the chronically failing or cardiomyopathic heart muscle. However, high-resolution image cytometry may be useful in beginning to understand the stereotypic and idiosyncratic responses of cardiac myocytes to injury, and the relationship between distinctive nuclear features and measures of cardiac myocyte function, gene expression or metabolism [12]. Myofibrillar and cytoskeletal changes in cardiac myocytes are also being studied in depth [13]. An understanding of these relationships represents one of the remaining important goals in myocardial diseases.

The myocardial interstitium has an intimate relationship with the cardiac myocytes that it envelopes. Much work has been done to define the evolution of aberrant cardiac matrix and the impact of such matrix on cardiac diastolic and systolic function [14]. Undoubtedly, an increased amount or altered nature of cardiac connective tissue impacts on the relationship between individual and groups of myocytes, thereby changing the electrical and physiologic status of the myocardium in question. Appreciation of interactions between matrix and myocytes will be linked to the emergent understanding of cardiac myocyte gap junctions [15,16].

Familial dilated cardiomyopathy

Idiopathic dilated cardiomyopathy (IDCM) is a primary myocardial disease, often of unknown cause, characterized by predominant left ventricular or biventricular dilatation and impaired myocardial contractility [17]. The pathophysiologic features of DCM include increased ventricular volume with ventricular wall thinning and moderate to severe reduction of contractile function [18]. In contrast, hypertrophic cardiomyopathy (HCM) is characterized by markedly thickened ventricular walls and distinctive histopathologic features, notably myocyte disarray with interstitial fibrosis, accompanied by hyperdynamic contractile function with reduced ventricular volumes. IDCM is the most common cause of congestive heart failure in the young, with an estimated prevalence of at least 36.5 per 100,000 in the USA. Among patients with IDCM, at least 30% of deaths are sudden, usually attributed to ventricular tachyarrhythmias [19]. The 5-year survival is <50% [20]. Approximately 50% of cardiac transplants are consequent to IDCM.

Familial DCM is defined as the presence of IDCM in two or more family members, as determined from a thorough family history. A positive clinical history of DCM in relatives of probands has yielded a prevalence of <10% of familial disease. However, the incidence of familial DCM is likely underestimated given the diversity of presentations and variability in penetrance leading to many early cases being unrecognized and virtually undetected clinically [18,21]. In fact, when first degree relatives of probands were assessed by physical examination, 12-lead electrocardiography and transthoracic echocardiography, regardless of the presence of symptoms, up to 35% were found to have DCM [18,22,23]. Overall, it is estimated that roughly 30% of individuals with IDCM have a significant familial, heritable component [24,25]. The incidence of IDCM has been estimated to be 5–8 cases per 100,000 per year [17,26–28], accounting for 10,000 deaths in the USA annually [29]. Affected individuals may have a relatively benign course, or develop progressive heart failure or experience sudden death from severe heart failure or ventricular arrhythmias. In general, the 5-year survival following a diagnosis of congestive cardiac failure in patients with familial DCM is 50% [17].

Molecular genetics, particularly linkage analysis, continues to facilitate the identification of genes and/or candidate genes responsible for the various

Table 4.1 Gene mutations and chromosomal locations in dilated cardiomyopathy (DCM).

Gene	Chromosomal location	Phenotype	Inheritance	Reference
Dystrophin	Xp21	Duchenne/Becker type muscular dystrophy, X-linked DCM	X-linked	61,64–72
Emerin	Xq28	Emery–Dreifuss muscular dystrophy	X-linked	124,125
Tafazin	Xq28	Barth syndrome	X-linked	37,73,75
Desmin	–	DCM	Autosomal dominant	35,36,48–52
Lamins A/C	1q11–q23	Emery–Dreifuss muscular dystrophy	Autosomal dominant	39
Lamins A/C	1p1–q21	DCM, conduction system disease	Autosomal dominant	40
α, β, γ, δ Sarcoglycans	–	DCM, limb girdle muscular dystrophy	Autosomal dominant	107–112
Actin	15q14	DCM	Autosomal dominant	94
Actin	2q31	DCM	Autosomal dominant	98
Actin	9q13–q22	DCM	Autosomal dominant	99
Actin	1q32	DCM	Autosomal dominant	97
Actin	10q21–q23	Mitral valve prolapse	Autosomal dominant	100
Actin	1p1–1q1	DCM, conduction system disease	Autosomal dominant	101
Actin	3p22–p25	DCM, conduction system disease	Autosomal dominant	102
Actin	2q14–q22	DCM, conduction system disease	Autosomal dominant	106
Actin	6q23	DCM, conduction system disease	Autosomal dominant	103
Actin	1q11–q21	DCM, limb girdle muscular dystrophy	Autosomal dominant	105
Actin	6q12–q16	DCM	Autosomal dominant	104
Actin	14q12	DCM	Autosomal dominant	81

forms of DCM [30,31]. The pattern of inheritance of familial DCM is variable and is most commonly autosomal dominant; but X-linked, autosomal recessive and mitochondrial inheritance also exist [32–34]. It has also become clear that these diseases are highly heterogeneous, with multiple genes identified for each of the major forms of cardiomyopathy. They exhibit marked clinical variability and can present alone, with conduction-system disease (sinus bradycardia, atrioventricular conduction block, atrial tachyarrhythmias) or with skeletal myopathy.

Grunig *et al.* [22] analysed the pedigrees of 445 consecutive patients with angiographically proven DCM and found that in 48 cases the patient (10.8%) had at least one additional first degree relative with DCM. The phenotypic groups identified in the 48 cases of familial DCM include DCM with muscular dystrophy, juvenile DCM with a rapidly progressive course in male relatives without muscular dystrophy, DCM with segmental hypokinesia of the left ventricle, DCM with conduction defects and DCM with sensorineural hearing loss. Gene mutations known to cause dilated cardiomyopathy include defects in the cytoskeletal proteins dystrophin [32], desmin [35,36], tafazzin [37,38] and lamins A/C [39,40], which cause ventricular dysfunction with conduction system diseases, as well as myocardial and skeletal muscle dysfunction. Table 4.1 summarizes mutations in known genes as well as published mutations in chromosomal loci that result in DCM.

Desminopathy

Desmin is the key intermediate-filament protein (type III) of cardiac and skeletal muscle supporting the structural arrangement of myofibrils by linking sarcomeres to the sarcolemma membrane [41,42]. In the heart, desmin is particularly abundant in the Purkinje fibers [43] and in cardiomyocytes, where it forms a double-banded structure at intercalated discs, at regular intervals along the sarcolemma [44,45]. Desmin filaments exist at the periphery of the Z-disc in striated muscle, where they keep adjacent myofibrils in lateral alignment [41]. It forms a three-dimensional scaffold around the myofibrillar Z-disc and interconnects the contractile apparatus with the subsarcolemmal cytoskeleton, the nuclei and other organelles. A single gene on human chromosome 2q35 encodes this muscle-specific protein [46].

Desminopathies present with heterogeneous clinical phenotypes, which may include cardiomyopathy, skeletal myopathy, respiratory insufficiency, neuropathy and smooth muscle disorders. The characteristic features of desminopathies are a loss of function caused by disorganization of the desmin filament network and the accumulation of misfolded desmin or alpha-B-crystallin to form toxic insoluble aggregates [47]. The observed clinical variation is in part dictated by type of desmin mutation. Indeed, 21 different mutations causing myopathy have been identified on the desmin gene. The most common mutations are those of helix 1B (encoded by exons 4–6) on residues 155–250. Severe disease, with a multisystem disorder involving skeletal, cardiac and smooth muscle, was found in a homozygous boy with a large deletion (7-amino acid, Arg173-Glu179) in the desmin gene [48]. In seven cases associated with skeletal and cardioskeletal myopathy, residues in helix 2B (337, 345, 357, 360, 370, 385 and 389) were replaced by proline, which disrupted the coiled-coil geometry [36,49–52]. Two mutations have been identified in the nonhelical tail domain, one at Lys449Thr and the other at Ile451Met in familial DCM [35,36].

Besides desmin, a gene implicated in desmin-related myopathy is the desmin-associated protein alpha-B-crystallin. Alpha-B-crystallin is a 22-kDa heat shock protein with molecular chaperone activity. A missense mutation in the alpha-B-crystallin chaperone gene on chromosome 11q22.3–q23.1 is known to cause hypertrophic cardiomyopathy [53].

Dystrophinopathies

Dystrophin, a member of the spectrin superfamily of proteins, is a key linker protein between the sarcolemma of the myocyte and the contractile apparatus, the sarcomere [54]. The dystrophin–glycoprotein complex (DGC) is a large multimeric complex specific to muscle cells [55]. In addition to providing a structural link between subcortical actin and the extracellular matrix, the DGC also appears to be involved in signaling with components such as neuronal nitric oxide synthase, dystrobrevin and syntrophin. In its full-length form, the dystrophin gene encodes a 427-kDa protein

that is composed of an amino-terminal globular domain, and 24 spectrin repeats that make up the rod region of the dystrophin molecule [54]. At its carboxyl terminus, dystrophin binds to dystroglycan and the syntrophins providing mechanical stability to the plasma membrane [56,57].

Dystrophinopathies include Duchenne muscular dystrophy, Becker muscular dystrophy and X-linked dilated cardiomyopathy caused by mutations in the destrophin gene on chromosome Xp21.1 [58,59]. All patients with dystrophin deficiency are at risk for the development of Duchenne muscular dystrophy and should be screened regularly. Dystrophin deficiency leads to a disruption of the transmembrane complex and loss of integrity at the sarcolemma.

Duchenne and Becker muscular dystrophy

Duchenne and Becker muscular dystrophies are severe X-linked genetic muscular diseases affecting approximately 1 in 3500 live male births and are the most common, childhood onset, muscular dystrophies [60]. Cardiac myopathy in Duchenne muscular dystrophy (DMD) may be severe and progressive, with life-threatening heart failure and/or arrhythmias, or may be asymptomatic for many years without clinical features. Mutations that cause DMD are typically deletions that are "out of frame" resulting in a dysfunctional protein [61]. The standard diagnostic evaluation for cardiac involvement, the echocardiogram, is a noninvasive method of evaluating left ventricular (LV) size and systolic function, allowing one to analyze a variety of more subtle parameters [62].

Becker muscular dystrophy (BMD) is also an X-linked DCM characterized with later onset and a slower progression of skeletal and cardiac myopathy. The prevalence of BMD is almost one-tenth of the frequency of DMD [59]. There are reports that a more severe form of BMD is associated with mutations in the promoter region (5′ amino-terminal end) of the dystrophin gene [59]. Thus far, several mutations in the dystrophin gene have been identified at the mRNA level in BMD patients [63–72].

Barth syndrome

The cardiac manifestations of Barth syndrome include left ventricular dilatation, endocardial fibroelastosis or a dilated hypertrophied left ventricle caused by mutations in the gene G4.5 on chromosome Xq28, which encodes the protein tafazzin [37]. The function of the tafazzin protein is unknown, but mutations in the G4.5 gene appear to be responsible for a diverse spectrum of cardiac disease with unique clinical phenotypes, including classic DCM, endocardial fibroelastosis and left ventricular noncompaction (LVNC) with or without clinical features of Barth syndrome (isolated LVNC). Recently, an autosomal recessive mutation at 3q26.33 in the DNAJC19 gene was found to cause a Barth-like syndrome, which includes early onset dilated cardiomyopathy with conduction defects, in the Canadian Dariusleut Hutterite population [73].

X-linked dilated cardiomyopathy

The X-linked DCM (X-LDCM) is a heterogeneous inherited cardiomyopathy and has been shown to be an allelic disorder to DMD, BMD and Barth syndrome [32,74,75]. X-LDCM occurs mostly in males during adolescence or early adulthood, with a rapidly progressive clinical course. The infantile form of X-LDCM and Barth syndrome typically presents in male infants and is characterized by neutropenia, 3-methylglutaconic aciduria, mitochondrial dysfunction and growth retardation [76]. X-LDCM is associated with increased serum creatine kinase (CK), cardiac phenotype, without clinical signs of skeletal myopathy and is caused by mutations in the dystrophin gene [77,78]. Several types of mutations including the deletion of the dystrophin muscle-promotor region [75], a stop mutation and alternative splicing of exon 29 [79] and a point mutation in the 5′ splice site of dystrophin gene have been reported in Barth syndrome [80].

Troponin mutations

Troponin is a sarcomeric protein with a central role in calcium regulation during contraction. The troponin complex consists of T (TnT), I (TnI) and C (TnC) subunits. Evidence exists for mutations in all three subunits being causative for cardiomyopathies. Although mutations in the troponin gene are mainly associated with HCM, mutations for TnT, TnI and TnC have been also shown to cause DCM [81–84]. The TnT mutation at ΔK210 was first shown to cause primary familial DCM, which was subsequently confirmed in another family [81,85]. The autosomal recessive TnI A2V mutation at the N terminus has been also shown to cause DCM [82]. HCM resulting from a mutation in TnT has a distinctive phase of dilatation [86]. Troponin mutation R141W may cause calcium desensitization through stabilization of the troponin and T-tropomyosin interaction leading to DCM [87]. Recently, a TnC mutation at G159D was reported to trigger DCM [84].

The functional consequences in DCM are different and less well understood than in HCM. However, the sarcomere mutations that have been shown to cause dilated cardiomyopathy are likely to diminish the mechanical function of cardiac myocytes. TnT mutation ΔK210 is located in a domain responsible for calcium-sensitive TnC binding and in an experimental setting had a calcium desensitizing effect on force generation in isolated cardiomyocytes [87–89]. Three of these basic residues (lysine 208, 209 and 210) participate in forming a tight binary complex with TnC, possibly by means of complementary interactions with a ring of acidic residues located on the surface of TnC [90–92]. Loss of TnT lysine residue 210 should reduce these ionic interactions and diminish activation of calcium-stimulated actomyosin ATPase, just as occurs with mutagenesis of TnC acidic residues [92], leading to a significant reduction in contraction force. The Gly to Asp substitution (G159D) on TnC gene causes DCM and reduces the rate of force development [84]. Decreased calcium sensitivity leads to reduced force generation, impinging on systolic cardiac function and stroke volume, and ventricular dilatation. Notably, none of the mutations in cardiac TnT that cause hypertrophic cardiomyopathy alters lysine residues in the calcium-sensitive TnC-binding domain.

Myosin and actin mutations

The location of cardiac myosin mutations that cause DCM suggests that such mutations impair contractile function. In this regard, Ser532Pro maps within an α-helical structure of the lower 50 kDa domain in myosin that contributes to the tight binding of actin [93]. This mutation disrupts stereospecific interactions between myosin and actin critical for initiating the power stroke of contraction. Demonstration that mutations in cardiac β-myosin heavy chain and cardiac TnT, along with two previously reported mutations in cardiac actin, cause DCM implicates mutations in other sarcomeric genes in causing this disorder [94–96]. Mutations in the gene encoding cardiac actin (15q14) have been shown in two unrelated families diagnosed with familial DCM [94]. To date, disease loci have been defined on chromosomes 1p1–1q1, 6q23, 3p22–25, 6q12–16, 1q32, 2q11–22, 2q31, 9q13–22, 10q21–23 and 14q11.2–13 [81,94,97–106].

Laminin-2 (merosin) mutations

Laminin-α2 is an extracellular matrix protein that acts with dystrophin and dystroglycan to mechanically link the actin cytoskeleton and the extracellular matrix. Mutations in the gene encoding the α2 chain of laminin-2 produce congenital muscular dystrophy [107]. As this gene is also expressed in the nervous system, laminin-α2 mutations can produce central and peripheral nervous system defects in addition to a severe skeletal muscular dystrophy and a mild cardiomyopathic phenotype [108].

Sarcoglycanopathies

Sarcoglycans are dystrophin-associated glycoproteins, which function to stabilize the interaction between α- and β-dystroglycan [109]. There are six sarcoglycan genes described so far: α, β, γ, δ, ε and ζ. Of them, α, β, γ and δ form a subcomplex in striated muscle and mutations in either one of the sarcoglycan genes results in different forms of a limb-girdle muscular dystrophy (LGMD-2D, -2E, -2C and -2F, respectively) [110–114].

Titin mutations

Titin, an abundant and giant sarcomeric protein, spans a significant part of the sarcomere and its mutations underlie a form of autosomal dominant DCM. At its N terminus, titin binds α-actinin at the Z-band, while it interacts with myosin-binding protein C and myomesin in the M domain [115]. Titin is composed of a number of immunoglobulin domains that are thought to provide elastic recoil to the muscle and participate in passive tension [115]. There are at least two splice forms of titin, one of which (N2BA) provides greater stiffness. The two major splice forms, N2B and N2BA, have varying expression in pathologic states such as DCM and pressure-overload hypertrophy [116,117]. The differential ratio of titin isoforms likely results from adaptive or maladaptive changes in gene expression in response to increased load.

In addition to the changes in functional titin splice forms, inherited mutations in titin may lead to cardiac and skeletal muscle myopathies. Genetic mutations in titin have been reported in several cases of familial DCM [118,119]. A variety of mutations has been described in these families including missense and splice site mutations. Very recently, mutations in the titin gene have been reported in patients with tibial muscular dystrophy [120]. Mutations in titin affect the terminal feature of this giant protein and alter one of the predicted binding sites for the calcium-activated protease, calpain. Autosomal recessive mutations in calpain-3, a muscle-specific form of calpain, cause a relatively common form of muscular dystrophy [121]. Calpain-3-associated LGMD is not thought to be associated with cardiomyopathy and appears to be more involved in regulating the passive stiffness of the myocyte sarcomere. Siu *et al.* [98] mapped familial DCM in three generations with autosomal dominant transmissions of DCM to chromosome 2q31. Interestingly, although the titin gene resides in this region, it did not co-segregate with disease in affected individuals.

Emery–Dreifuss muscular dystrophies

Lamins A/C are intermediate filament proteins, located in inner nuclear membrane, with a central rod-like structure flanked by amino- and carboxyl-terminal globular domains. Missense mutations scattered along the length of the lamins A/C sequence have now been described in DCM subjects in the absence of skeletal muscle disease. Like other intermediate filament proteins, lamins A/C aggregate into parallel, coiled-coil structures that can assemble into a higher order, head-to-tail filament. Overall, lamins are thought to provide structure to the nuclear membrane and can directly bind chromatin [122].

Genetic defects in lamins A/C (1p1–q21) were first discovered in association with an autosomal dominantly inherited Emery–Dreifuss muscular dystrophy (EDMD) phenotype [39]. The autosomal dominant and recessive variants are presented with isolated DCM, conduction-system disease, lipodystrophy and other phenotypes.

EDMD is a genetically heterogeneous X-linked disorder characterized by muscle weakness, contractures, AV nodal heart block and cardiomyopathy [123,124]. X-linked EDMD is caused by mutations in the *STA* gene encoding the nuclear protein emerin, a 34-kDa protein of the inner nuclear membrane. Emerin associates with the membrane through its carboxyl terminus and contains an LEM (Lamina-associated polypeptide-2, Emerin, MAN1 antigen) domain that binds directly to chromatin. Thus, emerin participates directly in a link from the nuclear membrane cytoskeleton (nucleoskeleton) to chromatin itself suggesting a role in the regulation of DNA synthesis or gene expression [123].

Cardiac involvement in EDMD is characterized by AV conduction defects, which start with sinus bradycardia, prolongation of the PR interval on electrocardiography (ECG), and evolution to heart block, atrial flutter and complete atrial paralysis [123]. A recent report describes two brothers with a relatively severe variant of EDMD who developed cardiac symptoms in the first decade of life and evidence of left ventricular hypokinesia in the second decade of life [125]. Cardiac involvement in EDMD was described recently by Boriani *et al.* [126] who reported the long term follow-up of 10 EDMD patients.

Arrhythmogenic right ventricular cardiomyopathy

Arrhythmogenic right ventricular cardiomyopathy (ARVC), formally known as right ventricular dysplasia (ARVD), is a disease of myocardium associated with cardiac dysfunction with frequent familial occurrence and a manifestation of arrhythmic sudden death in young adolescents and athletes. The introduction of the condition dates back to 1960, when Dalla Volta *et al.* [127,128], for the first time, reported patients with "auricularization of the right ventricular pressure curve" and referred to the disease as "sclerosis of the right ventricle after a myocardial infarction without coronary obstruction." It was only later that Vedel *et al.* [129] established the term "arrhythmogenic right ventricular dysplasia" in 1978. Afterward, Marcus *et al.* [130] reported 24 adult cases of ARVC/D with sustained ventricular tachycardia of right ventricular origin.

Because of its nature as a progressive myocardial disease of unknown etiology, ARVC has been included among cardiomyopathies by the Task Force of WHO/ISFC [2,131,132]. The prevalence of disease is unknown because of asymptomatic, non-diagnosed or misdiagnosed cases. In the general population, the prevalence varies from 1 in 5000 to 1 in 10,000 people [133,134]. Several groups have reported an approximate male : female ratio of 3 : 1 in ARVC [135,136]. ARVC has been indicated as the cause of up to 20% of sudden death, particularly among young competitive athletes [137,138]. The incidence of familial cases is in the range 15–50% in the published literature [130,132,139–141].

The typical clinical presentation consists of ventricular arrhythmias with the left bundle branch block (LBBB) pattern, a morphology indicating right ventricular origin; and ECG depolarization/repolarization changes mostly in the right precordial lead [56,130,142]. Pathologically, ARVC is characterized by acquired progressive degeneration, fatty or fibro-fatty replacement and patchment-like thinning of right ventricular myocardium causing dilatation of the right ventricle and ventricular tachycardia (VT) or ventricular fibrillation (VF) [132,142]. Among other manifestations are subtle wall aneurysms of the infundibular, apical and sub-tricuspid areas known as "triangle of dysplasia" and, in advanced cases, bi-ventricular pump failure caused by left ventricular involvement [132,136, 143–145].

At the microscopic level, within the subepicardial and deeper layers of right ventricular myocardium, abnormal fatty or fibro-fatty tissues surround the normal strand of fibers causing the slow conduction and ventricular arrhythmias in ARVC patients. The fibro-fatty cardiomyopathic variant is associated with considerable thinning of right ventricular free wall and reaches from epicardium to the endocardium [136,146].

In order to establish a universal diagnostic guideline, the Study Group on ARVC of the Working Group Myocardial and Pericardial Disease of the European Society of Cardiology and of the Scientific Council on Cardiomyopathies of the ISFC proposed the standardized major and minor diagnostic criteria encompassing structural, histologic, electrocardiographic, arrhythmic and genetic factors. Diagnosis of ARVC must be based on the presence of two major criteria, or one major plus two minor criteria, or four minor criteria [135,147, 148]. Assessment of structural, functional and histologic alterations can be performed with echocardiography, magnetic resonance imaging (MRI), ultrafast computed tomography (CT), conventional or radionuclide angiography (right ventricular angiography as the gold standard), invasive electrophysiologic study (EPS) and endomyocardial biopsy [136,143,148]. The existence of adipose and fibrous tissue between or around surviving myocardial fibers is the gold standard for pathologic diagnosis of ARVC [148,149].

Therapy must be directed towards prevention of sudden cardiac death (SCD) and/or treatment of heart failure, if present. In patients with nonlife-threatening and well-tolerated ventricular arrhythmias, pharmacologic therapy could be the first choice. In case of life-threatening ventricular arrhythmias, nonpharmacologic treatment such as implantable cardioverter-defibrillator (ICD), radiofrequency ablation and surgery can be effective [150–157]. Nevertheless, heart transplantation is the final therapeutic alternative for unmanageable congestive heart failure or untreatable ventricular arrhythmias.

ARVC is considered among idiopathic cardiomyopathies because of unknown etiology and pathogenesis. For the last decade, the etiology of ARVD has been in the center of scientific debate and advanced research. Myocyte apoptosis, inflammatory response to viral myocarditis, genetically determined dystrophy (atrophy) and some metabolic or structural defects have been proposed as potential mechanisms for progressive loss of right ventricular myocardium and fibro-fatty replacement in ARVC [132,158–160]. Regardless of suggested mechanisms and apart from a few cases, familial occurrence and genetic defects have been reported with largely autosomal dominant traits, variable and incomplete penetrance, and polymorphic phenotypic expression in ARVC patients [142,161–163].

Based on several familial linkage studies, nine loci have been associated with ARVC/D. In recent years, loci for autosomal dominant forms have been identified and mapped to chromosomes 14q23–q24 (ARVC/D1), 1q42–q43 (ARVC/D2), 14q12–q22 (ARVC/D3), 2q32 (ARVC/D4), 3p23

(ARVC/D5), 10p12–p14 (ARVC/D6) and 12p11 (ARVC/D9) [133,157,164–169]. In addition, autosomal recessive forms of ARVC have been linked to chromosomes 10q22 (ARVC/D7) and 6p23–p24 (ARVC/D8) [170,171]. The existence of several loci for the autosomal dominant forms, and the absence of any linkage to above loci in almost 50% of familial cases of ARVC provide strong indications for higher genetic heterogeneity. Recent studies have identified some involved genes including desmoplakin, plakophilin-2 (pkp2), cardiac ryanodine receptor 2 (RyR2) and transforming growth factors β1, 2, 3 (TGF-β1, 2, 3). Genes for other loci are yet to be determined. Table 4.2 summarizes related loci and identified gene mutations in various types of ARVC.

Cardiac ryanodine receptor 2 mutations

Mutation in RyR2 has been associated with arrhythmogenic right ventricular cardiomyopathy type 2 (ARVC2, OMIM 600996), and was the first gene mutation identified through the mapping studies in patients with ARVC. The disease locus was mapped to chromosome 1q42–q43 [165,168]. The same mutation has been also reported in familial polymorphic ventricular tachycardia [172] and catecholaminergic ventricular tachycardia [153]. ARVC2 is distinctive by the presence of peculiar polymorphic effort-induced ventricular arrhythmias, a high penetrance, a 1 : 1 male : female ratio, and less pronounced fibro-fatty substitution than in other ARVDs [142]. RyR2 is the cardiac counterpart of RyR1, the ryanodine receptor expressed in skeletal muscle cells. In cardiomyocytes, RyR2 can be activated by calcium and has a pivotal role in electromechanical (excitation–contraction) coupling by controlling calcium release from sarcoplasmic reticulum (SR) into the cytosol. RyR2 has a tetrameric structure consisting of four identical 565 kDa monomers. Within the cell, RyR2 binds to four FK506 binding proteins (FKBP12.6, also known as calstabin-2), a necessary interaction for a stabilized and coordinated gating of calcium channel (coupled gating) [173–175].

Studies in four independent families with recurrence of ARVD2 cases have revealed single nucleotide polymorphisms (SNPs) in exons 15, 28, 37 and 59 of the RyR2 gene [172]. Using polymerase chain reaction (PCR), single-strand conformation

Table 4.2 Gene mutations and chromosomal locations in arrhythmogenic right ventricular cardiomyopathy (ARVC).

Gene	Chromosomal location	Phenotype	Inheritance	Reference
TGF-β3	14q23–q24	ARVC-1, excessive fibrosis	Autosomal dominant	133
RyR2	1q42–q43	ARVC-2, effort-induced arrhythmias	Autosomal dominant	165,168,177
–	14q12–q22	ARVC-3	Autosomal dominant	164
–	2q32.1–q32.3	ARVC-4	Autosomal dominant	166
–	3p23	ARVC-5	Autosomal dominant	167
–	10p12–p14	ARVC-6	Autosomal dominant	169
–	10q22	ARVC-7	Autosomal recessive	170
Desmoplakin	6p23–p24	ARVC-8, impaired filament interaction, dilation	Autosomal recessive	149,171,178,179,181
Plakophilin-2	12p11	ARVC-9, disarrayed cytoskeleton, rupture of cardiac walls	Autosomal dominant	157
Plakoglobin	17p21	Naxos disease, diffuse palmoplantar keratoderma, woolly hair, ventricular arrhythmias	Autosomal recessive	192,193

RyR2, ryanodine receptor 2; TGF-β3, transforming growth factor β3.

polymorphism (SSCP) analysis, denaturing high-performance liquid chromatography (dHPLC) and direct sequencing, Laitinen *et al.* [172] succeeded in identifying invariable transmission of four of these sequence changes (R176Q, T2504M, N2386I and L433P) along family generations. These mutations occur in highly conserved cytoplasmic domain of the RyR2 protein, which is crucial for FKBP12.6 binding and proper calcium gating, causing after-depolarization, leading to arrhythmias in ARVC patients. At the structural level, imbalanced intracellular calcium homeostasis and massive calcium release from SR could lead to apoptotic and/or necrotic myocardial cell death and degeneration of cardiac muscle, a common histologic feature of ARVC disease [160,172,176]. Recently, a missense mutation in exon 3 (230C→T, A77V) of the RyR2 gene has been associated with both ARVC2 and catecholaminergic polymorphic ventricular tachycardia in a family with a history of sudden death, a potential indication of differing phenotypic expressions of the same disease [177]. These findings underscore the importance of parallel application of tissue examination, electrophysiologic assessment and genetic examination among family members of affected ARVC patients.

Desmoplakin mutations

Desmoplakin (DSP) was the second gene identified in association with an autosomal recessive of ARVC type 8 [178]. DSP is the most abundant protein in desmosomes and a key constituent of the innermost desmosomal plaque. The C terminus of desmoplakin binds to intermediate filament desmin, whereas the N terminus contains the putative PKC phosphorylation site, as well as binding sites for another desmosomal protein, plakoglobin [179]. Desmosomes are highly organized cell–cell adhesion junctions that form mechanical coupling between cellular intermediate filaments and the membrane of neighboring cells. The adhesive function of desmosomes is highly dependent on the proper functioning of desmosomal components wherein mutations could cause cellular detachment resulting in cell death and impaired regenerative capacity of tissue in response to mechanical stresses [180].

Norgett *et al.* [171] reported the first recessive mutation in DSP in members of three Ecuadorian families with Carvajal disease, who were homozygous for a single nucleotide deletion (7901delG) mapped to chromosome 6p23–p24. All affected individuals presented with clinical manifestations of dilated left ventricular cardiomyopathy, woolly hair and keratoderma [171]. Thereafter, a missense mutation (C1176G; AGC→AGG; S299R) in exon 7 of the DSP gene was reported for the first time in a family of 26 members affected by an autosomal dominant form of ARVC (ARVC8) spanning four generations [178]. The study disclosed that the mutation affects the N terminal region of DSP. Notably, in those patients, the mutation was tolerated by epidermal tissue and left ventricular myocardium, because there was no report of skin disorders or left ventricular involvement [178]. However, Bauce *et al.* [181] reported on left ventricular involvement in half of the ARVC8 cases while studying 38 individuals belonging to four families carrying different DSP mutations including three missense and one mutation in intron–exon splicing region. Their findings indicate that ARVC8 disorders are highly associated with sudden death as the first clinical manifestation. Primary left-sided (left ventricular) variants of ARVC have been increasingly reported on postmortem examination with fibro-fatty replacement restricted to the left ventricle [182]. Recently, a heterozygous single adenine insertion (2034insA) in DSP gene has been shown in a large family with autosomal dominant left-sided ARVC [149].

Of note, mutations in DSP C terminus have also been associated with skin disorders such as autosomal dominant form of striate palmoplantar keratoderma II (SPPK2) and autosomal recessive form of skin fragility woolly hair syndrome (SFWHS) without any presentation of cardiac involvement [183,184].

Plakophilin-2 mutations
Plakophilins (pkp1, -2 and -3) are a family of 42-amino acid armadillo-repeat containing proteins that are located in the outer dense plaque of desmosomes and in the nucleoplasm [185]. Plakophilins have important roles as intermediate proteins linking cadherins with DSP and intermediate filaments. Plakophilin-2 (pkp2) is prominent in cardiac cells and exists in two splice forms, 2a and 2b [186]. Using co-immunoprecipitation and yeast two-hybrid assays, Chen *et al.* [187] have shown pkp2 interactions with multiple desmosomal components including DSP, plakoglobin, desmoglein 1 and 2, and desmocollin 1a and 2a. The role of plakophilin

in embryogenesis, heart morphogenesis and junctional architecture has also been demonstrated in studies using pkp2 null mice [157].

In a study by Gerull *et al.* [157], heterozygous mutations in the pkp2 gene were identified in 32 of 120 unrelated individuals of western European descent with ARVC type 9 (ARVC9). In total, 25 pkp2 mutations, including 12 insertion-deletion mutations, six nonsense mutations, four missense mutations and three splice site mutations were identified and mapped to chromosome 12p11 [157]. It has been shown that in the absence of pkp2, DSP dissociates from desmosomal plaque of cardiac cells and forms granular aggregates within the cytoplasm [157]. Mutant pkp2 protein impairs desmosomal assembly and cardiac cell–cell junction leading to intercellular disruption during mechanical stress or exercise. This probably explains the high prevalence of ARVC among competitive athletes [157].

Transforming growth factor-β3 mutations
Mutation in transforming growth factor-β (TGF-β3) gene has been associated with ARVC type 1. ARVC1 locus was the first one identified and mapped to chromosome 14q23–q24 [133]. TGF-β1, -2 and -3 are pleiotropic cytokines with crucial roles in the process of wound repair and tissue remodeling following injury. TGF-β3 is involved in cardiac morphogenesis, mesenchymal differentiation, fibrous skeleton development, angiogenesis and trophoblast differentiation during hypoxia [188]. The regulatory role of the TGF-β family of proteins in heart morphogenesis has been studied using various knockout mice models [188].

Beffagna *et al.* [189] have reported mutations in 5′ untranslated region (5′UTR) and 3′UTR regions of the TGF-β3 gene in familial ARVC1 leading to a twofold increase in TGF-β3 synthesis. TGF-β3 has been shown to induce fibrosis in various tissues by increasing the expression of components of extracellular matrix and suppression of expression of matrix metalloproteinases [28,190]. On the basis of these findings, it is suggested that mutations in UTR regions of TGF-β3 may lead to extensive replacement of cardiac cells by fibrous tissue leading to disruption of electrical and mechanical signals within the heart. Members of the TGF-β family have been also shown to regulate the expression of desmosomal proteins such as plakoglobin, indicating a role for these growth factors in cell–cell junction assembly and stability [191].

Plakoglobin mutations

Plakoglobin (JUP, DSPIII, γ-catenin) was initially identified during mapping studies for Naxos disease. Naxos disease was first described in the Greek island of Naxos by Coonar *et al.* [192] as a familial cardiocutaneous syndrome with clinical presentations of diffuse, nonepidermolytic palmoplantar keratoderma, woolly hair, a high incidence of ventricular arrhythmias and sudden death in children or young adults. Further studies in nine families with 21 positive cases of disease led to mapping of the gene to chromosome 17p21 [192]. A homozygous two base pair (TG) deletion in the plakoglobin gene causes a frame-shift and premature translation termination leading to expression of a truncated form of plakoglobin [193].

Plakoglobin is located in cytosol, desmosomes and adherens junctions and, like other desmosomal proteins, has a fundamental role in cell–cell junction assembly through its C terminal domain [194]. Within the cytoplasm, plakoglobin also interacts with cadherins in adherens junctions and actin in sarcomeres [195]. A hypothesis is that mutation in plakoglobin protein could cause defects in the intercellular junctions and the intracellular cytoskeleton network leading to remodeling of gap junctions and impaired electrical coupling and conduction within myocardium [196].

Restrictive cardiomyopathy

According to the revised WHO Task Force, restrictive cardiomyopathy (RCM) is a myocardial disorder characterized by restrictive filling and reduced diastolic volume, with rapid early filling and slow late filling of either or both ventricles, with normal or near-normal systolic function [2,197]. Wall thickness may remain normal or increase, depending on the precipitating condition. RCM is the least common type of cardiomyopathy and is classified as primary and secondary [198].

Primary RCM includes endomyocardial fibrosis, Löffler endocarditis and idiopathic RCM of unknown etiology [199]. Idiopathic RCM occurs in the absence of a precipitating condition and is often characterized by noninfiltrative interstitial fibrosis of myocardium, skeletal myopathy and autosomal dominant transmission. The secondary form of RCM is often caused by precipitating pathologic conditions classified as myocardial infiltrative diseases including amyloidosis, sarcoidosis, Gaucher disease and Hurler disease; myocardial storage diseases such as hemochromatosis, Fabry disease and glycogen storage disease; and endomyocardial complications including hypereosinophilic syndrome, endomyocardial fibrosis, carcinoid, metastatic malignancy, radiation damage and anthracycline toxicity. Cardiac amyloidosis is the most prevalent and most thoroughly investigated entity of the secondary form of RCM [200].

In RCM patients, increased stiffness of the myocardium causes excessive pressure within either or both ventricles in order to preserve cardiac output. When only the right ventricle is affected, common clinical profiles include peripheral edema, ascites, hepatomegaly and elevated jugular venous pressure with inspiration (Kussmaul sign). With involvement of the left ventricle, exertional dyspnea, fatigue, exercise intolerance and evidence of pulmonary edema are common presentations [197,199,201,202]. Idiopathic RCM is usually associated with distal skeletal myopathy and manifestations of AV block, skeletal muscle weakness and a moderate increase in heart weight [203,204]. On microscopic examination, patchy endocardial fibrosis and frequent fibrosis of sinoatrial and AV nodes may be observed [203,205].

In cardiac amyloidosis, clinical presentations consist of biatrial dilatation and enlargement, right and/or left ventricular hypertrophy and thrombi in the atrial appendages. On histologic examination, the myocardium may present a rubbery texture as well as a waxy appearance. Complete heart block resulting from fibrosis of the AV and sinoatrial nodes necessitating permanent pacing has been reported [203,206]. Patchy endocardial fibrosis and deposition of insoluble amyloid protein fibrils within the myocardial interstitium or vessel walls are also present [205]. There are reports of eosinophilic deposits in cardiac valves, intramyocardial coronary arteries and within myocardial interstitium [207].

Endomyocardial fibrosis and Löffler's endocarditis (eosinophilic cardiomyopathy) are considered with forms of RCM associated with eosinophilia

[208]. Studies in animal models indicate that parasitic infection could result in myocardial accumulation of eosinophils causing damage [54,209]. In patients diagnosed with Löffler disease, valve fibrosis may lead to valvular regurgitation and AV valve stenosis that requires valvular replacement [210,211].

The first diagnostic dilemma is the distinction between RCM and constrictive pericarditis, as they have similar clinical signs and symptoms but different treatment options [212,213]. Recent reports have suggested that tissue Doppler imaging (TDI) and pulsed wave Doppler echocardiography (PWD) are more reliable methods in early detection of cardiac dysfunction in patients with amyloidosis [214,215]. Noninvasive Tc-99m pyrophosphate myocardial single photon emission computed tomography (SPECT) has also been successfully used by Casset-Senon *et al.* [216]. Endomyocardial biopsy is a valuable tool for the diagnosis of cardiac amyloidosis, where cardiac myocytes are surrounded by amyloid deposits forming the so-called "honeycomb" pattern [217–219].

The prognosis of RCM is generally poor except for those with reversible precipitating conditions such as hemochromatosis. The majority of individuals affected with RCM develop progressive deterioration as a result of congestive heart failure [220,221]. Specific therapies are considered according to the underlying cause. In patients with cardiac sarcoidosis, cardiac transplantation is the ultimate therapeutic alternative. However, there are reports of recurrence of sarcoid granulomas in the transplanted heart [15].

Cumulative evidence confirms familial occurrence of both autosomal dominant and recessive forms of RCM [204,206,211,222]. Two small studies have shown increased prevalence among girls in childhood [223,224]. Fitzpatrick *et al.* [203] reported an autosomal dominant restrictive cardiomyopathy associated with skeletal myopathy in an Italian family spanning five generations. A familial cardiomyopathy with variable hypertrophic and restrictive presentations affecting three generations of family members with common human leukocyte antigen (HLA) haplotype was documented by Feld and Caspi [225]. There are also reports of RCM associated with desmin accumulation in families with evidence of autosomal dominant inheritance [226,227].

Cardiac troponin I mutations

The contractile structure of cardiomyocyte comprises highly organized arrangements of myosin and actin filaments and associated troponin–tropomyocin complexes, all which form sarcomere units. More than 200 mutations in genes for sarcomeric proteins such as α- and β-myosin heavy chains, myosin-binding protein C, cardiac TnT and TnI, α-cardiac actin, cardiac titin/connectin and α-tropomyosin have been reported in various forms of cardiac cardiomyopathies [228]. Troponin is a key player in cardiac contraction and relaxation by preventing the interaction between actin and myosin heads to guarantee muscle relaxation and encompasses three subunits, TnC, TnI and TnT, each with a distinct structure and function.

TnI blocks the contractile interaction between actin and myosin through its inhibitory region [229,230]. In the presence of sufficient amount of calcium, TnI regulatory domain binds to the N terminal domain of TnC, leading to removal of TnI inhibitory action of muscle contraction. Mutation in the TnI gene could result in cardiac malfunction, particularly diastolic dysfunction, and has been associated with both restrictive and hypertrophic forms of cardiomyopathy [84].

A linkage study in 33 members of a family presenting with RMC and hypertrophic cardiomyopathy revealed a disease-causing mutation in a highly conserved region of cardiac troponin I (*TNNI3*) gene [83]. Further mutational analysis identified nucleotide substitution (87A→G) on exon 8 (D190H; lod score 4.8). Additional studies by the same group in nine unrelated patients who had been diagnosed with idiopathic RCM revealed that six patients were carriers of the same *TNNI3* mutations [83]. These findings indicate that idiopathic RCM may be considered a clinical expression of hereditary sarcomeric contractile protein disease. To date, six novel missense mutations in human TnI have been identified in RCM patients [83,231].

Desmin mutations

Defects in desmin and other desmin-related filaments, such as alpha-B-crystallin and plectin, result in myofibril fragility and impaired contraction [41,42]. Mice lacking desmin develop cardiac and skeletal myopathy, indicating an important role for desmin in muscle function [232,233]. In humans, mutation in desmin has been associated with a distinct myopathy (desmin myopathy), which is often accompanied by cardiomyopathy. Desmin-related myopathy is considered a dominantly inherited familial disorder, which is characterized by aggregation of desmin in skeletal or cardiac muscle fibers, proximal muscle weakness and cardioskeletal myopathy associated with arrhythmias and restrictive heart failure [36]. Missense A337P mutation on exon 5 and missense A360P and N3931 mutations on exon 6 have been identified in two separate families [234]. A homozygous deletion of 21 nucleotides in the desmin gene has been also reported in a family with a history of cerebrovascular attacks [48].

Transthyretin (prealbumin) mutations

Among different types of amyloidosis, cardiac involvement is more common in primary amyloidosis. Primary amyloidosis is a hereditary disorder caused by the deposition of immunoglobulin light chains, while secondary amyloidosis is a result of deposition of proteins other than immunoglobulin. A familial pattern has been observed in both types. Inherited forms of cardiac amyloidosis, which is associated with cardiomyopathic features, may be caused by mutation in serum protein transthyretin (TTR, prealbumin), which is produced mainly in liver. To date, over 80 different mutations have been reported in association with amyloid disease [235,236]. Lafitte *et al.* [237] have reported the cardiac manifestations in a French family of five sisters and one brother, three of whom presented with amyloidosis with deposits of transthyretin and apolipoprotein A1 resulting from a genetic mutation. A novel variant of the transthyretin gene encoding 59Thr→Lys associated with autosomal dominant hereditary systemic amyloidosis has been reported in Italian kindred in whom cardiac involvement was the major feature [238].

Fabry disease

Fabry disease is an X-linked systemic lysosomal storage disorder caused by lysosomal α-galactosidase A deficiency resulting in accumulation of glycosphingolipids such as globotriaosylceramide within brain, heart, kidney, skin and vasculature [239]. Almost 60% of males with Fabry disease present with cardiac abnormalities such as valvular dysfunction, left ventricular hypertrophy and conduction blocks [230,240]. To date, more than 300 mutations in α-galactosidase gene on the long arm of the X chromosome (Xq22.1) have been reported, mostly in single families [240]. Polymorphism of interleukin-6, eNOS, methylenetetrahydrofolate reductase (MTHFR), prothrombin, protein Z and factor V have also been associated with various manifestations of Fabry disease [241–246].

Unclassified cardiomyopathies
Mitochondrial cardiomyopathies

Mitochondria serve important functions in energy production in the form of adenosine triphosphate (ATP) via the pyruvate dehydrogenase complex, citrate cycle, β-oxidation, respiratory chain and oxidative phosphorylation, as well as in the mediation of the endogenous pathway of apoptosis [247]. The mitochondrion is especially important to the proper function of the adult myocardium, which has a continuous and enormous demand for energy to fulfill its function as a circulatory pump and to maintain ion homeostasis. Thus, the heart is particularly susceptible to disorders of mitochondrial function. The constituents of mitochondria are encoded largely by nuclear genes but also by mitochondrial DNA (mtDNA), which follows a maternal inheritance. The mitochondrial genome encodes 13 protein subunits of the electron transport chain, 22 transfer RNAs and two ribosomal RNAs [247]. The first nuclear mutation leading to a mitochondrial cardiomyopathy was discovered in 1992 [248]. Since then, over 100 point mutations have been found [249]. Mitochondriopathies caused by nuclear DNA mutations follow a Mendelian pattern of inheritance, either autosomal recessive, autosomal dominant or X-linked.

Mitochondrial cardiomyopathies were first described in 1958 by Kearns and Sayre [250]. Mitochondrial myopathy is defined as "muscle disease characterized by structurally or numerically abnormal mitochondria and/or abnormally functioning mitochondria" [251–253]. Several mtDNA disorders, including Kearns–Sayre syndrome, Leber hereditary optic neuropathy and Leigh syndrome result in a global impairment of mitochondrial respiratory function. Proteins most frequently affected by mutations are those of the respiratory chain and oxidative phosphorylation.

Cardiac muscle involvement may be predominant or minor in the clinical disease spectrum. In early childhood, heart involvement may present with congestive failure, cardiomegaly and lactic acidosis. Hearts with mitochondriopathy typically show biventricular hypertrophy and increased weight. Occasionally, there may be ventricular dilatation or endocardial fibroelastosis [254,255].

Ultrastructural findings in clinically significant cases are generalized, and include qualitative as well as quantitative mitochondrial changes. These morphologic abnormalities of mitochondria are usually, if not always, present in affected tissues when the mitochondrial disease is caused by a defect of the respiratory chain (the large majority of which are from mutations of the mtDNA, rather than nuclear DNA) [34]. Defects of substrate transport or utilization, on the other hand, often lack abnormalities of mitochondrial morphology [256]. The striking pathologic finding is fusiform swelling of cardiac myocytes with perinuclear cytoplasmic granulation and clearing of myofibrils. The accumulation of the structurally and functionally aberrant mitochondria is thought to be responsible for the enlargement of the myocardium [257]. Increased numbers and pleomorphism of mitochondria are seen on electron microscopic examination. Myofibrils are displaced by the mitochondrial hyperplasia. Qualitative abnormalities of the mitochondria include an abundance of tubular and vesicular cristae, concentrically arranged ("fingerprint") cristae, and stacked arrays of cristae.

In a recent report, the incidence of mitochondrial cardiomyopathies was estimated to be 4–5 in 100,000 live births, but may be as high as 1 in 5000–10,000 live births [258,259]. More than 95% of the mitochondriopathies are caused by nuclear DNA mutations, of which only a few have been identified thus far. On the other hand, less than 5% of mitochondriopathies are caused by mutations in mtDNA. Mutations involving nuclear DNA are transmitted by traditional Mendelian modes and primarily cause defects in substrate transport and utilization [260,261]. The prevalence of mtDNA mutations in adults is estimated to be 1 in 50,000 [262]. Mutations in mtDNA can be germline or somatic and have a mutation rate of several times that of nuclear DNA, probably because of a failure of proofreading by mtDNA polymerases. Any

mutations in a 16.6-kb ring structure located in the matrix of the mitochondrion are transmitted by maternal inheritance [263]. Typically, "mitochondrial cardiomyopathy" refers to abnormalities resulting from mtDNA mutations. These include deletions, most often encompassing several electron transport chain subunit coding regions, duplications and point mutations involving transfer RNA coding regions [264].

Mutations of mtDNA, through the process of mitotic segregation of mitochondria during embryogenesis, can show selective organ or tissue distribution. When heart is a dominant focus, clinical presentation is usually with concentric less common asymmetric septal hypertrophic cardiomyopathy, with decreased systolic function [257]. The Kearns–Sayre syndrome prominently features conduction disturbances [264]. "Pure" cardiomyopathy most often presents in infancy, but initial presentation in older patients has been reported [265].

Mitochondriopathies are typically multisystem disorders, which predominantly manifest in childhood in tissues with high oxygen consumption such as skeletal muscle, brain and myocardium. Mitochondriopathies are extremely phenotypically heterogeneous with symptoms including myalgia, fatigue, weakness, muscle cramps, muscle stiffness, double vision, sensory disturbances, deafness, wasting and ptosis [266]. Combinations of clinical features such as deafness, cardiomyopathy and diabetes together with encephalopathy and myopathy are highly susceptible of mitochondriopathy [267]. The combination of myopathy and deafness is also highly suggestive of mitochondriopathy, wherein the patients present with short stature, deafness and ptosis [262]. Myopathy with or without lactic acidosis is also the most common presenting feature [266]. Mitochondriopathy should be also considered when dealing with an unexplained association of manifestations with progressive course, involving seemingly unrelated organs [258]. A known pathogenic mutation in a symptomatic individual is regarded as diagnostic. Beyond that, the diagnostic approach to a patient with suspected mitochondriopathy requires an integral approach, incorporating clinical, electrophysiologic, imaging, histologic, biochemical and genetic investigations [262,268].

Mitochondrial disorders most often result in hypertrophic cardiomypathy as in the case with such disorders as Leber hereditary optic neuropathy, Complex I deficiency, Leigh syndrome, Friedreich ataxia and fatty acid oxidation disorders. However, such diseases may also result in dilated cardiomyopathy. Cardiomyopathy, typically with symmetrical hypertrophy, occurs with deficiency of carnitine, an essential co-factor for mitochondrial fatty acid oxidation. Most forms of systemic carnitine deficiency relate to defects in various enzymes of beta-oxidation. Primary systemic carnitine deficiency, a rare autosomal recessive disorder, causes a dilated cardiomyopathy that reverses with carnitine administration [262,268]. It is noteworthy that mitochondrial ultrastructure in this and other disorders of beta-oxidation, such as carnitine palmitoyltransferase II deficiency and long chain acyl coenzyme A dehydrogenase (LCAD) deficiency, is reported as normal [269].

Mutations in the mitochondrial genome frequently result in skeletal myopathy coincident with cardiac defects, including DCM, HCM and conduction defects. Similarly, inborn errors in several steps in the mitochondrial fatty acid oxidation pathway often manifest as a hypertrophic cardiomyopathic phenotype [270–272]. Cardiomyopathy in these patients usually appears during childhood and often presents as sudden onset heart failure, pulmonary edema and ventricular arrhythmia brought on by metabolic stress such as periods of fasting during infectious illness. A chronic cardiomyopathic phenotype may also develop. Together, these inherited metabolic cardiomyopathic disorders highlight the sensitivity of the heart to alterations of mitochondrial function.

Leigh syndrome

Leigh syndrome, also known as subacute necrotizing encephalomyopathy is a neuropathologically defined multisystem disorder of infancy [267]. Leigh syndrome is caused by mutations in mtDNA or nuclear DNA genes encoding for pyruvate dehydrogenase complex or respiratory chain components [267,268]. Complex IV (COX) is composed of 13 subunits and functions as the terminal complex of the electron-transport chain. COX-deficient Leigh syndrome is caused by mutations in the *SURF1*, *NDUFS* or *SDHA* genes [267]. Several distinct autosomal recessive mutations in COX have been associated with Leigh syndrome. The genes responsible are the COX assembly factors *SURF1*, *SCO1*, *SCO2* and *COX10* [273,274]. In addition, Leigh syndrome, French-Canadian type, is a human cytochrome c-oxidase deficiency, which was recently mapped to chromosome 2p16-21 [275]. Although most often associated with hypertrophic cardiomyopathy, Leigh syndrome may also result in a DCM phenotype.

Kearns–Sayre syndrome

Kearns–Sayre syndrome (KSS) is caused by sporadic, single large deletions or duplications in the mtDNA leading to a global impairment of mitochondrial respiratory function resulting in DCM [252,256]. KSS is a subtype of chronic progressive external ophthalmoplegia (CPEO) characterized by pigmentary retinopathy, cardiac conduction defects, cerebellar ataxia and an onset of less than 20 years [250]. The prognosis of KSS is rather poor as patients rarely survive beyond 30 years of age.

Myopathy, encephalopathy, lactacidosis and stroke-like episodes (MELAS)

MELAS is an early onset encephalomyopathy [276]. Typical features comprise stroke-like episodes with hemiparesis, hemianopsia, migraine, nausea and vomiting. Additional features are deafness, diabetes, seizures, dementia, ataxia, cortical blindness, optic atrophy, pigmentary retinopathy, dilative cardiomyopathy, myopathy (87% of cases), exercise intolerance, lactic acidosis and short stature (55% of cases) [259,277].

Several hypotheses exist for the mechanisms relating mitochondrial dysfunction to cardiomyopathy. First, reduced ATP production caused by mitochondrial oxidative deficit could lead to a state of energy deprivation, especially relevant in genetic forms of mitochondrial myopathy [278]. Second, production of reactive oxygen species (ROS) may be causative in the development of mitochondrial-induced metabolic cardiomyopathy [279]. Third, unabated mitochondrial proliferation could alter expression of structural and sarcomeric proteins [280,281].

Glycogen storage diseases

Disorders of metabolism account for a significant proportion of "idiopathic" cardiomyopathy in children [282]. Certain inherited disorders of metabolism present with a clinical picture predominated by heart disease, while in other forms extracardiac involvement, especially nervous system and liver, dominates cardiac consequences. Conditions significantly affecting the heart may involve endocardium, connective tissues and myocardium; heart dysfunction may result from valvular, coronary artery or myocardial disease. The morphologic pattern of cardiomyopathy resulting from inherited metabolic diseases can be hypertrophic, dilated or restrictive, with or without superimposed endocardial fibroelastosis [283].

In terms of cardiomyopathies, inherited metabolic disorders weigh more importantly in infants and children than in adults [284]. Although this group of disorders predominantly affects central nervous system and hepatic function, expression may primarily involve the heart. Servidei *et al.* [283] classified the hereditary metabolic cardiomyopathies into four groups: glycogen storage diseases, disorders of lipid metabolism, disorders of mitochondrial metabolism and "others" (including mucopolysaccharidoses and other storage disorders).

The gross morphology of cardiomyopathies in storage disorders can be dilated or restrictive, although hypertrophic is the most common pattern [285]. The classic infantile form (GSD IIa) of Pompe disease, (acid α-1,4-glucosidase, or acid maltase, deficiency) characteristically causes biventricular cardiomegaly and leads to cardiorespiratory failure before 1 year of age [286]. The autosomal recessive inherited lack of lysosomal acid α-1,4-glucosidase results in lysosomal accumulation of morphologically normal glycogen, mainly as β-particles, in numerous cell types, but chiefly in cardiac myocytes, skeletal muscle cells and hepatocytes. Excessive cytoplasmic and nuclear accumulations of glycogen also occur. Cardiomegaly, macroglossia, muscular weakness and hypotonia develop in early infancy [286]. The heart in a patient with infantile Pompe disease shows marked biventricular hypertrophy, with weight up to 6 times normal. Cardiomegaly may be evident at birth, and is progressive. Endocardial fibroelastosis of the left ventricle may be prominent [287]. The left ventricle and papillary muscles are most strikingly thickened. Light microscopy shows swelling of cardiac myocytes with central vacuolation and peripheral displacement of myofilaments.

Isolated left ventricular noncompaction cardiomyopathy

Left ventricular noncompaction cardiomyopathy (LVNC) is a rare congenital myocardial disorder resulting in multiple trabeculations in the left ventricular myocardium caused by disruption or failure of the compaction process of the myocardial trabeculae during early fetal development [288,289]. LVNC is generally characterized by hypertrophic and dilated left ventricle with a clinical presentation of reduced systolic function associated with a poor prognosis [290,291].

There is evidence that LVNC is a heterogeneous genetic disorder. A point mutation in the gene *G4.5*, which encodes the tafazzin protein, has been identified in a family with X-linked isolated LVNC, suggesting that LVNC is likely allelic with Barth syndrome [292]. Mutations in *ZASP* gene, a novel cardiac and skeletal muscle specific Z-line protein and α-dystrobrevin have been also reported in families with LVNC, left ventricular dysfunction and variable forms of congenital heart diseases [293,294]. Deletion of the gene encoding FK binding potein 12 (FKBP12) has been shown in a noncompaction disorder in mice [295,296]. However, to date there is no report of the same mutation in human cases of LVNC [296]. Recently, Sasse-Klaassen *et al.* [297] proposed that isolated LVNC in the adult is an autosomal dominant disorder in the majority of patients. An autosomal dominant familial transmission of LVNC has been reported and mapped to chromosome 11p15 by the same group [298].

Conclusions

New advances in molecular genetics have provided effective tools for identification of single gene defects in various forms of cardiomyopathy complications [299,300]. Understanding how mutations in different genes alter cardiac function may further our search for potential targets for therapeutic interventions. In general, DCM disorders are mostly caused by defects in the cytoskeleton components causing disruption in muscle force transmission [299]. RCM cases, on the other hand, are frequently linked to abnormal sarcomere proteins impairing force production [83], while ARVC abnormalities are brought about by impaired cytoskeletal proteins implicated in cell–cell junction [178,193].

Thus far, idiopathic cardiomyopathy has been the diagnosis of exclusion, and the classification of inherited cardiomyopathies is mainly based on phenotypic manifestations. However, as we gain a better understanding of the complex genetic etiology of such diseases, we may have better classification parameters by taking into account the underlying gene mutations.

Current challenges that limit the value of genetic testing for cardiomyopathies include the vast number of genes and mutations causing clinical disease. This situation is compounded by the fact that the majority of patients have mutations, which are yet undefined. In addition, sensitive, reproducible, rapid, inexpensive and high throughput DNA testing is required. Nevertheless, as we increase our understanding of genes and diseases, genetic testing will begin to take a more central role in the management of patients. Genetic counseling is a communication process that includes education for medical and psychosocial issues, as well as therapy for patients and their families. Currently, no established guidelines exist for when to refer a patient for genetic counseling. Thus, there is a great need for cardiologists who have expertise in genetics and vice versa who can become guiding experts in the genetics and genomics of cardiomyopathy. The horizon still beckons.

References

1 Kapoor AS. The spectrum of cardiomyopathies. In: Kapoor AS, Schroeder J, Yacoub M, eds. *Cardiomyopathies and Heart-Lung Transplantation.* McGraw-Hill, Inc., New York, 1991, 3–28.

2 Richardson P, McKenna W, Bristow M *et al.* Report of the 1995 World Health Organization/International Society and Federation of Cardiology Task Force on the definition and classification of cardiomyopathies. *Circulation* 1996; 93: 841–842.

3 Silver M, Silver M. Examination of the heart and of cardiovascular specimens in surgical pathology. In: Silver M, Gotlieb A, Schoen F, eds. *Cardiovascular Pathology.* Churchill Livingstone, New York, 2001, 1–29.

4 Fox CC, Hutchins GM. The architecture of the human ventricular myocardium. *Johns Hopkins Med J* 1972; 130: 289–299.

5 Veinot J, Ghadially F, Walley V. Light microscopy and ultrastructure of the blood vessels and heart. In: Silver M, Gotlieb A, Schoen F, eds. *Cardiovascular Pathology.* Churchill Livingstone, New York, 2001, 30–53.

6 Davies MJ, Pomerance A. Quantitative study of ageing changes in the human sinoatrial node and internodal tracts. *Br Heart J* 1972; 34: 150–152.

7 Lester W. Age-related cardiovascular changes. In: Silver M, Gotlieb A, Schoen F, eds. *Cardiovascular Pathology.* Churchill Livingstone, New York, 2001, 54–67.

8 Lenkiewicz JE, Davies MJ, Rosen D. Collagen in human myocardium as a function of age. *Cardiovasc Res* 1972; 6: 549–555.

9 Campbell SE, Korecky B, Rakusan K. Remodeling of myocyte dimensions in hypertrophic and atrophic rat hearts. *Circ Res* 1991; 68: 984–996.

10 Baroldi G. Anatomy and quantification of myocardial cell death. *Methods Achiev Exp Pathol* 1988; 13: 87–113.

11 Itoh S, Yanagishita T, Mukae S, Konno N, Katagiri T. Study on reperfusion injury on sarcoplasmic reticulum in acute myocardial ischemia. *Jpn Circ J* 1992; 56: 384–391.

12 Huhn KM, Palcic B, Wilson JE, McManus BM. Cytometric analysis of ventricular myocyte nuclei in idiopathic dilated cardiomyopathy: a tool for evaluation of disease progression? *Eur Heart J* 1995; 16 (Supplement O): 97–99.

13 Simpson DG, Sharp WW, Borg TK, Price RL, Samarel AM, Terracio L. Mechanical regulation of cardiac myofibrillar structure. *Ann N Y Acad Sci* 1995; 752: 131–140.

14 Weber KT, Sun Y, Katwa LC. Local regulation of extracellular matrix structure. *Herz* 1995; 20: 81–88.

15 Kanter HL, Saffitz JE, Beyer EC. Cardiac myocytes express multiple gap junction proteins. *Circ Res* 1992; 70: 438–444.

16 Davis LM, Kanter HL, Beyer EC, Saffitz JE. Distinct gap junction protein phenotypes in cardiac tissues with disparate conduction properties. *J Am Coll Cardiol* 1994; 24: 1124–1132.

17 Dec GW, Fuster V. Idiopathic dilated cardiomyopathy. *N Engl J Med* 1994; 331: 1564–1575.

18 Michels VV, Moll PP, Miller FA *et al.* The frequency of familial dilated cardiomyopathy in a series of patients with idiopathic dilated cardiomyopathy. *N Engl J Med* 1992; 326: 77–82.

19 Tamburro P, Wilber D: Sudden death in idiopathic dilated cardiomyopathy. *Am Heart J* 1992; 124: 1035–1045.

20 Fuster V, Gersh BJ, Giuliani ER, Tajik AJ, Brandenburg RO, Frye RL. The natural history of idiopathic dilated cardiomyopathy. *Am J Cardiol* 1981; 47: 525–531.

21 Redfield MM, Gersh BJ, Bailey KR, Ballard DJ, Rodeheffer RJ. Natural history of idiopathic dilated cardiomyopathy: effect of referral bias and secular trend. *J Am Coll Cardiol* 1993; 22: 1921–1926.

22 Grunig E, Tasman JA, Kucherer H, Franz W, Kubler W, Katus HA. Frequency and phenotypes of familial dilated cardiomyopathy. *J Am Coll Cardiol* 1998; 31: 186–194.

23 Keeling PJ, Gang Y, Smith G *et al.* Familial dilated cardiomyopathy in the United Kingdom. *Br Heart J* 1995; 73: 417–421.

24 Felker GM, Thompson RE, Hare JM *et al.* Underlying causes and long-term survival in patients with initially unexplained cardiomyopathy. *N Engl J Med* 2000; 342: 1077–1084.

25 Towbin JA, Bowles NE. The failing heart. *Nature* 2002; 415: 227–233.

26 Bagger JP, Baandrup U, Rasmussen K, Moller M, Vesterlund T. Cardiomyopathy in western Denmark. *Br Heart J* 1984; 52: 327–331.

27 Williams DG, Olsen EG. Prevalence of overt dilated cardiomyopathy in two regions of England. *Br Heart J* 1985; 54: 153–155.

28 Codd MB, Sugrue DD, Gersh BJ, Melton LJ. Epidemiology of idiopathic dilated and hypertrophic cardiomyopathy. A population-based study in Olmsted County, Minnesota, 1975–1984. *Circulation* 1989; 80: 564–572.

29 Gillum RF. Idiopathic cardiomyopathy in the United States, 1970–1982. *Am Heart J* 1986; 111: 752–755.

30 Mestroni L, Rocco C, Vatta M, Miocic S, Giacca M. Advances in molecular genetics of dilated cardiomyopathy. The Heart Muscle Disease Study Group. *Cardiol Clin* 1998; 16: 611–621, vii.

31 Bachinski LL, Roberts R. New theories. Causes of dilated cardiomyopathy. *Cardiol Clin* 1998; 16: 603–610, vii.

32 Towbin JA, Hejtmancik JF, Brink P *et al.* X-linked dilated cardiomyopathy. Molecular genetic evidence of linkage to the Duchenne muscular dystrophy (dystrophin) gene at the Xp21 locus. *Circulation* 1993; 87: 1854–1865.

33 Suomalainen A, Paetau A, Leinonen H, Majander A, Peltonen L, Somer H. Inherited idiopathic dilated cardiomyopathy with multiple deletions of mitochondrial DNA. *Lancet* 1992; 340: 1319–1320.

34 Kelly DP, Strauss AW: Inherited cardiomyopathies. *N Engl J Med* 1994; 330: 913–919.

35 Li D, Tapscoft T, Gonzalez O *et al.* Desmin mutation responsible for idiopathic dilated cardiomyopathy. *Circulation* 1999; 100: 461–464.

36 Dalakas MC, Park KY, Semino-Mora C, Lee HS, Sivakumar K, Goldfarb LG. Desmin myopathy, a skeletal myopathy with cardiomyopathy caused by mutations in the desmin gene. *N Engl J Med* 2000; 342: 770–780.

37 Bione S, D'Adamo P, Maestrini E, Gedeon AK, Bolhuis PA, Toniolo D. A novel X-linked gene, G4.5 is responsible for Barth syndrome. *Nat Genet* 1996; 12: 385–389.

38 D'Adamo P, Fassone L, Gedeon A *et al.* The X-linked gene G4.5 is responsible for different infantile dilated cardiomyopathies. *Am J Hum Genet* 1997; 61: 862–867.

39 Bonne G, Di Barletta MR, Varnous S *et al.* Mutations in the gene encoding lamin A/C cause autosomal dominant Emery–Dreifuss muscular dystrophy. *Nat Genet* 1999; 21: 285–288.

40 Fatkin D, MacRae C, Sasaki T *et al.* Missense mutations in the rod domain of the lamin A/C gene as causes of dilated cardiomyopathy and conduction-system disease. *N Engl J Med* 1999; 341: 1715–1724.

41 Lazarides E, Hubbard BD: Immunological characterization of the subunit of the 100 A filaments from muscle cells. *Proc Natl Acad Sci USA* 1976; 73: 4344–4348.

42 Gard DL, Lazarides E. The synthesis and distribution of desmin and vimentin during myogenesis in vitro. *Cell* 1980; 19: 263–275.

43 Thornell LE, Eriksson A. Filament systems in the Purkinje fibers of the heart. *Am J Physiol* 1981; 241: H291–H305.

44 Ferrans VJ, Roberts WC, Shugoll GI, Massumi RA, Ali N. Plasma membrane extensions in intercalated discs of human myocardium and their relationship to partial dissociations of the discs. *J Mol Cell Cardiol* 1973; 5: 161–169.

45 Tokuyasu KT. Visualization of longitudinally-oriented intermediate filaments in frozen sections of chicken cardiac muscle by a new staining method. *J Cell Biol* 1983; 97: 562–565.

46 Viegas-Pequignot E, Li ZL, Dutrillaux B, Apiou F, Paulin D. Assignment of human desmin gene to band 2q35 by nonradioactive *in situ* hybridization. *Hum Genet* 1989; 83: 33–36.

47 Goebel HH. Desmin-related neuromuscular disorders. *Muscle Nerve* 1995; 18: 1306–1320.

48 Munoz-Marmol AM, Strasser G, Isamat M *et al.* A dysfunctional desmin mutation in a patient with severe generalized myopathy. *Proc Natl Acad Sci USA* 1998; 95: 11312–11317.

49 Dagvadorj A, Goudeau B, Hilton-Jones D *et al.* Respiratory insufficiency in desminopathy patients caused by introduction of proline residues in desmin c-terminal alpha-helical segment. *Muscle Nerve* 2003; 27: 669–675.

50 Goudeau B, Dagvadorj A, Rodrigues-Lima F *et al.* Structural and functional analysis of a new desmin variant causing desmin-related myopathy. *Hum Mutat* 2001; 18: 388–396.

51 Sjoberg G, Saavedra-Matiz CA, Rosen DR *et al.* A missense mutation in the desmin rod domain is associated with autosomal dominant distal myopathy, and exerts a dominant negative effect on filament formation. *Hum Mol Genet* 1999; 8: 2191–2198.

52 Sugawara M, Kato K, Komatsu M *et al.* A novel *de novo* mutation in the desmin gene causes desmin myopathy with toxic aggregates. *Neurology* 2000; 55: 986–990.

53 Vicart P, Caron A, Guicheney P *et al.* A missense mutation in the alpha-B-crystallin chaperone gene

causes a desmin-related myopathy. *Nat Genet* 1998; 20: 92–95.

54 Hoffman EP, Brown RH Jr, Kunkel LM. Dystrophin: the protein product of the Duchenne muscular dystrophy locus. *Cell* 1987; 51: 919–928.

55 Rando TA. The dystrophin–glycoprotein complex, cellular signaling, and the regulation of cell survival in the muscular dystrophies. *Muscle Nerve* 2001; 24: 1575–1594.

56 Chamberlain JS, Corrado K, Rafael JA, Cox GA, Hauser M, Lumeng C. Interactions between dystrophin and the sarcolemma membrane. *Soc Gen Physiol Ser* 1997; 52: 19–29.

57 Lapidos KA, Kakkar R, McNally EM. The dystrophin glycoprotein complex: signaling strength and integrity for the sarcolemma. *Circ Res* 2004; 94: 1023–1031.

58 Cox GF, Kunkel LM. Dystrophies and heart disease. *Curr Opin Cardiol* 1997; 12: 329–343.

59 Finsterer J, Stollberger C. The heart in human dystrophinopathies. *Cardiology* 2003; 99: 1–19.

60 Beggs AH. Dystrophinopathy, the expanding phenotype. Dystrophin abnormalities in X-linked dilated cardiomyopathy. *Circulation* 1997; 95: 2344–2347.

61 Monaco AP, Bertelson CJ, Liechti-Gallati S, Moser H, Kunkel LM. An explanation for the phenotypic differences between patients bearing partial deletions of the DMD locus. *Genomics* 1988; 2: 90–95.

62 Mori K, Manabe T, Nii M, Hayabuchi Y, Kuroda Y, Tatara K. Myocardial integrated ultrasound backscatter in patients with Duchenne's progressive muscular dystrophy. *Heart* 2001; 86: 341–342.

63 Roberts RG, Passos-Bueno MR, Bobrow M, Vainzof M, Zatz M. Point mutation in a Becker muscular dystrophy patient. *Hum Mol Genet* 1993; 2: 75–77.

64 Hagiwara Y, Nishio H, Kitoh Y *et al*. A novel point mutation (G-1 to T) in a 5′ splice donor site of intron 13 of the dystrophin gene results in exon skipping and is responsible for Becker muscular dystrophy. *Am J Hum Genet* 1994; 54: 53–61.

65 Bartolo C, Papp AC, Snyder PJ *et al*. A novel splice site mutation in a Becker muscular dystrophy patient. *J Med Genet* 1996; 33: 324–327.

66 Patria SY, Alimsardjono H, Nishio H, Takeshima Y, Nakamura H, Matsuo M. A case of Becker muscular dystrophy resulting from the skipping of four contiguous exons (71–74) of the dystrophin gene during mRNA maturation. *Proc Assoc Am Physicians* 1996; 108: 308–314.

67 Shiga N, Takeshima Y, Sakamoto H *et al*. Disruption of the splicing enhancer sequence within exon 27 of the dystrophin gene by a nonsense mutation induces partial skipping of the exon and is responsible for Becker muscular dystrophy. *J Clin Invest* 1997; 100: 2204–2210.

68 Fajkusova L, Lukas Z, Tvrdikova M, Kuhrova V, Hajek J, Fajkus J. Novel dystrophin mutations revealed by analysis of dystrophin mRNA: alternative splicing suppresses the phenotypic effect of a nonsense mutation. *Neuromusc Disord* 2001; 11: 133–138.

69 Cagliani R, Fortunato F, Giorda R *et al*. Molecular analysis of LGMD-2B and MM patients: identification of novel DYSF mutations and possible founder effect in the Italian population. *Neuromusc Disord* 2003; 13: 788–795.

70 Tuffery-Giraud S, Saquet C, Chambert S, Claustres M. Pseudoexon activation in the DMD gene as a novel mechanism for Becker muscular dystrophy. *Hum Mutat* 2003; 21: 608–614.

71 Adachi K, Takeshima Y, Wada H, Yagi M, Nakamura H, Matsuo M. Heterogous dystrophin mRNA produced by a novel splice acceptor site mutation in intermediate dystrophinopathy. *Pediatr Res* 2003; 53: 125–131.

72 Beroud C, Carrie A, Beldjord C *et al*. Dystrophinopathy caused by mid-intronic substitutions activating cryptic exons in the DMD gene. *Neuromusc Disord* 2004; 14: 10–18.

73 Davey KM, Parboosingh JS, McLeod DR *et al*. Mutation of DNAJC19, a human homolog of yeast inner mitochondrial membrane co-chaperones, causes DCMA syndrome, a novel autosomal recessive Barth syndrome-like condition. *J Med Genet* 2006; 43: 385–393.

74 Berko BA, Swift M. X-linked dilated cardiomyopathy. *N Engl J Med* 1987; 316: 1186–1191.

75 Muntoni F, Cau M, Ganau A *et al*. Brief report: deletion of the dystrophin muscle-promoter region associated with X-linked dilated cardiomyopathy. *N Engl J Med* 1993; 329: 921–925.

76 Barth PG, Valianpour F, Bowen VM, *et al*. X-linked cardioskeletal myopathy and neutropenia (Barth syndrome): an update. *Am J Med Genet A* 2004; 126: 349–354.

77 Mestroni L, Giacca M. Molecular genetics of dilated cardiomyopathy. *Curr Opin Cardiol* 1997; 12: 303–309.

78 Ferlini A, Sewry C, Melis MA, Mateddu A, Muntoni F. X-linked dilated cardiomyopathy and the dystrophin gene. *Neuromusc Disord* 1999; 9: 339–346.

79 Franz W, Hermann R, Cremer M *et al*. Novel stop mutation and alternative splicing of exon 29 in th dystrophin gene associated with rapid progressive familial dilated cardiomyopathy. Abstract at the European Society of Human Gentics. *Med Genetik* 1995; 2: 192.

80 Milasin J, Muntoni F, Severini GM *et al*. A point mutation in the 5′ splice site of the dystrophin gene first intron responsible for X-linked dilated cardiomyopathy. *Hum Mol Genet* 1996; 5: 73–79.

81 Kamisago M, Sharma SD, DePalma SR *et al*. Mutations in sarcomere protein genes as a cause of dilated cardiomyopathy. *N Engl J Med* 2000; 343: 1688–1696.

82 Murphy RT, Mogensen J, Shaw A, Kubo T, Hughes S, McKenna WJ. Novel mutation in cardiac troponin I in recessive idiopathic dilated cardiomyopathy. *Lancet* 2004; 363: 371–372.

83 Mogensen J, Kubo T, Duque M *et al.* Idiopathic restrictive cardiomyopathy is part of the clinical expression of cardiac troponin I mutations. *J Clin Invest* 2003; 111: 209–216.

84 Mogensen J, Murphy RT, Shaw T *et al.* Severe disease expression of cardiac troponin C and T mutations in patients with idiopathic dilated cardiomyopathy. *J Am Coll Cardiol* 2004; 44: 2033–2040.

85 Hanson EL, Jakobs PM, Keegan H *et al.* Cardiac troponin T lysine 210 deletion in a family with dilated cardiomyopathy. *J Card Fail* 2002; 8: 28–32.

86 Fujino N, Shimizu M, Ino H *et al.* A novel mutation Lys273Glu in the cardiac troponin T gene shows high degree of penetrance and transition from hypertrophic to dilated cardiomyopathy. *Am J Cardiol* 2002; 89: 29–33.

87 Lu QW, Morimoto S, Harada K *et al.* Cardiac troponin T mutation R141W found in dilated cardiomyopathy stabilizes the troponin T-tropomyosin interaction and causes a Ca^{2+} desensitization. *J Mol Cell Cardiol* 2003; 35: 1421–1427.

88 Morimoto S, Lu QW, Harada K *et al.* Ca^{2+}-desensitizing effect of a deletion mutation Delta K210 in cardiac troponin T that causes familial dilated cardiomyopathy. *Proc Natl Acad Sci USA* 2002; 99: 913–918.

89 Jha PK, Mao C, Sarkar S. Photo-cross-linking of rabbit skeletal troponin I deletion mutants with troponin C and its thiol mutants: the inhibitory region enhances binding of troponin I fragments to troponin C. *Biochemistry* 1996; 35: 11026–11035.

90 Leszyk J, Collins JH, Leavis PC, Tao T. Cross-linking of rabbit skeletal muscle troponin subunits: labeling of cysteine-98 of troponin C with 4-maleimidobenzophenone and analysis of products formed in the binary complex with troponin T and the ternary complex with troponins I and T. *Biochemistry* 1988; 27: 6983–6987.

91 Leszyk J, Dumaswala R, Potter JD, Collins JH. Amino acid sequence of bovine cardiac troponin I. *Biochemistry* 1988; 27: 2821–2827.

92 Houdusse A, Love ML, Dominguez R, Grabarek Z, Cohen C. Structures of four Ca^{2+}-bound troponin C at 2.0 A resolution: further insights into the Ca^{2+}-switch in the calmodulin superfamily. *Structure* 1997; 5: 1695–1711.

93 Rayment I, Holden HM, Whittaker M *et al.* Structure of the actin-myosin complex and its implications for muscle contraction. *Science* 1993; 261: 58–65.

94 Olson TM, Michels VV, Thibodeau SN, Tai YS, Keating MT. Actin mutations in dilated cardiomyopathy, a heritable form of heart failure. *Science* 1998; 280: 750–752.

95 Richard P, Isnard R, Carrier L *et al.* Double heterozygosity for mutations in the beta-myosin heavy chain and in the cardiac myosin binding protein C genes in a family with hypertrophic cardiomyopathy. *J Med Genet* 1999; 36: 542–545.

96 Fatkin D, Christe ME, Aristizabal O *et al.* Neonatal cardiomyopathy in mice homozygous for the Arg403Gln mutation in the alpha cardiac myosin heavy chain gene. *J Clin Invest* 1999; 103: 147–153.

97 Durand JB, Bachinski LL, Bieling LC *et al.* Localization of a gene responsible for familial dilated cardiomyopathy to chromosome 1q32. *Circulation* 1995; 92: 3387–3389.

98 Siu BL, Niimura H, Osborne JA *et al.* Familial dilated cardiomyopathy locus maps to chromosome 2q31. *Circulation* 1999; 99: 1022–1026.

99 Krajinovic M, Pinamonti B, Sinagra G *et al.* Linkage of familial dilated cardiomyopathy to chromosome 9. Heart Muscle Disease Study Group. *Am J Hum Genet* 1995; 57: 846–852.

100 Bowles KR, Gajarski R, Porter P *et al.* Gene mapping of familial autosomal dominant dilated cardiomyopathy to chromosome 10q21–23. *J Clin Invest* 1996; 98: 1355–1360.

101 Kass S, MacRae C, Graber HL *et al.* A gene defect that causes conduction system disease and dilated cardiomyopathy maps to chromosome 1p1–1q1. *Nat Genet* 1994; 7: 546–551.

102 Olson TM, Keating MT. Mapping a cardiomyopathy locus to chromosome 3p22–p25. *J Clin Invest* 1996; 97: 528–532.

103 Messina DN, Speer MC, Pericak-Vance MA, McNally EM. Linkage of familial dilated cardiomyopathy with conduction defect and muscular dystrophy to chromosome 6q23. *Am J Hum Genet* 1997; 61: 909–917.

104 Sylvius N, Tesson F, Gayet C *et al.* A new locus for autosomal dominant dilated cardiomyopathy identified on chromosome 6q12–q16. *Am J Hum Genet* 2001; 68: 241–246.

105 van der Kooi AJ, van Meegen M, Ledderhof TM, McNally EM, de Visser M, Bolhuis PA. Genetic localization of a newly recognized autosomal dominant limb-girdle muscular dystrophy with cardiac involvement (LGMD1B) to chromosome 1q11–21. *Am J Hum Genet* 1997; 60: 891–895.

106 Jung M, Poepping I, Perrot A *et al.* Investigation of a family with autosomal dominant dilated cardiomyopathy defines a novel locus on chromosome 2q14–q22. *Am J Hum Genet* 1999; 65: 1068–1077.

107 Miyagoe-Suzuki Y, Nakagawa M, Takeda S. Merosin and congenital muscular dystrophy. *Microsc Res Tech* 2000; 48: 181–191.

108 Gilhuis HJ, ten Donkelaar HJ, Tanke RB *et al.* Non-muscular involvement in merosin-negative congenital

muscular dystrophy. *Pediatr Neurol* 2002; 26: 30–36.

109 Straub V, Duclos F, Venzke DP *et al.* Molecular pathogenesis of muscle degeneration in the delta-sarcoglycan-deficient hamster. *Am J Pathol* 1998; 153: 1623–1630.

110 Ben Hamida M, Ben Hamida C, Zouari M, Belal S, Hentati F. Limb-girdle muscular dystrophy 2C: clinical aspects. *Neuromuscul Disord* 1996; 6: 493–494.

111 Lim LE, Duclos F, Broux O *et al.* Beta-sarcoglycan: characterization and role in limb-girdle muscular dystrophy linked to 4q12. *Nat Genet* 1995; 11: 257–265.

112 Bonnemann CG, Modi R, Noguchi S *et al.* Beta-sarcoglycan (A3b) mutations cause autosomal recessive muscular dystrophy with loss of the sarcoglycan complex. *Nat Genet* 1995; 11: 266–273.

113 Duclos F, Broux O, Bourg N *et al.* Beta-sarcoglycan: genomic analysis and identification of a novel missense mutation in the LGMD2E Amish isolate. *Neuromuscul Disord* 1998; 8: 30–38.

114 Noguchi S, McNally EM, Ben Othmane K *et al.* Mutations in the dystrophin-associated protein gamma-sarcoglycan in chromosome 13 muscular dystrophy. *Science* 1995; 270: 819–822.

115 Li H, Linke WA, Oberhauser AF *et al.* Reverse engineering of the giant muscle protein titin. *Nature* 2002; 418: 998–1002.

116 Neagoe C, Kulke M, del Monte F *et al.* Titin isoform switch in ischemic human heart disease. *Circulation* 2002; 106: 1333–1341.

117 Wu Y, Bell SP, Trombitas K *et al.* Changes in titin isoform expression in pacing-induced cardiac failure give rise to increased passive muscle stiffness. *Circulation* 2002; 106: 1384–1389.

118 Gerull B, Gramlich M, Atherton J *et al.* Mutations of TTN, encoding the giant muscle filament titin, cause familial dilated cardiomyopathy. *Nat Genet* 2002; 30: 201–204.

119 Itoh-Satoh M, Hayashi T, Nishi H *et al.* Titin mutations as the molecular basis for dilated cardiomyopathy. *Biochem Biophys Res Commun* 2002; 291: 385–393.

120 Van den Bergh PY, Bouquiaux O, Verellen C *et al.* Tibial muscular dystrophy in a Belgian family. *Ann Neurol* 2003; 54: 248–251.

121 Jia Z, Petrounevitch V, Wong A *et al.* Mutations in calpain 3 associated with limb girdle muscular dystrophy: analysis by molecular modeling and by mutation in m-calpain. *Biophys J* 2001; 80: 2590–2596.

122 Goldman RD, Gruenbaum Y, Moir RD, Shumaker DK, Spann TP. Nuclear lamins: building blocks of nuclear architecture. *Genes Dev* 2002; 16: 533–547.

123 Helbling-Leclerc A, Bonne G, Schwartz K. Emery–Dreifuss muscular dystrophy. *Eur J Hum Genet* 2002; 10: 157–161.

124 Bione S, Maestrini E, Rivella S *et al.* Identification of a novel X-linked gene responsible for Emery–Dreifuss muscular dystrophy. *Nat Genet* 1994; 8: 323–327.

125 Talkop UA, Talvik I, Sonajalg M *et al.* Early onset of cardiomyopathy in two brothers with X-linked Emery–Dreifuss muscular dystrophy. *Neuromuscul Disord* 2002; 12: 878–881.

126 Boriani G, Gallina M, Merlini L *et al.* Clinical relevance of atrial fibrillation/flutter, stroke, pacemaker implant, and heart failure in Emery–Dreifuss muscular dystrophy: a long-term longitudinal study. *Stroke* 2003; 34: 901–908.

127 Dalla Volta S, Battaglia G, Zerbini E: "Auricularization" of right ventricular pressure curve. *Am Heart J* 1961; 61: 25–33.

128 Corrado D, Basso C, Thiene G *et al.* Spectrum of clinicopathologic manifestations of arrhythmogenic right ventricular cardiomyopathy/dysplasia: a multicenter study. *J Am Coll Cardiol* 1997; 30: 1512–1520.

129 Vedel J, Frank R, Fontaine G *et al.* Recurrent ventricular tachycardia and parchment right ventricle in the adult. Anatomical and clinical report of 2 cases. *Arch Mal Coeur Vaiss* 1978; 71: 973–981.

130 Marcus FI, Fontaine GH, Guiraudon G *et al.* Right ventricular dysplasia: a report of 24 adult cases. *Circulation* 1982; 65: 384–398.

131 Giles TD. New WHO/ISFC classification of cardiomyopathies: a task not completed. *Circulation* 1997; 96: 2081–2082.

132 Basso C, Thiene G, Corrado D, Angelini A, Nava A, Valente M. Arrhythmogenic right ventricular cardiomyopathy. Dysplasia, dystrophy, or myocarditis? *Circulation* 1996; 94: 983–991.

133 Rampazzo A, Nava A, Danieli GA *et al.* The gene for arrhythmogenic right ventricular cardiomyopathy maps to chromosome 14q23–q24. *Hum Mol Genet* 1994; 3: 959–962.

134 Norman MW, McKenna WJ. Arrhythmogenic right ventricular cardiomyopathy: perspectives on disease. *Z Kardiol* 1999; 88: 550–554.

135 Fontaine G, Fontaliran F, Hebert JL *et al.* Arrhythmogenic right ventricular dysplasia. *Annu Rev Med* 1999; 50: 17–35.

136 Kayser HW, van der Wall EE, Sivananthan MU, Plein S, Bloomer TN, de Roos A. Diagnosis of arrhythmogenic right ventricular dysplasia: a review. *Radiographics* 2002; 22: 639–648; discussion 649–650.

137 Shen WK, Edwards WD, Hammill SC, Bailey KR, Ballard DJ, Gersh BJ. Sudden unexpected nontraumatic death in 54 young adults: a 30-year population-based study. *Am J Cardiol* 1995; 76: 148–152.

138 Corrado D, Thiene G, Nava A, Rossi L, Pennelli N. Sudden death in young competitive athletes: clinico-

pathologic correlations in 22 cases. *Am J Med* 1990; 89: 588–596.

139 Laurent M, Descaves C, Biron Y, Deplace C, Almange C, Daubert JC. Familial form of arrhythmogenic right ventricular dysplasia. *Am Heart J* 1987; 113: 827–829.

140 Buja GF, Nava A, Martini B, Canciani B, Thiene G. Right ventricular dysplasia: a familial cardiomyopathy? *Eur Heart J* 1989, 10 (Supplement D): 13–15.

141 Wlodarska EK, Konka M, Kepski R *et al.* Familial form of arrhythmogenic right ventricular cardiomyopathy. *Kardiol Pol* 2004; 60: 1–14.

142 Nava A, Canciani B, Daliento L *et al.* Juvenile sudden death and effort ventricular tachycardias in a family with right ventricular cardiomyopathy. *Int J Cardiol* 1988; 21: 111–126.

143 Frances RJ. Arrhythmogenic right ventricular dysplasia/cardiomyopathy. A review and update. *Int J Cardiol* 2006; 110: 279–287.

144 Dokuparti MV, Pamuru PR, Thakkar B, Tanjore RR, Nallari P. Etiopathogenesis of arrhythmogenic right ventricular cardiomyopathy. *J Hum Genet* 2005; 50: 375–381.

145 Lindstrom L, Nylander E, Larsson H, Wranne B. Left ventricular involvement in arrhythmogenic right ventricular cardiomyopathy: a scintigraphic and echocardiographic study. *Clin Physiol Funct Imaging* 2005; 25: 171–177.

146 Fontaine G, Fontaliran F, Frank R. Arrhythmogenic right ventricular cardiomyopathies: clinical forms and main differential diagnoses. *Circulation* 1998; 97: 1532–1535.

147 McKenna WJ, Thiene G, Nava A *et al.* Diagnosis of arrhythmogenic right ventricular dysplasia/cardiomyopathy. Task Force of the Working Group Myocardial and Pericardial Disease of the European Society of Cardiology and of the Scientific Council on Cardiomyopathies of the International Society and Federation of Cardiology. *Br Heart J* 1994; 71: 215–218.

148 Corrado D, Basso C, Nava A, Thiene G. Arrhythmogenic right ventricular cardiomyopathy: current diagnostic and management strategies. *Cardiol Rev* 2001; 9: 259–265.

149 Norman M, Simpson M, Mogensen J *et al.* Novel mutation in desmoplakin causes arrhythmogenic left ventricular cardiomyopathy. *Circulation* 2005; 112: 636–642.

150 Leclercq JF, Coumel P. Characteristics, prognosis and treatment of the ventricular arrhythmias of right ventricular dysplasia. *Eur Heart J* 1989; 10 (Supplement D): 61–67.

151 Movsowitz C, Callans DJ, Schwartzman D, Gottlieb C, Marchlinski FE. The results of atrial flutter ablation in patients with and without a history of atrial fibrillation. *Am J Cardiol* 1996; 78: 93–96.

152 Wichter T, Borggrefe M, Haverkamp W, Chen X, Breithardt G. Efficacy of antiarrhythmic drugs in patients with arrhythmogenic right ventricular disease. Results in patients with inducible and noninducible ventricular tachycardia. *Circulation* 1992; 86: 29–37.

153 Priori SG, Aliot E, Blomstrom-Lundqvist C *et al.* Task Force on Sudden Cardiac Death of the European Society of Cardiology. *Eur Heart J* 2001; 22: 1374–1450.

154 Fontaine G, Tonet J, Gallais Y *et al.* Ventricular tachycardia catheter ablation in arrhythmogenic right ventricular dysplasia: a 16-year experience. *Curr Cardiol Rep* 2000; 2: 498–506.

155 Fontaine G, Prost-Squarcioni C. Implantable cardioverter defibrillator in arrhythmogenic right ventricular cardiomyopathies. *Circulation* 2004; 109: 1445–1447.

156 Guiraudon GM, Klein GJ, Sharma AD, Yee R, Guiraudon CM. Surgical therapy for arrhythmogenic right ventricular adiposis. *Eur Heart J* 1989, 10 (Supplement D): 82–83.

157 Gerull B, Heuser A, Wichter T *et al.* Mutations in the desmosomal protein plakophilin-2 are common in arrhythmogenic right ventricular cardiomyopathy. *Nat Genet* 2004; 36: 1162–1164.

158 James TN. Normal and abnormal consequences of apoptosis in the human heart. From postnatal morphogenesis to paroxysmal arrhythmias. *Circulation* 1994; 90: 556–573.

159 Hofmann R, Trappe HJ, Klein H, Kemnitz J. Chronic (or healed) myocarditis mimicking arrhythmogenic right ventricular dysplasia. *Eur Heart J* 1993; 14: 717–720.

160 Valente M, Calabrese F, Thiene G *et al.* In vivo evidence of apoptosis in arrhythmogenic right ventricular cardiomyopathy. *Am J Pathol* 1998; 152: 479–484.

161 Rakovec P, Rossi L, Fontaine G, Sasel B, Markez J, Voncina D. Familial arrhythmogenic right ventricular disease. *Am J Cardiol* 1986; 58: 377–378.

162 Nava A, Scognamiglio R, Thiene G *et al.* A polymorphic form of familial arrhythmogenic right ventricular dysplasia. *Am J Cardiol* 1987; 59: 1405–1409.

163 Canciani B, Nava A, Toso V, Martini B, Thiene G. A casual spontaneous mutation as possible cause of the familial form of arrhythmogenic right ventricular cardiomyopathy (arrhythmogenic right ventricular dysplasia). *Clin Cardiol* 1992; 15: 217–219.

164 Severini GM, Krajinovic M, Pinamonti B *et al.* A new locus for arrhythmogenic right ventricular dysplasia on the long arm of chromosome 14. *Genomics* 1996; 31: 193–200.

165 Rampazzo A, Nava A, Erne P *et al.* A new locus for arrhythmogenic right ventricular cardiomyopathy (ARVD2) maps to chromosome 1q42–q43. *Hum Mol Genet* 1995; 4: 2151–2154.

166 Rampazzo A, Nava A, Miorin M *et al.* ARVD4, a new locus for arrhythmogenic right ventricular cardiomyopathy, maps to chromosome 2 long arm. *Genomics* 1997; 45: 259–263.

167 Ahmad F, Li D, Karibe A *et al.* Localization of a gene responsible for arrhythmogenic right ventricular dysplasia to chromosome 3p23. *Circulation* 1998; 98: 2791–2795.

168 Bauce B, Nava A, Rampazzo A *et al.* Familial effort polymorphic ventricular arrhythmias in arrhythmogenic right ventricular cardiomyopathy map to chromosome 1q42–43. *Am J Cardiol* 2000; 85: 573–579.

169 Li D, Ahmad F, Gardner MJ *et al.* The locus of a novel gene responsible for arrhythmogenic right-ventricular dysplasia characterized by early onset and high penetrance maps to chromosome 10p12–p14. *Am J Hum Genet* 2000; 66: 148–156.

170 Melberg A, Oldfors A, Blomstrom-Lundqvist C *et al.* Autosomal dominant myofibrillar myopathy with arrhythmogenic right ventricular cardiomyopathy linked to chromosome 10q. *Ann Neurol* 1999; 46: 684–692.

171 Norgett EE, Hatsell SJ, Carvajal-Huerta L *et al.* Recessive mutation in desmoplakin disrupts desmoplakin-intermediate filament interactions and causes dilated cardiomyopathy, woolly hair and keratoderma. *Hum Mol Genet* 2000; 9: 2761–2766.

172 Laitinen PJ, Brown KM, Piippo K *et al.* Mutations of the cardiac ryanodine receptor (RyR2) gene in familial polymorphic ventricular tachycardia. *Circulation* 2001; 103: 485–490.

173 Marks AR: Ryanodine receptors, FKBP12, and heart failure. *Front Biosci* 2002; 7: D970–977.

174 Marx SO, Ondrias K, Marks AR. Coupled gating between individual skeletal muscle Ca^{2+} release channels (ryanodine receptors). *Science* 1998; 281: 818–821.

175 Brillantes AB, Ondrias K, Scott A *et al.* Stabilization of calcium release channel (ryanodine receptor) function by FK506-binding protein. *Cell* 1994; 77: 513–523.

176 Mallat Z, Tedgui A, Fontaliran F, Frank R, Durigon M, Fontaine G. Evidence of apoptosis in arrhythmogenic right ventricular dysplasia. *N Engl J Med* 1996; 335: 1190–1196.

177 d'Amati G, Bagattin A, Bauce B *et al.* Juvenile sudden death in a family with polymorphic ventricular arrhythmias caused by a novel RyR2 gene mutation: evidence of specific morphological substrates. *Hum Pathol* 2005; 36: 761–767.

178 Rampazzo A, Nava A, Malacrida S *et al.* Mutation in human desmoplakin domain binding to plakoglobin causes a dominant form of arrhythmogenic right ventricular cardiomyopathy. *Am J Hum Genet* 2002; 71: 1200–1206.

179 Cheong JE, Wessagowit V, McGrath JA. Molecular abnormalities of the desmosomal protein desmoplakin in human disease. *Clin Exp Dermatol* 2005; 30: 261–266.

180 Payne AS, Hanakawa Y, Amagai M, Stanley JR. Desmosomes and disease: pemphigus and bullous impetigo. *Curr Opin Cell Biol* 2004; 16: 536–543.

181 Bauce B, Basso C, Rampazzo A *et al.* Clinical profile of four families with arrhythmogenic right ventricular cardiomyopathy caused by dominant desmoplakin mutations. *Eur Heart J* 2005; 26: 1666–1675.

182 Michalodimitrakis M, Papadomanolakis A, Stiakakis J, Kanaki K. Left side right ventricular cardiomyopathy. *Med Sci Law* 2002; 42: 313–317.

183 Armstrong DK, McKenna KE, Purkis PE *et al.* Haploinsufficiency of desmoplakin causes a striate subtype of palmoplantar keratoderma. *Hum Mol Genet* 1999; 8: 143–148.

184 Whittock NV, Wan H, Morley SM *et al.* Compound heterozygosity for non-sense and mis-sense mutations in desmoplakin underlies skin fragility/woolly hair syndrome. *J Invest Dermatol* 2002; 118: 232–238.

185 Mertens C, Hofmann I, Wang Z *et al.* Nuclear particles containing RNA polymerase III complexes associated with the junctional plaque protein plakophilin 2. *Proc Natl Acad Sci USA* 2001; 98: 7795–7800.

186 Mertens C, Kuhn C, Franke WW. Plakophilins 2a and 2b: constitutive proteins of dual location in the karyoplasm and the desmosomal plaque. *J Cell Biol* 1996; 135: 1009–1025.

187 Chen X, Bonne S, Hatzfeld M, van Roy F, Green KJ. Protein binding and functional characterization of plakophilin 2. Evidence for its diverse roles in desmosomes and beta-catenin signaling. *J Biol Chem* 2002; 277: 10512–10522.

188 Azhar M, Schultz Jel J, Grupp I *et al.* Transforming growth factor beta in cardiovascular development and function. *Cytokine Growth Factor Rev* 2003; 14: 391–407.

189 Beffagna G, Occhi G, Nava A *et al.* Regulatory mutations in transforming growth factor-beta3 gene cause arrhythmogenic right ventricular cardiomyopathy type 1. *Cardiovasc Res* 2005; 65: 366–373.

190 Leask A, Abraham DJ: TGF-beta signaling and the fibrotic response. *FASEB J* 2004; 18: 816–827.

191 Kapoun AM, Liang F, O'Young G *et al.* B-type natriuretic peptide exerts broad functional opposition to transforming growth factor-beta in primary human cardiac fibroblasts: fibrosis, myofibroblast conversion, proliferation, and inflammation. *Circ Res* 2004; 94: 453–461.

192 Coonar AS, Protonotarios N, Tsatsopoulou A *et al.* Gene for arrhythmogenic right ventricular cardiomyopathy with diffuse nonepidermolytic palmoplantar keratoderma and woolly hair (Naxos disease) maps to 17q21. *Circulation* 1998; 97: 2049–2058.

193 McKoy G, Protonotarios N, Crosby A *et al.* Identification of a deletion in plakoglobin in arrhythmogenic right ventricular cardiomyopathy with palmoplantar keratoderma and woolly hair (Naxos disease). *Lancet* 2000; 355: 2119–2124.

194 Sacco PA, McGranahan TM, Wheelock MJ, Johnson KR. Identification of plakoglobin domains required for association with N-cadherin and alpha-catenin. *J Biol Chem* 1995; 270: 20201–20206.

195 Knudsen KA, Wheelock MJ. Plakoglobin, or an 83-kD homologue distinct from beta-catenin, interacts with E-cadherin and N-cadherin. *J Cell Biol* 1992; 118: 671–679.

196 Saffitz JE. Dependence of electrical coupling on mechanical coupling in cardiac myocytes: insights gained from cardiomyopathies caused by defects in cell-cell connections. *Ann N Y Acad Sci* 2005; 1047: 336–344.

197 Fatkin D, Graham RM. Molecular mechanisms of inherited cardiomyopathies. *Physiol Rev* 2002; 82: 945–980.

198 Abelmann WH. Classification and natural history of primary myocardial disease. *Prog Cardiovasc Dis* 1984; 27: 73–94.

199 Kushwaha SS, Fallon JT, Fuster V. Restrictive cardiomyopathy. *N Engl J Med* 1997; 336: 267–276.

200 Asher CR, Klein AL. Diastolic heart failure: restrictive cardiomyopathy, constrictive pericarditis, and cardiac tamponade: clinical and echocardiographic evaluation. *Cardiol Rev* 2002; 10: 218–229.

201 Child JS, Perloff JK. The restrictive cardiomyopathies. *Cardiol Clin* 1988; 6: 289–316.

202 Benotti JR, Grossman W, Cohn PF. Clinical profile of restrictive cardiomyopathy. *Circulation* 1980; 61: 1206–1212.

203 Fitzpatrick AP, Shapiro LM, Rickards AF, Poole-Wilson PA. Familial restrictive cardiomyopathy with atrioventricular block and skeletal myopathy. *Br Heart J* 1990; 63: 114–118.

204 Ishiwata S, Nishiyama S, Seki A, Kojima S. Restrictive cardiomyopathy with complete atrioventricular block and distal myopathy with rimmed vacuoles. *Jpn Circ J* 1993; 57: 928–933.

205 Siegel RJ, Shah PK, Fishbein MC. Idiopathic restrictive cardiomyopathy. *Circulation* 1984; 70: 165–169.

206 Katritsis D, Wilmshurst PT, Wendon JA, Davies MJ, Webb-Peploe MM. Primary restrictive cardiomyopathy: clinical and pathologic characteristics. *J Am Coll Cardiol* 1991; 18: 1230–1235.

207 Hughes SE, McKenna WJ. New insights into the pathology of inherited cardiomyopathy. *Heart* 2005; 91: 257–264.

208 Fauci AS, Harley JB, Roberts WC, Ferrans VJ, Gralnick HR, Bjornson BH. NIH conference. The idiopathic hypereosinophilic syndrome. Clinical, pathophysiologic, and therapeutic considerations. *Ann Intern Med* 1982; 97: 78–92.

209 Schaffer SW, Dimayuga ER, Kayes SG. Development and characterization of a model of eosinophil-mediated cardiomyopathy in rats infected with Toxocara canis. *Am J Physiol* 1992; 262: H1428–1434.

210 Boustany CW Jr, Murphy GW, Hicks GL Jr. Mitral valve replacement in idiopathic hypereosinophilic syndrome. *Ann Thorac Surg* 1991; 51: 1007–1009.

211 Gudmundsson GS, Ohr J, Leya F, Jacobs WR, Godwin JE, Schwartz J. An unusual case of recurrent Loffler endomyocarditis of the aortic valve. *Arch Pathol Lab Med* 2003; 127: 606–609.

212 Yazdani K, Maraj S, Amanullah AM. Differentiating constrictive pericarditis from restrictive cardiomyopathy. *Rev Cardiovasc Med* 2005; 6: 61–71.

213 Chinnaiyan KM, Leff CB, Marsalese DL. Constrictive pericarditis versus restrictive cardiomyopathy: challenges in diagnosis and management. *Cardiol Rev* 2004; 12: 314–320.

214 Demir M, Paydas S, Cayli M, Akpinar O, Balal M, Acarturk E. Tissue Doppler is a more reliable method in early detection of cardiac dysfunction in patients with AA amyloidosis. *Ren Fail* 2005; 27: 415–420.

215 Hancock EW. Differential diagnosis of restrictive cardiomyopathy and constrictive pericarditis. *Heart* 2001; 86: 343–349.

216 Casset-Senon D, Secchi V, Arbeille P, Cosnay P. Localization of myocardial amyloid deposits in cardiac amyloidosis by Tc-99m pyrophosphate myocardial SPECT: implication for medical treatment. *Clin Nucl Med* 2005; 30: 496–497.

217 Hesse A, Altland K, Linke RP *et al.* Cardiac amyloidosis: a review and report of a new transthyretin (prealbumin) variant. *Br Heart J* 1993; 70: 111–115.

218 Olson LJ, Gertz MA, Edwards WD *et al.* Senile cardiac amyloidosis with myocardial dysfunction. Diagnosis by endomyocardial biopsy and immunohistochemistry. *N Engl J Med* 1987; 317: 738–742.

219 Arbustini E, Merlini G, Gavazzi A *et al.* Cardiac immunocyte-derived (AL) amyloidosis: an endomyocardial biopsy study in 11 patients. *Am Heart J* 1995; 130: 528–536.

220 Huang XP, Du JF. Troponin I, cardiac diastolic dysfunction and restrictive cardiomyopathy. *Acta Pharmacol Sin* 2004; 25: 1569–1575.

221 Rivenes SM, Kearney DL, Smith EO, Towbin JA, Denfield SW. Sudden death and cardiovascular collapse in children with restrictive cardiomyopathy. *Circulation* 2000; 102: 876–882.

222 Rapezzi C, Ortolani P, Binetti G, Picchio FM, Magnani B. Idiopathic restrictive cardiomyopathy in the young: report of two cases. *Int J Cardiol* 1990; 29: 121–126.

223 Cetta F, O'Leary PW, Seward JB, Driscoll DJ. Idiopathic restrictive cardiomyopathy in childhood: diagnostic features and clinical course. *Mayo Clin Proc* 1995; 70: 634–640.

224 Lewis AB. Clinical profile and outcome of restrictive cardiomyopathy in children. *Am Heart J* 1992; 123: 1589–1593.

225 Feld S, Caspi A. Familial cardiomyopathy with variable hypertrophic and restrictive features and common HLA haplotype. *Isr J Med Sci* 1992; 28: 277–280.

226 Zachara E, Bertini E, Lioy E, Boldrini R, Prati PL, Bosman C. Restrictive cardiomyopathy due to desmin accumulation in a family with evidence of autosomal dominant inheritance. *G Ital Cardiol* 1997; 27: 436–442.

227 Zhang J, Kumar A, Stalker HJ *et al.* Clinical and molecular studies of a large family with desmin-associated restrictive cardiomyopathy. *Clin Genet* 2001; 59: 248–256.

228 Alpert NR, Mulieri LA, Warshaw D: The failing human heart. *Cardiovasc Res* 2002; 54: 1–10.

229 Wilkinson JM, Perry SV, Cole HA, Trayer IP: The regulatory proteins of the myofibril. Separation and biological activity of the components of inhibitory-factor preparations. *Biochem J* 1972; 127: 215–228.

230 Layland J, Solaro RJ, Shah AM. Regulation of cardiac contractile function by troponin I phosphorylation. *Cardiovasc Res* 2005; 66: 12–21.

231 Harada K, Morimoto S; Inherited cardiomyopathies as a troponin disease. *Jpn J Physiol* 2004; 54: 307–318.

232 Milner DJ, Weitzer G, Tran D, Bradley A, Capetanaki Y. Disruption of muscle architecture and myocardial degeneration in mice lacking desmin. *J Cell Biol* 1996; 134: 1255–1270.

233 Capetanaki Y, Milner DJ, Weitzer G. Desmin in muscle formation and maintenance: knockouts and consequences. *Cell Struct Funct* 1997; 22: 103–116.

234 Goldfarb LG, Park KY, Cervenakova L *et al.* Missense mutations in desmin associated with familial cardiac and skeletal myopathy. *Nat Genet* 1998; 19: 402–403.

235 Saraiva MJ: Transthyretin mutations in health and disease. *Hum Mutat* 1995; 5: 191–196.

236 Frigerio R, Fabrizi GM, Ferrarini M *et al.* An unusual transthyretin gene missense mutation (TTR Phe33Val) linked to familial amyloidotic polyneuropathy. *Amyloid* 2004; 11: 121–124.

237 Lafitte S, Lafitte M, Perron JM, Ennouchi D, Vital C, Roudaut R. [Cardiac manifestations of amyloidosis by deposits of transthyretin and apolipoprotein A1. Report of 3 families]. *Arch Mal Coeur Vaiss* 2003; 96: 631–635.

238 Booth DR, Tan SY, Hawkins PN, Pepys MB, Frustaci A. A novel variant of transthyretin, 59Thr⇒Lys, associated with autosomal dominant cardiac amyloidosis in an Italian family. *Circulation* 1995; 91: 962–967.

239 Moore DF, Scott LT, Gladwin MT *et al.* Regional cerebral hyperperfusion and nitric oxide pathway dysregulation in Fabry disease: reversal by enzyme replacement therapy. *Circulation* 2001; 104: 1506–1512.

240 Masson C, Cisse I, Simon V, Insalaco P, Audran M. Fabry disease: a review. *Joint Bone Spine* 2004; 71: 381–383.

241 Revilla M, Obach V, Cervera A, Davalos A, Castillo J, Chamorro A. A-174G/C polymorphism of the interleukin-6 gene in patients with lacunar infarction. *Neurosci Lett* 2002; 324: 29–32.

242 Paradossi U, Ciofini E, Clerico A, Botto N, Biagini A, Colombo MG. Endothelial function and carotid intima-media thickness in young healthy subjects among endothelial nitric oxide synthase Glu298⇒Asp and T-786⇒C polymorphisms. *Stroke* 2004; 35: 1305–1309.

243 Bertina RM, Koeleman BP, Koster T *et al.* Mutation in blood coagulation factor V associated with resistance to activated protein C. *Nature* 1994; 369: 64–67.

244 Frosst P, Blom HJ, Milos R *et al.* A candidate genetic risk factor for vascular disease: a common mutation in methylenetetrahydrofolate reductase. *Nat Genet* 1995; 10: 111–113.

245 Lichy C, Kropp S, Dong-Si T *et al.* A common polymorphism of the protein Z gene is associated with protein Z plasma levels and with risk of cerebral ischemia in the young. *Stroke* 2004; 35: 40–45.

246 Longstreth WT Jr, Rosendaal FR, Siscovick DS *et al.* Risk of stroke in young women and two prothrombotic mutations: factor V Leiden and prothrombin gene variant (G20210A). *Stroke* 1998; 29: 577–580.

247 DiMauro S, Schon EA. Mitochondrial respiratory-chain diseases. *N Engl J Med* 2003; 348: 2656–2668.

248 Burgeois M, Goutieres F, Chretien D, Rustin P, Munnich A, Aicardi J: Deficiency in complex II of the respiratory chain, presenting as a leukodystrophy in two sisters with Leigh syndrome. *Brain Dev* 1992; 14: 404–408.

249 Finsterer J: Mitochondriopathies. *Eur J Neurol* 2004; 11: 163–186.

250 Kearns TP, Sayre GP. Retinitis pigmentosa, external ophthalmophegia, and complete heart block: unusual syndrome with histologic study in one of two cases. *AMA Arch Ophthalmol* 1958; 60: 280–289.

251 Sengers RC, Stadhouders AM, Trijbels JM. Mitochondrial myopathies. Clinical, morphological and biochemical aspects. *Eur J Pediatr* 1984; 141: 192–207.

252 DiMauro S, Schon EA. Mitochondrial DNA mutations in human disease. *Am J Med Genet* 2001; 106: 18–26.

253 Smeitink J, van den Heuvel L, DiMauro S. The genetics and pathology of oxidative phosphorylation. *Nat Rev Genet* 2001; 2: 342–352.

254 Hubner G, Grantzow R. Mitochondrial cardiomyopathy with involvement of skeletal muscles. *Virchows Arch A Pathol Anat Histopathol* 1983; 399: 115–125.

255 Hodgson S, Child A, Dyson M. Endocardial fibroelastosis: possible X linked inheritance. *J Med Genet* 1987; 24: 210–214.

256 DiMauro S, Bonilla E, Zeviani M, Nakagawa M, DeVivo DC. Mitochondrial myopathies. *Ann Neurol* 1985; 17: 521–538.

257 Guenthard J, Wyler F, Fowler B, Baumgartner R. Cardiomyopathy in respiratory chain disorders. *Arch Dis Child* 1995; 72: 223–226.

258 Gillis L, Kaye E. Diagnosis and management of mitochondrial diseases. *Pediatr Clin North Am* 2002; 49: 203–219.

259 McFarland R, Taylor RW, Turnbull DM. The neurology of mitochondrial DNA disease. *Lancet Neurol* 2002; 1: 343–351.

260 Orstavik KH, Skjorten F, Hellebostad M, Haga P, Langslet A. Possible X linked congenital mitochondrial cardiomyopathy in three families. *J Med Genet* 1993; 30: 269–272.

261 Clarke L, Dimmick J, Da A. Pathology of inherited metabolic diseases. In: Dimmick J, Kalousek D, eds. *Developmental Pathology of the Embryo and Fetus.* JB Lippincott, Philadelphia, 1992, 199–234.

262 Chinnery PF, Turnbull DM. Clinical features, investigation, and management of patients with defects of mitochondrial DNA. *J Neurol Neurosurg Psychiatry* 1997; 63: 559–563.

263 Anderson S, Bankier AT, Barrell BG *et al.* Sequence and organization of the human mitochondrial genome. *Nature* 1981; 290: 457–465.

264 Anan R, Nakagawa M, Miyata M *et al.* Cardiac involvement in mitochondrial diseases. A study on 17 patients with documented mitochondrial DNA defects. *Circulation* 1995; 91: 955–961.

265 Laschi R, Govoni E, Cenacchi G, Trotta F. Calcium pyrophosphate dihydrate microcrystal-associated arthropathy. *Ultrastruct Pathol* 1986; 10: 395–400.

266 Finsterer J, Jarius C, Eichberger H. Phenotype variability in 130 adult patients with respiratory chain disorders. *J Inherit Metab Dis* 2001; 24: 560–576.

267 Leonard JV, Schapira AH. Mitochondrial respiratory chain disorders I: mitochondrial DNA defects. *Lancet* 2000; 355: 299–304.

268 Fadic R, Johns DR. Clinical spectrum of mitochondrial diseases. *Semin Neurol* 1996; 16: 11–20.

269 Hug G, Bove KE, Soukup S. Lethal neonatal multiorgan deficiency of carnitine palmitoyltransferase II. *N Engl J Med* 1991; 325: 1862–1864.

270 Rocchiccioli F, Wanders RJ, Aubourg P *et al.* Deficiency of long-chain 3-hydroxyacyl-CoA dehydrogenase: a cause of lethal myopathy and cardiomyopathy in early childhood. *Pediatr Res* 1990; 28: 657–662.

271 Kelly DP, Whelan AJ, Ogden ML *et al.* Molecular characterization of inherited medium-chain acyl-CoA dehydrogenase deficiency. *Proc Natl Acad Sci USA* 1990; 87: 9236–9240.

272 Bennett MJ, Rinaldo P, Strauss AW. Inborn errors of mitochondrial fatty acid oxidation. *Crit Rev Clin Lab Sci* 2000; 37: 1–44.

273 Shoubridge EA. Cytochrome c oxidase deficiency. *Am J Med Genet* 2001; 106: 46–52.

274 Robinson BH. Human cytochrome oxidase deficiency. *Pediatr Res* 2000; 48: 581–585.

275 Mootha VK, Lepage P, Miller K *et al.* Identification of a gene causing human cytochrome c oxidase deficiency by integrative genomics. *Proc Natl Acad Sci USA* 2003; 100: 605–610.

276 Pavlakis SG, Phillips PC, DiMauro S, De Vivo DC, Rowland LP. Mitochondrial myopathy, encephalopathy, lactic acidosis, and strokelike episodes: a distinctive clinical syndrome. *Ann Neurol* 1984; 16: 481–488.

277 Schapira AH, Cock HR. Mitochondrial myopathies and encephalomyopathies. *Eur J Clin Invest* 1999; 29: 886–898.

278 Ingwall JS, Weiss RG. Is the failing heart energy starved? On using chemical energy to support cardiac function. *Circ Res* 2004; 95: 135–145.

279 Ishikawa K, Kimura S, Kobayashi A *et al.* Increased reactive oxygen species and anti-oxidative response in mitochondrial cardiomyopathy. *Circ J* 2005; 69: 617–620.

280 Levak-Frank S, Radner H, Walsh A *et al.* Muscle-specific overexpression of lipoprotein lipase causes a severe myopathy characterized by proliferation of mitochondria and peroxisomes in transgenic mice. *J Clin Invest* 1995; 96: 976–986.

281 Tokunaga M, Mita S, Sakuta R, Nonaka I, Araki S. Increased mitochondrial DNA in blood vessels and ragged-red fibers in mitochondrial myopathy, encephalopathy, lactic acidosis, and stroke-like episodes (MELAS). *Ann Neurol* 1993; 33: 275–280.

282 Bonnet D, de Lonlay P, Gautier I *et al.* Efficiency of metabolic screening in childhood cardiomyopathies. *Eur Heart J* 1998; 19: 790–793.

283 Servidei S, Bertini E, DiMauro S: Hereditary metabolic cardiomyopathies. *Adv Pediatr* 1994; 41: 1–32.

284 Towbin JA, Lipshultz SE. Genetics of neonatal cardiomyopathy. *Curr Opin Cardiol* 1999; 14: 250–262.

285 Kohlschutter A, Hausdorf G. Primary (genetic) cardiomyopathies in infancy. A survey of possible disorders and guidelines for diagnosis. *Eur J Pediatr* 1986; 145: 454–459.

286 Hirschhorn R. storage disease type II: Acid alpha-glucosidase (acid maltase) deficiency. In: Scriver C,

Beaudet A, Sly W, Valle D, eds. *The Metabolic and Molecular Bases of Inherited Disease*. McGraw Hill, Inc., New York, 1995, 2443–2464.

287 Dincsoy MY, Dincsoy HP, Kessler AD, Jackson MA, Sidbury JB Jr. Generalized glycogenosis and associated endocardial fibroelastosis. Report of 3 cases with biochemical studies. *J Pediatr* 1965; 67: 728–740.

288 Chin TK, Perloff JK, Williams RG, Jue K, Mohrmann R. Isolated noncompaction of left ventricular myocardium. A study of eight cases. *Circulation* 1990; 82: 507–513.

289 Murphy RT, Thaman R, Blanes JG *et al.* Natural history and familial characteristics of isolated left ventricular non-compaction. *Eur Heart J* 2005; 26: 187–192.

290 Ichida F, Hamamichi Y, Miyawaki T *et al.* Clinical features of isolated noncompaction of the ventricular myocardium: long-term clinical course, hemodynamic properties, and genetic background. *J Am Coll Cardiol* 1999; 34: 233–240.

291 Jenni R, Wyss CA, Oechslin EN, Kaufmann PA. Isolated ventricular noncompaction is associated with coronary microcirculatory dysfunction. *J Am Coll Cardiol* 2002; 39: 450–454.

292 Bleyl SB, Mumford BR, Thompson V *et al.* Neonatal, lethal noncompaction of the left ventricular myocardium is allelic with Barth syndrome. *Am J Hum Genet* 1997; 61: 868–872.

293 Ichida F, Tsubata S, Bowles KR *et al.* Novel gene mutations in patients with left ventricular noncompaction or Barth syndrome. *Circulation* 2001; 103: 1256–1263.

294 Vatta M, Mohapatra B, Jimenez S *et al.* Mutations in Cypher/ZASP in patients with dilated cardiomyopathy and left ventricular non-compaction. *J Am Coll Cardiol* 2003; 42: 2014–2027.

295 Shou W, Aghdasi B, Armstrong DL *et al.* Cardiac defects and altered ryanodine receptor function in mice lacking FKBP12. *Nature* 1998; 391: 489–492.

296 Kenton AB, Sanchez X, Coveler KJ *et al.* Isolated left ventricular noncompaction is rarely caused by mutations in G4.5, alpha-dystrobrevin and FK Binding Protein-12. *Mol Genet Metab* 2004; 82: 162–166.

297 Sasse-Klaassen S, Gerull B, Oechslin E, Jenni R, Thierfelder L. Isolated noncompaction of the left ventricular myocardium in the adult is an autosomal dominant disorder in the majority of patients. *Am J Med Genet A* 2003; 119: 162–167.

298 Sasse-Klaassen S, Probst S, Gerull B *et al.* Novel gene locus for autosomal dominant left ventricular noncompaction maps to chromosome 11p15. *Circulation* 2004; 109: 2720–2723.

299 Watkins H: Genetic clues to disease pathways in hypertrophic and dilated cardiomyopathies. *Circulation* 2003; 107: 1344–1346.

300 Seidman JG, Seidman C. The genetic basis for cardiomyopathy: from mutation identification to mechanistic paradigms. *Cell* 2001; 104: 557–567.

5

CHAPTER 5

The long QT syndrome

Sabina Kupershmidt, PhD, *Kamilla Kelemen,* MD,
& Tadashi Nakajima, MD, PhD

Introduction

Cardiac arrhythmias are a major cause of morbidity and mortality throughout the world. On average, every day over 1000 individuals in the USA die suddenly because of fatal ventricular arrhythmias, most often caused by underlying heart disease in middle-aged to elderly patients. This chapter deals with the long QT syndrome (LQTS), which has evolved into a paradigm for arrhythmia studies, where basic science and clinical research have danced a well-choreographed pas de deux to yield fundamental insights into arrhythmia mechanisms and cardiac electrophysiology. We hope to demonstrate convincingly that these advances would have been unthinkable without information gleaned from the study of rare, hereditary ion-channel diseases, often called "channelopathies."

In particular, recent advances in human molecular genetic research prepared the way for the identification of genes responsible for a number of inherited, potentially lethal arrhythmia syndromes, including the LQTS [1–3], the Brugada syndrome [4], catecholaminergic polymorphic ventricular tachycardias [5], progressive cardiac conduction defects [6], familial atrial fibrillation [7] and familial sick sinus syndrome [8]. As we shall discuss in greater detail, the underlying cause of all of these conditions was eventually traced to defects in ion channels or channel-associated proteins. Among those, the LQTS remains the best studied, and identification and functional analysis of genes responsible for LQTS form the basis for a better understanding of the pathophysiology of arrhythmia syndromes in general.

By definition, LQTS is characterized by a prolongation of the QT interval on the surface electrocardiogram (ECG; Figs 5.1a and 5.2a). The QT interval extends from the QRS complex, to the end of the T wave representing ventricular depolarization and repolarization, respectively (Fig. 5.2a). A rate-corrected QT interval (QTc) longer than 440 ms is considered prolonged, although there is considerable variability due to age and sex (see "Diagnosis"). In addition to a prolonged QT interval, other ECG parameters may also be affected. Clinical manifestations of LQTS includes syncope and a characteristic polymorphic ventricular tachycardia whose ECG tracing twists around the iso-electric line, called torsade de pointes (TdP), which is often precipitated by physical or emotional stress (Figs 5.1b and 5.2c) and can deteriorate into life-threatening ventricular fibrillation. Significantly, it occurs mostly in young individuals without underlying structural heart disease.

The LQTS exists in two forms, congenital and acquired LQTS. Congenital LQTS is defined as an inherited disorder, caused by gene mutations and can be further subdivided according to locus-specific characteristics (Table 5.1). The various congenital LQT syndromes show different phenotypes. Gene-specific differences have been reported in terms of morphology of the ST-T wave complex [9], triggers for cardiac events [10–12] and risk of cardiac events [13]. In the past few years, genotype–phenotype studies have become one of the most active areas of LQTS research.

The acquired LQTS is precipitated by co-factors such as exposure to certain drugs (Table 5.2) and electrolyte abnormalities, but it presents with

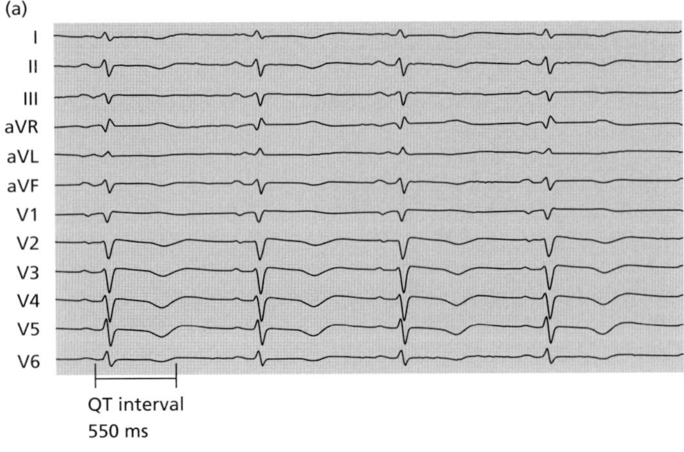

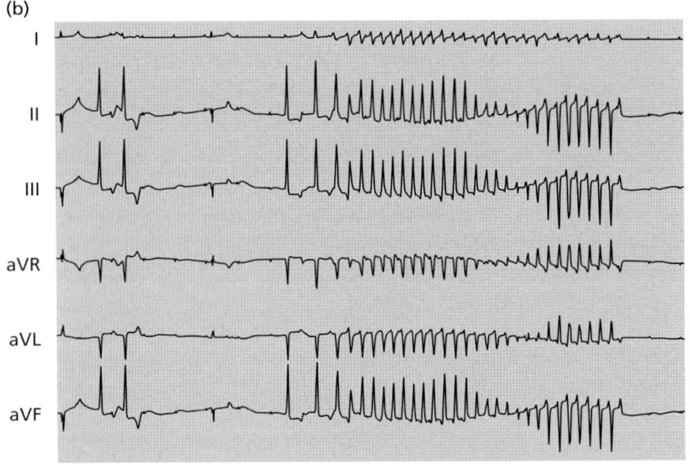

Figure 5.1 (a) Surface ECG of a patient with the long QT syndrome (LQTS). The QT interval measures 550 ms. (b) Drug-induced TdP tachyarrhythmia in a chronic AV block dog model. The class III anti-arrhythmic agent almokalant was applied to induce torsade de pointes (TdP).

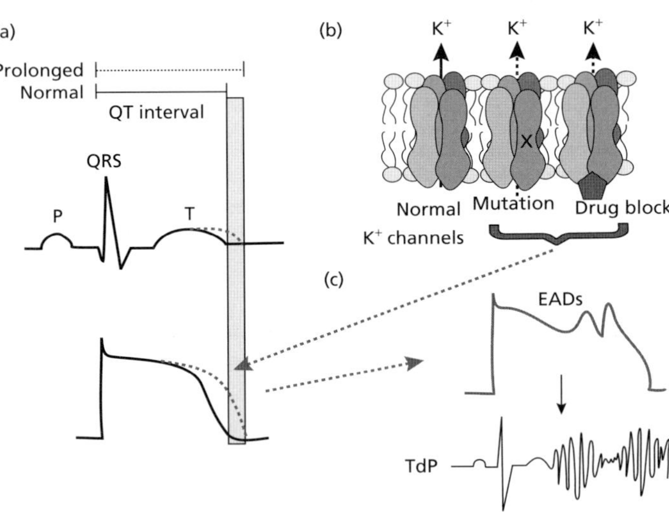

Figure 5.2 (a) Schematic depiction of a surface electrocardiograph (ECG) trace. The QT interval is outlined. Below: idealized depiction of a ventricular action potential. The gray-shaded area and the stippled lines denote a prolongation of the QT interval. (b) Tetrameric K^+ channels are depicted embedded within the lipid bilayer of the membrane. K^+ flows from the inside to the outside of the cell. Normal: a wild-type channel conducts unimpeded current. Mutants: a mutation within a channel monomer prevents current flow. Drug block: an anti-arrhythmic drug or an unrelated drug (in the case of the acquired long QT syndrome [LQTS]) blocks the flow of K^+ by binding to the inside of the channel pore. (c) Mutations or drug block can lead to prolongation of the QT interval, which, in turn, leads to early afterdepolarizations (EADs) and torsade de pointes (TdP).

Table 5.1 Genetics of the long QT syndrome (LQTS).

Phenotype	Genotype (protein)	Current	Localization	Inherited form	LQTS (%)
LQT1	KCNQ1 (KvLQT1)	I_{Ks} α-subunit	Chromosome 11p15.5	Autosomal dominant	60
LQT2	KCNH2 (HERG)	I_{Kr} α-subunit	Chromosome 7q35–36	Autosomal dominant	35
LQT3	SCN5A (Nav1.5)	I_{Na}	Chromosome 3p21–23	Autosomal dominant	3–4
LQT4	ANK2 (Ankyrin B)	–	Chromosome 4q25–27	Autosomal dominant	<1
LQT5	KCNE1 (minK)	I_{Ks} β-subunit	Chromosome 21q22.1	Autosomal dominant	<1
LQT6	KCNE2 (MiRP1)	I_{Kr} β-subunit	Chromosome 21q22.1	Autosomal dominant	<1
LQT7 (Andersen syndrome)	KCNJ2 (Kir2.1)	I_{K1}	Chromosome 17q23–24	Autosomal dominant	<1
Timothy syndrome (LQT8)	CACNA1C (Ca$_v$1.2)	I_{Ca-L}	?	Sporadic? Autosomal recessive?	<1
JLN1	KCNQ1 (KvLQT1)	I_{Ks} α-subunit	Chromosome 11p15.5	Autosomal recessive	<1
JLN2	KCNE1 (minK)	I_{Ks} β-subunit	Chromosome 21q22.1–22.2	Autosomal recessive	<1

JLN1, Jervell and Lange–Nielson syndrome 1; JLN2, Jervell and Lange–Nielson syndrome 2.

Table 5.2 Examples of drugs causing QT prolongation and torsade de pointes (TdP).

Anti-arrhythmics	Class IA
	Quinidine, disopyramide, procainamide
	Class III
	Sotalol, amiodarone, almokalant, dofetilide
	Class IV
	Bepridil
Anti-infectious agents	Erythromycin, clarithromycin, clindamycin, quinolone, amantadine, pentamidine, imidazole, chloroquine, quinine, ketoconazole, halofantrine
Histamine antagonists	Terfenadine, astemizole, fexofenadine
Serotonine antagonists	Fluoxetine, zimelidine
Diuretics	Indapamide, triamterene
Antipsychotics	Haloperidol, droperidol, chlorpromazine, desipramine, doxepin, lithium, maprotiline, sertinole, imipramine
Anticholinergics	Cisapride, acetylcholine, terodiline
Inotropics	Amrinone, milrinone

symptoms similar to those of the congenital LQTS. The acquired LQTS represents a significant clinical problem because its manifestation is unpredictable and it has been estimated that up to 3% of all therapeutic drug prescriptions may result in TdP as a side effect. Indeed, the acquired LQTS has become the leading cause for drug withdrawal from the market [14,15].

Public and scientific awareness of the importance of the LQTS are reflected in the International LQTS Registry, which was established in 1979 [16,17]. The registry constitutes a long-term project intended to promote a better understanding and management of the LQTS. It encompasses clinical phenotypes and family pedigrees, and has played an important part in investigating this arrhythmia syndrome.

In this chapter, we describe the latest molecular and pathophysiologic insights into LQTS mechanisms.

Historical development

Historically, the LQTS has been divided into two groups, the Jervell and Lange-Nielsen syndrome (JLN syndrome) and the Romano–Ward syndrome

(RW syndrome). In 1957, Jervell and Lange-Nielsen formally reported a rare, inherited autosomal recessive disorder defined by prolongation of the QT interval, congenital bilateral neural deafness, syncopal episodes and sudden death [18]. In the early 1960s, Romano and Ward independently reported a similar familial disorder, also associated with QT prolongation but not with associated deafness and inherited in an autosomal dominant fashion [19,20]. "Autosomal recessive" indicates that both copies of a gene (alleles) must be altered for a person to be affected. In this case, it is likely that the patients' parents are merely carriers of the mutant gene and are themselves unaffected, whereas in the "dominant" form, the mutation of a single allele is sufficient to cause disease.

It is difficult to find exact figures for the incidence of the RW syndrome but estimates range from 1 in 10,000 to 1 in 5000 [21,22] people. The JLN syndrome is less common, and thought to occur in approximately 1.6–6 in 1 million children. Yet it is important to recognize that, in general, the 10-year mortality rate is very high (71%) in patients who are left untreated. Precise estimates for the incidence of the acquired LQTS are also not available, although it is known to be much more common than the congenital form.

Until the first genetic loci responsible for the syndrome were identified in the 1990s, clinical and experimental studies suggested that LQTS might be caused by an imbalance of sympathetic nerve activity or by defects in cardiac ion channels. The "sympathetic imbalance" theory was initially favored because of reports that the QT interval could be prolonged by right stellate ganglionectomy or by stimulating the left stellate ganglion [21,23], and that left stellate ganglionectomy was an effective therapy for patients unresponsive to drug treatment [21,24] (see "Clinical treatment").

This hypothesis has been revised during the last decade, when mutations in eight genes were identified that can cause the congenital LQTS. These included the genes for the cardiac sodium current, I_{Na} [1], multiple genes contributing to the rapid and slow components of the delayed rectifier potassium currents, I_{Kr} and I_{Ks}, respectively [2,3,25,26], the Kir2.1 channel (I_{K1}) [27] and the L-type calcium channel (I_{Ca-L}) (Fig. 5.3) [28]. Providing a novel twist, ankyrin B, a protein acting primarily as an adapter between membrane proteins and the cytoskeleton, was discovered to be the first non-ion channel gene responsible for one form of the LQTS [29].

Clinically silent ion channel gene mutations have been characterized in families with cases of the acquired LQTS, indicating that this form of the disorder also carries a genetic component [30–34]. Additional risk factors for the acquired LQTS include individual differences in drug metabolism caused by genetic mutations or polymorphisms that alter the function of drug-metabolizing enzymes (CYP2D6, CYP3A) [34].

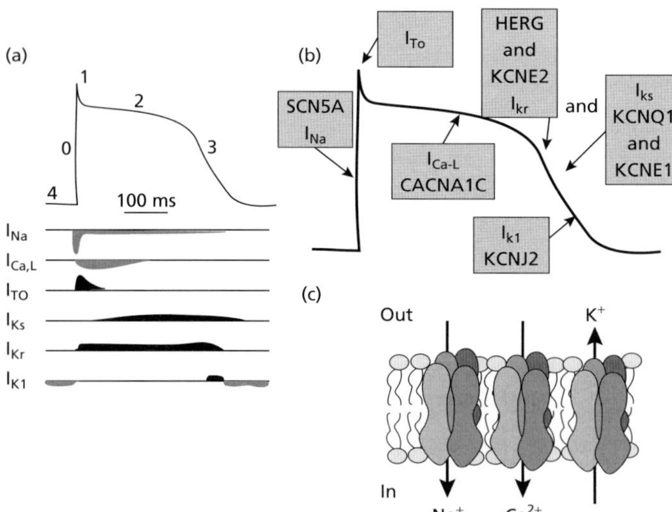

Figure 5.3 (a) Top: The five phases of an idealized action potential (AP). Bottom: Schematized activation patterns of the currents discussed in this chapter. Inward currents are shown below the solid line in gray, outward currents are above the solid line, in black. (b) Shows the currents associated with the various phases of the AP. The gene names corresponding to the currents are denoted in capital letters. (c) Tetrameric K+ channels are embedded within the lipid bilayer of the membrane. Na+ and Ca2+ are depolarizing and flow into the cell, while K+ repolarizes and thus flows to the outside of the cell.

In the early 2000s, a molecular link between sudden infant death syndrome (SIDS) and LQTS was also proposed [35,36]. SIDS is a frequent cause of unexplained death among otherwise healthy infants. The pathophysiologic mechanisms responsible for SIDS may be multifactorial and remain poorly understood. However, postmortem genetic testing in some patients identified mutations/polymorphisms in LQT-associated genes (SCN5A, KCNQ1 and KCNH2) [35,37,38].

Molecular mechanisms

Ionic currents involved in shaping the cardiac action potential

The cardiac action potential (AP) reflects the integrated electrical activity of many ionic currents across the cell membrane through voltage-gated ion channels, ionic pumps and ionic exchangers. Depolarizing currents convey positively charged ions into the cell, whereas the repolarizing currents ferry positively charged ions out of the cell (Fig. 5.3c). This is illustrated with the help of an idealized AP, which can be divided into five phases (Fig. 5.3a):

Phase 0 The cardiomyocyte membrane is depolarized by a rapid, transient influx of Na^+ ions through voltage-gated sodium channels (I_{Na}). Because depolarization makes the cytoplasmic side of the plasma membrane more positive, this is reflected in electrophysiologic recordings as the upstroke of the action potential.

Phase 1 During early repolarization, the transient efflux of K^+ through transient outward (I_{to}) channels (Fig. 5.3b), which inactivate rapidly, terminates the AP upstroke.

Phase 2 The prolonged plateau phase of the cardiac action potential is maintained by a balance between inward Ca^{2+} through L-type Ca^{2+} channels and efflux of repolarizing K^+ (Fig. 5.3a,b).

Phase 3 During late repolarization the Ca^{2+} channels have inactivated and the outward K^+ currents are unopposed to sustain the downward stroke of the action potential through the delayed rectifier K^+ channels (I_{Kr}, I_{Ks}) (Fig. 5.3a,b).

Phase 4 The resting membrane potential is maintained between the end of one AP and the beginning of the next by a balance between Na^+ and Ca^{2+} leak currents and the inward rectifier (I_{K1}) current (Fig. 5.3a,b). At rest, there is no substantial electrochemical gradient for K^+ to enter or exit the cell because the reversal potential for K^+ is close to the resting membrane potential (-85 mV).

In healthy people, the AP progresses through its five phases within 200–300 ms but in LQTS patients, it takes in excess of 440–460 ms. This implies an imbalance of depolarizing and repolarizing currents which may, in theory, be brought about by too much of the inward Na^+ current or by inhibition of one or more of the outward K^+ currents, resulting in a delay in repolarization. Indeed, this theory was proven correct by a series of seminal discoveries made in the 1990s which rapidly led to the development of LQTS as a molecular paradigm for arrhythmia development [39]. An important factor in determining the success of these studies was the availability of genetic material collected from families afflicted with the LQTS, by researchers working in Salt Lake City, Utah, who had the foresight to establish such DNA banks and use the power of human genetics to usher in a new era of arrhythmia research [40–42].

Cardiac ion channels

At this point, a short discussion of the profound impact on human genetic diseases made by the study of the model organism *Drosophila* is warranted. Voltage-gated potassium channels had first been cloned from the central nervous system of fruit flies in the 1980s and their structure and functional relationships had been worked out in some detail [43,44]. The *Shaker* potassium channels were the first to be cloned [45–47] and received their moniker because flies carrying channel mutations shake their legs under ether anesthesia [48]. A Ca^{2+} modulated, voltage-dependent potassium channel, called the *ether-a-go-go-gene* (eag) for similar reasons, was also first cloned from fruit flies and its biophysical behavior was studied in *Xenopus* oocytes [49]. The mechanistic insights gained from studying ion channels isolated from *Drosophila* eventually converged with powerful genetic techniques applied to large families, to enable rapid progress in the identification of the molecular mechanisms underlying arrhythmias.

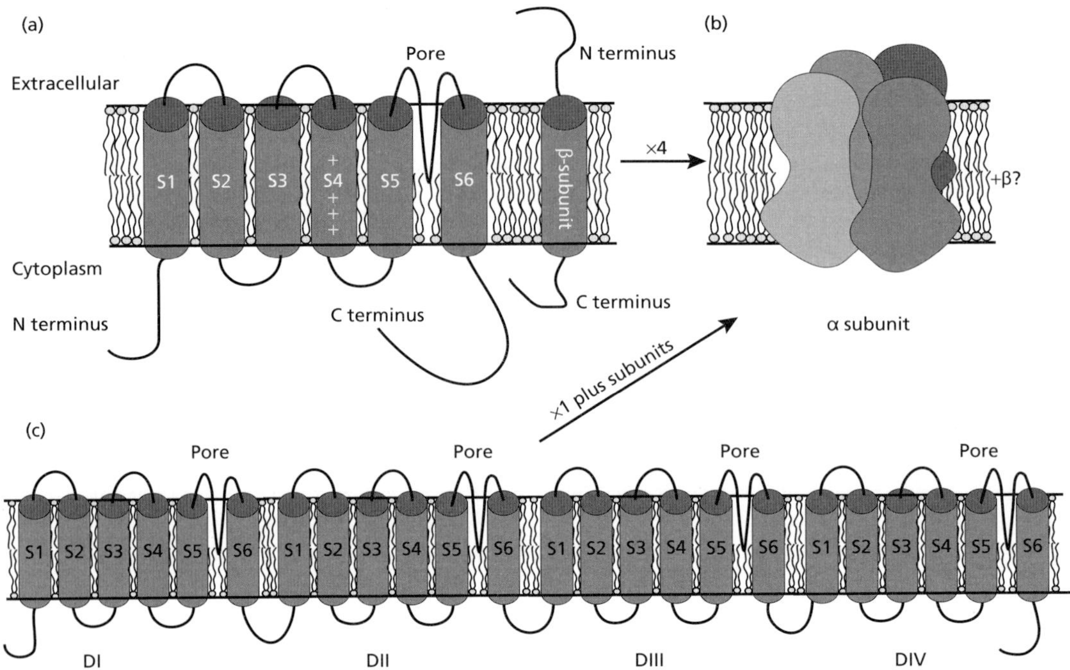

Figure 5.4 (a) Schematic depiction of a voltage-gated K$^+$ channel embedded in the lipid bilayer. The cylinders denote the membrane-spanning regions (S1–S6). The S4 domain contains a periodic array of positively charged amino acid residue that act as the voltage sensor. A function-modulating β subunit often co-assembles with the pore forming α subunit. (b) Four individual monomers form a functional pore-forming tetramer. In some cases, modulatory β-subunits co-assemble at various and often undetermined stoichiometries. (c) The Na$^+$ and Ca^{2+} channels consist of a linked repeat of four individual K$^+$ channel-like modules. The individual modules are denoted domains I–IV (DI–DIV). Accessory subunits also assemble with Na$^+$ and Ca^{2+} channels, although they are not shown in this figure.

Ion channel structure–function relationships

The predicted structure of voltage-gated potassium channels includes six membrane-spanning regions termed S1–S6, a voltage sensing domain (S4) and a pore region located between S5 and S6 (Fig. 5.4). The S4 domain features a regular array of positively charged amino acids at every third position, which move across the membrane in response to depolarization [50]. The N and C terminal regions are cytoplasmically located (Fig. 5.4a) and are involved in regulating the channel's response to the environment. Four monomers assemble to yield a tetrameric structure (also called the α subunit; Fig. 5.4) forming a hydrophobic envelope which surrounds a central cavity or pore, after traversing various membranous intracellular checkpoints on their way from the endoplasmic reticulum (ER) to the plasma membrane [51,52]. The pore-forming α subunits are known to be modulated in their biophysical properties, pharmacologic responses, tissue distribution and intracellular trafficking by smaller accessory (β) subunits which often have a single membrane-spanning domain [53–57] (Fig. 5.4a). The differential assortment of α–β associations may also contribute to greater combinatorial diversity of ion channel function [58]. It is not entirely clear at which point function-modulating β subunits assemble with their corresponding α subunits, or whether such assemblies can be transient, although work carried out in heterologous systems indicates that this occurs early in the biosynthesis pathway during transit through the ER [59].

The structure of the cardiac sodium and L-type calcium channels, SCN5A and CACNA1C, is a bit more complex than that of the K$^+$ channels. These channels consist of four homologous domains, each of which contain six transmembrane domains, similar to the architecture of four linked potassium channel modules (Fig. 5.4c).

Important aspects of voltage-gated K^+ channel structure have recently been worked out to considerably facilitate the investigation of structure–function relationship of ion channels [60–69]. Dr. Roderick MacKinnon's group achieved crucial breakthroughs in this area by producing and purifying relatively large amounts of a bacterial potassium channel (KcsA, isolated from *Streptomyces lividans*). They subsequently devised a method to turn the protein into well-ordered crystals, a prerequisite for determining a molecule's structure. This led to an X-ray crystallographic structure of the pore-forming region of a K^+ channel for which Dr. MacKinnon shared the 2003 Nobel Prize in Chemistry with Dr. Peter Agre. The crystal structure showed that the four subunits are arranged in the shape of an inverted teepee whose wide opening contains the selectivity filter or pore. The ions must fit precisely into the pore and must pass through it in single file. Ions other than K^+ are either too large or too small to align properly with the sides of the pore, making the pore selective for K^+ [70]. Recently, the first structure of a mammalian *Shaker*-type voltage-gated ion channel was published [71].

The congenital LQTS: Affected genes

A combination of two basic approaches was used to identify LQTS-related genes:

1 The candidate gene approach, which posits a likely mechanistic hypothesis based on existing physiologic evidence; and
2 Positional cloning, where a disease-causing gene is identified based on its relative chromosomal position with respect to previously defined DNA markers. In addition to genetic methods, the molecular basis of delayed cardiac repolarization was subsequently worked out through a combination of electrophysiologic, molecular biologic and biochemical methods.

Initially, the analysis of a large kindred in Utah who inherited the LQTS in an autosomal dominant fashion led to the linkage of the LQT phenotype to the Harvey ras-1 locus on chromosome 11 by positional cloning [40,41]. The locus was denoted as LQT1, and although the Harvey ras-1 gene thus became a candidate for LQT, it was subsequently

demonstrated by direct sequence analysis and further linkage studies that it was not, in fact, the culprit gene [72]. Additional LQT loci were quickly identified on other chromosomes: LQT2 was localized to chromosome 7, LQT3 to chromosome 3 and LQT4 to chromosome 4 (Table 5.1) [3,42,73,74].

Potassium channels

The first potassium channel gene shown to underlie the LQTS was the human *ether-a-go-go-related gene* (abbreviated as HERG), whose more recently assigned gene name is KCNH2 [2]. Initially, five unrelated kindreds of patients suffering from a prolonged QT interval, syncope, seizures and aborted sudden cardiac death in an autosomal dominant manner, were genetically linked to polymorphic markers on chromosome 7q35–36 (Table 5.1), the known site of the LQT2 locus. Subsequently, KCNH2 was mapped to the same chromosomal location, shown to harbor mutations in the affected families and its mRNA was detected in human heart. This led to the conclusion that KCNH2 mutations caused the LQT2 syndrome. Molecular cloning of the KCNH2 cDNA soon permitted functional studies and the channel was expressed in *Xenopus* oocytes and characterized biophysically with the two-electrode voltage clamp technique [75,76]. These studies revealed that KCNH2 encoded a channel protein that produced a voltage-sensitive potassium-selective current nearly identical to the cardiac delayed rectifier I_{Kr}. At the functional level, many LQT2 mutations result in reduced current levels or in dominant negative suppression of I_{Kr} in heterologous expression systems. This implied that the molecular mechanism of chromosome 7-linked LQT2 syndrome was a malfunctioning I_{Kr} channel.

Within the same year, KCNQ1 (Fig. 5.3b) was implicated as the underlying cause for LQT1. Again, positional cloning was used to map the offending gene to chromosome interval 11p15.5. When a gene with a high degree of sequence similarity to K^+ channels was identified in that position, it was initially named KvLQT1, for "voltage-gated K^+ channel associated with the LQT1 locus" (the gene name is now KCNQ1; Table 5.1) and examined for mutations in DNA samples from patients with LQT1. Mutations in KCNQ1 were found to co-segregate with the disease and the associated mRNA was found to be expressed in human heart [3]. Similar to KCNH2, *in vitro* expression studies of mutated KCNQ1 proteins suggested multiple biophysical consequences to K^+ current through the channel, all of them ultimately inducing a reduction of function [77,78]. At the time, it was uncertain which specific cardiac ionic current was affected in LQT1 patients. However, in the following year it became clear that a modulatory β subunit, variably called IsK, minK and, most recently, KCNE1 (gene name), when co-expressed with KCNQ1 in heterologous systems, it generated a current corresponding to the cardiac delayed rectifier, I_{Ks} (Fig. 5.3b) [79,80].

Compared with α subunits, KCNE1 is a very small protein with only a single membrane-spanning domain and lacking the classic pore region characteristic of voltage-gated potassium channels (Fig. 5.4a). However, it dramatically influences the biophysical and pharmacologic properties of the current mediated by its KCNQ1 α subunit to resemble those of the cardiac myocyte current I_{Ks}. In light of this information, a candidate gene approach was used to sequence the KCNE1 gene in LQTS families that had not yet been linked to any other loci, and yielded missense mutations in KCNE1 in affected members of two different families [81]. This finding was especially significant because it lent credibility to the notion that the genetics of the syndrome was partly determined by factors that modulate ion channel proteins rather than being limited to the pore-forming subunits themselves. Since then, the search for genetic mutations in similar proteins that might also cause arrhythmias has been ongoing.

In 1999, as the Human Genome Project-driven deposition in public databases of human and mouse genomic sequences neared completion, amino acid sequence alignments revealed that the KCNE protein family was, in fact, comprised of five members [26,82]. The second member to be cloned, KCNE2 (formerly called MiRP1, the MinK-related protein 1) was found to influence the biophysical and pharmacologic characteristics of the HERG-mediated I_{Kr} current [26]. In the ensuing years, KCNE2 was found to affect a number of other ionic currents [83–89] making it difficult to determine its true physiologic role [90]. Despite this uncertainty, KCNE2 was found to carry mutations or polymorphisms in subjects with the congenital and acquired (drug-associated) QT prolongation. One of the KCNE2 variants, associated with clarithromycin-induced arrhythmia, rendered the I_{Kr} channel more susceptible to blockade by the antibiotic. This revealed a mechanism for the acquired form of the LQTS: genetic variants in ion channel associated proteins can remain clinically silent until additional stressors such as drug challenge uncover the disease phenotype [26,32]. When tested in heterologous expression systems, most mutations in K^+ channel or subunits that cause the LQTS in patients result in decreased K^+ current, which is also called a "loss-of-function" phenotype (although a "reduction of function" might be the more appropriate terminology) [25,75,91,92]. Several molecular mechanisms are now known to result in decreased current levels:

1 Decreased channel number at the plasma membrane because of a mutation in one allele (haploinsufficiency).

2 Alterations in the gating process caused by a change in the permeation pathway, which hinders the movement of ions through the open pore, or changes in the process whereby the channel opens or inactivates [91,93–96].

3 Defective 'trafficking' of the channel through the obligate cellular compartments to the cell surface [97–100].

At the time of this writing, the Internet-based *Inherited Arrhythmias Database* maintained by the *European Society of Cardiology* listed 180 LQTS-associated mutations of the KCNQ1 and 197 of the KCNH2 potassium channels and the list is steadily growing. We have therefore confined ourselves to a general discussion of their effects on repolarization.

The sodium channel

When the LQT3 locus was initially localized to chromosome 3p21–24 by positional cloning, it was discovered that this area of the genome comprised several genes. However, the cardiac voltage-gated sodium channel, SCN5A (Table 5.1; Fig. 5.4c) at that location, had previously been cloned and characterized [1,101,102] and became a strong candidate because of its known functional properties. Na^+ channels open quickly in response to depolarization but within a few milliseconds the channels cease to conduct via a process termed "fast inactivation" (Fig. 5.3a,b). During prolonged depolarizations, however, a small fraction of channels do not fully inactivate and still conduct Na^+. Functional analysis of LQT3-associated SCN5A mutations revealed that most mutations produce "gain of function" defects by slightly disrupting Na^+ channel fast inactivation thereby causing a small but persistent I_{Na} during the AP plateau, a property sufficient to delay repolarization and increase vulnerability of the heart to arrhythmias [103–106].

Interestingly, LQT3 patients carrying the SCN5A 1795insD mutation (i.e., the amino acid aspartate [D] is inserted following amino acid 1795) exhibit ECG features of both LQTS and Brugada syndrome: QT interval prolongation at slow heart rates and ST segment elevations with exercise, respectively. Functional analysis revealed that this mutation disrupts fast inactivation, causing persistent I_{Na} throughout the action potential plateau and prolonging cardiac repolarization at slow heart rates, while at the same time augmenting slow inactivation, delaying recovery of Na^+ channel availability between stimuli and reducing I_{Na} at rapid heart rates [107]. Thus, although sodium channel mutations causing the LQTS generally result in a "gain of function," and potassium channel mutations result in "loss of function" phenotype, the cellular consequences are comparable because both types of defects lead to delayed repolarization and increased cellular excitability. Although some experts do not consider LQT4 or LQT7 and LQT8 to be representatives of the classic LQTS [17,108], they are nevertheless associated with QT prolongation, and we will include them in our discussion.

Ankyrin mutations and LQT4

In 1995, Schott *et al.* [109] reported a large French kindred with LQTS, sinus node dysfunction with bradycardia, atrial fibrillation, polymorphic ventricular tachycardia, syncope and sudden cardiac death. This was the first description of LQT4, an autosomal dominant condition that mapped to chromosome 4q25–27. Cardiac reoplarization defects, as measured by QTc, in patients of this genotype were not as severe as in other families with previously described forms of the LQTS [21,29,109,110]. Because of the large size of the gene, it took several more years to pinpoint the genetic defect of LQT4 to a loss of function mutation in ankyrin B, also known as ankyrin 2 [111,112]. Ankyrins were originally identified in erythrocytes and, in addition to the heart, they are also expressed in brain, kidney, skeletal muscle, liver and lung. Because the cellular function of ankyrin B is to link membrane proteins to the cytoskeleton, LQT4 became the first form of the LQTS not caused by a defect in an ion channel protein. A total of three ankyrins (ankyrins R, B and G) are encoded in the mammalian genome. Cardiomyocytes express the 220-kDa ankyrin B and the 190-kDa ankyrin G [29,113,114]. In LQT4 patients, the E1425G mutation in ankyrin B was found to be responsible for disease and a mouse model heterozygous for the E1425G mutation, the ankyrin $B^{+/-}$ mouse, was generated. The homozygous mutation is lethal but haploinsufficient ankyrin $B^{+/-}$ mice live to adulthood and display 50% reduction of ankyrin B in the heart. The mutant phenotype included abnormal calcium homeostasis such as elevated calcium transients and loss of cellular targeting of the Na^+/K^+ ATPase, Na^+/Ca^+ exchanger and inositol 1,4,5 triphosphate receptor, all of them ankyrin-binding proteins, to the cell membrane. Thus, in this model, calcium overload in the sarcoplasmic reticulum indirectly caused by a Na^+/K^+ ATPase deficiency may be the final trigger causing depolarizations that can initiate the arrhythmias. Like the LQT4 kindred, the ankyrin B mutant mice show catecholamine-induced polymorphic ventricular tachycardia and sudden death.

Ankyrin G is associated with the principal cardiac voltage-gated sodium channel, SCN5A (Fig. 5.3b). As described above, gain of function mutations in the SCN5A gene lead to long LQT3 [1]. In addition, SCN5A loss of function mutation may also lead to the Brugada syndrome with right bundle branch block, ventricular arrhythmia and the risk of sudden cardiac death [107,115,116]. Wild-type SCN5A and ankyrin G interact directly in biochemical assays, whereas the E1053K SCN5A mutation isolated from a patient with Brugada syndrome is no longer able to do so. The failure to interact with ankyrin G leads to reduced membrane expression of the sodium channel in cardiac myocytes (albeit not other cell types) and its accumulation in intracellular compartments, resulting in the Brugada syndrome phenotype [113]. Although the Brugada syndrome is independent of the LQTS, it is of interest in this context because it illustrates how mutations in ankyrin, which link to the cytoskeleton, can influence the function of ion channels at the cell surface. It is to be expected that other proteins that regulate ion channel trafficking and intracellular processing will also be found to cause LQTS, considering that 30–35% of patients with a definite clinical diagnosis of LQTS remain to be genotyped [117].

The Andersen syndrome (LQT7)

In 1963, Klein *et al.* [259] reported two girls with cardiac arrhythmia and periodic paralysis. This was the first description of Andersen syndrome, a rare disease with potassium-sensitive periodic paralysis, prolongation of the QT interval with ventricular arrhythmias, clinodactyly, micrognathia and low-set ears [118–120]. The mean QTc was found to be 479–493 ms [121].

Andersen syndrome is inherited as an autosomal dominant disorder with variable penetrance, exhibiting cardiac arrhythmias as a primary manifestation [117,120,122,123]. In 2002, the Andersen syndrome locus was mapped to chromosome 17q23 [27], and subsequently a missense mutation (D71V) in the potassium channel gene KCNJ2 was shown to cause the phenotype. KCNJ2 encodes the inward rectifier K^+ channel Kir2.1, which is an important contributor to I_{K1}. Kir 2.1 is expressed both in skeletal and cardiac muscle [124] where it is a key determinant of the cardiac resting potential and the terminal phase of repolarization (Fig. 5.3a,b).

A major difference between LQT7 (Andersen syndrome) and the other forms of the LQTS is the low incidence of syncope and cardiac arrest. LQT7-associated cardiac arrhythmias degenerate only very rarely into TdP or ventricular fibrillation [110], suggesting that the substrate for arrhythmia susceptibility here differs from the other forms of the LQTS. Reduced Kir2.1 prolongs the terminal phase of the cardiac action potential and, in the setting of hypokalemia, induces Na^+/Ca^{2+} exchanger-dependent delayed afterdepolarizations and spontaneous arrhythmias. In patients with Andersen syndrome a unique form of arrhythmia, called bidirectional tachycardia, which is characterized by alternating polarity of the QRS axis, has been reported [125,126]. Bidirectional tachycardia is typically associated with digitalis toxicity and is also characteristic for familial catecholaminergic polymorphic ventricular tachycardia, which is not related to LQTS [127]. In contrast, other forms of LQTS are more commonly characterized by early

afterdepolarization (EAD)-induced TdP (see "Cellular mechanisms").

Timothy syndrome (LQT8)

The first case of Timothy syndrome was described in 1992 as a sporadic case of congenital heart disease, LQTS and syndactyly [128]. Several other cases followed that presented with various forms of arrhythmias including bradycardia, AV block, TdP, ventricular fibrillation and sudden death in early childhood [28,129]. Recently, a gain of function missense mutation (G406R) in exon 8A of the cardiac L-type calcium channel CACNA1C (Ca_v1.2) was found to be responsible for the diverse physiologic and developmental defects in Timothy syndrome (Fig. 5.3a,b and Fig. 5.4c). Because the CACNA1C channel is widely expressed in organs such as the heart, pancreas, brain, ectodermal cells of developing digits, bladder, prostate, uterus, stomach and smooth muscle, it is not surprising that Timothy syndrome is a multisystem disorder affecting heart (congenital heart disease, arrhythmia), skin (syndactyly), eyes (myopia), teeth (cavities, small teeth), immune system (immunodeficiency with recurrent infections) and brain (autism, mental retardation). The cardiac phenotype known to be associated with the syndrome includes patent ductus arteriosus, patent foramen ovale, ventricular septal defects and tetralogy of Fallot. The average QTc in Timothy syndrome is 580 ms, but in sporadic cases QTc ranges from 620–730 ms [28]. In addition, hypocalcemia and hypoglycemia are also observed. The mechanism underlying the arrhythmia phenotype is reduced CACNA1C channel inactivation, which results in maintained depolarizing Ca^{2+} currents during the plateau phase of the cardiac action potential. Even modest changes in inward calcium current can lead to significant QT interval prolongation and may result in life-threatening arrhythmia and sudden cardiac death. Because the Timothy syndrome is the most recently discovered syndrome associated with QT interval prolongation it has been designated as LQT8.

Cellular mechanisms

The interaction of several membrane currents is responsible for cardiac excitation and repolarization (Fig. 5.3) and the morphology of the cardiac action potential depends on the balance between depolarizing and repolarizing currents which varies in different regions of cardiac tissue. On the surface ECG the QRS complex reflects the depolarization, and the T wave reflects the repolarization phase of the ventricle (Fig. 5.2). Typically, the T wave is much longer than the QRS complex because repolarization takes longer than the fast upstroke of the action potential. Inherent in the T wave is the electrical inhomogeneity of the ventricle during the repolarization phase. This inhomogeneity is based on the fact that the ventricular wall consists of three different layers: epicardium, midmyocardium (or M cell layer) and endocardium. Action potential duration and configuration are different in each of these layers [130,131] (Fig. 5.5). In the dog model, Antzelevitch *et al.* [132–134] described that action potential durations (APD) are longest in the midmyocardium, because of a low density of I_{Ks}, whereas the endo- and epicardium show shorter APD. Voltage gradients between the endocardium and the M-cell layer and between the epicardium and the M region during phase 2 and 3 of the ventricular AP determine the height and shape of the T wave. However, to date, a functional role of M cells in transmural dispersion of repolarization has not been demonstrated in the human heart [135].

Electrophysiologically, a distinction is made between global and local dispersion of refractoriness. In the ventricle, "global dispersion" denotes dispersion in refractoriness over the whole ventricle, whereas "local dispersion" describes the inhomogeneity in refractoriness in a smaller, local area of the ventricle. Local dispersion of refractoriness is most arrhythmogenic in animal models [136,137] as well as in humans. For example, a study by Misier

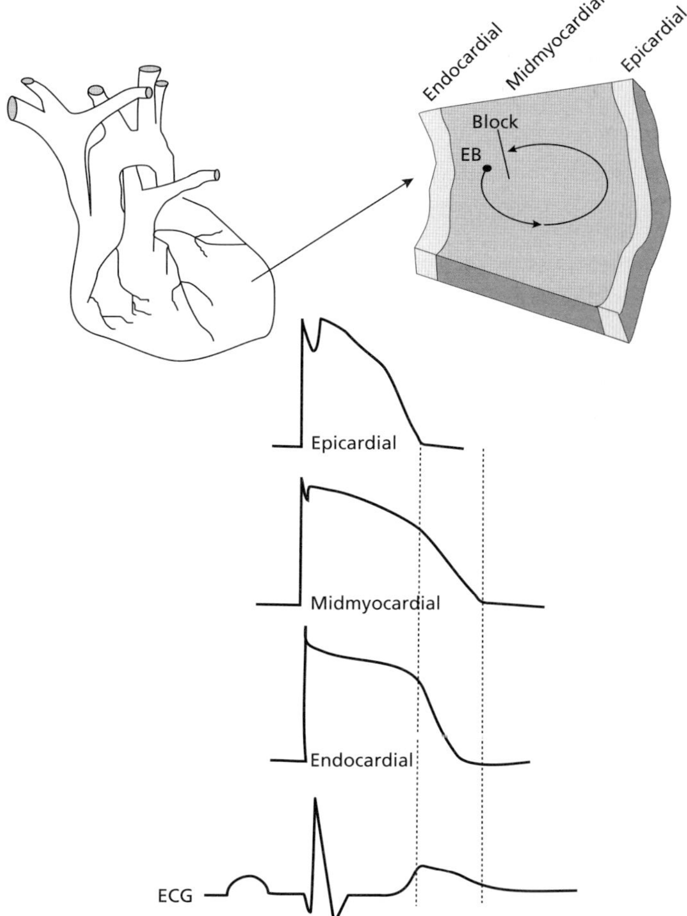

Figure 5.5 Temporal relationship between action potentials (APs) recorded from the three different layers of the ventricular wall and the T wave of the surface ECG. The three different cardiac layers and the corresponding APs are shown within a wedge of ventricular tissue. APDs are shortest in the epicardial or outer layer, longest in the midmyocardium (M cells) and of intermediate duration in the endocardium. Repolarization of the epicardium, midmyocardium and endocardium coincides with the peak, the end and the latter half of the T wave, respectively. A re-entry circuit is shown as a result of an early beat (EB) and unidirectional block within the wedge (see Fig. 5.6 for greater detail).

et al. [138] showed that patients with ventricular fibrillation after myocardial infarction have greater local dispersion of refractoriness in the nonischemic part of the ventricle than patients with ventricular tachycardia that did not degenerate into ventricular fibrillation.

Dispersion in APD has been related to dispersion in QT interval duration [139,140]. In a clinical study, Priori *et al.* [141] showed that LQTS patients, especially RW syndrome patients who were treated with beta-blockade, and patients who underwent left cardiac sympathectomy had similar levels of QT dispersion, indicating that QT dispersion may be a useful index for therapeutic efficacy.

A prolonged QT interval may give rise to life-threatening ventricular tachyarrhythmias, especially TdP. Re-entry and triggered activity are the two major mechanisms used to explain how the TdP arrhythmia arises. In brief, dispersion of repolarization can lead to unidirectional block of a wave of electrical excitation [137] and cause re-entry. Conduction is temporarily blocked within an area of refractory tissue. If conduction velocity is slowed and re-entrant circuits occur in the refractory area, this can create a substrate for arrhythmias (Fig. 5.6). However, unidirectional block is often not sufficient to initiate life-threatening arrhythmias and a triggering mechanism is still required. The trigger for the LQTS seem to be EADs (Fig. 5.2c), which set the stage for ectopic beats if the afterdepolarization is large enough to reach the threshold potential for activation (Fig. 5.6).

The typical initiation of TdP is a "long–short" sequence, where a premature ventricular beat is followed by a pause, then a supraventricular beat followed by a premature ventricular beat at short cycle length, starting the ventricular tachyarrhythmia that twists around the isoelectric line (Figs 5.1b and 5.2c) [142,143].

A more detailed look at the possible mechanisms of TdP reveals the likelihood that re-entry and triggered activity are not completely separable and that both mechanisms have a role. According to this theory, EADs occur in specific areas of the myocardium, prolonging the cardiac action potential and favouring the development of re-entry (Fig. 5.6). As a consequence, inhomogeneous repolarization and afterdepolarizations are both required for the development and maintenance of TdP

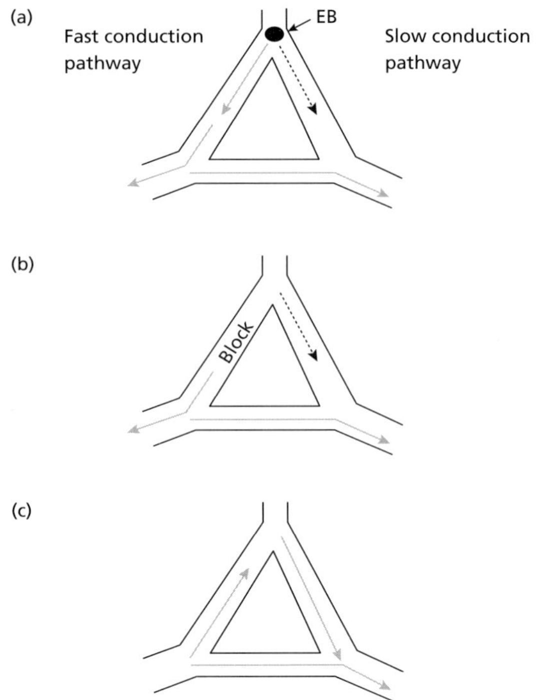

Figure 5.6 Three conditions conspire to cause re-entry: (a) A combination of slow and fast conducting pathways, shown on the left and right sides of the figure, respectively. (b) A unidirectional block in the fast conducting pathway. (c) The conduction velocity in the slow conducting pathway is slow enough to allow recovery of the refractory tissue in the fast conducting pathway. The re-entry mechanism is initiated by an extra beat (EB, part a) that meets a refractory tissue where conduction is temporarily blocked (unidirectional block, part b). Because it cannot enter the fast conducting pathway, it advances along the slow conducting pathway (b). When the EB arrives at the distal end of the fast conducting pathway, which has in the meantime recovered from refractoriness, it enters the fast conduction pathway in a retrograde fashion, thus creating the circuit movements that are called re-entry.

[144]. In most cases TdP is self-limiting, but it may also deteriorate into life-threatening ventricular fibrillation with the risk of sudden cardiac death.

Acquired LQTS

While drugs designed to treat arrhythmias can prevent disease and sudden cardiac death in some patients, they can also unpredictably provoke serious adverse events, including fatal arrhythmias in others [145]. The SWORD and CAST clinical trials were conducted to evaluate the efficacy of antiarrhythmic drugs based on the hypothesis that altering either the refractory period (by using class III drugs that block potassium currents) or the conduction velocity (by using class I drugs that block sodium currents), respectively, would prevent tachycardia. Unfortunately, the trials became prime examples of how an incomplete understanding of drug effects on wave propagation can cost human lives. Indeed, more patients died after taking the drugs than after taking placebo and the trials had to be discontinued [146–148].

Class III anti-arrhythmics (e.g. amiodarone, sotalol and bretylium) were designed to delay cardiac repolarization and TdP is not an unexpected side effect during treatment. However, for many other therapeutic agents, AP prolongation is an undesired side effect, which seriously diminishes their intended use. The problem is so serious that half of all drug withdrawals since 1998 were in response to potential proarrhythmic effects [149]. The most common underlying mechanism is I_{Kr} block, which causes the acquired or drug-induced LQTS (e.g., the drugs that cause most acquired LQTS such as quinidine, sotalol, erythromycin, terfenadine and astemizole all block HERG).

Examples of drugs that were withdrawn because of HERG toxicity are Propulsid, a heart burn medication that was prescribed to 30 million US residents since 1993 [150], and Seldane, an antihistamine which also totaled millions of dollars in US sales [151]. Many of these drugs were prescribed for relatively benign or low-risk conditions and for most of them, the potential to prolong the QT interval and cause arrhythmias was not recognized until after approval and clinical use. To complicate matters, the metabolites of certain drugs may also lead to QT prolongation and TdP. Typical examples include procainamide [152] and astemizole [153].

Genetic factors also appear to contribute to the acquired LQTS: a number of "silent" mutations and functional polymorphisms in LQT-associated genes (SCN5A, KCNQ1, KCNE1, KCNE2) have been associated with an increased vulnerability for the disease [15,30–34,154,155]. Some of the mutations identified, resulted in reduced current levels when tested in *in vitro* assays, which could put the patient at increased risk for the acquired LQTS [155]. Additional risk factors for the acquired LQTS include underlying heart disease such as cardiac hypertrophy or congestive heart failure [154,156–164] where ion channel genes are no longer expressed at normal levels thereby causing prolonged myocardial repolarization, as well as individual differences in drug metabolism caused by genetic mutations or polymorphisms that alter the function of drug-metabolizing enzymes (CYP2D6, CYP3A) [34].

Experimental models for the LQTS

Numerous studies have shown that there is not always a correlation between prolongation of repolarization and the potentially lethal TdP arrhythmia [165–172]. The FDA now requires that any new drug needs to be investigated with respect to its potential to delay repolarization and proarrhythmic properties [173]. This can be difficult to demonstrate because of a relatively low incidence of provoked TdP *in vivo* because healthy animals usually have adequate repolarization reserve, a term that denotes a redundancy of repolarizing currents or the abiltity of cardiomyocytes to maintain efficient repolarization in the face of reduced net outward current. For example, I_{Ks} can function as a reserve reservoir of repolarizing current if I_{Kr} is impaired [174,175].

The dog model

Nevertheless, experimental animal models frequently used to study susceptibility for acquired TdP are dogs with chronic complete atrioventricular (AV) nodal block created by radiofrequency ablation [176–180]. The model is readily adapted to study drug-induced arrhythmias because the electrophysiologic alterations show striking similarities with those of humans [181].

The principle behind this model is that complete AV block results in bradycardia-induced hemodynamic overload of the ventricles, thus initiating a cascade of compensatory processes for decreased cardiac output and increased end-diastolic pressure. This includes the development of hypertrophy [182], contractile alterations to preserve cardiac output [183], prolongation of ventricular repolarization and increased dispersion of repolarization (see "Cellular mechanisms") [181] as well as EAD-triggered arrhythmias and drug-induced TdP

[184]. In this model, the blockade of I_{Ks}, in addition to drug-induced I_{Kr} inhibition, seems to constitute the underlying mechanism, because a downregulation of KCNQ1 and KCNE1 was demonstrated [185]. Additionally, the canine wedge preparation model was used to demonstrate that bifurcated T waves and a U wave can occur in conjunction with QT interval prolongation [133].

The rabbit model

A second animal model used to study the acquired LQTS is anesthetized rabbits in which TdP is evoked by administration of a test agent [165, 180,186], and the effects of test drugs on APD are recorded from Purkinje fibers and ventricular cells. Carlsson *et al.* [165] showed that almokalant and other QT-prolonging agents, such as sematilide, and cesium chloride induced a prolongation of the QT interval, as well as EADs and TdP in a dose-dependent manner.

The arterially perfused canine or rabbit transmural ventricular wedge preparation was initially developed to study transmural repolarization heterogeneity across the ventricular wall (Fig. 5.5) [133,187] but is also useful for the assessment of cardiac safety of new compounds [188,189]. In this model, transmural wedges are excised from the left ventricle and the tissue is cannulated and arterially perfused. The wedges are stimulated with bipolar silver electrodes and transmural ECGs are recorded. In addition, intracellular floating electrodes can be used to measure transmembrane APs from endocardial, midmyocardial and epicardial layers of the ventricular wall. With this technique, Shimizu *et al.* [189–191] were able to mimic LQT1 using the I_{Ks}-blocker chromanol, LQT2 using the I_{Kr}-blocker D-sotalol, and LQT3 using ATX-II, which increases sodium currents. To a large extent, the insights about the role M cells have in setting up the development of arrhythmias were generated using this model (see "Cellular mechanisms").

The mouse model

The drawback of the dog and rabbit models is that they are not genetically tractable, a property that, in addition to experimental accessibility, is highly desirable in an ideal model organism. Over the last two decades the mouse has grown into that particular role for several reasons:

1 The field of mouse genetics is comparatively well developed because spontaneous and induced mouse mutations have been studied, maintained, inbred, genetically mapped and described at the Jackson Laboratory in Bar Harbor, Maine, since 1929.
2 Techniques for manipulating gene function, such as inactivating genes ("knockouts") or turning them on (transgenics) exist, which are not available in any other model vertebrate.
3 The mouse genome is fully sequenced and 99% of mouse genes have a direct human counterpart.

Genetically modified mouse models of ion channel diseases have been derived both in the form of transgenics, where genetic coding material is added to an otherwise (at least theoretically) undisturbed genetic background, as well as in the form of "knockouts," where the function of the target gene is disrupted by genetic means. As the salient characteristics of engineered arrhythmia mouse models were recently summarized in a comprehensive review [192], we will restrict ourselves to a short discussion. Several genetically modified mouse models were constructed before it was fully appreciated that the mouse is a useful but not ideal model to study cardiac arrhythmias. Specialized techniques had to be developed to investigate mouse cardiac electrophysiology [193–198]. For example, changes in mouse ECGs can now be analyzed in conscious animals with the help of implantable telemetry devices and signal averaging to avoid the confounding effects of anesthetic agents [199]. These investigations uncovered fundamental differences between the electrical activity of the hearts of mice and humans. At rest, the mouse heart beats at a rate approximately 10-fold that of humans (600–700 beats/min) and there are clear differences in AP morphology. In adult mouse ventricular myocytes, a distinct plateau phase is missing and repolarization is rapid, carried mainly by I_{To} [198,200,201]. I_{Kr} and I_{Ks}, the currents most often affected in the congenital LQTS, are generally thought to be lacking in adult ventricular myocytes, although they may simply be exceedingly rare, occurring only in specialized subpopulations of mouse cardiocytes. For example, KCNE1, the required β subunit for I_{Ks} formation, is restricted to cells from the cardiac conduction system in the adult mouse heart and one would not expect to record I_{Ks} from a randomly selected mouse myocyte because conduction system-derived cells make up an exceedingly small proportion of cardiocytes [202]. Spontaneous ventricular arrythmias are rare in either wild-type mice, or in mice with genetically altered levels of LQTS-associated genes. However, this is not simply because of the small size of the mouse heart, because both atrial and ventricular arrhythmias can be generated in mice [203–208].

Lower organisms

Some intrepid researchers are exploring simpler, more genetically tractable model organisms such as *Caenorhabditis elegans*, taking advantage of the rhythmic contractility of the worm pharynx to identify proteins that affect electrical excitability in that organism and then relating it back to the mammalian channels [209]. The zebrafish model is discussed below. However, it is evident that the ideal animal model in which to study integrated cardiac myocyte biology is still lacking.

Theoretical models

Data derived from *in vivo* and *in vitro* experimental models are now being incorporated systematically into theoretical computer models of cardiac electrophysiology that are uniquely capable of simulating cardiac electrical function in a way that is not possible experimentally. These models are described by systems of ordinary differential equations approximating the dynamical behavior of ion channels, pumps, exchangers, as well as Na^+, Ca^{2+} and K^+ concentrations during the AP [210,211]. Pioneering work in this respect was carried out in the early 1990s by Luo and Rudy [211,212] who established a mathematical model of the membrane action potential of mammalian ventricular cells. Henry and Rappel [213] later extended the dynamic Luo–Rudy model to investigate the role of M cells with a special focus on LQT3.

Modifying factors

Adrenergic stimulation

Sympathetic stimulation is one of the most important factors in cardiac repolarization. Experimental data reveal that β-adrenergic stimulation augments a number of ionic currents in cardiac myocytes that contribute to repolarization, including the delayed rectifier, I_{Ks}, the Ca^{2+} activated chloride current ($I_{Cl(Ca)}$), the L-type Ca^{2+} current (I_{Ca}), and the Na^+/Ca^{2+} exchange current (I_{Na-Ca}), thus impacting the balance of net outward (I_{Ks}, $I_{Cl(Ca)}$) and inward (I_{Ca}) current [214]. Indeed, sympathetic stimulation is often the trigger for arrhythmic events, especially in LQT1 patients [12] (see "Symptoms"). Clinical studies of LQTS patients infused with epinephrine suggest that sympathetic stimulation produces genotype-specific responses of the QT interval [215]. Sympathetic stimulation remarkably prolongs the QTc interval at fast heart rates and peak levels of epinephrine in LQT1 and LQT2 patients. The QTc continues to be prolonged in LQT1 patients even at steady state conditions of epinephrine, whereas in LQT2 patients it then shortens to baseline levels. In LQT3 patients, the QTc is less prolonged at peak levels and also returns to baseline at steady state. Thus, in LQT1 patients who have suffered a loss of function in I_{Ks}, β-adrenergic stimulation does not sufficiently augment the current to hasten repolarization, resulting in persistent QTc interval prolongation. In contrast, in LQT3 patients with a gain-of-function in I_{Na}, β-adrenergic stimulation does not markedly prolong the QTc, which is probably because of an increase in I_{Ks} as well as a reduction in the I_{Na-Ca} because of persistent I_{Na}.

Thus, patients with different genotypes carry varying degrees of risk for arrhythmias in response to adrenergic stimulation, which has raised the possibility of distinct therapeutic strategies in the management of patients with these LQTS variants (see "Therapy").

Electrolytes

Another factor impacting cardiac repolarization is electrolyte imbalance. Hypokalemia is known to prolong QT interval in healthy populations [34,216], and hypokalemia, as well as severe hypomagnesemia constitute risk factors in the LQTS [28]. The underlying reason appears to be, paradoxically, that the magnitude of I_{Kr} decreases with reduced extracellular K^+, despite an increase in the chemical driving force [217,218]. This can be explained by a relief of extracellular Na^+-mediated I_{Kr} block under high extracellular $[K^+]$ conditions [219]. Thus, decreasing serum K^+ prolongs repolarization by decreasing the repolarizing current, whereas an increase of serum K^+ augments I_{Kr} magnitude. Indeed, it has been reported that long-term oral potassium administration may be effective therapy for LQT2 (see "Therapy") [220,221].

Gender and sex hormones

The QTc interval is known to be age- and sex-dependent in the normal population, with lower values in adult males [222]. Male LQTS patients have an earlier onset of adverse cardiac events and a higher risk of first events in childhood than females, with decreasing risk after puberty. In contrast, females have a lower incidence of cardiac events in childhood and higher incidence of events in adulthood; their risk of first cardiac events does not decrease in adulthood [223]. The lower incidence of cardiac events among adult males may be because of a shortening of the QTc interval post puberty, which is more prominent in males than in females. Analyses of genotyped families show that LQT1 and LQT2 syndrome males have shorter QTc intervals than females. In contrast, LQT3 syndrome males have longer QTc intervals than females [224].

In the drug-induced LQTS, female gender is associated with a greater incidence of cardiac events than male gender [225]. Drug-induced QT prolongation in females occurs more often during menstruation and the ovulatory phase of the menstrual cycle than during the luteal phase [226]. Although sex hormones are probably implicated, the precise mechanisms responsible for age and sex differences in QTc interval are still unknown.

Symptoms

The major symptoms of LQTS patients are syncope, seizure-like activity and sudden cardiac arrest resulting from TdP and ventricular fibrillation. Tragically, death is the first symptom in 10–15% of patients who die of complications, because LQTS patients rarely feel palpitations.

Most clinical and basic studies on genotype–phenotype correlation have been performed on patients with either the LQT1, LQT2 or LQT3, because this group comprises approximately 60%, 35% and 3–4% of all patients, respectively, compared with all other genotypes which represent less than 1% of the total (Table 5.1) [22]. It was found that triggers of cardiac events differ among the genotypes: LQT1 patients experience the majority of their events during exercise or conditions associated with elevated sympathetic activity but rarely during rest or sleep. In contrast, this situation is reversed in LQT2 and LQT3 patients. Swimming as a trigger is particularly frequent in LQT1 patients, while auditory stimuli are relatively frequent in LQT2 patients [12]. These characteristics can mostly be explained through the differences in response to adrenergic stimulation.

The incidence of first cardiac event (syncope, cardiac arrest and sudden death) before the age of 40, and prior to initiation of therapy, also differs among the genotypes. It is lower in LQT1 (30%) than for LQT2 (46%) or LQT3 (42%). Moreover, the genetic locus affects not only the clinical course of the LQTS but also modulates the effects of the QTc interval and gender on clinical manifestation. A risk stratification scheme for LQTS patients, according to length of QTc interval, genotype and gender, has been proposed [13] and is depicted in Table 5.3.

Diagnosis

The classic ECG feature of LQTS patients is prolongation of the rate-corrected QT interval (QTc), as measured by Bazett's formula (QTc = QT/√RR). The benchmark QTc value of 440 ms was often used in the past as an index for diagnosing LQTS. However, the ECG changes in LQTS are not merely limited to a prolongation of the QTc interval, but may also include factors such as T and U wave abnormalities, bradycardia and episodic polymorphic ventricular tachycardia. As a consequence, diagnostic criteria for LQTS were refined based on a scoring system of clinical characteristics including ECG findings, clinical history and family history [227] (Table 5.4).

Because the symptoms of LQTS patients often look like those of patients with neurologic disorders, clinical suspicion is a crucial element in diagnosing the LQTS. If healthy people experience unexplained syncope, cardiac arrest and sudden death, LQTS should be on the list of differential diagnosis. However, the diagnosis of LQTS can be quite difficult in some "borderline" patients. In addition to 12-lead ECGs and Holter monitoring, provocative tests such as epinephrine infusion or treadmill

Table 5.3 Risk stratification of the long QT syndrome (LQTS).

Low risk for cardiac event (<30%)	Intermediate risk for cardiac event (30–50%)	High risk for cardiac event (>50%)
QTc <500 ms	QTc <500 ms	QTc >500 ms
Male sex	Female sex, LQT2	Female sex, LQT
LQT1, LQT2	Female sex, LQT3	
	Male sex, LQT3	
	QTc >500 ms	QTc >500 ms
	Female sex, LQT3	LQT1
		LQT2
		Male sex, LQT3

Table 5.4 Diagnostic criteria.

Electrocardiographic findings	Points
QTc (calculated by Bazett formula*)	
>480 ms	3
460–470 ms	2
450 ms in males	1
Tosade de pointes	2
T-wave alternans	1
Notched T-wave in 3 leads of the surface ECG	1
Low heart rate for age (resting heart rate below the second percentile for age)	0.5
Clinical history	
Syncope	
with stress	2
without stress	1
Congenital deafness	0.5
Family history	
Family members with definite LQTS	1
Unexplained sudden cardiac death before the age of 30 among immediate family members	0.5

ECG, electrocardiogram; LQTS, long QT syndrome.
A score of 4 or more defines LQTS.
* Bazett formula: QTc = QT/√RR. Scoring: <1 point, low probability; 2–3 points, intermediate probability; >4 points: high probability.

exercises may be useful [215,228], especially in latent KCNQ1 mutation carriers [229]. The results may differ among LQTS genotypes (see "Modifying factors"). In the absence of other indicators, invasive electrophysiologic study can be useful for diagnosing LQTS. If LQTS is clinically diagnosed or suspected, family members should also be screened by ECGs. Genetic testing is available and recommended as a final diagnosis (success rate, 50–70%) and to direct genotype-specific therapy.

Clinical treatment

Clinical tratment of patients with congenital LQTS is most effective if it is genotype-specific. Several studies on drug-induced LQTS based on the wedge preparation model support genotype-specific treatments for the different types of LQTS, with LQT3 being the best studied [189–191,230].

Treatment of LQT1

The first choice of therapy for LQT1 are beta-blockers. Schwartz *et al.* [12] reported that suppression of cardiac events by beta-blockers is more frequent in patients with LQT1 (81%) than in LQT2 (59%) and LQT3 (50%). However, if beta-blockade is not sufficient to suppress cardiac events, mexiletine, a class IB anti-arrhythmic (sodium-channel blocker) should be considered for treatment. It has been shown experimentally that mexiletine is able to suppress TdP [190,191]. An adjunct therapy may be a calcium-channel blocker such as verapamil. A

study by Aiba *et al.* [231] showed positive effects of verapamil in shortening the QT interval and decreasing the risk for TdP. On the other hand, potassium-channel openers such as nicorandil, available in Europe and Japan, are of lesser value, because it has been demonstrated that a high intravenous dosage of nicorandil is required to abolish EADs and TdP [230]. LQT1 patients are most sensitive to sympathetic stimulation, and left cardiac sympathetic denervation is thought to be most effective in LQT1 patients, if beta-blocker therapy fails [232]. LQT1 patients should avoid exercise or do so only under careful supervision, because cardiac events in LQT1 patients are most frequently exercise-induced, and are especially associated with swimming.

Treatment of LQT2

As for LQT1 patients, beta-blockers are also the first choice of treatment for LQT2 patients.

Additional therapies such as mexiletine and/or verapamil might have a higher impact in LQT2 patients because they seem to have a higher recurrence rate of cardiac events when they are exclusively on beta-blocker therapy [12,233]. It is important to consider serum potassium levels in LQT2 patients, because I_{Kr}, the defective current in LQT2, is very sensitive to extracellular potassium levels [133,217–219]. Etheridge *et al.* [234] found that long-term oral administration of potassium has beneficial effects, improving disturbances in repolarization. Additionally, restricted exercise is meaningful in LQT2 patients [235].

Treatment of LQT3

Because a gain of function defect in the sodium-channel gene, SCN5A, is responsible for LQT3, sodium-channel blockade with mexiletine is of special interest. Indeed, preliminary clinical and basic experimental data suggest that the class Ib anti-arrhythmic drug, mexiletine, is more effective in abbreviating the QT interval in LQT3 than in LQT1 or LQT2 [191,236,237], possibly because of the block of SCN5A mutation-associated persistent I_{Na} by mexiletine [238].

In this context it is interesting to note that the SCN5A D1790G mutation does not produce persistent I_{Na} during depolarization, but alters the kinetics and voltage dependence of the inactivated state, which can prolong ventricular repolarization [239]. In this case, administration of the class Ic anti-arrhythmic flecainide is more effective than the class Ib mexiletine, possibly because of the high sensitivity of this mutation to the use-dependent block of I_{Na} by flecainide [240]. Based on preliminary clinical data, mexiletine should be applied with beta-blockers or co-administered with an implantable cardioverter-defibrillator (ICD). LQT3 patients are not recommended for single beta-blocker therapy because experimental studies show no protective or even harmful effects [189,235]. Pacemakers seem to be of specific benefit for LQT3 patients. Schwartz *et al.* [241] showed that an increase of heart rate may effectively abbreviate the QT interval.

Treatment of LQT4–8

At present, genotype-specific treatment of LQT4–8 has not been determined because of the small number of cases. In general, beta-blockers should be the first therapy of choice. In LQT5 the responsible defect is a mutation in KCNE1, which is responsible for I_{Ks}, thus the recommended therapy for LQT1 patients also applies to LQT5. Similarly, the therapy scheme for LQT2 should be beneficial for LQT6 [242]. In all cases, additional prospective clinical studies are needed to provide a better assessment of LQT treatment other than beta-blockers.

ICD therapy

In general, ICD therapy is indicated in LQTS patients who have experienced cardiac arrest or repetitive episodes of syncope. In cases involving young LQTS patients who have not themselves had cardiac events but whose family members have experienced episodes of syncope, cardiac arrest or even sudden cardiac death, an ICD implantation should be considered early in therapy. In an 8-year follow-up study of 125 LQT patients, Zareba *et al.* [243] and Welde [244] showed that incidence of mortality was higher in those patients who did not receive an ICD.

Pacemaker therapy

A study by Eldar *et al.* [245] showed that, in general, a combination of beta-blocker and cardiac pacing appears to be highly advantageous for LQTS patients. Especially patients with LQT3 derive great benefit from pacemaker therapy [241]. Currently, ICD and pacemaker are combined in one device, which facilitates the use of a pacemaker as adjunct therapy.

Because a large number of drugs have the potential to induce TdP tachyarrhythmias with the risk of deteriorating into life-threatening ventricular fibrillation, patients with congenital LQTS should avoid taking those drugs. Examples of drugs with torsadogenic potential are listed in Table 5.2. A more comprehensive list of drugs that can cause TdP can be found on the Internet at http://www.qtdrugs.org [246].

A challenge for the pharmaceutical industry

A major problem facing the pharmaceutical industry is the identification of potentially toxic effects of drugs before those drugs reach the market and endanger the health of patients. This is associated with an ever-escalating cost for drug development which was estimated to be US$1.9 billion in 2005 [247]. In addition, only 21% of drugs entering phase I clinical trials actually reach the market [248]. Thus, the pharmaceutical industry considers it increasingly important to identify potential toxicities early in the drug discovery process and the demand for assays that can accurately predict the effect of compounds on the QT interval is great [249].

A number of screens exist for detecting drug-induced repolarization abnormalities, although most are not high-throughput and costly (i.e., of limited use early in the drug discovery process). Typical assays used to determine the QT prolonging potential of test compounds include the *in vivo* dog model, as well as *in vitro* models including the analysis of APD from freshly isolated rabbit Purkinje fibers, or the Langendorff perfused *ex vivo* heart model [249,250] (see "Experimental models for the LQTS"). *In vitro* assays, such as the measurement of HERG channel activity in a cell culture system, can be adapted to high throughput formats but are limited by their inherent biologic simplicity, high rate of false positives and the inability to detect drug interactions [251].

An exciting new model that holds promise for high throughput screening applications is the zebrafish, *Danio rerio*. Zebrafish embryos are transparent which permits convenient monitoring of the heart rate, "gene knockdown" approaches via morpholinos [252] allow for easy gene manipulation, the culture of the embryos is easy and adaptable to high throughput formats, small amounts of drugs are sufficient for *in vivo* testing, and drugs can be applied to the bath where they are taken up through the embryo's skin. Moreover, it is possible to stop the heart beat in zebrafish embryos for days while they survive by diffusion [253]. Although zebrafish have two-chambered hearts, human and zebrafish cardiac structure and function appear largely conserved. For example, hereditary cardiomyopathy in humans has been linked to mutations in several genes that also affect the heartbeat in zebrafish [254–256]. A zebrafish ortholog of HERG/KCNH2, zERG, was cloned and antisense knockdown of this gene was shown to produce bradycardia and arrhythmia [257]. The usefulness of zebrafish in screening for QT prolonging drugs was demonstrated, when compounds associated with long

QT and TdP in humans, such as terfenadine and cisapride reduced heart rate and caused dose-dependent AV nodal block in this model [257,258].

Conclusions

We hope that at this point the reader has developed an appreciation that even a relatively rare genetic syndrome can provide an impressive example of the power of translational research. The LQTS posed a challenge to basic scientists, clinicians and industry alike. They responded by combining basic science techniques including population genetics, biochemistry, molecular biology, electrophysiology, with clinical studies to elucidate general arrhythmia mechanisms and to improve therapies for the betterment of public health.

References

1 Wang Q, Shen J, Splawski I *et al.* SCN5A mutations associated with an inherited cardiac arrhythmia, long QT syndrome. *Cell* 1995; 80: 805–811.

2 Curran ME, Splawski I, Timothy KW *et al.* A molecular basis for cardiac arrhythmia: HERG mutations cause long QT syndrome. *Cell* 1995; 80: 795–803.

3 Wang Q, Curran ME, Splawski I *et al.* Positional cloning of a novel potassium channel gene: KVLQT1 mutations cause cardiac arrhythmias. *Nat Genet* 1996; 12: 17–23.

4 Chen Q, Kirsch GE, Zhang D *et al.* Genetic basis and molecular mechanism for idiopathic ventricular fibrillation. *Nature* 1998; 392: 293–296.

5 Priori SG, Napolitano C, Tiso N *et al.* Mutations in the cardiac ryanodine receptor gene (hRyR2) underlie catecholaminergic polymorphic ventricular tachycardia. *Circulation* 2001; 103: 196–200.

6 Tan HL, Bink-Boelkens MT, Bezzina CR *et al.* A sodium-channel mutation causes isolated cardiac conduction disease. *Nature* 2001; 409: 1043–1047.

7 Chen YH, Xu SJ, Bendahhou S *et al.* KCNQ1 gain-of-function mutation in familial atrial fibrillation. *Science* 2003; 299: 251–254.

8 Benson DW, Wang DW, Dyment M *et al.* Congenital sick sinus syndrome caused by recessive mutations in the cardiac sodium channel gene (SCN5A). *J Clin Invest* 2003; 112: 1019–1028.

9 Moss AJ. T-wave patterns associated with the hereditary long QT syndrome. *Cardiac Electrophysiol Rev* 2002; 6: 311–315.

10 Ackerman MJ, Tester DJ, Porter CJ. Swimming, a gene-specific arrhythmogenic trigger for inherited long QT syndrome. *Mayo Clin Proc* 1999; 74: 1088–1094.

11 Moss AJ, Robinson JL, Gessman L *et al*. Comparison of clinical and genetic variables of cardiac events associated with loud noise versus swimming among subjects with the long QT syndrome. *Am J Cardiol* 1999; 84: 876–879.

12 Schwartz PJ, Priori SG, Spazzolini C *et al*. Genotype–phenotype correlation in the long-QT syndrome: gene-specific triggers for life-threatening arrhythmias. *Circulation* 2001; 103: 89–95.

13 Priori SG, Schwartz PJ, Napolitano C *et al*. Risk stratification in the long-QT syndrome. *N Engl J Med* 2003; 348: 1866–1874.

14 Chiang CE. Congenital and acquired long QT syndrome. Current concepts and management. *Cardiol Rev* 2004; 12: 222–234.

15 Roden DM. Drug-induced prolongation of the QT interval. *N Engl J Med* 2004; 350: 1013–1022.

16 Moss AJ, Schwartz PJ. 25th Anniversary of the International Long-QT Syndrome Registry: An ongoing quest to uncover the secrets of long-QT syndrome. *Circulation* 2005; 111: 1199–1201.

17 Schwartz PJ. Management of long QT syndrome. *Nature Clin Pract Cardiovasc Med* 2005; 2: 346–351.

18 Jervell A, Lange-Nielsen F. Congenital deaf-mutism, functional heart disease with prolongation of the Q-T interval and sudden death. *Am Heart J* 1957; 54: 59–68.

19 Romano C, Gemme G, Pongiglione R. [Rare cardiac arrythmias of the pediatric age. II. Syncopal attacks due to paroxysmal ventricular fibrillation. (Presentation of 1st Case in Italian Pediatric Literature)]. *Clin Pediatr (Bologna)* 1963; 45: 656–683.

20 Ward OC. A new familial cardiac syndrome in children. *J Ir Med Assoc* 1964; 54: 103–106.

21 Chiang CE, Roden DM. The long QT syndromes: genetic basis and clinical implications. *J Am Coll Cardiol* 2000; 36: 1–12.

22 Vincent GM. Romano-Ward syndrome. GeneReviews at GeneTests: Medical Genetics Information Resource (database online). Copyright, University of Washington, Seattle. 1997–2005. Available at http://www.genetests.org

23 Yanowitz F, Preston JB, Abildskov JA. Functional distribution of right and left stellate innervation to the ventricles. Production of neurogenic electrocardiographic changes by unilateral alteration of sympathetic tone. *Circ Res* 1966; 18: 416–428.

24 Schwartz PJ, Periti M, Malliani A. The long Q-T syndrome. *Am Heart J* 1975; 89: 378–90.

25 Splawski I, Tristani-Firouzi M, Lehmann MH *et al*. Mutations in the hminK gene cause long QT syndrome and suppress IKs function. *Nat Genet* 1997; 17: 338–340.

26 Abbott GW, Sesti F, Splawski I *et al*. MiRP1 forms IKr potassium channels with HERG and is associated with cardiac arrhythmia. *Cell* 1999; 97: 175–187.

27 Plaster NM, Tawil R, Tristani-Firouzi M *et al*. Mutations in Kir2.1 cause the developmental and episodic electrical phenotypes of Andersen's syndrome. *Cell* 2001; 105: 511–519.

28 Splawski I, Timothy KW, Sharpe LM *et al*. Ca(V)1.2 calcium channel dysfunction causes a multisystem disorder including arrhythmia and autism. *Cell* 2004; 119: 19–31.

29 Mohler PJ, Schott JJ, Gramolini AO *et al*. Ankyrin-B mutation causes type 4 long-QT cardiac arrhythmia and sudden cardiac death. *Nature* 2003; 421: 634–639.

30 Splawski I, Timothy KW, Tateyama M *et al*. Variant of SCN5A sodium channel implicated in risk of cardiac arrhythmia. *Science* 2002; 297: 1333–1336.

31 Paulussen AD, Gilissen RA, Armstrong M *et al*. Genetic variations of KCNQ1, KCNH2, SCN5A, KCNE1, and KCNE2 in drug-induced long QT syndrome patients. *J Mol Med* 2004; 82: 182–188.

32 Sesti F, Abbott GW, Wei J *et al*. A common polymorphism associated with antibiotic-induced cardiac arrhythmia. *Proc Natl Acad Sci USA* 2000; 97: 10613–10618.

33 Kubota T, Horie M, Takano M *et al*. Evidence for a single nucleotide polymorphism in the KCNQ1 potassium channel that underlies susceptibility to life-threatening arrhythmias. *J Cardiovasc Electrophysiol* 2001; 12: 1223–1229.

34 Aerssens J, Paulussen AD. Pharmacogenomics and acquired long QT syndrome. *Pharmacogenomics* 2005; 6: 259–270.

35 Schwartz PJ, Priori SG, Dumaine R *et al*. A molecular link between the sudden infant death syndrome and the long-QT syndrome. *N Engl J Med* 2000; 343: 262–267.

36 Tester DJ, Ackerman MJ. Sudden infant death syndrome: How significant are the cardiac channelopathies? *Cardiovasc Res* 2005; 67: 388–396.

37 Schwartz PJ, Priori SG, Bloise R *et al*. Molecular diagnosis in a child with sudden infant death syndrome. *Lancet* 2001; 358: 1342–1343.

38 Christiansen M, Tonder N, Larsen LA *et al*. Mutations in the HERG K+-ion channel: a novel link between long QT syndrome and sudden infant death syndrome. *Am J Cardiol* 2005; 95: 433–434.

39 Roden DM, Spooner PM. Inherited long QT syndromes: a paradigm for understanding arrhythmogenesis. *J Cardiovasc Electrophysiol* 1999; 10: 1664–1683.

40 Keating M, Atkinson D, Dunn C *et al*. Linkage of a cardiac arrhythmia, the long QT syndrome, and the Harvey ras-1 gene. *Science* 1991; 252: 704–706.

41 Keating M, Dunn C, Atkinson D *et al*. Consistent linkage of the long-QT syndrome to the Harvey ras-1 locus on chromosome 11. *Am J Hum Genet* 1991; 49: 1335–1339.

42 Jiang C, Atkinson D, Towbin JA *et al.* Two long QT syndrome loci map to chromosomes 3 and 7 with evidence for further heterogeneity. *Nat Genet* 1994; 8: 141–147.

43 Jan LY, Jan YN. Cloned potassium channels from eukaryotes and prokaryotes. *Ann Rev Neurosci* 1997; 20: 91–123.

44 Jan LY, Jan YN. Tracing the roots of ion channels. *Cell* 1992; 69: 715–718.

45 Papazian DM, Schwarz TL, Tempel BL *et al.* Cloning of genomic and complementary DNA from Shaker, a putative potassium channel gene from *Drosophila*. *Science* 1987; 237: 749–753.

46 Pongs O, Kecskemethy N, Muller R *et al.* Shaker encodes a family of putative potassium channel proteins in the nervous system of *Drosophila*. *EMBO J* 1988; 7: 1087–1096.

47 Iverson LE, Tanouye MA, Lester HA *et al.* A-type potassium channels expressed from Shaker locus cDNA. *Proc Natl Acad Sci USA* 1988; 85: 5723–5727.

48 Kaplan WD, Trout WE III. The behavior of four neurological mutants of *Drosophila*. *Genetics* 1969; 61: 399–409.

49 Bruggemann A, Pardo LA, Stuhmer W *et al.* Ether-a-go-go encodes a voltage-gated channel permeable to K$^+$ and Ca^{2+} and modulated by cAMP. *Nature* 1993; 365: 445–448.

50 Jan LY, Jan YN. Structural elements involved in specific K$^+$ channel functions. *Annu Rev Physiol* 1992; 54: 537–555.

51 Roden DM, Kupershmidt S. From genes to channels: normal mechanisms. *Cardiovasc Res* 1999; 42: 318–326.

52 Deutsch C. The birth of a channel. *Neuron* 2003; 40: 265–276.

53 Rhodes KJ, Strassle BW, Monaghan MM *et al.* Association and colocalization of the Kvbeta1 and Kvbeta2 beta-subunits with Kv1 alpha-subunits in mammalian brain K$^+$ channel complexes. *J Neurosci* 1997; 17: 8246–8258.

54 Takumi T, Ohkubo H, Nakanishi S. Cloning of a membrane protein that induces a slow voltage-gated potassium current. *Science* 1988; 242: 1042–1045.

55 Trimmer JS, Rhodes KJ. Building (potassium) channels to the 21st century. *Trends Neurosci* 1997; 20: 99–100.

56 England SK, Uebele VN, Shear H *et al.* Characterization of a voltage-gated K$^+$ channel beta subunit expressed in human heart. *Proc Natl Acad Sci USA* 1995; 92: 6309–6313.

57 Nuss HB, Chiamvimonvat N, Perez Garcia MT *et al.* Functional association of the beta 1 subunit with human cardiac (hH1) and rat skeletal muscle (mu 1) sodium channel alpha subunits expressed in Xenopus oocytes. *J Gen Physiol* 1995; 106: 1171–1191.

58 Pourrier M, Schram G, Nattel S. Properties, expression and potential roles of cardiac K$^+$ channel accessory subunits: MinK, MiRPs, KChIP, and KChAP. *J Membr Biol* 2003; 194: 141–152.

59 Nagaya N, Papazian DM. Potassium channel alpha and beta subunits assemble in the endoplasmic reticulum. *J Biol Chem* 1997; 272: 3022–3027.

60 Jiang Y, Lee A, Chen J *et al.* X-ray structure of a voltage-dependent K$^+$ channel. *Nature* 2003; 423: 33–41.

61 Jiang Y, Lee A, Chen J *et al.* The open pore conformation of potassium channels. *Nature* 2002; 417: 523–526.

62 Jiang Y, Lee A, Chen J *et al.* Crystal structure and mechanism of a calcium-gated potassium channel. *Nature* 2002; 417: 515–522.

63 Kreusch A, Pfaffinger PJ, Stevens CF *et al.* Crystal structure of the tetramerization domain of the Shaker potassium channel. *Nature* 1998; 392: 945–948.

64 Bezanilla F. The voltage-sensor structure in a voltage-gated channel. *Trends Biochem Sci* 2005; 30: 166–168.

65 Long SB, Campbell EB, MacKinnon R. Voltage sensor of Kv1.2: structural basis of electromechanical coupling. *Science* 2005; 309: 903–908.

66 Cuello LG, Cortes DM, Perozo E. Molecular architecture of the KvAP voltage-dependent K$^+$ channel in a lipid bilayer. *Science* 2004; 306: 491–495.

67 Perozo E, Cortes DM, Cuello LG. Three-dimensional architecture and gating mechanism of a K$^+$ channel studied by EPR spectroscopy. *Nat Struct Biol* 1998; 5: 459–469.

68 Gulbis JM, Zhou M, Mann S *et al.* Structure of the cytoplasmic beta subunit-T1 assembly of voltage-dependent K$^+$ channels. *Science* 2000; 289: 123–127.

69 Perozo E, Cortes DM, Cuello LG. Structural rearrangements underlying K$^+$-channel activation gating. *Science* 1999; 285: 73–78.

70 Doyle DA, Cabral JM, Pfuetzner RA *et al.* The structure of the potassium channel: molecular basis of K$^+$ conduction and selectivity. *Science* 1998; 280: 69–77.

71 Long SB, Campbell EB, MacKinnon R. Crystal structure of a mammalian voltage-dependent Shaker family K$^+$ channel. *Science* 2005; 309: 897–903.

72 Roy N, Kahlem P, Dausse E *et al.* Exclusion of HRAS from long QT locus. *Nat Genet* 1994; 8: 113–114.

73 Curran M, Atkinson D, Timothy K *et al.* Locus heterogeneity of autosomal dominant long QT syndrome. *J Clin Invest* 1993; 92: 799–803.

74 Benhorin J, Kalman YM, Medina A *et al.* Evidence of genetic heterogeneity in the long QT syndrome. *Science* 1993; 260: 1960–1962.

75 Sanguinetti MC, Jiang C, Curran ME *et al.* A mechanistic link between an inherited and an acquired cardiac arrhythmia: HERG encodes the IKr potassium channel. *Cell* 1995; 81: 299–307.

76 Trudeau MC, Warmke JW, Ganetzky B *et al.* HERG, a human inward rectifier in the voltage-gated potassium channel family. *Science* 1995; 269: 92–95.

77 Huang LQ, Bitner-Glindzicz M, Tranebjaerg L *et al.* A spectrum of functional effects for disease causing mutations in the Jervell and Lange-Nielsen syndrome. *Cardiovasc Res* 2001; 51: 670–680.

78 Li RA, Miake J, Hoppe UC *et al.* Functional consequences of the arrhythmogenic G306R KvLQT1 K$^+$ channel mutant probed by viral gene transfer in cardiomyocytes. *J Physiol (Lond)* 2001; 533: 127–133.

79 Barhanin J, Lesage F, Guillemare E *et al.* K(V)LQT1 and IsK (minK) proteins associate to form the I(Ks) cardiac potassium current. *Nature* 1996; 384: 78–80.

80 Sanguinetti MC, Curran ME, Zou A *et al.* Coassembly of K(V)LQT1 and minK (IsK) proteins to form cardiac I(Ks) potassium channel. *Nature* 1996; 384: 80–83.

81 Splawski I, Tristani-Firouzi M, Lehmann MH *et al.* Mutations in the hminK gene cause long QT syndrome and suppress IKs function. *Nat Genet* 1997; 17: 338–340.

82 Piccini M, Vitelli F, Seri M *et al.* KCNE1-like gene is deleted in AMME contiguous gene syndrome: identification and characterization of the human and mouse homologs. *Genomics* 1999; 60: 251–257.

83 Lewis A, McCrossan ZA, Abbott GW. MinK, MiRP1 and MiRP2 diversify Kv3.1 and Kv3.2 potassium channel gating. *J Biol Chem* 2003; 279: 7884–7892.

84 Decher N, Bundis F, Vajna R *et al.* KCNE2 modulates current amplitudes and activation kinetics of HCN4: influence of KCNE family members on HCN4 currents. *Pflugers Arch* 2003; 446: 633–640.

85 Mazhari R, Greenstein JL, Winslow RL *et al.* Molecular interactions between two long-QT syndrome gene products, HERG and KCNE2, rationalized by *in vitro* and *in silico* analysis. *Circ Res* 2001; 89: 33–38.

86 Deschenes I, Tomaselli GF. Modulation of Kv4.3 current by accessory subunits. *FEBS Lett* 2002; 528: 183–188.

87 Zhang M, Jiang M, Tseng GN. minK-related peptide 1 associates with Kv4.2 and modulates its gating function: potential role as beta subunit of cardiac transient outward channel? *Circ Res* 2001; 88: 1012–1019.

88 Yu H, Wu J, Potapova I *et al.* MinK-related peptide 1: A beta subunit for the HCN ion channel subunit family enhances expression and speeds activation. *Circ Res* 2001; 88: E84–E87.

89 Abbott GW, Goldstein SA. Disease-associated mutations in KCNE potassium channel subunits (MiRPs) reveal promiscuous disruption of multiple currents and conservation of mechanism. *FASEB J* 2002; 16: 390–400.

90 Weerapura M, Nattel S, Chartier D *et al.* A comparison of currents carried by HERG, with and without coexpression of MiRP1, and the native rapid delayed rectifier current. Is MiRP1 the missing link? *J Physiol* 2002; 540: 15–27.

91 Sanguinetti MC, Curran ME, Spector PS *et al.* Spectrum of HERG K$^+$-channel dysfunction in an inherited cardiac arrhythmia. *Proc Natl Acad Sci USA* 1996; 93: 2208–2212.

92 Shalaby FY, Levesque PC, Yang WP *et al.* Dominant-negative KvLQT1 mutations underlie the LQT1 form of long QT syndrome. *Circulation* 1997; 96: 1733–1736.

93 Lees-Miller JP, Duan Y, Teng GQ *et al.* Novel gain-of-function mechanism in K(+) channel-related long-QT syndrome: altered gating and selectivity in the HERG1 N629D mutant. *Circ Res* 2000; 86: 507–513.

94 Nakajima T, Kurabayashi M, Ohyama Y *et al.* Characterization of S818L mutation in HERG C-terminus in LQT2: Modification of activation–deactivation gating properties. *FEBS Letters* 2000; 481: 197–203.

95 Chen J, Zou A, Splawski I *et al.* Long QT syndrome-associated mutations in the Per-Arnt-Sim (PAS) domain of HERG potassium channels accelerate channel deactivation. *J Biol Chem* 1999; 274: 10113–10118.

96 Sanguinetti MC. Dysfunction of delayed rectifier potassium channels in an inherited cardiac arrhythmi. *Ann N Y Acad Sci* 1999; 868: 406–413.

97 Delisle BP, Anson BD, Rajamani S *et al.* Biology of cardiac arrhythmias: Ion channel protein trafficking. *Circ Res* 2004; 94: 1418–1428.

98 January CT, Gong QM, Zhou ZF. Long QT syndrome: Cellular basis and arrhythmia mechanism in LQT2. *J Cardiovasc Electrophysiol* 2000; 11: 1413–1418.

99 Kupershmidt S, Yang T, Chanthaphaychith S *et al.* Defective human ether-a-go-go related gene trafficking linked to an endoplasmic reticulum retention signal in the carboxy terminus. *J Biol Chem* 2002; 277: 27442–27448.

100 Kanki H, Kupershmidt S, Yang T *et al.* A structural requirement for processing the cardiac K$^+$ channel KCNQ1. *J Biol Chem* 2004; 279: 33976–33983.

101 George AL Jr, Varkony TA, Drabkin HA *et al.* Assignment of the human heart tetrodotoxin-resistant voltage-gated Na$^+$ channel alpha-subunit gene (SCN5A) to band 3p21. *Cytogenet Cell Genet* 1995; 68: 67–70.

102 Gellens ME, George AL Jr, Chen L *et al.* Primary structure and functional expression of the human cardiac tetrodotoxin-insensitive voltage-dependent sodium channel. *Proc Natl Acad Sci USA* 1992; 89: 554–558.

103 Bennett PB, Yazawa K, Makita N *et al.* Molecular mechanism for an inherited cardiac arrhythmia. *Nature* 1995; 376: 683–685.

104 Dumaine R, Wang Q, Keating MT *et al.* Multiple mechanisms of Na$^+$ channel-linked long-QT syndrome. *Circ Res* 1996; 78: 916–924.

105 Makita N, Shirai N, Nagashima M *et al.* A *de novo* missense mutation of human cardiac Na$^+$ channel exhibiting novel molecular mechanisms of long QT syndrome. *FEBS Lett* 1998; 423: 5–9.

106 Kambouris NG, Nuss HB, Johns DC *et al.* A revised view of cardiac sodium channel "blockade" in the long-QT syndrome. *J Clin Invest* 2000; 105: 1133–1140.

107 Veldkamp MW, Viswanathan PC, Bezzina C *et al.* Two distinct congenital arrhythmias evoked by a multidysfunctional Na⁺ channel. *Circ Res* 2000; 86: 91–97.

108 Zhang L, Benson DW, Tristani-Firouzi M *et al.* Electrocardiographic features in Andersen–Tawil syndrome patients with KCNJ2 mutations: Characteristic T-U-wave patterns predict the KCNJ2 genotype. *Circulation* 2005; 111: 2720–2726.

109 Schott JJ, Charpentier F, Peltier S *et al.* Mapping of a gene for long QT syndrome to chromosome 4q25–27. *Am J Hum Genet* 1995; 57: 1114–1122.

110 Zareba W, Moss AJ, Schwartz PJ *et al.* Influence of genotype on the clinical course of the long-QT syndrome. International Long-QT Syndrome Registry Research Group. *N Engl J Med* 1998; 339: 960–965.

111 Mohler PJ, Schott JJ, Gramolini AO *et al.* Ankyrin-B mutation causes type 4 long-QT cardiac arrhythmia and sudden cardiac death. *Nature* 2003; 421: 634–639.

112 Mohler PJ, Splawski I, Napolitano C *et al.* A cardiac arrhythmia syndrome caused by loss of ankyrin-B function. *Proc Natl Acad Sci USA* 2004; 101: 9137–9142.

113 Mohler PJ, Rivolta I, Napolitano C *et al.* Nav1.5 E1053K mutation causing Brugada syndrome blocks binding to ankyrin-G and expression of Nav1.5 on the surface of cardiomyocytes. *Proc Natl Acad Sci USA* 2004; 101: 17533–17538.

114 Tuvia S, Buhusi M, Davis L *et al.* Ankyrin-B is required for intracellular sorting of structurally diverse Ca²⁺ homeostasis proteins. *J Cell Biol* 1999; 147: 995–1008.

115 Baroudi G, Napolitano C, Priori SG *et al.* Loss of function associated with novel mutations of the SCN5A gene in patients with Brugada syndrome. *Can J Cardiol* 2004; 20: 425–430.

116 Mok NS, Priori SG, Napolitano C *et al.* A newly characterized SCN5A mutation underlying Brugada syndrome unmasked by hyperthermia. *J Cardiovasc Electrophysiol* 2003; 14: 407–411.

117 Ganelin R, Marks JF, Usher P *et al.* J Periodic paralysis with cardiac arrhythmia. *J Pediatr* 1963; 62: 371–385.

118 Andersen ED, Krasilnikoff PA, Overvad H. Intermittent muscular weakness, extrasystoles, and multiple developmental anomalies. A new syndrome? *Acta Paediatr Scand* 1971; 60: 559–564.

119 Tawil R, Ptacek LJ, Pavlakis SG *et al.* Andersen's syndrome: potassium-sensitive periodic paralysis, ventricular ectopy, and dysmorphic features. *Ann Neurol* 1994; 35: 326–330.

120 Sansone V, Griggs RC, Meola G *et al.* Andersen's syndrome: a distinct periodic paralysis. *Ann Neurol* 1997; 42: 305–312.

121 Kimbrough J, Moss AJ, Zareba W *et al.* Clinical implications for affected parents and siblings of probands with long-QT syndrome. *Circulation* 2001; 104: 557–562.

122 Canun S, Perez N, Beirana LG. Andersen syndrome autosomal dominant in three generations. *Am J Med Genet* 1999; 85: 147–156.

123 Plaster NM, Tawil R, Tristani-Firouzi M *et al.* Mutations in Kir2.1 Cause the developmental and episodic electrical phenotypes of Andersen's syndrome. *Cell* 2001; 105: 511–519.

124 Kubo Y, Baldwin TJ, Jan YN *et al.* Primary structure and functional expression of a mouse inward rectifier potassium channel. *Nature* 1993; 362: 127–133.

125 Stubbs WA. Bidirectional ventricular tachycardia in familial hypokalaemic periodic paralysis. *Proc R Soc Med* 1976; 69: 223–224.

126 Fukuda K, Ogawa S, Yokozuka H *et al.* Long-standing bidirectional tachycardia in a patient with hypokalemic periodic paralysis. *J Electrocardiol* 1988; 21: 71–75.

127 Grimm W, Ritter M, Alter P *et al.* Bidirectional ventricular tachycardia due to digitalis intoxication. *Z Kardiol* 2005; 94: 79–80.

128 Reichenbach H, Meister EM, Theile H. [The heart-hand syndrome. A new variant of disorders of heart conduction and syndactylia including osseous changes in hands and feet]. *Kinderarztl Prax* 1992; 60: 54–56.

129 Marks ML, Trippel DL, Keating MT. Long QT syndrome associated with syndactyly identified in females. *Am J Cardiol* 1995; 76: 744–745.

130 Antzelevitch C. Transmural dispersion of repolarization and the T wave. *Cardiovasc Res* 2001; 50: 426–431.

131 Belardinelli L, Antzelevitch C, Vos MA. Assessing predictors of drug-induced torsade de pointes. *Trends Pharmacol Sci* 2003; 24: 619–625.

132 Antzelevitch C, Sicouri S, Litovsky SH *et al.* Heterogeneity within the ventricular wall. Electrophysiology and pharmacology of epicardial, endocardial, and M cells. *Circ Res* 1991; 69: 1427–1449.

133 Yan GX, Antzelevitch C. Cellular basis for the normal T wave and the electrocardiographic manifestations of the long-QT syndrome. *Circulation* 1998; 98: 1928–1936.

134 Liu DW, Antzelevitch C. Characteristics of the delayed rectifier current (IKr and IKs) in canine ventricular epicardial, midmyocardial, and endocardial myocytes. A weaker IKs contributes to the longer action potential of the M cell. *Circ Res* 1995; 76: 351–365.

135 Taggart P, Sutton PM, Opthof T *et al.* Transmural repolarisation in the left ventricle in humans during normoxia and ischaemia. *Cardiovasc Res* 2001; 50: 454–462.

136 Allessie MA, Bonke FI, Schopman FJ. Circus movement in rabbit atrial muscle as a mechanism of tachycardia. II. The role of nonuniform recovery of excitability in the

occurrence of unidirectional block, as studied with multiple microelectrodes. *Circ Res* 1976; 39: 168–177.

137　Kuo CS, Munakata K, Reddy CP *et al.* Characteristics and possible mechanism of ventricular arrhythmia dependent on the dispersion of action potential durations. *Circulation* 1983; 67: 1356–1367.

138　Misier AR, Opthof T, van Hemel NM *et al.* Dispersion of "refractoriness" in noninfarcted myocardium of patients with ventricular tachycardia or ventricular fibrillation after myocardial infarction. *Circulation* 1995; 91: 2566–2572.

139　Zabel M, Portnoy S, Franz MR. Electrocardiographic indexes of dispersion of ventricular repolarization: an isolated heart validation study. *J Am Coll Cardiol* 1995; 25: 746–752.

140　Zabel M, Franz MR, Siedow A *et al.* QT dispersion as a marker of risk in patients awaiting heart transplantation? *J Am Coll Cardiol* 1998; 31: 1442–1443.

141　Priori SG, Napolitano C, Diehl L *et al.* Dispersion of the QT interval. A marker of therapeutic efficacy in the idiopathic long QT syndrome. *Circulation* 1994; 89: 1681–1689.

142　Kay GN, Plumb VJ, Arciniegas JG *et al.* Torsade de pointes: the long-short initiating sequence and other clinical features: observations in 32 patients. *J Am Coll Cardiol* 1983; 2: 806–817.

143　Roden DM, Woosley RL, Primm RK. Incidence and clinical features of the quinidine-associated long QT syndrome: implications for patient care. *Am Heart J* 1986; 111: 1088–1093.

144　Brugada P, Wellens HJ. Early afterdepolarizations: role in conduction block, "prolonged repolarization-dependent reexcitation," and tachyarrhythmias in the human heart. *Pacing Clin Electrophysiol* 1985; 8: 889–896.

145　Roden DM. Risks and benefits of antiarrhythmic therapy. *N Engl J Med* 1994; 331: 785–791.

146　Cardiac Arrhythmia Suppression Trial (CAST) Investigators. Preliminary report: effect of encainide and flecainide on mortality in a randomized trial of arrhythmia suppression after myocardial infarction. *N Engl J Med* 1989; 321: 406–412.

147　Waldo AL, Camm AJ, deRuyter H *et al.* Survival with oral d-sotalol in patients with left ventricular dysfunction after myocardial infarction: Rationale, design, and methods (the SWORD trial). *Am J Cardiol* 1995; 75: 1023–1027.

148　Tan LB. SWORD trial of d-sotalol. *Lancet* 1996; 348: 827–828.

149　Noble D. Modeling the heart: from genes to cells to the whole organ. *Science* 2002; 295: 1678–1682.

150　Heartburn drug Propulsid pulled from US market. *Washington Post* Friday, March 24, 2000; 2005, A11.

151　Defective Drugs. adrugrecall.com. 2005.

152　Lazzara R. Antiarrhythmic drugs and torsade de pointes. *Eur Heart J* 1993; 14 (Supplement H): 88–92.

153　Zhou Z, Vorperian VR, Gong Q *et al.* Block of HERG potassium channels by the antihistamine astemizole and its metabolites desmethylastemizole and norastemizole. *J Cardiovasc Electrophysiol* 1999; 10: 836–843.

154　Schulze-Bahr E, Haverkamp W, Eckardt L *et al.* Genetic aspects in acquired long QT syndrome: a piece in the puzzle. *Eur Heart J Suppl* 2001; 3: K48–K52.

155　Yang P, Kanki H, Drolet B *et al.* Allelic variants in long-QT disease genes in patients with drug-associated torsades de pointes. *Circulation* 2002; 105: 1943–1948.

156　Nattel S, Khairy P, Schram G. Arrhythmogenic ionic remodeling. Adaptive responses with maladaptive consequences. *Trends Cardiovasc Med* 2001; 11: 295–301.

157　Swynghedauw B, Baillard C, Milliez P. The long QT interval is not only inherited but is also linked to cardiac hypertrophy. *J Mol Med* 2003; 81: 336–345.

158　Kawasaki T, Azuma A, Kuribayashi T *et al.* Determinant of QT dispersion in patients with hypertrophic cardiomyopathy. *Pacing Clin Electrophysiol* 2003; 26: 819–826.

159　Ramakers C, Vos MA, Doevendans PA *et al.* Coordinated down-regulation of KCNQ1 and KCNE1 expression contributes to reduction of I(Ks) in canine hypertrophied hearts. *Cardiovasc Res* 2003; 57: 486–496.

160　Dong D, Duan Y, Guo J *et al.* Overexpression of calcineurin in mouse causes sudden cardiac death associated with decreased density of K^+ channels. *Cardiovasc Res* 2003; 57: 320–332.

161　Ji S, Cesario D, Valderrabano M *et al.* The molecular basis of cardiac arrhythmias in patients with cardiomyopathy. *Curr Heart Fail Rep* 2004; 1: 98–103.

162　McNair WP, Ku L, Taylor MR *et al.* SCN5A mutation associated with dilated cardiomyopathy, conduction disorder, and arrhythmia. *Circulation* 2004; 110: 2163–2167.

163　Antzelevitch C. Molecular genetics of arrhythmias and cardiovascular conditions associated with arrhythmias. *J Cardiovasc Electrophysiol* 2003; 14: 1259–1272.

164　Boccalandro F, Velasco A, Thomas C *et al.* Relations among heart failure severity, left ventricular loading conditions, and repolarization length in advanced heart failure secondary to ischemic or idiopathic dilated cardiomyopathy. *Am J Cardiol* 2003; 92: 544–547.

165　Carlsson L, Almgren O, Duker G. QTU-prolongation and torsades de pointes induced by putative class III antiarrhythmic agents in the rabbit: etiology and interventions. *J Cardiovasc Pharmacol* 1990; 16: 276–285.

166　Hondeghem LM, Carlsson L, Duker G. Instability and triangulation of the action potential predict serious proarrhythmia, but action potential duration prolongation is antiarrhythmic. *Circulation* 2001; 103: 2004–2013.

167 Hondeghem LM, Dujardin K, De Clerck F. Phase 2 prolongation, in the absence of instability and triangulation, antagonizes class III proarrhythmia. *Cardiovasc Res* 2001; 50: 345–353.

168 van Opstal JM, Verduyn SC, Leunissen HDM *et al.* Electrophysiological parameters indicative of sudden cardiac death in the dog with chronic complete AV-block. *Cardiovasc Res* 2001; 50: 354–361.

169 Antzelevitch C, Belardinelli L, Zygmunt AC *et al.* Electrophysiological effects of ranolazine, a novel anti-anginal agent with antiarrhythmic properties. *Circulation* 2004; 110: 904–910.

170 Milberg P, Eckardt L, Bruns HJ *et al.* Divergent proarrhythmic potential of macrolide antibiotics despite similar QT prolongation: Fast phase 3 repolarization prevents early afterdepolarizations and torsade de pointes. *J Pharmacol Exp Ther* 2002; 303: 218–225.

171 Thomsen MB, Verduyn SC, Stengl M *et al.* Increased short-term variability of repolarization predicts d-sotalol-induced torsades de pointes in dogs. *Circulation* 2004; 110: 2453–2459.

172 Thomsen MB, Volders PGA, Stengl M *et al.* Electrophysiological safety of sertindole in dogs with normal and remodeled hearts. *J Pharmacol Exp Ther* 2003; 307: 776–784.

173 Morganroth J. A definitive or thorough phase 1 QT ECG trial as a requirement for drug safety assessment. *J Electrocardiol* 2004; 37: 25–29.

174 Jost N, Virag L, Bitay M *et al.* Restricting excessive cardiac action potential and QT prolongation. A vital role for IKs in human ventricular muscle. *Circulation* 2005; 112: 1392–1399.

175 Silva J, Rudy Y. Subunit interaction determines IKs participation in cardiac repolarization and repolarization reserve. *Circulation* 2005; 112: 1384–1391.

176 Belardinelli L, Antzelevitch C, Vos MA. Assessing predictors of drug-induced torsade de pointes. *Trends Pharmacol Sci* 2003; 24: 619–625.

177 Weissenburger J, Davy JM, Chezalviel F *et al.* Arrhythmogenic activities of antiarrhythmic drugs in conscious hypokalemic dogs with atrioventricular block: comparison between quinidine, lidocaine, flecainide, propranolol and sotalol. *J Pharmacol Exp Ther* 1991; 259: 871–883.

178 Vos MA, de Groot SHM, Verduyn SC *et al.* Enhanced susceptibility for acquired torsade de pointes arrhythmias in the dog with chronic, complete AV block is related to cardiac hypertrophy and electrical remodeling. *Circulation* 1998; 98: 1125–1135.

179 Schreiner KD, Kelemen K, Zehelein J *et al.* Biventricular hypertrophy in dogs with chronic AV block: effects of cyclosporin A on morphology and electrophysiology. *Am J Physiol Heart Circ Physiol* 2004; 287: H2891–H2898.

180 Chiba K, Sugiyama A, Hagiwara T *et al. In vivo* experimental approach for the risk assessment of fluoroquinolone antibacterial agents-induced long QT syndrome. *Eur J Pharmacol* 2004; 486: 189–200.

181 Vos MA. Preclinical evaluation of antiarrhythmic drugs: new drugs should be safe to be successful. *J Cardiovasc Electrophysiol* 2001; 12: 1034–1036.

182 Vos MA, de Groot SH, Verduyn SC *et al.* Enhanced susceptibility for acquired torsade de pointes arrhythmias in the dog with chronic, complete AV block is related to cardiac hypertrophy and electrical remodeling. *Circulation* 1998; 98: 1125–1135.

183 de Groot SH, Schoenmakers M, Molenschot MM *et al.* Contractile adaptations preserving cardiac output predispose the hypertrophied canine heart to delayed afterdepolarization-dependent ventricular arrhythmias. *Circulation* 2000; 102: 2145–2151.

184 Verduyn SC, Vos MA, van der ZJ *et al.* Further observations to elucidate the role of interventricular dispersion of repolarization and early afterdepolarizations in the genesis of acquired torsade de pointes arrhythmias: a comparison between almokalant and d-sotalol using the dog as its own control. *J Am Coll Cardiol* 1997; 30: 1575–1584.

185 Ramakers C, Vos MA, Doevendans PA *et al.* Coordinated down-regulation of KCNQ1 and KCNE1 expression contributes to reduction of I(Ks) in canine hypertrophied hearts. *Cardiovasc Res* 2003; 57: 486–496.

186 Carlsson L, Abrahamsson C, Andersson B *et al.* Proarrhythmic effects of the class III agent almokalant: importance of infusion rate, QT dispersion, and early afterdepolarisations. *Cardiovasc Res* 1993; 27: 2186–2193.

187 Yan GX, Shimizu W, Antzelevitch C. Characteristics and distribution of M cells in arterially perfused canine left ventricular wedge preparations. *Circulation* 1998; 98: 1921–1927.

188 Yan GX, Wu Y, Liu T *et al.* Phase 2 early afterdepolarization as a trigger of polymorphic ventricular tachycardia in acquired long-QT syndrome: direct evidence from intracellular recordings in the intact left ventricular wall. *Circulation* 2001; 103: 2851–2856.

189 Shimizu W, Antzelevitch C. Differential effects of beta-adrenergic agonists and antagonists in LQT1, LQT2 and LQT3 models of the long QT syndrome. *J Am Coll Cardiol* 2000; 35: 778–786.

190 Shimizu W, Antzelevitch C. Sodium channel block with mexiletine is effective in reducing dispersion of repolarization and preventing torsade des pointes in LQT2 and LQT3 models of the long-QT syndrome. *Circulation* 1997; 96: 2038–2047.

191 Shimizu W, Antzelevitch C. Cellular basis for the ECG features of the LQT1 form of the long-QT syndrome:

effects of beta-adrenergic agonists and antagonists and sodium channel blockers on transmural dispersion of repolarization and torsade de pointes. *Circulation* 1998; 98: 2314–2322.

192 Nerbonne JM. Studying cardiac arrhythmias in the mouse: A reasonable model for probing mechanisms? *Trends Cardiovasc Med* 2004; 14: 83–93.

193 Berul CI. Electrophysiological phenotyping in genetically engineered mice. *Physiol Genomics* 2003; 13: 207–216.

194 Gehrmann J, Berul CI. Cardiac electrophysiology in genetically engineered mice. *J Cardiovasc Electrophysiol* 2000; 11: 354–368.

195 Berul CI, Aronovitz MJ, Wang PJ *et al. In vivo* cardiac electrophysiology studies in the mouse. *Circulation* 1996; 94: 2641–2648.

196 London B. Cardiac arrhythmias: from (transgenic) mice to men. *J Cardiovasc Electrophysiol* 2001; 12: 1089–1091.

197 Papadatos GA, Wallerstein PM, Head CE *et al.* Slowed conduction and ventricular tachycardia after targeted disruption of the cardiac sodium channel gene Scn5a. *Proc Natl Acad Sci USA* 2002; 99: 6210–6215.

198 London B. Cardiac arrhythmias: from (transgenic) mice to men. *J Cardiovasc Electrophysiol* 2001; 12: 1089–1091.

199 Mitchell GF, Jeron A, Koren G. Measurement of heart rate and Q-T interval in the conscious mouse. *Am J Physiol* 1998; 274: H747–H751.

200 Wang L, Duff HJ. Developmental changes in transient outward current in mouse ventricle. *Circ Res* 1997; 81: 120–127.

201 Wang L, Feng ZP, Kondo CS *et al.* Developmental changes in the delayed rectifier K$^+$ channels in mouse heart. *Circ Res* 1996; 79: 79–85.

202 Kupershmidt S, Yang T, Anderson ME *et al.* Replacement by homologous recombination of the minK gene with lacZ reveals restriction of minK expression to the mouse cardiac conduction system. *Circ Res* 1999; 84: 146–152.

203 Temple J, Frias P, Rottman J *et al.* Atrial fibrillation in KCNE1-null mice. *Circ Res* 2005; 97: 62–69.

204 Cerrone M, Colombi B, Santoro M *et al.* Bidirectional ventricular tachycardia and fibrillation elicited in a knock-in mouse model carrier of a mutation in the cardiac ryanodine receptor. *Circ Res* 2005; 96: e77–e82.

205 Tanaka M, Berul CI, Ishii M *et al.* A mouse model of congenital heart disease: cardiac arrhythmias and atrial septal defect caused by haploinsufficiency of the cardiac transcription factor Csx/Nkx2.5. *Cold Spring Harb Symp Quant Biol* 2002; 67: 317–325.

206 Xiao HD, Fuchs S, Campbell DJ *et al.* Mice with cardiac-restricted angiotensin-converting enzyme (ACE) have atrial enlargement, cardiac arrhythmia, and sudden death. *Am J Pathol* 2004; 165: 1019–1032.

207 Gutstein DE, Morley GE, Vaidya D *et al.* Heterogeneous expression of Gap junction channels in the heart leads to conduction defects and ventricular dysfunction. *Circulation* 2001; 104: 1194–1199.

208 Gutstein DE, Morley GE, Tamaddon H *et al.* Conduction slowing and sudden arrhythmic death in mice with cardiac-restricted inactivation of connexin 43. *Circ Res* 2001; 88: 333–339.

209 Petersen CI, McFarland TR, Stepanovic SZ *et al. In vivo* identification of genes that modify ether-a-go-go-related gene activity in *Caenorhabditis elegans* may also affect human cardiac arrhythmia. *Proc Natl Acad Sci USA* 2004; 101: 11773–11778.

210 Luo CH, Rudy Y. A dynamic model of the cardiac ventricular action potential. II. Afterdepolarizations, triggered activity, and potentiation. *Circ Res* 1994; 74: 1097–1113.

211 Luo CH, Rudy Y. A dynamic model of the cardiac ventricular action potential. I. Simulations of ionic currents and concentration changes. *Circ Res* 1994; 74: 1071–1096.

212 Luo CH, Rudy Y. A model of the ventricular cardiac action potential. Depolarization, repolarization, and their interaction. *Circ Res* 1991; 68: 1501–1526.

213 Henry H, Rappel WJ. The role of M cells and the long QT syndrome in cardiac arrhythmias: simulation studies of reentrant excitations using a detailed electrophysiological model. *Chaos* 2004; 14: 172–182.

214 Roden DM, Lazzara R, Rosen M *et al.* Multiple mechanisms in the long-QT syndrome. Current knowledge, gaps, and future directions. The SADS Foundation Task Force on LQTS. *Circulation* 1996; 94: 1996–2012.

215 Noda T, Takaki H, Kurita T *et al.* Gene-specific response of dynamic ventricular repolarization to sympathetic stimulation in LQT1, LQT2 and LQT3 forms of congenital long QT syndrome. *Eur Heart J* 2002; 23: 975–983.

216 Nosworthy A. Images in clinical medicine. Hypokalemia. *N Engl J Med* 2003; 349: 2116.

217 Yang T, Roden DM. Extracellular potassium modulation of drug block of IKr. Implications for torsade de pointes and reverse use-dependence. *Circulation* 1996; 93: 407–411.

218 Yang T, Snyders DJ, Roden DM. Rapid inactivation determines the rectification and [K$^+$]o dependence of the rapid component of the delayed rectifier K$^+$ current in cardiac cells. *Circ Res* 1997; 80: 782–789.

219 Numaguchi H, Johnson JP Jr, Petersen CI *et al.* A sensitive mechanism for cation modulation of potassium current. *Nat Neurosci* 2000; 3: 429–430.

220 Tan HL, Alings M, Van Olden RW *et al.* Long-term (subacute) potassium treatment in congenital HERG-related long QT syndrome (LQTS2). *J Cardiovasc Electrophysiol* 1999; 10: 229–233.

221 Etheridge SP, Compton SJ, Tristani-Firouzi M *et al.* A new oral therapy for long QT syndrome: long-term oral potassium improves repolarization in patients with HERG mutations. *J Am Coll Cardiol* 2003; 42: 1777–1782.

222 Pham TV, Rosen MR. Sex, hormones, and repolarization. *Cardiovasc Res* 2002; 53: 740–751.

223 Locati EH, Zareba W, Moss AJ *et al.* Age- and sex-related differences in clinical manifestations in patients with congenital long-QT syndrome: findings from the International LQTS Registry. *Circulation* 1998; 97: 2237–2244.

224 Lehmann MH, Timothy KW, Frankovich D *et al.* Age-gender influence on the rate-corrected QT interval and the QT-heart rate relation in families with genotypically characterized long QT syndrome. *J Am Coll Cardiol* 1997; 29: 93–99.

225 Lehmann MH, Hardy S, Archibald D *et al.* Sex difference in risk of torsade de pointes with d,l-sotalol. *Circulation* 1996; 94: 2535–2541.

226 Rodriguez I, Kilborn MJ, Liu XK *et al.* Drug-induced QT prolongation in women during the menstrual cycle. *JAMA* 2001; 285: 1322–1326.

227 Schwartz PJ, Moss AJ, Vincent GM *et al.* Diagnostic criteria for the long QT syndrome. An update. *Circulation* 1993; 88: 782–784.

228 Takenaka K, Ai T, Shimizu W *et al.* Exercise stress test amplifies genotype–phenotype correlation in the LQT1 and LQT2 forms of the long-QT syndrome. *Circulation* 2003; 107: 838–44.

229 Shimizu W, Noda T, Takaki H *et al.* Epinephrine unmasks latent mutation carriers with LQT1 form of congenital long-QT syndrome. *J Am Coll Cardiol* 2003; 41: 633–642.

230 Shimizu W, Antzelevitch C. Effects of a K(+) channel opener to reduce transmural dispersion of repolarization and prevent torsade de pointes in LQT1, LQT2, and LQT3 models of the long-QT syndrome. *Circulation* 2000; 102: 706–712.

231 Aiba T, Shimizu W, Inagaki M *et al.* Cellular and ionic mechanism for drug-induced long QT syndrome and effectiveness of verapamil. *J Am Coll Cardiol* 2005; 45: 300–307.

232 Schwartz PJ, Priori SG, Cerrone M *et al.* Left cardiac sympathetic denervation in the management of high-risk patients affected by the long-QT syndrome. *Circulation* 2004; 109: 1826–1833.

233 Priori SG, Napolitano C, Schwartz PJ *et al.* Association of long QT syndrome loci and cardiac events among patients treated with beta-blockers. *JAMA* 2004; 292: 1341–1344.

234 Etheridge SP, Compton SJ, Tristani-Firouzi M *et al.* A new oral therapy for long QT syndrome: long-term oral potassium improves repolarization in patients with HERG mutations. *J Am Coll Cardiol* 2003; 42: 1777–1782.

235 Shimizu W. The long QT syndrome: Therapeutic implications of a genetic diagnosis. *Cardiovasc Res* 2005; 67: 347–356.

236 Schwartz PJ, Priori SG, Locati EH *et al.* Long QT syndrome patients with mutations of the SCN5A and HERG genes have differential responses to Na$^+$ channel blockade and to increases in heart rate. Implications for gene-specific therapy [see comments]. *Circulation* 1995; 92: 3381–3386.

237 Shimizu W, Antzelevitch C. Sodium channel block with mexiletine is effective in reducing dispersion of repolarization and preventing torsade des pointes in LQT2 and LQT3 models of the long-QT syndrome. *Circulation* 1997; 96: 2038–2047.

238 Wang DW, Yazawa K, Makita N *et al.* Pharmacological targeting of long QT mutant sodium channels. *J Clin Invest* 1997; 99: 1714–1720.

239 Abriel H, Wehrens XH, Benhorin J *et al.* Molecular pharmacology of the sodium channel mutation D1790G linked to the long-QT syndrome. *Circulation* 2000; 102: 921–925.

240 Benhorin J, Taub R, Goldmit M *et al.* Effects of flecainide in patients with new SCN5A mutation: mutation-specific therapy for long-QT syndrome? *Circulation* 2000; 101: 1698–1706.

241 Schwartz PJ, Priori SG, Locati EH *et al.* Long QT syndrome patients with mutations of the SCN5A and HERG genes have differential responses to Na$^+$ channel blockade and to increases in heart rate. Implications for gene-specific therapy. *Circulation* 1995; 92: 3381–3386.

242 Shimizu W. The long QT syndrome: Therapeutic implications of a genetic diagnosis. *Cardiovasc Res* 2005; 67: 347–356.

243 Zareba W, Moss AJ, Daubert JP *et al.* Implantable cardioverter defibrillator in high-risk long QT syndrome patients. *J Cardiovasc Electrophysiol* 2003; 14: 337–341.

244 Welde AA. Is there a role for implantable cardioverter defibrillators in long QT syndrome? *J Cardiovasc Electrophysiol* 2002; 13: S110–S113.

245 Eldar M, Griffin JC, Van Hare GF *et al.* Combined use of beta-adrenergic blocking agents and long-term cardiac pacing for patients with the long QT syndrome. *J Am Coll Cardiol* 1992; 20: 830–837.

246 Woosley RL. ArizonaCERT. University of Arizona Health Sciences Center.

247 Mervis J. Productivity counts—but the definition is key. *Science* 2005; 309: 726.

248 Kaitin KI. Post-approval R&D raises total drug development costs to $897 million. *Tufts Center Study Drug Dev Impact Rep* 2003; 5: 1.

249 Netzer R, Ebneth A, Bischoff U *et al.* Screening lead compounds for QT interval prolongation. *Drug Discov Today* 2001; 6: 78–84.

250 Padrini R, Speranza G, Nollo G *et al.* Adaptation of the QT interval to heart rate changes in isolated perfused guinea pig heart: influence of amiodarone and D-sotalol. *Pharmacol Res* 1997; 35: 409–416.

251 Brown AM. Drugs, hERG and sudden death. *Cell Calcium* 2004; 35: 543–547.

252 Summerton J, Weller D. Morpholino antisense oligomers: design, preparation, and properties. *Antisense Nucleic Acid Drug Dev* 1997; 7: 187–195.

253 Langheinrich U. Zebrafish: a new model on the pharmaceutical catwalk. *Bioessays* 2003; 25: 904–912.

254 Sehnert AJ, Stainier DY. A window to the heart: can zebrafish mutants help us understand heart disease in humans? *Trends Genet* 2002; 18: 491–494.

255 Sehnert AJ, Huq A, Weinstein BM *et al.* Cardiac troponin T is essential in sarcomere assembly and cardiac contractility. *Nat Genet* 2002; 31: 106–110.

256 Chen JN, Fishman MC. Genetics of heart development. *Trends Genet* 2000; 16: 383–388.

257 Langheinrich U, Vacun G, Wagner T. Zebrafish embryos express an orthologue of HERG and are sensitive toward a range of QT-prolonging drugs inducing severe arrhythmia small star, filled. *Toxicol Appl Pharmacol* 2003; 193: 370–382.

258 Milan DJ, Peterson TA, Ruskin JN *et al.* Drugs that induce repolarization abnormalities cause bradycardia in zebrafish. *Circulation* 2003; 107: 1355–1358.

259 Klein R, Gamelin R, Marks JF et al. Periodic paralysis with cardiac arrhythmia. *J Pediatr* 1963; 62: 371–385.

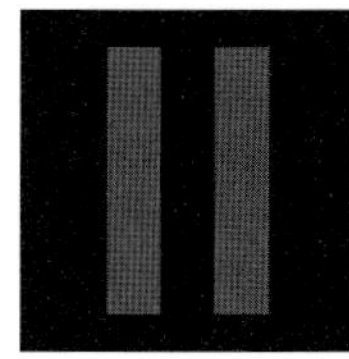

PART II
Cardiovascular polygenic disorders

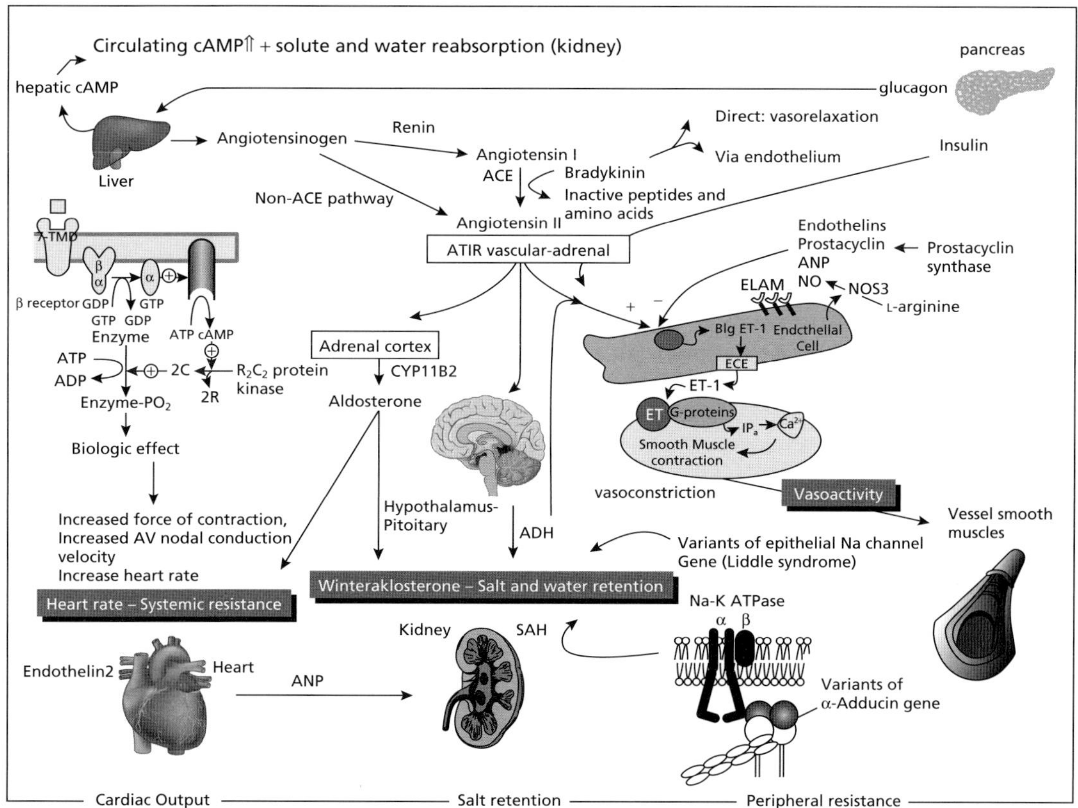

Network of pathways and genes postulated to be associated with blood pressure regulation.
ACE, angiotensin-converting enzyme; ADH, antidiuretlc hormone (vasopressin); ANP, atrial natriuretic peptide; AT1R, angiotensin II type 1 receptor; AV, atrioventricular; ECE, endothelin-converting enzyme ELAM, endothelial leukocyte adhesion molecule 1 (E-selectin); ET-1, endothelin-1; IP3, inositol tris-phosphate; NO, nitric oxide; NOS, nitric oxide synthase; SAH, SA hypertension-associated homolog (rat); 7-TMD, seven-transmembrane domain. Reprinted from Marteau J-B, Zaiou M, Siest G, Visvikis-Siest S. Genetic determinants of blood pressure regulation. *J Hypertens* 2005; 23: 2127–2143 with permission from Lippincott Williams and Wilkins.

CHAPTER 6

Atherosclerosis

Päivi Pajukanta, MD, PhD, *Kiat Tsong Tan,* MD, MRCP, FRCR
& Choong-Chin Liew, PhD

Introduction

For hundreds of years observers have been interested in the unusual lesions of atherosclerosis. Renaissance artist Leonardo da Vinci complained of the "waxy fat" that made some arteries difficult to draw and provided some early descriptions of arteriosclerosis in elderly men. The first complete and accurate description of intimal atheromatous lesions is that of the eighteenth century Italian anatomist Antonio Scarpa who wrote graphically of the "slow, morbid ulcerated, steatomatous, fungus, squamous degeneration of the internal coat of the artery." The word "atheroma" derives from the Greek for gruel or porridge and was first used with reference to human arteries by Swiss physiologist Albrecht von Haller in 1755; the term "atherosclerosis" as we use it today to describe disease of the coronary artery intima was coined in the first years of the twentieth century by Leipzig pathologist, Felix Marchand [1,2].

Data from previous clinical and epidemiologic studies have shown that several risk factors, including age, male sex, family history of myocardial infarction (MI), increased serum total and low density lipoprotein cholesterol (LDL-C), decreased serum high density lipoprotein cholesterol (HDL-C), smoking and diabetes mellitus, predict the risk for atherogenesis [3–11]. More recently, inflammation linked with disadvantageous plasma lipoprotein profile and chronic infections were suggested as risk factors for coronary artery disease (CAD) [12–14]. It has become evident that atherosclerosis is a complex multifactorial phenomenon. Furthermore, risk factors appear to cluster and interact in individuals and families, making it challenging to

determine the level of risk. Evaluation of risk is further hampered by the largely unknown relationship and interactions among the underlying genetic and environmental factors. Currently, atherosclerosis is considered to be a highly complex heterogeneous disease involving the actions of more than 400 genes [15] and an ever increasing number of genetic, environmental and endogenous risk factors continue to be identified as acting singly and in combination to modify gene expression to contribute to or to protect against the development of CAD. Today atherosclerosis, presenting as CAD, stroke and peripheral artery disease, is the most common cause of morbidity and mortality in western and westernizing societies. About half of all people in the USA die from atherosclerosis-related complications, and cardiovascular disease worldwide is expected to increase significantly over the next 20 years [16].

Up until about 20 years ago, the lesions of atherosclerosis were mainly regarded as degenerative by-products of the atherosclerotic process; atherosclerosis itself was considered to be a build up of bland degenerative lipid products, and angina and thrombosis were thought to be the consequence of narrowing of the artery to occlusion as a result of lipid accumulation. However, this view has drastically changed. For the past 20 years researchers have been dissecting out the intricate cellular and molecular signaling and communication pathways, the genetic, molecular and cellular activity that drives atherosclerotic lesion formation. Thanks in large part to this work, atherosclerosis today is regarded as a complex, ongoing inflammatory process. The lesions of atherosclerosis are considered to have important roles in driving the atheromatous process from initial endothelial injury to final plaque disruption and disease manifestations such as stroke and MI [17].

In this chapter we explore first the cellular and tissue changes that initiate the process of lesion formation in atherosclerosis and, second, the molecular and gene level changes as the disease advances from benign to increasingly more dangerous. Studies in the molecular biology of atherosclerosis over the past two decades have provided numerous clues to novel diagnostic, prognostic and therapeutic approaches to atherosclerosis and in the final section we review how molecular biologic insights can be translated into clinical applications for the future.

Endothelial dysfunction and lesion formation: A general view

Theories of the pathogenesis of atherosclerosis have undergone considerable changes in the past few decades. Earlier conceptualizations of atherosclerosis as for the most part a disorder of lipid storage have given way to the "response to injury" hypothesis originally described by Ross and Glomset [18]. In this model, atherosclerosis is considered to be essentially an inflammatory and immune response process, triggered and maintained as a response to ongoing systemic biochemical injury (reviewed in [19,20]). This hypothesis focuses on the cardiovascular system not as a set of passive mechanical structures but at the molecular biologic level as active players in atherosclerotic events. Gimbrone and Topper [21] have well described the blood vessel as a "community of cells." It is by exploring the microcomponents of this community and their inter- and intracellular interactions and signals as they work together to maintain homeostasis in health and become maladaptive and ultimately self-destructive in disease that we can best understand atherosclerosis.

The lesions of atherosclerosis start with vascular endothelial dysfunction. The vascular endothelium is the 700 m^2, single-cell-thick luminal lining of the vascular system. The healthy functioning of the vascular endothelium is the first line of defense against atherosclerosis and this highly interesting tissue has lately received a great deal of research attention [21–23].

The endothelium is most obviously a tissue of structural importance. It is the innermost layer of the artery and acts as a barrier between the blood flowing in the intravascular space and the wall of the artery itself. The endothelium is also a complex and active tissue, even a cardiovascular organ in its own right, with paracrine, endocrine and autocrine functions [21]. The endothelial cells synthesize and release vasoactive substances and have a number of important functional properties. First, healthy endothelial cells regulate the vascular tone of the cardiovascular system; second, they are antithrombotic, inhibiting platelet aggregation and coagulation so that blood circulates through the arterial vessels without clotting; and third, the endothelium is nonadhesive. Endothelial cells are able to sense changes in their microclimate; to signal these changes to other cells and to respond to alterations in order to maintain vascular homeostasis.

It is thus in specific areas of the vascular system where the endothelium is functioning less well that the first changes leading to atherosclerosis become evident. In areas of endothelial dysfunction the endothelial cells lose some protective function, becoming pro- rather than anti-atherogenic. In particular, in areas of altered function the endothelium becomes more permeable to plasma lipoproteins, shows increased monocyte adhesion, altered vasoreactivity and other signs of inflammatory changes begin to occur in the vessel wall.

It is at these areas of endothelial dysfunction that the first lesions associated with atherosclerosis develop, the so-called fatty streaks. Fatty streaks are yellowish inflammatory lesions that develop in the intima within the artery wall. Varying from the size of a pinhead to covering large areas, fatty streaks are to be found in the arteries of very young children [24] and even in premature fetuses [25], and by the age of 10–14 years some 50% of childhood autopsy specimens show evidence of fatty streak lesions [26]. In humans, fatty streaks can be found in the aorta in the first few years of life; the coronary arteries in the teens; and cerebral arteries in third to fourth decades of life [16].

Analysis of characteristic fatty streak lesions finds them to contain inflammatory monocytes and T lymphocyte cells as well as oxidized lipids such as low density lipoprotein (LDL). Intralesional monocytes become macrophages, able to digest large amounts of the surrounding oxidated lipoprotein particles. Such lipid laden macrophages, or foam cells, together with T cells form the bulk of fatty streaks. It is these lipid filled foam cells that account for the gruel- or cereal-like yellowish substance that is observed in atheroma.

The next stage in development of atherosclerosis is the formation of "intermediate or fibrofatty lesions." Similar in appearance to the fatty streaks, intermediate lesions are more complex in composition. Ongoing inflammation stimulates smooth muscle cells to migrate to the area of injury. Contractile smooth muscle cells undergo phenotypic changes, becoming noncontractile and then fibrous.

Neither the fatty streaks nor the more complex intermediate lesions are immediately harmful. Such lesions can in fact be viewed as protective responses to insult, as is wound healing in general in the body. If the injury were a one time or occasional event, then these changes would be benign and reversible. However, as Ross [26] observes, in the atherogenic milieu the biochemical injury tends to be constant and chronic, such as smoking, diabetes or hypercholesterolemia. Such ongoing insult prevents the inflammatory process from ceasing, becoming maladaptive and harmful as the lesions continue to develop.

The next stage of atherogenesis is the formation of arterial plaques. These are well-defined lesions containing a lipid core, collagen, elastic fibres and proteoglycans and covered with cap of fibrous tissue, composed of smooth muscle cells and connective tissue. Patients with early atherosclerotic plaques often do not exhibit symptoms of chronic ischemia as the arterial lumen can be remarkably normal at this stage, because of the phenomenon of positive remodeling. However, the absence of symptoms does not mean that these patients are not at risk of acute atherothrombosis, as even the smallest of plaques is liable to rupture. As the plaque grows, the lesions may obtrude into the artery lumen and it was long thought that it was this gradual and insidious narrowing of the lumen over time that was the causal event in angina and coronary thrombosis. However, it is now believed that although occlusion may have a role in some cases, it is the physical disruption of plaques that is the cause of acute coronary events. That is, the lesions become unstable and either stimulate *in situ* thrombosis or, more rarely, break off to form distal emboli. Either of these events can cause obstruction of the coronary arterial tree, which ultimately results in the manifestation of acute cardiac ischemia.

Steps in lesion formation: Molecular and gene levels

The challenge for the molecular biologist is to identify the genes and gene expression changes that drive the pathologic process of lesion formation from dysfunction of the normal healthy endothelium to fatty intimal deposits, plaque formation, rupture and thrombosis [23].

The initiating events in endothelial dysfunction have long been a subject of investigation. If endothelial dysfunction were solely a consequence of biochemical risk factors such as smoking or high cholesterol or homocysteine levels then we would expect that blood vessels would be uniformly prone to disease: that atheroma would be found throughout the length of the arteries. However, it has long been known that atherosclerotic lesions do not occur uniformly or randomly throughout the vascular system. On the contrary, atheroma is more likely to be found in specific locations in the vasculature: in arterial branches and bifurcations; straight vessels, by contrast, are less likely to develop atherogenic lesions [27].

The most relevant geographical differences between vascular regions where atheromatous lesions occur compared with where they do not occur is that blood flow patterns are significantly altered in arterial branches compared to straight arterial regions. Hemodynamic shear stress forces such as turbulence tends to affect curvatures in the arterial landscape; straight vessels experience uniform, relatively constant laminar shear stress. Thus, shear stress may act as a "local" risk factor, contributing to endothelial dysfunction.

Are there genes whose expression may be altered under conditions of altered shear stresses? One such molecule is the nitric oxide synthase 3 (*NOS3*) gene encoding the enzyme endothelial cell nitric oxide synthase (eNOS). The *NOS3* gene is a key factor linking shear stress and endothelial dysfunction.

The biologically ubiquitous gas nitric oxide was named molecule of the year by *Science* magazine in 1992 [28]. Nitric oxide was identified as the original endothelium derived relaxant factor [29] and is of crucial importance in maintaining healthy endothelium [30]. In vascular tissue, nitric oxide is atheroprotective: it regulates vascular tone and vasomotor function, it counteracts leukocyte adhesion to the endothelium, opposes vascular smooth muscle proliferation and inhibits platelet aggregation. Nitric oxide is thus a major component of defense against vascular injury, inflammation and thrombosis [22].

Nitric oxide is synthesized in endothelial tissue by eNOS encoded by *NOS3*. The *NOS3* gene expression is influenced by biomechanical fluid shear stresses generated by local conditions of blood flow. Cultured human endothelial cells undergo gene expression changes under conditions of altered shear stress. For example, Topper *et al.* [27] showed that steady laminar blood flow, mimicking conditions occurring as blood flows through straight vessels, upregulates NOS3, as well as manganese superoxide dismutase and other atheroprotective genes. By upregulating genes that are anti-oxidant, anti-thrombotic and anti-adhesive, laminar blood flow creates conditions within the artery that are atheroprotective. Turbulent blood flow, by contrast, which might occur at artery branches, does not appear to upregulate *NOS3*. Thus, specific areas within the artery may have decreased local nitric oxide production, leading to endothelial dysfunction at these vulnerable areas.

Areas of altered endothelial function are also characterized by increasing "stickiness" of the endothelial cells to circulating monocytes. Normally, mononuclear leukocytes circulate freely through the blood and do not adhere to the vascular endothelium. In conditions conducive to atherosclerosis such as hypercholesterolemia, however, large numbers of mononuclear cells can be found attaching to the endothelium in the specific areas prone to atheroma [31,32].

Abnormal localized stickiness is a result of endothelial cell activation or overexpression of specific leukocyte adhesion molecules. Leukocyte adhesion molecules are proteins which, when activated on the surface of the endothelial cells, increase adherence of monocytes and T cells to the endothelial surface [33]. Adhesion molecule expression occurs at the specific focal sites that are prone to develop atherosclerosis. Normal endothe-

lium shows little to no such expression of adhesion molecules.

One of the first endothelial adhesion molecules to be identified was vascular cell adhesion molecule 1 (VCAM-1). VCAM-1 is specialized to recruit circulating leukocytes, specifically monocytes and T lymphocytes. Early experiments showed that in response to cholesterol feeding in rabbits, endothelial cells express mononuclear leukocyte selective VCAM-1 in localized areas of the aorta close to developing atheroma [34]. VCAM-1 expression is increased in early stage disease [35] and is also expressed in advanced atherogenic plaques [36]. Mice genetically engineered with defective VCAM-1 expression showed reduced foam cell lesion development [37]. Other adhesion molecules upregulated in atherosclerosis include P selectin, E selectin and intercellular adhesion molecule 1 (ICAM-1) [33].

Once monocytes and T cells attach to activated endothelial cells via adhesion molecules, these cells are able to effect passage through the single cell layer of endothelium, between the endothelial cells and into the coronary vessel intima. Little is known about the process of transmigration. Specific proinflammatory chemokines are expressed in atheroma and have been identified, including monocyte chemoattractant protein 1 (MCP-1) [38] and various T-cell chemoattractants [39].

Endothelial dysfunction is also characterized by increased permeability to lipoproteins [21]. It is now firmly established that cholesterol levels are important contributors to atherogenesis, in particular levels of LDL. In endothelial dysfunction, circulating LDL attaches itself to the wall of the artery in the areas that endothelial cells have become altered because of shear stress. Via transcellular or pericellular mechanisms, LDL transmigrates through the normally impermeable endothelial barrier into the intima.

Although LDL circulating in plasma is nontoxic, once trapped in the subendothelial matrix, LDL seems to become more susceptible to enzymatic induced oxidative changes, becoming oxidized LDL (oxLDL), a proinflammatory substance [20]. For example, oxLDL contains bioactive lysophosphatidyl choline and other phospholipids, which act to upregulate several genes including the adhesion molecules VCAM-1 and ICAM-1 and growth factors such as platelet derived growth factor (PDGF) and heparin epidermal binding growth factor-like protein (HB-EGF) [40,41]. Growth factors are important mediators of smooth muscle cell and fibroblast migration and proliferation. The exact role of oxLDL and the relevance of this modified lipoprotein to atherosclerosis continue to be under investigation.

The combination of monocytes and oxLDL within the arterial intima initiates a number of gene changes driving the activation of monocytes. This transformation involves the activation of molecular scavenger receptors on mononuclear cells. Scavenger receptors are proteins structured to be able to recognize and rapidly to accumulate oxLDL [42]. Several molecules of this class have been identified including scavenger receptors of the SRA series and CD36 (reviewed in [20]). Genetically engineered ApoE deficient mice lacking scavenger receptor expression are less likely than control mice to develop atherosclerosis [20].

Macrophages contribute to host cell defense by acting as phagocytes, recognizing and removing foreign or noxious substances, and it is thought that initial macrophage removal of cytotoxic and inflammatory oxLDL is a protective process [20,43]. Macrophages also act as signaling cells in the inflammatory cascade and release cytokines such as tumor necrosis factor α (TNF-α) [44,45]. Macrophage derived TNF-α has important proinflammatory autocrine and paracrine effects. Cell stimulation by TNF-α leads to the downstream activation of the proinflammatory transcription factor nuclear factor κB (NF-κB) [46]. Activation of NF-κB leads to the upregulation of various adhesion molecules (e.g. VCAM-1, E selectin), the production and release of proinflammatory cytokines (e.g. interleukin-1β [IL-1β]) and the induction of molecules that favor a prothrombotic state (e.g. tissue factor, plasminogen activator inhibitor 1) [47]. Therefore, TNF-α has a net effect of propagating endothelial dysfunction as well as promoting the influx of more inflammatory cells.

TNF-α increases expression of macrophage-produced iNOS. We have earlier spoken of NO as atheroprotective when produced in small amounts by endothelial cells. When produced in high levels by macrophages, inducible NO (iNOS) is antimicrobial. However, antimicrobial NO is also athero-genic. The high levels of NO produced by iNOS damages proteins and DNA [48].

Macrophages can also secrete IL-1β which is another potent proinflammatory cytokine [49]. IL-1β has similar effects to TNF-α as it also induces NF-κB activation [47]. Indeed, the importance of IL-1β in atherogenesis was illustrated by a study that showed that the blockade of IL-1β in an ApoE $-/-$ mouse model of atherosclerosis attenuated the formation of atherosclerotic plaques [50].

Macrophages that have taken up great quantities of lipids are called foam cells, and lesions full of foam cells are called fatty streaks. The initial pathologic changes in the fatty streak are fully reversible. However, with persistence of the atherogenic stimuli, the fatty streak may progress to become an atherosclerotic plaque. The conversion of the fatty streak to the atherosclerotic plaque occurs as a result of the death of foam cells and the influx of vascular smooth muscle (SMC) cells to form the characteristic lipid core and fibrous cap, respectively. The origin of the SMCs in atherosclerotic plaques is currently a matter of debate. Classically, plaque SMCs are believed to arise from the pre-existing medial SMCs [51].

More recently, it has been proposed that at least some of the plaque SMCs are derived from stem cells [52]. Whatever the origin of the SMC in the atherosclerotic plaque, their phenotypic characteristics are different from those of SMCs found in healthy blood vessels. Normally, SMCs are quiescent, contractile and nonproliferative and act to control artery tonus. Within the atherosclerotic environment SMCs become noncontractile and proliferative. SMCs become altered under the influence of cytokines and growth factors from leukocytes, platelets and endothelial cells. γ-Interferon (IFN-γ), which is secreted by CD4$^+$ lymphocytes of the Th1 phenotype, has been shown to promote vascular SMC proliferation [53,54]. Indeed, this effect has been suggested to be caused by the induction of the PDGF receptor in SMCs [54]. Another molecule, macrophage inhibitory factor (MIF), a potent proinflammatory cytokine secreted by SMCs, endothelium and macrophages, has been shown to promote SMC proliferation [55,56]. Indeed, MIF immunoreactivity has been demonstrated to co-localize to areas of atherosclerosis [55]. In addition, MIF $-/-$ mice are more resistant to atherosclerosis than their MIF +/+ littermates [53]. Analysis of atherosclerotic lesions from MIF −/− mice has shown decreased lipid deposition and reduced smooth muscle proliferation and intimal thickening.

Conversely, some cytokines have been demonstrated to suppress SMC proliferation. The anti-inflammatory cytokine, IL-10, has been observed to inhibit the proliferation of SMC after vascular injury [57]. In addition, IL-10 hyperexpression in a murine model of atherosclerosis has been shown to reduce plaque formation [58].

Many of the cytokine-induced SMC changes are mediated by modification of gene expression. Indeed, *in vitro* stimulation of SMCs by TNF-α induced the upregulation of more than 50 genes and the downregulation of around 20 others [59]. The genes whose expression are affected by TNF-α treatment are diverse, and they range from those involved in the immune response through those encoding structural proteins to those involved in cellular metabolism [59].

The next stage in atherosclerosis is the formation of plaque [60]. It is now thought that it is the rupture of the plaque leading to thrombosis rather than stenosis leading to occlusion that is the precipitating factor in the majority of acute atherosclerotic events [61,62]. Plaque rupture is now suspected to be the cause of approximately 70% of fatal MI and sudden coronary deaths [63].

The atherosclerotic lesion is composed of a mass of fatty material and inflammatory cells overlaid with a fibrous cap. The major question in considering plaque lesions is: What causes some lesions to suddenly rupture and cause life-threatening thrombosis? [64]. Gene expression changes and other factors that contribute either to plaque stability or to plaque instability are subjects of increasing interest in this regard (reviewed in [60,65]).

Stable and unstable plaques show striking dissimilarities in architecture and histology. The unstable vulnerable plaque lesion has a large lipid core, which can comprise some 40% and upwards of the lesion, and a thin fibrous cap [66]. Stable plaque comprises about 14% lipid and the stable cap, which can occupy more than 70% of the lesion, acts to protect against rupture [61]. The unstable lesion also contains large numbers of inflammatory macrophages and T cells [22]. By contrast, the stable cap is relatively nonactivated and noninflammatory [60].

The stability of the fibrous cap is largely determined by its collagen content. Collagen and other cap proteins such as elastin and glycosaminoglycans are in turn synthesized by the SMCs. In the unstable inflammatory lesion, fibrous cap thinning is caused, on the one hand, by increased collagen breakdown by proteases and, on the other, by decreased collagen synthesis by SMCs.

The formation of the calcified plaques is also dependent on gene expression. Genes that are active in bone formation are also activated during plaque formation [67,68]. This may also contribute to plaque stability.

It is now recognized that collagen and elastin breakdown are affected by activated macrophages that overproduce matrix metalloproteinases, in particular collagenases and gelatinases, enzymes that have a key role in breaking down collagen and elastin [16]. Advanced plaque is also made up of about 20% T cells, which produce the lymphokine IFN-γ that inhibits the production of collagen matrix by SMCs. Macrophage content in unstable plaque is high and, in turn, macrophage expression of these proteins is strongly induced by inflammatory cytokines such as TNF-α, PDGF and IL-1 [69–71].

The other important process occurring in plaque destabilization is a reduction in SMCs. Cytokines such as TNF-α and IFN-γ induce apoptosis of SMCs. As a consequence, SMC synthesized collagen, elastin and glycosaminoglycans decrease and large areas of necrotic and apoptotic SMCs begin to accumulate in unstable plaques [72–74]. Increased apoptosis of SMCs brings into play the Fas death pathways and other cell death pathways. The lipid cores of unstable plaque lesions have been described as a cemetery of cells and cell types [74]. Extracellular protein tenascin C, which regulates cell adhesion, both induces metalloproteinase expression and causes SMC apoptosis. It is not present in normal vessel but is expressed in unstable plaque [75,76].

Plaque rupture usually occurs specifically at the edges of the lesion. It is here that inflammatory activity is the most intense, T lymphocytes and lipid-filled macrophages predominate and SMCs are less common. When the cap ruptures the contents of the lesion plaque lipids, tissue factor, collagen and other materials come into contact with blood components, initiating platelet activation, coagulation and thrombosis [64].

In addition to being prone to rupture, the inflamed plaque has a prothrombotic effect. This finding is not surprising, given that the thrombotic and inflammatory pathways are intimately linked [77]. Indeed, it is now believed that thrombosis can play a vital part in the initiation of atherosclerosis. The adhesion of platelets to seemingly normal endothelium has been shown to promote the formation of the atherosclerotic plaque in the apoE −/− mouse [78]. The initial adhesion of the platelet to the vessel wall has been shown to be mediated by a member of the integrin family, GPIb/IX/V [78]. Indeed, this adhesion molecule may represent a potential pharmacologic target as its inhibition can attenuate the formation of the atherosclerotic plaque in the mouse model of atherosclerosis [78]. GPIb/IX/V can bind von Willebrand factor (vWF) adherent to the damaged vessel wall, thus allowing the platelet to roll on the endoluminal surface of the vasculature [79]. This interaction slows the passage of the platelet in the blood stream, thus allowing for other platelet adhesion molecules, such as GPIIb/IIIa, to establish more permanent links to the vessel wall. GPIIb/IIIa is another member of the integrin family, whose main physiologic ligands are fibrinogen and vWF [80]. The presence of multiple GPIIb/IIIa binding sites on fibrinogen allows this molecule to act as a cross-link between different platelets, thus facilitating platelet aggregation [77]. GPIIb/IIIa binding to fibrinogen also mediates various physiologic changes within the platelet leading to shape change and platelet granular release.

Thrombosis also has a vital role in the progression of atherosclerosis. Exposure of the subintima to blood, through plaque rupture, disruption of plaque microvessels or superficial endothelial disruption, leads to activation of the coagulation cascade as well as circulating platelets [17]. The resulting thrombus can be integrated into the atherosclerotic plaque, thus contributing to its size. In addition, thrombus constituents, such as erythrocyte membrane cholesterol, may provide additional antigenic stimuli for plaque growth [81]. Platelets are also a rich source of proinflammatory cytokines, such as RANTES, CD40L and IL-1β,

which further promote inflammation [77]. CD40L has an important role in plaque destabilization by inducing the production of various metalloproteinases (MMPs) responsible for the breakdown of the extracellular matrix in the plaque [82]. Indeed, CD40L levels are associated with the presence of lipid-rich plaques in humans [83]. It is therefore not difficult to visualize a vicious cycle whereby the inflamed plaque initiates local thrombosis, which then promotes further inflammation and plaque destabilization.

Genetic background of atherosclerosis: Atherosclerosis is a complex trait

Common DNA sequence variants which each may have a small to moderate phenotypic effect have been suggested to determine genetic susceptibility to common complex traits such as atherosclerosis [84–86]. Currently, however, the DNA sequence variants, whether rare or common, conferring the susceptibility to atherosclerosis in the general population are still largely unknown.

Atherosclerosis and other complex traits do not follow a simple Mendelian mode of inheritance. Instead, relatives of an affected individual are likely to have disease-predisposing alleles, making the disease more common among the first degree relatives of the proband and less common in less closely related relatives, resulting in a familial aggregation of the complex trait. However, the observed familial aggregation does not necessarily mean a strong genetic contribution. It may be by chance alone because atherosclerosis is common at the population levels or to great extent explained by shared environmental factors but previous studies have shown that family history of coronary heart disease (CHD) significantly increases the risk for CHD [3,87,88]. Heritability is also used to estimate the degree of genetic involvement. Heritability is the fraction of the total phenotypic variance of a trait caused by genes. It is worth noting that heritability does not reveal how many genes are involved or how the different genes interact. For CHD the heritability has been estimated to be 56–63% [89].

Currently, a gene for a monogenic disease can be mapped and identified, providing that there are enough informative families available for analysis.

Unknown allelic spectra	Unknown mode of inheritance	
	Unknown allele frequencies of the disease-predisposing alleles	
	Unknown risk associated with the disease	
	Epistasis	
	Locus and allelic heterogeneity	
Phenotype	Difficulties in assessing the complex phenotype (lack of unequivocal diagnostic criteria)	
	Late onset of the disease	
	Pleiotrophy	
	Phenocopies	
	Incomplete penetrance	
Technical issues	Affordable large-scale genotyping methods	
Statistical analyses	Multiple testing	
	Limited statistical power	

Table 6.1 Gene identification in atherosclerosis is hampered by several factors.

However, the success in identifying genes for complex diseases including atherosclerosis has been relatively modest (reviewed in [90]), and most progress has been made with rare familial (Mendelian) forms of these complex traits [91]. Table 6.1 shows some of the factors hampering gene identification in atherosclerosis.

To increase the impact of genetic involvement and thus the possibilities of identifying contributing genes for complex diseases, investigation of cases with a likely familial component; families with multiple affected individuals; subjects with an early onset disease; and extreme phenotypes can be utilized [92]. Importantly, study samples originating from genetically isolated populations such as Sardinians and Finns, where genetic and environmental heterogeneity are reduced, have been very successfully used to identify genes for rare monogenic diseases (reviewed in [93,94]). These populations may provide some advantages into gene identification of complex traits as well, although there are likely to be multiple predisposing alleles for complex traits even in these population isolates. An important advantage of population isolates may turn out to be the relatively low environmental and lifestyle variability that can be expected among these populations. These are important factors because they present less confounding factors into the statistical analyses.

Alleles contributing to complex traits are also suggested to have only a minor to moderate effect on the phenotype [84–86]. Furthermore, in com-

plex traits such as atherosclerosis the underlying DNA sequence variants may differ from the ones identified for typical monogenic diseases [94], and they may mostly even represent noncoding variants, residing in conserved regulatory regions. For associated single nucleotide polymorphisms (SNPs) in noncoding regions, the functional analyses are challenging, and currently they focus on investigation of cross-species conservation and/or identification of regulatory elements such as transcription factor, RNA splicing factor and microRNA binding sites using approaches of bioinformatics [95].

Two main projects facilitating gene identification in atherosclerosis: the Human Genome Project and the International HapMap Project

The Human Genome Project (HGP) was started in 1988 first to map and then to sequence the human genes. The initial emphasis was in building both physical and genetic linkage maps of 22 human autosomal chromosomes in order to provide dense maps of microsatellites, expressed sequence tags and sequence-tagged sites for mapping purposes. The ultimate goal was to sequence the human genome. The first version was available in 2001 [96,97] and, in 2003, the HGP announced the completion of the DNA reference sequence of *Homo sapiens*. The HGP provided the essential tools for gene identification of complex traits, including atherosclerosis. As a result of the successful HGP, human genetics is expected to be one of the key players in providing new insights and better understanding of diseases, not only of the rare monogenic disorders but also of common complex diseases such as CAD, and other atherosclerosis-related disorders including stroke, diabetes and hypertension.

Meaningful analysis of the enormous amount of data that the HGP produced is crucial for successful genetic analysis of atherosclerosis and other complex traits [90]. To tackle the millions of SNPs available for association tests, approaches identifying the causal variants among those in an associated haplotype were recently developed [98–101]. The important message of these studies was that most of the human genome may consist of blocks of variable length over which only a few common haplotypes are detected. The mean size of blocks was initially estimated to be 22 kb and ~80% of the genome in blocks >10 kb in populations of European ancestry [102]. In African-Americans, the mean size of blocks was initially estimated to be 11 kb and ~60% of the genome in blocks >10 kb [102]. These observations led to the establishment of the International HapMap Project which is currently determining the linkage disequilibrium (LD) patterns across the human genome in blood samples taken from people in Japan, Nigeria and China as well as from people of northern and western European ancestry in the USA in order to allow the efficient selection of SNPs for regional and genome-wide association studies [103]. The overall aim is to provide a restricted number of tag SNPs for genotyping to cover most of the common variation in the human genome without genotyping redundant SNPs. The ultimate success of this project is to great extent dependent on the hypothesis of common variants underlying common disorders [84,86]. Using these data generated by the HapMap project, the tag SNPs capturing most of the genetic variation can be selected for regional and genome-wide association analyses, hopefully providing a powerful shortcut to gene identification of atherosclerosis among other complex traits. For example, currently, the genotypes of millions of SNPs for 30 trios from Centre d'Etude du Polymorphisme Humain (CEPH) subjects of north European ancestry are available online (http://www.hapmap.org/).

Previous data show that some genomic regions fit better to the block theory than others [104]. Thus, the actual practical usefulness of the haplotype method will depend on the specific patterns of LD in the region of interest [104] as well as on the underlying LD structure of the study population. Furthermore, the haplotype-block strategy is agnostic about types and location of functional SNPs. However, in most Mendelian diseases the identified causative mutations have turned out to be coding variants (reviewed in [94]). Thus, an alternative strategy, focusing on identification and testing for association of SNPs in coding and regulative regions, has been proposed for association testing of complex traits [94]. In this sequence based approach, about 10 times smaller number of SNPs need to be genotyped than when using the haplotype-block strategy [94]. In addition, low frequency disease alleles could also be detected. This may be of importance because rare DNA sequence variants may have a role in individual families, and because both rare and common variants seem to confer for instance the susceptibility to low plasma levels of HDL-C in the general population [105].

Candidate genes contributing to the development of atherosclerosis

Genetic components of atherosclerosis can be investigated using several strategies, including genome-wide scans for novel genes (Table 6.2) and candidate gene approaches (reviewed in [90]). Candidate gene studies of known genes, mainly using case–control study samples have been the method of choice for several decades. In these studies, alleles of unrelated affected subjects are compared with alleles of unrelated unaffected subjects. When an association is detected, however, it may be difficult to demonstrate the direct causality. Differences in age, sex or ethnicity between case and control groups can contribute to the observed association and cause stratification bias resulting in false positive associations [106]. Therefore, careful selection of a control group is crucial for a meaningful case–control study. Multiple testing can also lead to false positive results. These difficulties partly explain the small number of findings replicated in several study samples and populations. Replication in independent study samples is of utmost

Table 6.2 Genome-wide screens for myocardial infarction (MI), coronary artery disease (CAD) or coronary artery calcification using linkage or association analysis [90]. After Lusis *et al.* 2004 [90].

Trait	Chromosome (gene)	Lod score or P value	Population (number of subjects)	Method	Reference
MI	6p21 (LTA)	0.0000003	Japanese (94 cases/ 658 controls)	Association	Ozaki *et al.* 2002 [135]
MI	13q12–13 (ALOX5AP)	2.9 0.000002	Icelandic (multiple study samples; 779 cases and 6624 controls for ALOX5AP)	Linkage: Allele sharing (Allegro); and association	Helgadottir *et al.* 2004 [114]
MI	14qter	3.9	German (1406)	Linkage: Variance component (SOLAR)	Broeckel *et al.* 2002 [132]
MI	1p34–36	11	USA (1163)	Linkage: Allele sharing (SAGE)	Wang *et al.* 2004 [213]
CAD	2q21–22, Xq23–26	3.2 3.5	Finnish (364)	Linkage: Allele sharing (MAPMAKER/SIBS)	Pajukanta *et al.* 2000 [129]
CAD	3q13	3.3	USA (1168)	Linkage: Allele sharing and parametric (Genehunter Plus)	Hauser *et al.* 2004 [133]
CAD	16p13-pter	3	North-Eastern Indian (535)	Linkage: Allele sharing (MAPMAKER/SIBS)	Francke *et al.* 2001 [130]
Coronary artery calcification	10q21.3	3.2	USA (94)	Linkage: Allele sharing (MLS)	Lange *et al.* 2002 [134]

LTA indicates lymphotoxin-alpha and ALOX5AP arachidonate 5-lipoxygenase-activating protein gene.

importance when evaluating the significance of association results.

Genes suggested to contribute to CAD risk include apolipoprotein E, apolipoprotein (a), methyl-tetrahydrofolate reductase, angiotensin-converting enzyme and *NOS3* genes [107–111]. The size and nature of these and other CAD susceptibility genes are largely unknown. An autosomal dominant form of CAD was recently shown to be caused by a mutation in the myocyte enhancer factor-2 (MEF2A) transcription factor gene [91], implicating the MEF2A signaling pathway in the pathogenesis of myocardial infarction (MI). However, subsequent studies showed that mutations in this gene are not a common cause of CAD at the population level [112,113]. Recently, the gene encoding 5-lipoxygenase activating protein was also shown to confer risk of MI and stroke in subjects from Iceland and UK [114], and the finding was replicated in Japanese [115]. Table 6.3 shows the currently known common DNA variations contributing to CAD and its risk factors (reviewed in [90]).

Example of a candidate gene for atherosclerosis: *NOS3*

Identifying polymorphisms that may indicate increased susceptibility to atherosclerosis and heart disease is an area of active research interest. A major contender for prognostic polymorphisms is the *NOS3* gene. *NOS3* (or the endothelial isoform of nitric oxide synthase) is expressed primarily in vascular endothelium [116]. The nitric oxide synthesized by this enzyme has antiplatelet effects as well as promoting smooth muscle relaxation. Indeed, nitric oxide is the predominant vasodilator found in the healthy vasculature. Therefore, loss of nitric oxide activity would promote vasoconstriction and platelet activation. Impairment of nitric oxide activity has been described in many conditions that predispose to atherosclerosis, including hypertension, hypercholesterolemia and diabetes [117–119].

Theoretically, genetically controlled subtle disturbances in nitric oxide synthesis caused by perturbations in the *NOS3* gene might predispose

Table 6.3 Common DNA sequence variations contributing to coronary heart disease (CHD) and its risk factors. After Lusis *et al.* 2004 [90]. Only genes showing evidence of linkage or association in multiple studies are cited.

Trait	Gene	Variation	Reference
LDL/VLDL	LDL receptor	Many mutations	Goldstein *et al.* 1995 [192]
	PCSK9	Many mutations	Abifadel *et al.* 2003 [193]
	ApoE	Three common missense alleles explain ~5% of variance of cholesterol	Sing *et al.* 1985 [194]
	ApoAI-CIII-AIV-AV cluster	Multiple polymorphisms	Talmud *et al.* 2002 [195]
HDL levels	Hepatic lipase	Promoter polymorphism	Shohet *et al.* 2002 [196]
	ABCA1	Many polymorphisms	Frikke-Schmidt *et al.* 2004 [197]
FCHL	Upstream transcription factor 1	Intronic and 3'UTR polymorphims	Pajukanta *et al.* 2004 [167]
Lp(a)	Apo(a)	Many alleles of apo(a) explain >90% variance	Boerwinkle *et al.* 1992 [198]
Homocysteine	Methylene tetrahydrofolate reductase	Missense polymorphism	Kang *et al.* 1993 [199] Ma *et al.* 1996 [200]
Coagulation	Fibrinogen B	Promoter polymorphism	Hamsten *et al.* 1993 [201]
	Plasminogen activator inhibitor type 1	Promoter polymorphism	Hamsten *et al.* 1993 [201]
	Factor VIII	Common missense	Hamsten *et al.* 1993 [201] Thomas *et al.* 1995 [202]
Blood pressure	Angiotensinogen	Missense and promoter polymorphisms	Caulfield *et al.* 1995 [203]
	β_2-Adrenergic receptor	Missense polymorphism	Lusis *et al.* 2002 [204]
	Alpha-adducin	Missense polymorphism; support from studies in rats	Lusis *et al.* 2002 [204]
CAD	Angiotensin converting enzyme	Insertion-deletion polymorphism	Staessen *et al.* 1997 [205]
	Serum paraoxonase	Missense polymorphism affecting enzymatic activity; animal studies support	Shih *et al.* 2001 [206] Tward *et al.* 2002 [207]
	Toll-like receptor 4	Missense polymorphism	Kiechl *et al.* 2002 [208]
	Arachidonate 5-lipoxygenase-activating protein	Haplotype of 4 SNPs	Helgadottir *et al.* 2004 [114]
Stroke	Phosphodiesterase 4D	Regulatory polymorphism	Gretarsdottir *et al.* 2003 [209]
Diabetes, obesity and insulin resistance	PPARγ	Missense polymorphism	Altshuler *et al.* 2000 [210]
	Calpain 10		Horikawa *et al.* 2000 [211]
	Hepatocyte nuclear factor-4α	Promoter polymorphism	Silander *et al.* 2004 [212]
	Transcription factor 7-like 2	Intronic	Grant *et al.* 2006 [214]

CAD, coronary artery disease; FCHL, familial combined hyperlipidemia; HDL, high density lipoprotein; LDL, low density lipoprotein; PPAR, peroxisome proliferator activated receptor; SNP, single nucleotide polymorphism; VLDL, very low density lipoprotein.

individuals to develop atherosclerosis, given some environmental or endogenous upset to uncover the effects of the gene alteration [111]. There is some evidence that certain polymorphisms in this gene may affect atherosclerosis development. Among several polymorphisms found in the *NOS3* gene (reviewed in [111]), the Glu298Asp polymorphism is of especial interest. This polymorphism, which is located in a coding region of the gene, exon 7, could alter mature protein activity, thus affecting enzyme activity and reducing local nitric oxide synthesis. Hingorani *et al.* [120] investigated Glu298Asp in an

English study sample. They found a strong association between the Glu298Asp polymorphism and the risk for coronary heart disease. Studies have also linked Glu298Asp to essential hypertension, resistance to therapy [121] and to coronary spasms [122]. However, the relevance of the data remains uncertain, as other investigators have not been able to replicate these findings [123].

Other polymorphisms have also been found in the *NOS3* gene. Polymorphisms in the promoter may influence mRNA transcription; intronic polymorphisms are less likely to have functional roles. In Japanese, the promoter polymorphism T-786C was shown to be linked to vasospasm, which in turn is linked to low endothelial NO [124]. T-786C is also associated with MI [125] and diabetes [126]. Wang *et al.* [127] reported that risk of CAD was increased in smokers with a 27 base pair repeat polymorphism in intron 4 of *NOS3*.

The evidence linking *NOS3* polymorphisms to clinical states is still inconsistent [30]. With increasing availability of large scale genotyping techniques such as microarray technology, the search for interesting diagnostic or prognostic polymorphisms will be made easier, as thousands of gene variants can be searched simultaneously.

Polymorphisms related to identifying individuals at risk of unstable plaque and plaque rupture are also under investigation. For example, MMP gene expression is regulated at the transcriptional level and responds to variety of factors such as TNF-α. Genetic variations in the MMP promoter regions may act to affect extracellular matrix degradation, thereby increasing susceptibility to atherosclerosis [74,128]. A polymorphism in MMP-3 promoter was also linked to MI independently of other risk factors, suggesting a susceptibility to plaque rupture (reviewed in [128]). Several SNPs in thrombospondin genes, deficiencies of which are associated with increases in MMP-2, have been associated with premature familial MI [74].

Genomic approaches to identify genes for atherosclerosis

During the last decade, a genome-wide scan has become a popular approach to identify novel genes for complex traits, and several scans have been performed for MI, CAD [129–133] as well as for coron-

ary artery calcification [134], mainly using linkage analysis of families or affected sib-pairs (Table 6.2) (reviewed in [90]). For this approach no a priori knowledge of disease pathophysiology is required, thus enabling identification of novel genes and pathways for atherosclerosis. In 1996 the idea of genome-wide association studies was introduced [85]. Association analysis has been shown to be more powerful than linkage analysis for detecting alleles of complex traits with only modest effects [85]. However, searching the whole genome using an association approach requires genotyping of hundreds of thousands of SNPs [94]. So far, only a few such studies have been completed for complex traits, because of the technical demands related to genotyping of such a large number of SNPs [135,136]. In one such study, analyses of 92,788 gene-based SNPs showed that functional SNPs in the lymphotoxin-alpha gene are associated with susceptibility to MI [135]. In another study, over 100,000 SNPs were genotyped and the human complement factor H gene was identified to be associated with age-related macular degeneration [136]. Advances in genotyping technologies have recently made this approach both more affordable and feasible. However, genome-wide association analyses are still facing a number of challenges, including problems related to multiple testing, suitable study design, SNP selection and interactions between polymorphisms [137,138]. Selecting the tag SNPs, produced by the HapMap Project, instead of random, evenly spaced SNPs may turn out to be the most effective way to cover most of the genetic variation in the human genome. However, it is difficult to detect rare causative SNPs using this approach, suggesting that substantial resequencing of genes is warranted to identify rare causative variants.

DNA microarrays provide a practical and economic tool for studying gene expression in atherosclerosis on a genomic scale [90]. Complementing classic linkage and association studies, expression arrays can relate changes in gene expression to atherosclerosis and have great potential to identify novel genes, their pathways and networks associated with atherosclerosis. Recently, when atherosclerotic plaques of patients with stable and unstable angina were investigated for gene expression differences, several genes previously linked to hemostasis, such as the protein S (PROS1) gene, the cyclo-oxygenase

1 (COX-1) gene, the IL-7 gene, and the MCP-1 and MCP-2 genes, were shown to be expressed at significantly lower levels in samples from unstable angina patients [139]. These findings suggest that these genes may have a role in plaque rupture.

The limitations of expression arrays are often related to the relevant tissues and small sample sizes available for investigation in humans. Small sample sizes typically available for microarrays as well as increased genetic heterogeneity can make distinguishing statistically significant differential expression between cases and controls especially challenging. Furthermore, atherosclerosis is likely to result from small quantitative differences in multiple genes, rather than major expression changes in a few genes. This phenomenon is exemplified by a recent study of type 2 diabetic males [140], where analysis of co-regulated sets of genes rather than individual genes identified metabolic pathways that are altered in diabetic individuals. In more detail, Mootha *et al.* [140] used the Gene Set Enrichment Analysis (GSEA) to identify a set of PGC-1α responsive genes involved in oxidative phosphorylation that were coordinately downregulated by approximately 20% in the muscle of diabetic individuals, with no single gene showing significant differential expression between diagnostic categories.

As a result of the successful HGP, recent progress of the HapMap Project and advancing genotyping and microarray technologies, it is finally possible to identify the DNA sequence variants conferring susceptibility to atherosclerosis using whole-genome approaches where information obtained from linkage, association, gene expression and functional analyses are combined to verify the signals.

Genes, lipids and atherosclerosis

Stein *et al.* [141] point out that some people are unlikely to develop atherosclerosis and CHD even in the face of high dietary cholesterol intake or frank hypercholesterolemia. Individuals show a wide variety of responses to dietary cholesterol, about 9% of populations studied are hyper-responders and 9% are hypo-responders. For example of the latter, a case was reported of one man who ate about 25 eggs a day (about 6 g cholesterol) and yet remained normocholesterolemic at 88 years of age [142]. It has been known for some time that genetic variation can have dramatic effects in causing inter-individual variation in cholesterol levels. Some of the genes implicated in lipid metabolism are discussed below.

Familial hypercholesterolemia is one of the most common inborn errors of metabolism and is most often a result of mutations of the low density lipoprotein receptor (LDL-R) gene [143]. Although countless loss of function mutations have been described in the LDL-R gene (these can be viewed at http://www.ucl.ac.uk/fh), these can generally be classified into one of five categories:

1 Those that do not produce a detectable LDL-R (e.g. 'null alleles');

2 Those that code for a protein that cannot be transported from the endoplasmic reticulum to the Golgi body and hence to the cell surface;

3 LDL-R that cannot bind the corresponding ligand;

4 LDL-R that binds LDL normally but cannot be internalized; and

5 LDL-R that cannot be recycled to the cell surface after transporting cholesterol into the cell [143]. Individuals who are heterozygous for a mutant allele have a two- to threefold increase in LDL-C and develop premature CHD after the age of 35. Homozygotes exhibit 6–8 times above normal levels of cholesterol and develop ischemic heart disease in their teenage years. Therefore, it is important to identify individuals who suffer from familial hypercholesterolemia in order to institute early lipid lowering treatment.

Apolipoprotein E (ApoE), found in various classes of lipoproteins, binds to LDL-R and mediates the uptake of the lipoprotein by the cell [144]. ApoE is believed to have a protective effect against the development of atherosclerosis. Mice that are −/− for the ApoE gene are severely hypercholesterolemic and are prone to early onset atherosclerosis [145]. Numerous polymorphisms of the gene coding for ApoE have been described [144]. Of the three common alleles, ε2, ε3 and ε4 at a single locus, ε3 and ε2 are the most and least common alleles, respectively [144,146]. ε2 homozygosity can give rise to type III hyperlipoproteinemia [146], a condition that is associated with premature atherosclerosis brought about by the defective cellular uptake of lipoprotein remnants [147,148]. Based on the finding that the majority of cases of type III hyperlipoproteinemia are homozygous for ε2, it is perhaps surprising to find that ε2 heterozygosity is not associated with CHD [149]. However, ε3/4 heterozygosity is associated with an increased risk of ischemic heart disease when compared with ε3/3 homozygotes, with an odds ratio of 1.30 (95% confidence interval [CI], 1.18–1.44) [149]. The mechanism by which ε4 heterozygosity affects atherogenesis is unclear at present.

Lipoprotein lipase is important in the regulation of triglyceride-rich lipoproteins such as VLDL and chylomicrons. Impairment of lipoprotein lipase activity may delay the clearance of these lipoproteins from the circulation [150]. An Asp9Asn mutation in the lipoprotein lipase gene may be associated with disease progression while some other variations are believed have a protective effect against MI [150,151].

Lipoprotein (a) levels appear to be a marker of the risk of disease progression in atherosclerosis and cardiovascular risk, although not all studies have shown consistent results [152–154]. Genetic variation seems to have an effect on plasma lipoprotein (a) levels, although the significance of this remains to be determined [155].

Cholesteryl ester transfer protein (CETP) mediates the transfer of cholesteryl esters from HDL to LDL and VLDL, therefore promoting the transport of cholesterol to the hepatocyte [156]. Inhibition of CETP has been shown to protect against atheroma formation in rabbits [157]. A recent meta-analysis of over 13,000 patients has shown that the so-called Taq1B polymorphism of the CETP gene can be related to cardiovascular risk [158].

Sitosterolemia is an autosomal recessive condition in which there is excessive intestinal uptake but decreased biliary secretion of plant sterols and cholesterol [159]. Patients with this condition usually have hypercholesterolemia and are at risk of developing premature atherosclerosis. Sitosterolemia has been found to be caused by mutations in the genes coding for ABCG5 and ABCG8 [160]. These genes belong to the ATP-binding cassette (ABC) superfamily of transmembrane transporters that are responsible for the movement of a diverse range of substances [161]. ABCG5 and ABCG8 are expressed in the liver and intestines and have been shown to be responsible for the secretion of cholesterol into the bile [162].

Another member of the ABC superfamily, ABCA1, controls the efflux of intracellular cholesterol to lipid-poor ApoA-I, the major apolipoprotein of HDL [161]. Impairment of ABCA1 activity leads to an autosomal recessive condition known as Tangier disease [163]. Patients with this condition have accumulation of cholesterol in various tissues, leading to a multisystem disorder featuring premature atherosclerosis, hepatosplenomegaly, polyneuropathy and epidermal lesions. In a mouse model of atherosclerosis, overexpression of the ABCA1 gene attenuated atherogenesis and increased HDL levels [164]. These findings suggest that HDL levels may reflect ABCA1 activity [165]. Indeed, certain mutations in the ABCA1 genes have been shown to be associated with reduced HDL levels and an increased risk of developing CHD [165,166].

Upstream transcription factor 1 (USF1), residing on chromosome 1q21, regulates several genes of lipid and glucose metabolism. Variants of USF1 were recently associated with a common familial dyslipidemia, familial combined hyperlipidemia (FCHL) in Finnish [167], Mexican [168] and Caucasian families [169]. As FCHL predisposes the affected individuals to CAD, it is of importance that a recent study implicated USF1 variants in CAD [170] and that the 1q21 region has also been linked to MI [133]. Taken together these studies suggest that USF1 should be further investigated as a potential candidate gene conferring the susceptibility to CAD.

Genetic polymorphisms, inflammation and atherosclerosis

There is great interest in elucidating the role played by polymorphisms of the genes regulating inflammation. Early studies on a rodent model have shown that a genetic predisposition to inflammation co-segregates with the tendency to form atherosclerotic plaques [171].

5-Lipoxygenase is an enzyme involved in the production of the leukotrienes, and thus has a major role in promoting inflammation. A recent study showed that 6% of an American population sample possess two minor variant alleles in the promoter polymorphism of the 5-lipoxygenase gene [172]. It was found that individuals with the variant alleles have higher intima-media thickness measurements and a greater degree of systemic inflammation (as measured by high-sensitivity C-reactive protein [CRP]). Indeed, the 5-lipoxygenase gene has been shown to be associated with adverse cardiovascular events in both Icelandic and Scottish subjects [114,173]. In addition, LDLR −/− mice that had one copy of their 5-lipoxygenase gene removed showed a dramatic reduction in plaque formation when compared with "normal" LDLR −/− controls [174]. However, this finding has not been replicated in LDL −/− 5-LO −/− and ApoE −/− 5-LO −/− mice [175].

A genome-wide association study of Japanese subjects has found a polymorphism in the lymphotoxin-α (also known as TNF-β) gene to be associated with susceptibility to MI [135]. However, the result of this study was not replicated in another publication from another Japanese group [176]. A study performed in a German population similarly failed to reveal an association between the lymphotoxin-α polymorphism and CHD [177]. Similarly, studies on polymorphisms of TNF-α gene provided conflicting results [177,178].

Genes, coagulation, fibrinolysis and atherosclerosis

Variations in genes controlling the coagulation pathway may also play a part in influencing the natural history of atherosclerosis. A high plasma level of fibrinogen has been associated with an increased risk of developing adverse cardiovascular events [179]. Fibrinogen is the precursor of fibrin. In addition, its numerous binding sites for the platelet GPIIb/IIIa receptor enables it to act as a cross-link during platelet aggregation [77]. A recent study has suggested that genetic variation may play a part in determining plasma fibrinogen levels and that the genes responsible for this are located on chromosomes 2 and 10 [180]. A -455G/A polymorphism in the β-fibrinogen gene has also been shown to affect plasma fibrinogen levels [181]. However, at present, it is unclear if this polymorphism may influence the risk of atherothrombosis, as some studies have reported a positive association between the two while others have not [181,182]. At present, there is no strong evidence to suggest that genetic polymorphisms in genes coding for the other components of the coagulation pathway would influence atherothrombosis.

Plasminogen activator inhibitor-1 (PAI-1) is the major circulating inhibitor of tissue-type plasminogen activator (t-PA). PAI-1 acts to regulate thrombolytic activity by preventing excessive systemic fibrinolysis. Therefore, PAI-1 can be considered to be a procoagulant molecule. Indeed, plasma levels of PAI-1 can be related to the risk of developing adverse cardiovascular events [183]. However, it is unclear if elevated PAI-1 levels are a cause or effect of atherosclerosis. In addition, the fact that the elevations of PAI-1 levels are accompanied by rises in t-PA further complicates matters. Most studies on the PAI-1 gene have focused on the 4G/5G insertion/deletion polymorphism at position -675 in the promoter region of the gene [184]. The 4G (deletion) allele is associated with higher PAI-1 levels and it is conceivable that patients with this allele may be more prone to thrombosis. Because studies on the 4G/5G polymorphism have yielded conflicting results, a meta-analysis was performed which showed a weak association between the 4G allele and the risk of atherothrombosis [185].

Other genes

An autosomal dominant form of CHD has been linked to a 21-bp deletion in the transcription factor MEF2A [91]. MEF2A is known to be involved in vasculogenesis but the pathophysiologic pathway by which the MEF2A polymorphism may influence the atherogenesis is uncertain at present. Some of the other candidate genes that may be related to the development of atherosclerotic disease are summarized in Table 6.3.

The future in atherosclerosis

Atherosclerosis is no longer thought of as a lipid accumulation or passive degeneration but as an active process of cells signaling and actively participating in atherosclerotic remodeling. Hundreds of genes are likely to be involved in this process as well as alterations in cell–cell communication [186]. Increased understanding of the molecular and cell biologic events underlying the atherosclerotic process has led to new concepts to develop the necessary prognostic indicators, diagnostic tests and targeted therapies for atherosclerosis and CHD. Plaque stabilization and identification of vulnerable patients is also emerging as an important goal to prevent complications of atherosclerosis [63].

The concept of atherosclerosis as inflammation has led to several new insights into novel, clinically applicable, risk factor markers. For example, measurement of inflammatory markers such as plasma CRP or serum amyloid A could provide a noninvasive method for assessing cardiovascular risk. Large increases in levels of these proteins can be discerned in plasma following inflammatory stimuli. CRP, produced in the liver and possibly in cells in atheromatous plaque in response to upstream proinflammatory cytokines, is being developed as a potentially important risk factor with predictive power for cardiovascular events [187].

In addition, because early endothelial changes may herald the development of later disease development, tests to assess the health of endothelium may serve as markers for disease and as targets for therapy [21,33]. For example, methods to detect marker "activation antigens" such as the endothelial leukocyte adhesion molecules, E selectin, ICAM-1 and VCAM-1, could be developed as biomarkers in tissues or circulating blood to identify inflammatory endothelial cell changes for early diagnosis [21]. Because VCAM-1 is an important early step in atherosclerotic processes, targeting this protein might be of therapeutic value, and several animal studies have suggested that blocking VCAM-1 acts to reduce neointimal hyperplasia [33].

Novel therapies for suppressing endothelial activation could provide new means of preventing atherosclerotic changes. For example, the peroxisome proliferator activated receptor (PPAR) subfamily appears to have anti-inflammatory properties and can reduce VCAM-1 and tissue factor gene expression by cells in atheroma [19].

The atherosclerosis as inflammation hypothesis has also led to new understanding of the mechanisms by which several existing pharmaceuticals are effective in reducing the complications of CAD. Angiotensin-converting enzyme (ACE) inhibitors promote bradykinin a nitric oxide stimulus, and decreased angiotensin increases nitric oxide bioavailability [22]. Thus, these agents appear to have effects on improving endothelial function beyond blood pressure decreases. The important role of the unstable plaque and plaque rupture in thrombosis has also increased interest in mechanisms that may effect plaque stabilization, by lipid lowering, ACE inhibition and antibiotics (reviewed in [60,188]).

Gene therapy

Gene therapeutic technologies might be used to remedy eNOS deficits (reviewed in [189]). Rabbits that are fed a high cholesterol diet are prone to develop endothelial dysfunction and atherosclerosis [189]. Impairment of nitric oxide activity has been shown to be a key factor behind this observation. Enhancement of NOS activity in these rabbits by the use of an adenovirus-mediated NOS gene transfer restored endothelial derived relaxing factor activity [190]. The use of the adenovirus-mediated NOS gene therapy was also shown to reduce macrophage infiltration of the carotid arteries of cholesterol-fed animals [191]. In addition, the expression of adhesion molecules such as ICAM-1 and VCAM-1 are significantly downregulated in the carotid arteries of the treated rabbit [191]. Therefore, it is possible that *NOS3* gene therapy may be useful in the treatment of human atherosclerosis.

Other gene therapy targets include MMP inhibitors (reviewed in [128]). Tissue inhibitors of MMPs might be increased at the local tissue level by administration of exogenous recombinant tissue inhibitors or by stimulating increased endogenous expression via gene therapy. Synthetic MMP inhibitors have also been investigated, including antibiotics such as doxycycline [128]. In addition, the apoptosis of fibroblasts on the surface of plaques would increase the likelihood of plaque rupture, and blocking apoptosis would be another approach to preventing complications of atherosclerosis [15]. It might also be possible to interfere with important transcription factors that regulate groups of genes involved in atherosclerosis. For example, NF κB, a key inflammation regulator in the vessel wall, PPARs and Sp/XKLF family of zinc finger genes may all prove to be important targets (reviewed in [186]).

Conclusions

Many highly intriguing genes have been discovered to be active in the progress of atherosclerosis. Furthermore, identification of genes contributing to susceptibility to atherosclerosis and other complex traits is expected to accelerate rapidly because the Human Genome Project and the HapMap Project have made the sequence of the human genome and human genome sequence variation data publicly available. Even so, we need more detailed knowledge about the numerous genes and gene targets involved in various stages of development of complex atherosclerotic lesions. New genomic technologies such as microarray chip technologies are beginning to be used to identify novel genes and pathways as well as to understand gene–gene and gene–environment interactions. For example, recent studies comparing gene expression in stable and ruptured plaque, as well as various responses of macrophages to oxLDL are pioneering the way to the fuller utilization of genomic technology in exploring the complexities of atherosclerosis.

References

1 Schwartz CJ, Mitchell JR. The morphology, terminology and pathogenesis of arterial plaques. *Postgrad Med J* 1962; 38: 25–34.

2 Gotto AM Jr. Some reflections on arteriosclerosis past, present and future. *Circulation* 1985; 72: 8–17.

3 Rose G. Familial patterns in ischaemic heart disease. *Br J Prev Soc Med* 1964; 18: 75–80.

4 Kannel WB, Castelli WP, Gordon T, McNamara PM. Serum cholesterol, lipoproteins, and the risk of coronary heart disease: the Framingham study. *Ann Intern Med* 1971; 74: 1–12.

5 Pooling Project Research Group. The relationship of blood pressure, serum cholesterol, smoking habit, relative weight and ECG abnormalities to incidence of major coronary events. 1978; Monograph 60, American Heart Association, Dallas.

6 Keys A. *Seven Countries: A Multivariate Analysis of Death and Coronary Heart Disease*. Harvard University Press, Cambridge, 1980.

7 Pyörälä K, Laakso M, Uusitupa M. Diabetes and atherosclerosis: an epidemiologic view. *Diabetes Metab Rev* 1987; 3: 463–542.

8 Grundy SM, Wilhelmsen L, Rose G *et al.* Coronary heart disease in high risk populations: lessons from Finland. *Eur Heart J* 1990; 11: 462–471.

9 Grundy SM, Balady GJ, Criqui MH *et al.* Primary prevention of coronary heart disease: guidance from Framingham. A statement for healthcare professionals from the AHA Task Force on Risk Reduction. *Circulation* 1998; 97: 1876–1887.

10 Genest J Jr, Cohn JS. Clustering of cardiovascular risk factors: targeting high-risk individuals. *Am J Cardiol* 1995; Supplement 76: 8A–20A.

11 Wilson PW, D'Agostino RB, Levy D *et al.* Prediction of coronary heart disease using risk factor categories. *Circulation* 1998; 97: 1837–1847.

12 Ridker PM, Buring JE, Shih J *et al.* Prospective study of C-reactive protein and the risk of future cardiovascular events among apparently healthy women. *Circulation* 1998; 98: 731–733.

13 Saikku P, Leinonen M, Mattila K *et al.* Serological evidence of an association of a novel chlamydia, TWAR, with chronic coronary heart disease and acute myocardial infarction. *Lancet* 1998; 2: 983–986.

14 Noll G. Pathogenesis of atherosclerosis: a possible relation to infection. *Atherosclerosis* 1998; 140 (Supplement 1): S3–S9.

15 Doevendans PA, Jukema W, Spiering W *et al.* Molecular genetics and gene expression in atherosclerosis. *Int J Cardiol* 2001; 80: 161–172.

16 Lusis AJ. Atherosclerosis. *Nature* 2000; 407: 233–241.

17 Libby P. Inflammation in atherosclerosis. *Nature* 2002; 420: 868–874.

18 Ross R, Glomset JA. The pathogenesis of atherosclerosis. *N Engl J Med* 1976; 295: 369–377.

19 Libby P, Ridker PM, Maseri A. Inflammation and atherosclerosis. *Circulation* 2002; 105: 1135–1143.

20 Glass CK, Witztum JL. Atherosclerosis: the road ahead. *Cell* 2001; 104: 503–516.

21 Gimbrone MA, Topper JN. Biology of the vessel wall: endothelium. In: Chien KR, ed. *Molecular Basis of Cardiovascular Disease*. WB Saunders, Philadelphia, 1999: 331–348.

22 Behrendt D, Ganz P. Endothelial function: from vascular biology to clinical applications. *Am J Cardiol* 2002; 90: 40L–48L.

23 Bonetti PO, Lerman LO, Lerman A. Endothelial dysfunction, a marker of atherosclerotic risk. *Arterioscler Thromb Vasc Biol* 2003; 23: 168–175.

24 Stary HC. Evolution and progression of atherosclerotic lesions in coronary arteries of children and young adults. *Arteriosclerosis* 1989; 9: 19–32.

25 Napoli C, Lerman L. Involvement of oxidation-sensitive mechanisms in the cardiovascular effects of hypercholesterolemia. *Mayo Clinic Proc* 2001; 76: 619–631.

26 Ross R. The pathogenesis of atherosclerosis a perspective for the 1990s. *N Engl J Med* 1993; 362: 801–809.

27 Topper JN, Cai J, Falb D, Gimbrone MA Jr. Identification of vascular endothelial genes differentially responsive to fluid mechanical stimuli: cyclooxygenase-2, manganese superoxide dismutase, and endothelial cell nitric oxide synthase are selectively up-regulated by

steady laminar shear stress. *Proc Nat Acad Sci USA* 1996; 93: 10417–10422.

28 Humphries SE, Ordovas JM. Genetics and atherosclerosis: broadening the horizon. *Atherosclerosis* 2001; 154: 517–519.

29 Henderson AH. It all used to be so simple in the old days. *Eur Heart J* 2001; 22: 648–653.

30 Albrecht EWJA, Stegeman CA, Heeringa P *et al.* Protective role of endothelial nitric oxide synthase. *J Pathol* 2003; 199: 8–17.

31 Gerrity RG, Naito HK, Richardson M, Schwartz CJ. Dietary induced atherogenesis in swine. Morphology of the intima in prelesion stages. *Am J Pathol* 1979; 95: 775–792.

32 Joris I, Zand T, Nunnari JJ. Studies on the pathogenesis of atherosclerosis. I. Adhesion and emigration of mononuclear cells in the aorta of hypercholesterolemic rats. *Am J Pathol* 1983; 113: 341–358.

33 Blankenberg S, Barbaux S, Tiret L. Adhesion molecules and atherosclerosis. *Atherosclerosis* 2003; 170: 191–203.

34 Cybulsky M, Gimbrone MA. Endothelial expression of a mononuclear adhesion molecule during atherogenesis. *Science* 1991; 251: 788–791.

35 Li H, Cybulsky MI, Gimbrone MA *et al.* An atherogenic diet rapidly induces VCAM-1 a cytokine regulatable mononuclear leukocyte adhesion molecule in rabbit aortic endothelium *Arterioscler Thromb* 1993; 13: 197–204.

36 O'Brien KD, Allen MD, McDonald TO *et al.* Vascular cell adhesion molecule 1 is expressed in human coronary atherosclerotic plaques. *J Clin Invest* 1993; 92: 945–951.

37 Cybulsky MI, Iiyama K, Li H, *et al.* A major role for VCAM-1 but not ICAM-1 in early atherogenesis *J Clin Invest* 2001; 107: 1255–1262.

38 Boring L, Gosling J, Cleary M, Charo IF. Decreased lesion formation in CCR2–/– mice reveals a role for chemokines in the initiation of atherosclerosis. *Nature* 1998; 394: 894–897.

39 Mach F, Sauty A, Iarossi AS *et al.* Differential expression of three T lymphocyte activating CXC chemokines by human atheroma-associated cells. *J Clin Invest* 1999; 104: 1041–1050.

40 Kume N, Cybulsky MI, Gimbrone MA Jr. Lysophosphatidylcholine, a component of atherogenic lipoproteins, induces mononuclear leukocyte adhesion molecules in cultured human and rabbit arterial endothelial cells. *J Clin Invest* 1992; 90: 1138–1144.

41 Kume N, Gimbrone MA Jr. Lysophosphatidylcholine transcriptionally induces growth factor gene expression in cultured human endothelial cells. *J Clin Invest* 1994; 93: 907–911.

42 Kodama T, Freeman M, Rohrer L *et al.* Type 1 macrophage scavenger receptor contains alpha helical and collaged like coiled cells. *Nature* 1990; 343: 531–535.

43 Krieger M, Action A, Ashkenas J *et al.* Molecular flypaper, host defense, and atherosclerosis. *J Biol Chem* 1993; 68: 4569–4572.

44 Hayes AL, Smith C, Foxwell BM *et al.* CD45-induced tumor necrosis factor alpha production in monocytes is phosphatidylinositol 3-kinase-dependent and nuclear factor kappa-B independent. *J Biol Chem* 1999; 274: 33455–33461.

45 Koike J, Nagata K, Kudo S *et al.* Density-dependent induction of TNF-alpha release from human monocytes by immobilised P-selectin. *FEBS Lett* 2000; 477: 84–88.

46 Thommesen L, Sjursen W, Gasvik K *et al.* Selective inhibitors of cytosolic or secretory phospholipase A2 blocks TNF induced activation of transcription factor nuclear factor-kappaB and expression of ICAM-1. *J Immunol* 1998; 161: 3421–3430.

47 de Martin R, Hoeth M, Hofer-Warbinek R *et al.* The transcription factor NFkB and the regulation of vascular cell function. *Arterioscler Thromb Vasc Biol* 2000; 20: e83–88.

48 Beckman JS, Koppenol WH. Nitric oxide, superoxide, and peroxynitrite: the good, the bad, and ugly. *Am J Physiol Cell Physiol* 1996; 271: C1424–C1437.

49 Dinarello CA, JG Cannon, Wolff SM *et al.* Tumour necrosis factor is an endogenous pyrogen and induces production of interleukin 1. *J Exp Med* 1986; 163: 1433–1450.

50 Elhage R, Maret A, Pieraggi MT *et al.* Differential effects of interleukin-1 receptor antagonist and tumour necrosis factor binding protein on fatty streak formation in apolipoprotein E-deficient mice. *Circulation* 1998; 97: 242–244.

51 Campbell GR, Campbell JH, Manderson JA *et al.* Arterial smooth muscle. A multifunctional mesenchymal cell. *Arch Pathol Lab Med* 1988; 112: 977–986.

52 Religa P, Bojakowski K, Maksymowicz M *et al.* Smooth muscle progenitor cells of bone marrow origin contribute to the development of neointimal thickenings in rat aortic allografts and injured rat carotid arteries. *Transplantation* 2002; 74: 1310–1315.

53 Buono C, Come G, Stavrakis GF *et al.* Influence of interferon-gamma on the extent and phenotype of diet induced atherosclerosis in the LDLR-deficient mouse. *Arterioscler Thromb Vasc Biol* 2003; 23: 454–460.

54 Tellides GD, Tereb A, Kirkiles-Smith NC *et al.* Interferon-gamma elicits arteriosclerosis in the absence of leukocytes. *Nature* 2000; 403: 207–211.

55 Burger-Kentischer A, Goebel H, Seiler R *et al.* Expression of macrophage migration inhibitory factor in different stages of human atherosclerosis. *Circulation* 2002; 105: 1561–1566.

56 Pan JH, Sukhova GK, Yang JT *et al.* Macrophage migration inhibitory factor deficiency impairs atherosclerosis

in low density lipoprotein receptor deficient mice. *Circulation* 2004; 109: 3149–3153.

57 Feldman LJ, Aguirre L, Ziol M *et al.* Interleukin-10 inhibits intimal hyperplasia after angioplasty or stent implantation in hypercholesterolaemic rabbits. *Circulation* 2000; 101: 908–916.

58 Pinderski LJ, Fischbein MP, Subbanagounder G *et al.* Overexpression of interleukin-10 by activated T lymphocytes inhibits atherosclerosis in LDL receptor deficient mice by altering lymphocyte and macrophage phenotypes. *Circ Res* 2002; 90: 1064–1071.

59 Jang WG, Kim HS, Park KG *et al.* Analysis of proteome and transcriptome of tumor necrosis factor alpha stimulated vascular smooth muscle cells with or without alpha lipoic acid. *Proteomics* 2004; 4: 3383–3393.

60 Forrester JS. Prevention of plaque rupture: a new paradigm of therapy. *Ann Intern Med* 2002; 137: 823–833.

61 Falk E. Plaque rupture with severe pre-existing stenosis precipitating coronary thrombosis. Characteristics of coronary atherosclerotic plaques underlying fatal occlusive thrombi. *Br Heart J* 1983; 50: 127–134.

62 Davies MJ, Thomas A. Plaque fissuring: the cause of acute myocardial infarction, sudden ischaemic death, and crescendo angina. *Br Heart J* 1985; 53: 363–373.

63 Naghavi M, Libby P, Falk E *et al.* From vulnerable plaque to vulnerable patient. *Circulation* 2003; 108: 1664–1672.

64 Falk E, Shah PK, Fuster V. Coronary plaque disruption. *Circulation* 1995; 92: 657–671.

65 Davies MJ. The composition of coronary artery plaques. *N Engl J Med* 1997; 336: 1312–1314.

66 Felton CV, Crook D, Davies MJ, Oliver MF. Relation of plaque lipid composition and morphology to the stability of human aortic plaques. *Arterioscler Thromb Vasc Biol* 1997; 17: 1337–1345.

67 Davies PF. Endothelium as a signal transduction interface for flow forces: cell surface dynamics. *Thromb Haemost* 1993; 70: 124–128.

68 Watson KE, Demer LL. The atherosclerosis-calcification link? *Curr Opin Lipid* 1996; 7: 101–104.

69 Saren P, Welgus HG, Kovanen P. TNF-alpha and IL-1beta selectively induce expression of 92-kDa gelatinase by human macrophages. *J Immunol* 1996; 157: 4159–4165.

70 Kol A, Sukhova GK, Lichtman AH, Libby P. Chlamydial heat shock protein 60 localizes in human atheroma and regulates macrophage tumor necrosis factor-alpha and matrix metalloproteinase expression. *Circulation* 1998; 98: 300–307.

71 Rajavashisth TB, Xu TP, Jovinge S *et al.* Membrane type 1 matrix metalloproteinase expression in human atherosclerotic plaques. *Circulation* 1999; 99: 3103–3109.

72 Bennett MR, Evan GI, Schwartz SM. Apoptosis of human vascular smooth muscle cells derived from normal vessels and coronary atherosclerotic plaques. *J Clin Invest* 1995; 95: 2266–2274.

73 Henderson EL, Geng YJ, Sukhova GK *et al.* Death of smooth muscle cells and expression of mediators of apoptosis by T lymphocytes in human abdominal aortic aneurysms. *Circulation* 1999; 99: 96–104.

74 Lopes N, Vasudevan SS, Alvarez RJ *et al.* Pathophysiology of plaque instability insights at the genomic level. *Prog Cardiovasc Dis* 2002; 44: 323–328.

75 Wallner K, Li C, Shah PK MD *et al.* Tenascin-C is expressed in macrophage-rich human coronary atherosclerotic plaque *Circulation* 1999; 99: 1284–1289.

76 LaFleur DW, Chiang J, Fagin JA. Aortic smooth muscle cells interact with tenascin C through its fibrinogen like domain. *J Biol Chem* 1997; 52: 32798–32803.

77 Tan KT, Lip GY. Platelets, atherosclerosis and the endothelium: new therapeutic targets? *Expert Opin Invest Drugs* 2003; 12: 1765–1776.

78 Massberg S, Brand K, Gruner S *et al.* A critical role of platelet adhesion in the initiation of atherosclerotic lesion formation. *J Exp Med* 2002; 196: 887–896.

79 Savage B, Almus-Jacobs F, Ruggeri ZM. Specific synergy of multiple substrate-receptor interactions in platelet thrombus formation under flow. *Cell* 1998; 94: 657–666.

80 Hynes RO. Integrins: Versatility, modulation, and signalling in cell adhesion. *Cell* 1992; 69: 11–25.

81 Kolodgie FD, Gold HK, Burke AP *et al.* Intraplaque haemorrhage and progression of coronary atheroma. *N Engl J Med* 2003; 349: 2316–2325.

82 Schonbeck U, Mach F, Sukhova GK *et al.* Regulation of matrix metalloproteinase in human smooth muscle cells by T lymphocytes: A role for CD40L in plaque rupture? *Circ Res* 1997; 81: 448–451.

83 Blake GJ, Ostfeld RJ, Yucel EK *et al.* Soluble CD40 ligand levels indicate lipid accumulation in carotid atheroma: an *in vivo* study with high resolution MRI. *Arterioscler Thromb Vasc Biol* 2003; 23: E11–E14.

84 Lander ES. The new genomics: global views of biology. *Science* 1996; 274: 536–539.

85 Risch N, Merikangas K. The future of genetic studies of complex human diseases. *Science* 1996; 273: 1516–1517.

86 Collins FS, Guyer MS, Chakravarti A. Variations on a theme: cataloging human DNA sequence variation. *Science* 1997; 278: 1580–1581.

87 Sholtz RI, Rosenma RH, Brand RJ. The relationship of reported parental history to the incidence of coronary heart disease in Western Collaborative Group Study. *Am J Epidemiol* 1975; 102: 350–356.

88 Jousilahti P, Puska P, Vartiainen E, Pekkanen J, Tuomilehto J. Parental history of premature coronary heart disease: an independent risk factor of myocardial infarction. *J Clin Epidemiol* 1996; 49: 497–503.

89 Nora JJ, Lortscher RH, Spangler RD, Nora AH, Kimberling WJ. Genetic-epidemiologic study of early-onset

ischemic heart disease. *Circulation* 1980; 61: 503–508.

90 Lusis AJ, Mar R, Pajukanta P. Genetics of atherosclerosis. *Annu Rev Genomics Hum Genet* 2004; 5: 189–218.

91 Wang L, Fan C, Topol SE *et al.* Mutation of MEF2A in an inherited disorder with features of coronary artery disease. *Science* 2003; 302: 1578–1581.

92 Lander ES, Schork NJ. Genetic dissection of complex traits. *Science* 1994; 265: 2037–2048.

93 Peltonen L, Palotie A, Lange K. Use of population isolates for mapping complex traits. *Nat Rev Genet* 2000; 1: 182–190.

94 Botstein D, Risch N. Discovering genotypes underlying human phenotypes: past successes for mendelian disease, future approaches for complex disease. *Nat Genet* 2003; Supplement: 228–237.

95 Xie X, Lu J, Kulbokas EJ *et al.* Systematic discovery of regulatory motifs in human promoters and 3′ UTRs by comparison of several mammals *Nature* 2005; 434: 338–345.

96 Lander ES, Linton LM, Birren B *et al.* Initial sequencing and analysis of the human genome. *Nature* 2001; 409: 860–921.

97 Venter JC, Adams MD, Myers EW *et al.* The sequence of the human genome. *Science* 2001; 291: 1304–1351.

98 Daly MJ, Rioux JD, Schaffner SF *et al.* High-resolution haplotype structure in the human genome. *Nat Genet* 2001; 29: 229–232.

99 Patil N, Berno AJ, Hinds DA *et al.* Blocks of limited haplotype diversity revealed by high-resolution scanning of human chromosome 21. *Science* 2001; 294: 1719–1723.

100 Reich DE, Cargill M, Bolk S *et al.* Linkage disequilibrium in the human genome. *Nature* 2001; 411: 199–204.

101 Rioux JD, Daly MJ, Silverberg MS *et al.* Genetic variation in the 5q31 cytokine gene cluster confers susceptibility to Crohn disease. *Nat Genet* 2001; 29: 223–228.

102 Gabriel SB, Schaffner SF, Nguyen H *et al.* The structure of haplotype blocks in the human genome. *Science* 2002; 296: 2225–2229.

103 The International HapMap Project. *Nature* 2003; 426: 789–796.

104 Wall JD, Pritchard JK. Haplotype blocks and linkage disequilibrium in the human genome. *Nat Rev Genet* 2003; 4: 587–597.

105 Cohen JC, Kiss RS, Pertsemlidis A *et al.* Multiple rare alleles contribute to low plasma levels of HDL cholesterol. *Science* 2004; 305: 869–872.

106 Weeks DE, Lathrop GM. Polygenic disease: methods for mapping complex disease traits. *Trends Genet* 1995; 11: 513–519.

107 Kraft HG, Lingenhel A, Kochl S *et al.* Apolipoprotein(a) kringle IV repeat number predicts risk for coronary heart disease. *Arterioscler Thromb Vasc Biol* 1996; 16: 713–719.

108 Cambien F, Poirier O, Lecerf L *et al.* Deletion polymorphism in the gene for angiotensin-converting enzyme is a potent risk factor for myocardial infarction. *Nature* 1992; 359: 641–644.

109 Boushey CJ, Beresford SA, Omenn GS, Motulsky AG. A quantitative assessment of plasma homocysteine as a risk factor for vascular disease: probable benefits of increasing folic acid intakes. *JAMA* 1995; 274: 1049–1057.

110 Wilson PW, Schaefer EJ, Larson MG, Ordovas JM. Apolipoprotein E alleles and risk of coronary disease A meta-analysis. *Arterioscler Thromb Vasc Biol* 1996; 16: 1250–1255.

111 Hingorani AD. Polymorphisms in endothelial nitric oxide synthase and atherogenesis: John French Lecture 2000. *Atherosclerosis* 2000; 154: 521–527.

112 Weng L, Kavaslar N, Ustaszewska A. Lack of MEF2A mutations in coronary artery disease. *J Clin Invest* 2005; 115: 1016–1020.

113 Kajimoto K, Shioji K, Tago N *et al.* Assessment of MEF2A mutations in myocardial infarction in Japanese patients. *Circ J* 2005; 69: 1192–1195.

114 Helgadottir A, Manolescu A, Thorleifsson G *et al.* The gene encoding 5-lipoxygenase activating protein confers risk of myocardial infarction and stroke. *Nat Genet* 2004; 36: 233–239.

115 Kajimoto K, Shioji K, Ishida C *et al.* Validation of the association between the gene encoding 5-lipoxygenase-activating protein and myocardial infarction in a Japanese population. *Circ J* 2005; 69: 1029–1034.

116 Moncada S. Nitric oxide: discovery and impact on clinical medicine. *J R Soc Med* 1999; 92: 164–169.

117 Cardillo C, Kilcoyne CM, Quyyumi AA *et al.* Selective defect in nitric oxide synthesis may explain the impaired endothelium-dependent vasodilation in patients with essential hypertension. *Circulation* 1998; 97: 851–856.

118 Zeiher AM, Drexler H, Saubier B, Just H. Endothelium-mediated coronary blood flow modulation in humans. Effects of age, atherosclerosis, hypercholesterolaemia, and hypertension. *J Clin Invest* 1993; 92: 652–662.

119 Saenz de Tejada I, Goldstein I, Azadzoi K *et al.* Impaired neurogenic and endothelium-mediated relaxation of penile smooth muscle from diabetic men with impotence. *N Engl J Med* 1989; 320: 1025–1030.

120 Hingorani AD, Liang CF, Fatibene J *et al.* A common variant of the endothelial nitric oxide synthase (Glu^{298}→Asp) is a major risk factor for coronary artery disease in the UK *Circulation* 1999; 100: 1515–1520.

121 Jachymova M, Horky K, Bultas J *et al.* Association of the Glu298Asp polymorphism in the endothelial nitric

oxide synthase gene with essential hypertension resistant to conventional therapy. *Biochem Biophys Res Commun* 2001; 284: 426–430.

122 Yoshimura M, Yasue H, Nakayama M *et al.* A missense Glu298Asp variant in the endothelial nitric oxide synthase gene is associated with coronary spasm in the Japanese. *Hum Genet* 1998; 103: 65–69.

123 Cai H, Wilcken DEL, Wang XL. The glu298→asp (894G-T) mutation at exon 7 of the endothelial nitric oxide synthase gene and coronary artery disease. *J Mol Med* 1999; 77: 511–514.

124 Nakayama M, Yasue H, Michihiro Y *et al.* T^{-786}→C Mutation in the 5′-flanking region of the endothelial nitric oxide synthase gene is associated with coronary spasm *Circulation* 1999; 99: 2864–2870.

125 Nakayama M, Yasue H, Yoshimura M *et al.* T^{-786}→C mutation in the 5′-flanking region of the endothelial nitric oxide synthase gene is associated with myocardial infarction, especially without coronary organic stenosis *Am J Cardiol* 2000; 86: 628–634.

126 Zanchi A, Moczulski DK, Hanna LS *et al.* Risk of advanced diabetic nephropathy in type 1 diabetes is associated with endothelial nitric oxide synthase gene polymorphism. *Kidney Int* 2000; 57: 405–413.

127 Wang XL, Sim AS, Badenhop RF *et al.* A smoking-dependent risk of coronary artery disease associated with a polymorphism of endothelial nitric oxide synthase gene. *Nat Med* 1996; 2: 41–45.

128 Loftus IM, Naylor AR, Bell PRF, Thompson MM. Matrix metalloproteinases and atherosclerotic plaque instability. *Br J Surg* 2002; 89: 680–694.

129 Pajukanta P, Cargill M, Viitanen L *et al.* Two loci on chromosomes 2 and X for premature coronary heart disease identified in early- and late-settlement populations of Finland. *Am J Hum Genet* 2000; 67: 1481–1493.

130 Francke S, Manraj M, Lacquemant C *et al.* A genome-wide scan for coronary heart disease suggests in Indo-Mauritians a susceptibility locus on chromosome 16p13 and replicates linkage with the metabolic syndrome on 3q27. *Hum Mol Genet* 2001; 10: 2751–2765.

131 Harrap SB, Zammit KS, Wong ZY *et al.* Genome-wide linkage analysis of the acute coronary syndrome suggests a locus on chromosome 2. *Arterioscler Thromb Vasc Biol* 2002; 22: 874–878.

132 Broeckel U, Hengstenberg C, Mayer B *et al.* A comprehensive linkage analysis for myocardial infarction and its related risk factors. *Nat Genet* 2002; 30: 210–214.

133 Hauser ER, Vrossman DC, Granger CB *et al.* A genomewide scan for early-onset coronary artery disease in 438 families: the GENECARD Study. *Am J Hum Genet* 2004; 75: 436–447.

134 Lange LA, Lange EM, Bielak LF *et al.* Autosomal genome-wide scan for coronary artery calcification loci in sibships at high risk for hypertension. *Arterioscler Thromb Vasc Biol* 2002; 22: 418–423.

135 Ozaki K, Ohnishi Y, Iida A *et al.* Functional SNPs in the lymphotoxin-alpha gene that are associated with susceptibility to myocardial infarction. *Nat Genet* 2002; 32: 650–654.

136 Klein RJ, Zeiss C, Chew EY *et al.* Complement factor H polymorphism in age-related macular degeneration. *Science* 2005; 308: 385–389.

137 Carlson CS, Eberle MA, Kruglyak L, Nickerson DA. Mapping complex disease loci in whole-genome association studies. *Nature* 2004; 429: 446–452.

138 Hirschhorn JN, Daly MJ. Genome-wide association studies for common diseases and complex traits. *Nat Rev Genet* 2005; 6: 95–108.

139 Randi AM, Biguzzi E, Falciani F *et al.* Identification of differentially expressed genes in coronary atherosclerotic plaques from patients with stable or unstable angina by cDNA array analysis. *J Thromb Haemost* 2003; 1: 829–835.

140 Mootha VK, Lindgren CM, Eriksson KF *et al.* PGC-1alpha-responsive genes involved in oxidative phosphorylation are coordinately downregulated in human diabetes. *Nat Genet* 2003; 34: 267–273.

141 Stein O, Thiery J, Stein Y. Is there a genetic resistance to atherosclerosis? *Atherosclerosis* 2002; 160: 1–10.

142 Kern F Jr. Normal plasma cholesterol in an 88-year-old man who eats 25 eggs a day. Mechanisms of adaptation. *N Engl J Med.* 1991; 28: 896–899.

143 Hobbs HH, Brown MS, Goldstein JL. Molecular genetics of LDL receptor gene in familial hypercholesterolaemia. *Hum Mutat* 1992; 1: 445–466.

144 Greenow K, Pearce NJ, Ramji D. The key role of apolipoprotein E in atherosclerosis. *J Mol Med* 2005; 83: 342–392.

145 Plump AS, Smith JD, Hayek T *et al.* Severe hypercholesterolemia and atherosclerosis in apolipoprotein E-deficient mice created by homologous recombination in ES cells. *Cell* 1992; 71: 343–353.

146 Mahley RW, Rall SC. Apolipoprotein E: far more than a transport protein. *Annu Rev Genomics Hum Genet* 2000; 1: 507–537.

147 Schaefer EJ, Gregg RE, Ghiselli G *et al.* Familial apolipoprotein E deficiency. *J Clin Invest* 1986; 78: 1206–1219.

148 Mahley RW, Huang Y, Rall SC. Pathogenenesis of type III hyperlipoproteinaemia. *J Lipid Res* 1999; 40: 1933–1949.

149 Song Y, Stampfer MJ, Liu S. Meta-analysis: apolipoprotein E genotypes and risk for coronary heart disease. *Ann Intern Med* 2004; 141: 137–147.

150 Jukema JW, van Boven AJ, Groenmeijer B *et al.* The Asp9 Asn mutation in the lipoprotein lipase gene is

associated with increased progression of coronary atherosclerosis. REGRESS Study Group, Interuniversity Cardiology Institute, Utrecht, the Netherlands. Regression Growth Evaluation Statin Study. *Circulation* 1996; 94: 1913–1918.

151 Humphries SE, Nicaud V, Margalef J. Lipoprotein lipase gene variation is associated with a paternal history of premature coronary artery disease and fasting and postprandial plasma triglycerides: the European Atherosclerosis Research Study (EARS). *Arterioscler Thromb Vasc Biol* 1998; 18: 526–534.

152 Kronenberg F, Kronenberg MF, Kiechl S *et al.* Role of lipoprotein(a) and apolipoprotein(a) phenotype in atherogenesis: prospective results from the Bruneck study. *Circulation* 1999; 100: 1154–1160.

153 Ridker PM, Hennekens CH, Stampfer MJ. A prospective study of lipoprotein(a) and the risk of myocardial infarction. *JAMA* 1993; 270: 2195–2199.

154 Rosengran A, Wilhelmsen L, Eriksen E *et al.* Lipoprotein(a) and coronary heart disease: a prospective case–control study in a general population sample of middle-aged men. *Br Med J* 1990; 301: 1248–1251.

155 Kim JH, Kim KH, Nam SM. The apolipoprotein(a) size, pentanucleotide repeat, C/T(+93) polymorphisms of apolipoprotein(a) gene, serum lipoprotein(a) concentrations and their relationship in a Korean population. *Clin Chim Acta* 2001; 314: 113–123.

156 Barter BJ, Brewer HB, Chapman J *et al.* Cholesteryl ester transfer protein. *Arterioscler Thromb Vasc Biol* 2003; 23: 160–167.

157 Okamoto H, Yonemori F, Wakitani K *et al.* A cholesteryl ester transfer protein inhibitor attenuates atherosclerosis in rabbits. *Nature* 2000; 406: 203–206.

158 Boekholdt S, Sacks FM, Jukema JW *et al.* Cholesteryl ester transfer protein TaqIB variant, high-density lipoprotein cholesterol levels, cardiovascular risk, and efficacy of pravastatin treatment. *Circulation* 2005; 111: 278–287.

159 Salen G, Shefer S, Nguyen L. Sitosterolemia. *J Lipid Res* 1992; 33: 945–955.

160 Berge KE, Tian H, Graf GA *et al.* Accumulation of dietary cholesterol in sitosterolemia caused by mutations in adjacent ABC transporters. *Science* 2000; 290: 1771–1775.

161 Stefkova J, Poledne R, Hubacek JA. ATP-binding cassette (ABC) transporters in human metabolism and diseases. *Physiol Res* 2004; 53: 235–243.

162 Graf GA, Li WP, Gerard RD *et al.* Coexpression of ATP-binding cassette proteins ABCG5 and ABCG8 permits their transport to the apical surface. *J Clin Invest* 2002; 110: 659–669.

163 Oram JF. Molecular basis of cholesterol homeostasis: lessons from Tangier disease and ABCA1. *Trends Mol Med* 2002; 8: 168–173.

164 Joyce CW, Amar MJ, Lambert G *et al.* The ATP binding cassette transporter A1 (ABCA1) modulates the development of aortic atherosclerosis in C57BL/6 and apoE-knockout mice. *Proc Natl Acad Sci USA* 2002; 99: 407–412.

165 Alrasadi K, Ruel IL, Marcil M *et al.* Functional mutations of the ABCA1 gene in subjects of French-Canadian descent with HDL deficiency. *Atherosclerosis* 2006; 188: 281–291.

166 Frikke-Schmidt R, Noordesgardt BG, Schnohr P *et al.* Mutation in ABCA1 predicted risk of ischemic heart disease in the Copenhagen City Heart Study Population. *J Am Coll Cardiol* 2005; 46: 1516–1520.

167 Pajukanta P, Lilja HE, Sinsheimer J *et al.* Familial combined hyperlipidemia is associated with upstream transcription factor 1 (USF1). *Nat Genet* 2004; 36: 371–376.

168 Huertas-Vazquez A, Aguilar-Salinas C, Lusis AJ. Familial combined hyperlipidemia in Mexicans: Association with upstream transcription factor 1 and linkage on chromosome 16q24.1. *Arterioscler Thromb Vasc Biol* 2005; 25: 1985–1991.

169 Coon H, Xin Y, Hopkins PN *et al.* Upstream stimulatory factor 1 associated with familial combined hyperlipidemia, LDL cholesterol, and triglycerides. *Hum Genet* 2005; 117: 444–451.

170 Komulainen K, Alanne M, Auro K *et al.* Risk alleles of USF1 gene predict cardiovascular disease of women in two prospective studies. *PLoS Genet* 2006; 2 : e69 [Epub ahead of print].

171 Liao F, Andalibi A, Qiao JH *et al.* Genetic evidence for a common pathway mediating oxidative stress, inflammatory gene induction, and aortic fatty streak formation in mice. *J Clin Invest* 1994; 94: 877–884.

172 Dwyer JH, Allayee H, Dwyer KM *et al.* Arachidonate 5-lipoxygenase promoter genotype, dietary arachidonic acid, and atherosclerosis. *N Engl J Med* 2004; 350: 29–37.

173 Helgadottir A, Gretarsdottir S, St Clair D *et al.* Association between the gene encoding 5-lipoxygenase-activating protein and stroke replicated in a Scottish population. *Am J Hum Genet* 2005; 76: 505–509.

174 Mehrabian M, Allayee H, Wong J *et al.* Identification of 5-lipoxygenase as a major gene contributing to atherosclerosis susceptibility in mice. *Circ Res* 2002; 91: 120–126.

175 Lotzer K, Funk CD, Habenicht AJ. The 5-lipoxygenase pathway in arterial wall biology and atherosclerosis. *Biochim Biophys Acta* 2005; 1736: 30–37.

176 Yamada A, Ichihara S, Murase Y *et al.* Lack of association of polymorphisms of the lymphotoxin alpha gene with myocardial infarction in Japanese. *J Mol Med* 2004; 82: 477–483.

177 Koch W, Kastrati A, Bottiger C *et al.* Interleukin-10 and tumor necrosis factor gene polymorphisms and risk of coronary artery disease and myocardial infarction. *Atherosclerosis* 2001; 159: 137–144.

178 Elghannam H, Tavackoli S, Ferlic L *et al.* A prospective study of genetic markers of susceptibility to infection and inflammation, and the severity, progression and regression of coronary atherosclerosis and its response to therapy. *J Mol Med* 2000; 78: 562–568.

179 Maresca G, DiBlasio A, Marchioli R *et al.* Measuring plasma fibrinogen to predict stroke and myocardial infarction: an update. *Arterioscler Thromb Vasc Biol* 1999; 19: 1368–1377.

180 Yang Q, Tofler GH, Cupples LA *et al.* A genome-wide search for genes affecting circulating fibrinogen levels in the Framingham Heart Study. *Thromb Res* 2003; 110: 57–64.

181 Folsom AR, Aleksic N, Ahn C *et al.* Beta-fibrinogen gene -455G/A polymorphism and coronary heart disease incidence: the Atherosclerosis Risk in Communities (ARIC) Study. *Ann Epidemiol* 2001; 11: 166–170.

182 Lee AJ, Fowkes FG, Lowe GD *et al.* Fibrinogen, factor VII and PAI-1 genotypes and the risk of coronary and peripheral atherosclerosis: Edinburgh Artery Study. *Thromb Haemost* 1999; 81: 553–560.

183 Thogersen AM, Jansson JH, Boman K *et al.* High plasminogen activator inhibitor and tissue plasminogen activator levels in plasma precede a first acute myocardial infarction in both men and women. *Circulation* 1998; 98: 2241–2247.

184 Dawson SJ, Wiman B, Hamsten A *et al.* The two allele sequences of a common polymorphism in the promoter of the plasminogen activator inhibitor-1 (PAI-1) gene respond differently to interleukin-1 in HepG2 cells. *J Biol Chem* 1993; 268: 10739–10745.

185 Iacoviello L, Burzotta F, Di Castelnuovo A *et al.* The 4G/5G polymorphism of PAI-1 promoter gene and the risk of myocardial infarction: a meta-analysis. *Thromb Haemost* 1998; 80: 1029–1030.

186 Monajemi H, Arkenbout EK, Pannekoek H. Gene expression in atherogenesis. *Thromb Haemost* 2001; 86: 404–412.

187 Libby P, Ridker PM. Inflammation and atherosclerosis: role of C-reactive protein in risk assessment. *Am J Med* 2004; 116 (Supplement 6A): 9S–16S.

188 Libby P. What have we learned about the biology of atherosclerosis? The role of inflammation. *Am J Cardiol* 2001; 88 (7B): 3J–6J.

189 Zuckerbraun BS, Tzeng E. Vascular gene therapy: a reality of the 21st century. *Arch Surg* 2002; 137: 854–861.

190 Channon KM, Qian H, Nephlioeva V *et al. In vivo* gene transfer of nitric oxide synthase enhances vasomotor function in carotid arteries from normal and cholesterol-fed rabbits. *Circulation* 1998; 98: 1905–1911.

191 Qian H, Nephlioeva V, Shetty GA *et al.* Nitric oxide synthase gene therapy rapidly reduces adhesion molecule expression and inflammatory cell infiltration in carotid arteries of cholesterol-fed rabbits. *Circulation* 1999; 99: 2979–2982.

192 Goldstein JL, Hobbs HH, Brown MS. Familial hypercholesterolemia. In: Scriver CR, Beaudet AL, Sly WS, Valle D, eds. *Metabolic Basis of Inherited Disease.* McGraw Hill, New York. 1995; 1267–1282.

193 Abifadel M, Varret M, Rabes JP *et al.* Mutations in PCSK9 cause autosomal dominant hypercholesterolemia. *Nat Genet* 2003; 34: 154–156.

194 Sing CF, Davignon J. Role of the apolipoprotein E polymorphism in determining normal plasma lipid and lipoprotein variation. *Am J Hum Genet* 1985; 37: 268–285.

195 Talmud PJ, Hawe E, Martin S *et al.* Relative contribution of variation within the APOC3/A4/A5 gene cluster in determining plasma triglycerides. *Hum Mol Genet* 2002; 11: 3039–3046.

196 Shohet RV, Vega GL, Bersot TP *et al.* Sources of variability in genetic association studies: insights from the analysis of hepatic lipase (LIPC). *Hum Mutat* 2002; 19: 536–542.

197 Frikke-Schmidt R, Nordestgaard BG, Jensen GB, Tybjaerg-Hansen A. Genetic variation in ABC transporter A1 contributes to HDL cholesterol in the general population. *J Clin Invest* 2004; 114: 1343–1353.

198 Boerwinkle E, Leffert CC, Lin J, Lackner C, Chiesa G, Hobbs HH. Apolipoprotein(a) gene accounts for greater than 90% of the variation in plasma lipoprotein(a) concentrations. *J Clin Invest* 1992; 90: 52–60.

199 Kang SS, Passen EL, Ruggie N *et al.* Thermolabile defect of methylenetetrahydrofolate reductase in coronary artery disease. *Circulation* 1993; 88: 1463–1469.

200 Ma J, Stampfer MJ, Hennekens CH *et al.* Methylenetetrahydrofolate reductase polymorphism, plasma folate, homocysteine, and risk of myocardial infarction in US physicians. *Circulation* 1996; 94: 2410–2416.

201 Hamsten A. The hemostatic system and coronary heart disease. *Thromb Res* 1993; 70: 1–38.

202 Thomas AE, Green FR, Lamlum H, Humphries SE. The association of combined alpha and beta fibrinogen genotype on plasma fibrinogen levels in smokers and non-smokers. *J Med Genet* 1995; 32: 585–589.

203 Caulfield M, Lavender P, Newell-Price J *et al.* Linkage of the angiotensinogen gene locus to human essential hypertension in African Caribbeans. *J Clin Invest* 1995; 96: 687–692.

204 Lusis AJ, Weinreb A, Drake TA, Allayee H. Genetics of atherosclerosis. In: Topol E, ed. *Textbook of Cardiovascular Medicine*, 2nd edn. Lippincott Williams and Wilkins, Philadelphia, 2002.

205 Staessen JA, Wang JG, Ginocchio G *et al.* The deletion/insertion polymorphism of the angiotensin converting enzyme gene and cardiovascular-renal risk. *J Hypertens* 1997; 15: 1579–1592.

206 Shih DM, Reddy S, Lusis AJ. CHD and atherosclerosis: human epidemiological studies and trangenic mouse models. In: Costa LG, Furlong C, eds. *Paraoxonase (PON1) in Health and Disease*. Kluwer, Boston, 2001.

207 Tward A, Xia YR, Wang XP *et al.* Decreased atherosclerotic lesion formation in human serum paraoxonase transgenic mice. *Circulation* 2002; 106: 484–490.

208 Kiechl S, Lorenz E, Reindl M *et al.* Toll-like receptor 4 polymorphisms and atherogenesis. *N Engl J Med* 2002; 347: 185–192.

209 Gretarsdottir S, Thorleifsson G, Reynisdottir ST *et al.* The gene encoding phosphodiesterase 4D confers risk of ischemic stroke. *Nat Genet* 2003; 35: 131–138.

210 Altshuler D, Hirschhorn JN, Klannemark M *et al.* The common PPARgamma Pro12Ala polymorphism is associated with decreased risk of type 2 diabetes. *Nat Genet* 2000; 26: 76–80.

211 Horikawa Y, Oda N, Cox NJ *et al.* Genetic variation in the gene encoding calpain-10 is associated with type 2 diabetes mellitus. *Nat Genet* 2000; 26: 163–175.

212 Silander K, Mohlke KL, Scott LJ *et al.* Genetic variation near the hepatocyte nuclear factor-4 alpha gene predicts susceptibility to type 2 diabetes. *Diabetes* 2004; 53: 1141–1149.

213 Wang Q, Rao S, Shen GQ *et al.* Premature myocardial infarction novel susceptibility locus on chromosome 1P34-36 identified by genomewide linkage analysis. *Am J Hum Genet* 2004; 74: 262–271.

214 Grant SFA, Thorleifsson G, Reynisdottir I *et al.* Variant of transcription factor 7-like 2 (TCF7L2) gene confers risk of type 2 diabetes. *Nat Genet* 2006; 38: 320–323.

CHAPTER 7

Heart failure

Markus Meyer, MD, *& Peter VanBuren,* MD

Overview

Introduction

Major advances in the treatment of cardiovascular diseases in general (e.g., coronary artery disease and sudden cardiac death) have resulted in improved overall mortality but ultimately contribute to the increasing incidence and prevalence of heart failure. Currently more than 550,000 new cases of heart failure are diagnosed in the USA each year [1]. Based on a 2002 estimate, the prevalence of heart failure in the US is approximately 5 million, with many more cases of preclinical heart failure uncounted in the population at large [2]. Once diagnosed with heart failure, an individual's 1 year mortality is approximately 20%, with 80% and 70% 8-year mortalities in men and women, respectively. Since 1992, the number of deaths from heart failure has increased by 35%. Nearly 1,000,000 patients were hospitalized in the USA with heart failure in 2002. Currently (2005), the estimated cost of heart failure in the USA is approximately 28 billion dollars [1].

These staggering statistics underscore the fact that this disease is a major public health burden which warrants active and broad based research to discover novel therapies. While some patients presenting with heart failure have inherited disorders where a single gene is the cause, the majority of heart failure cases are acquired and involve the interaction of multiple genes. Thus, by definition, acquired forms of heart failure are polygenic disorders. Gene regulation and expression are the cornerstones of understanding the processes that govern the pathophysiology of heart failure.

It is now well understood that heart failure is generally triggered by a direct insult to the myo-cardium. The source of the insult can be broadly categorized as ischemic/infarction, toxic metabolic (e.g., diabetes, alcohol, chemotherapy), inflammatory (e.g., myocarditis), hemodynamic (e.g., valvular heart disease, hypertension) or inherited (e.g., dilated cardiomyopathy, hypertrophic cardiomyopathy). Furthermore, the underlying conditions of diabetes, hypertension, renal disease, obesity and sleep apnea can contribute in an additive fashion to other myocardial insults. Following an insult, the heart hypertrophies seen as an increase in cardiac mass as well as an increase in myocyte length and width at the cellular level. In most cardiomyopathies, if the inciting insult (or its sequel) persists the left ventricle will dilate, which is then followed by further deterioration of systolic function (Fig. 7.1). In end-stage cardiomyopathy, the gross phenotype of the heart is strikingly similar irrespective of the initial etiology. The process of hypertrophy, dilation and dysfunction is termed ventricular remodeling or, more aptly, adverse remodeling. The progression of heart failure is a dynamic process where adverse remodeling can be partially reversed or at least halted with the initiation of therapies and/or removal of the insult. In the following pages, we review the stages of heart failure and probe known and probable changes in gene expression as a function of the progression of the disease.

Polygenic factors in the development of heart failure

Acquired heart failure is a polygenic disorder and several predisposing medical conditions increase the risk of developing heart failure. For example, hypertension, diabetes and renal disease are associated with altered hemodynamic stress, neurohormonal and inflammatory signaling which trigger

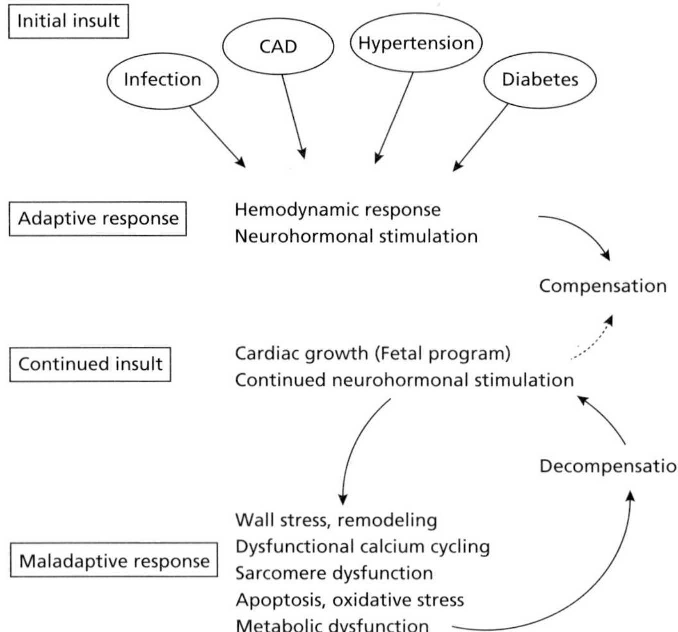

Figure 7.1 Polygenic diseases leading to heart failure. Intra- or extracardiac insults trigger hemodynamic and neurohormonal responses as an adaptive response. The intracellular pathways activated by prolonged mechanical or hormonal stimuli mediate cellular growth (hypertrophy), and have additional effects on the transcriptional control and post translational modification of proteins involved in contractility, metabolism, apoptosis and the extracellular matrix. These changes detrimental to cardiac function sustain a vicious cycle of decompensation and additional growth (maladaptive response). CAD, coronary artery disease.

the development of heart failure. These diseases increase the risk of developing heart failure, both as indirect sequelae of the disease process and as a result of direct modulation of cardiac gene expression.

A patient's race and sex are important factors in the pathogenesis of heart failure. The incidence of heart failure is currently declining in women but not in men [3]. After the onset of heart failure, women demonstrate better survival when compared with men [3,4]. The proportion of women with diastolic heart failure with preserved systolic function is greater than in men [2], which may be a factor in their improved prognosis. In response to pressure overload, the pattern of hypertrophy tends to be more concentric in woman and dilated in men [5], indicating that there is a sex difference in the remodeling of the left ventricle, a process that is ultimately tied to myocardial gene expression. Evidence of gender-specific responses to therapy is seen in the subset analyses of the SOLVD and CONSENSUS-1 studies which demonstrated that men achieved a greater beneficial response to angiotensin-converting enzyme inhibitors (ACE-I) than women [6]. In contrast, the response to beta-blocker therapy in heart failure is greater in women than in men [7,8].

Estrogen is likely a major (but not the sole) factor in the differential response of women to specific therapies compared with men [9]. Estrogen has shown attenuated hypertrophic response in animal models of myocardial failure [10]. Estrogen can directly affect transcription through two nuclear hormone receptors (α and β). In addition, estrogen appears to affect cell signaling through the phosphoinositide-3 kinase (PI3K)-Akt-dependent and mitogen activated protein kinase (MAPK) pathways which are discussed in more detail below [11,12]. In addition, estrogen is a potent vasodilator and increases nitric oxide and atrial natriuretic peptide production. Furthermore, testosterone can also affect cardiac function; thus, it may be the relative influence of sex hormones that is a determining factor in the therapeutic response of heart failure patients. However, more research needs to be conducted to better understand the effects of gender on the remodeling process in heart failure.

Black patients have a greater prevalence of heart failure and higher mortality compared with white patients [1]. These differences may be related, in part, to the increased incidence of hypertension and renal disease in black patients [13]. Differences in neurohormonal signaling may underlie the poor prognosis and altered response to heart failure therapy [14]. Specifically, response rate to ACE-I in patients with an African-American background is less pronounced than in patients with a Caucasian background [14,15], whereas the response to specific beta-blockers has been variable. Bucindolol conferred a benefit in white patients but was associated with a trend toward higher mortality in black patients [16]. In contrast, the benefit of carvedilol is similar for black and white patients [17]. These differences likely underlie the different mechanisms of action of specific beta-blockers. Taken together these results suggest that there are fundamental differences in signaling pathways and genetic regulation in black and white populations. This was the focus of the recently published African American Heart Failure Trial (A-HeFT) trial, which revisited the use of nitrates and hydralazine in black patients with heart failure. A 43% reduction in mortality was demonstrated in spite of the fact that many of the patients were already being treated with beta-blockers and ACE-I or angiotensin receptor blockers [18]. Given that black patients have been found to have lower bioavailability of nitric oxide than white patients [19], the increased nitric oxide level associated with the combined therapy of nitroglycerin and hydralazine may be uniquely suited to benefit black patients with heart failure.

It is now understood that polymorphisms in specific genes can predispose patients to heart failure. Gene polymorphisms are population variations in DNA composition that occur at a frequency >1%. In African-Americans these include polymorphisms in the α2c and β1 adrenergic receptors, endothelial nitric oxide synthase (eNOS), aldosterone synthase, transforming growth factor β1 (TGF-β1) and G proteins [20,21]. In addition, polymorphisms in the angiotensin-converting enzyme (deletion allele) has been reported in the general population with relatively high frequency [22]. All of these polymorphisms may predispose a carrier to the development of cardiomyopathy, impact prognosis or affect the response to a specific treatment for heart failure.

In summary, differences in genetic make up or alteration in genetic expression resulting from an underlying condition can affect the risk of developing heart failure, its prognosis and response to therapy. In the future, determining a patient's genetic profile may help tailor therapies in an effort to optimize response.

Humoral and hemodynamic factors in heart failure

Hemodynamic and mechanical considerations

During most of the last century, heart failure was understood as a disorder of cardiac ejection or systolic function. Only since the 1970s have we begun to understand that impaired diastolic function also contributes to most cases of heart failure. This is most obvious in end stage dilated cardiomyopathy where both systolic and diastolic function are impaired. In other words, a ventricle that ejects poorly in systole usually does not fill normally in diastole, and a heart that fills poorly in diastole usually does not eject normally in systole [23]. The initial hemodynamic, morphologic and molecular changes are acute adaptive responses that aim to preserve cardiac output after a myocardial insult. The earliest response involves adjustments of cardiac parameters including preload, afterload, heart rate and changes in the speed of contraction and relaxation. These early hemodynamic adaptations are supported by intrinsic myocardial mechanisms that control cardiac contractility and output. Such mechanisms include preload and heart rate dependent inotropy first described more than a century ago by Starling and Bowditch [24,25].

Many of these early adaptive processes are regulated by neurohormonal factors such as catacholamines and angiotensin II, whose levels are increased in the blood in an attempt to further stabilize cardiac output and blood pressure. This response is best exemplified in the activation of the renin–angiotensin–aldosterone system, where reduced renal perfusion leads to the renal release of renin, aldosterone and vasopressin. The end result is increased vasoconstriction and renal free water reabsorption. Norepinephrine, released primarily by the adrenal glands, increases heart rate and contractility through β_1-adrenergic stimulation whereas α-adrenergic stimulation in the peripheral vasculature leads to vasoconstriction. Thus, many effects of initial neurohormonal stimulation appear to be adaptive by attempting to maintain cardiac output and blood pressure. However, in the past 20 years we have come to understand that neurohormones, when chronically activated, lead to detrimental changes by creating unfavorable hemodynamic conditions, such as increased cardiac afterload, excessive preload and volume overload. Moreover, neurohormones can induce cardiac growth in the form of concentric and eccentric hypertrophy by directly activating intracellular signaling pathways that initiate a hypertrophy gene program, as discussed in greater detail below. This process of cardiac hypertrophy in the setting of myocardial stress and failure is maladaptive, as evidenced by clinical trials that demonstrate left ventricular hypertrophy is an independent predictor of mortality. Such maladaptive hypertrophy contrasts with adaptive hypertrophy (i.e., fetal cardiac development, cardiac growth into adulthood and cardiac growth as a response to exercise) which is governed by many of the same mechanisms. If cardiac remodeling results in ventricular dilatation, unfavorable chamber geometry and increased wall stress results. This is based on the Law of Laplace, where wall stress is proportional to (pressure × radius) / (2 × wall thickness). In concentric hypertrophy, where chamber dimensions are preserved and wall thickness is increased, wall stress is usually not increased (even with increased intracardiac pressures) and thus is likely not a contributing factor in the pathologic processes of heart failure. However, in dilated cardiomyopathy where chamber dimensions are increased and wall thickness is largely unchanged or even reduced, wall stress is markedly elevated and thus believed to be a contributing factor to heart failure.

Myocardial wall stress is now well recognized to have direct effects on cardiac gene expression by activating intracellular signaling cascades independent of neurohormonal stimulation [26,27]. The clinically best known genes activated by wall stress are the natriuretic hormones and in particular brain natriuretic peptide (BNP). The natriuretic peptides are known to decrease afterload and increase diuresis. However, despite the favorable hemodynamic effects of natriuretic proteins, myocardial wall stress is known to induce both physiologic and pathologic myocardial hypertrophy through several mechanisms.

Surprisingly, while baseline cardiac output is either unchanged or slightly reduced in most forms of systolic heart failure, the intrinsic mechanisms to control contractility in the myocardium are markedly disturbed [28]. The Frank–Starling curve

is shifted to higher preloads in order to maintain similar systolic pressures, and the heart rate dependent increase in cardiac output is blunted. This explains why a severely dilated heart with low ejection fraction can have a normal stroke volume but typically shows a reduced contractile reserve compared with a normal heart, especially when challenged by higher heart rates or circulatory demands. This illustrates that ejection fraction is a poor measure of contractile function which does not correlate well with cardiac output and symptoms but has repeatedly proven to be a good predictor of survival in systolic heart failure [29,30]. Cardiac contractility is a much more complex and poorly defined measure which integrates preload and afterload, chamber size and shape, inotropy, speed of cardiac relaxation and the speed and sequence of cardiac excitation. The interplay of these factors determines if cardiac output and blood pressure are sufficient to maintain organ perfusion and provide reserve capacity in times of higher circulatory needs. This supports the concept that even a minor disturbance in this system over time can induce cardiac hypertrophy as an adaptive process which itself can worsen the imbalance leading to a vicious cycle of insult and maladaptive changes which ultimately can result in symptoms and a diagnosis of heart failure.

Neurohormonal and cytokine signaling

While the depressed contractility of the heart is central to the clinical syndrome of heart failure, there are several circulating neurohormones and cytokines that, in addition to having early inotropic effects, can directly or indirectly affect cardiac gene regulation and thus alter cardiac function. The neurohormones include catacholamines, angiotensin II, endothelin, aldosterone, vasopressin, estrogen, testosterone and growth hormone.

Epinephrine and norepinephrine

Circulating catecholamine levels are markedly elevated in heart failure and positively correlate with mortality [31]. Adrenergic stimulation of the heart by norepinephrine and epinephrine occurs through α and β adrenergic receptors. The β-adrenergic receptor has three isoforms with the β_1 isoform demonstrating 75% predominance in the nonfailing human cardiac myocyte. Stimulation of the β-adrenergic receptor causes an acute increase in myocardial contractility, but chronic adrenergic stimulation results in contractile dysfunction, adverse ventricular remodeling and increased mortality. The β-adrenergic receptor is a seven trans-membrane, G protein coupled receptor (discussed in more detail below). When stimulated it activates adenylate cyclase through a Gαs protein coupled mechanism to generate cyclic AMP to ultimately activate protein kinase A (PKA) [32]. Many of the acute increases in contractile response to β-adrenergic stimulation are mediated through PKA. In contrast, chronic β-adrenergic receptor stimulation ($β_1$ receptor subtype) is associated with the induction of several signaling pathways (e.g. the c-Jun N terminal kinases [JNKs] and Ca^{2+} activated calmodulin [CAM] kinase II) that are involved in ventricular remodeling and apoptosis [33]. In chronic heart failure, the β-adrenergic receptor is downregulated largely by reduced gene transcription [32]. Moreover, the chronic adrenergic stimulation of heart failure results in desensitization of the β-adrenergic receptor [35]. Polymorphisms of the β-adrenergic receptor gene have been reported in patients with heart failure [20,36,37]. Many of these polymorphisms have been shown to alter function *in vitro* [38], and are implicated in the differential therapeutic responses of patients with heart failure [39]. Finally, these polymorphism have been associated with the development and progression of heart failure [20,36].

The α-adrenergic receptor is much less abundant than the β receptor (1 : 10 ratio). The α-adrenergic receptor is a G protein coupled receptor (Gαq) which ultimately mediates the activation of protein kinase C (PKC). Despite its lower abundance, chronic stimulation of the α-adrenergic receptor elicits a cardiac hypertrophic response through the MAPK and calcineurin mediated pathways.

Angiotensin II

Renin is released by the kidney largely with the activation of the sympathetic nervous system, reduced renal perfusion and vasopressin release. Renin converts angiotensinogen to angiotensin I. The major source of angiotensinogen is the liver but other organs such as the heart also produce angiotensinogen. Angiotensin I is converted to angiotensin II by angiotensin-converting enzyme which also

hydrolyzes bradykinin. Angiotensin II levels are markedly elevated in heart failure because of the activation of renin. Angiotensin II is a potent vasoconstrictor that also triggers aldosterone release and stimulates renal tubular retention of sodium and water and thus directly contributes to increased myocardial hemodynamic stress in heart failure. The heart contains two cell membrane-associated angiotensin receptors. Stimulation of the angiotensin II type 1 (AT1) receptor is associated with several pathologic changes in the myocardium. Stimulation of the AT1 receptor occurs through a Gαq protein mediated mechanism to activate PKC through the same pathways as for α-adrenergic stimulation. Angiotensin II infusion in rats induced a failing cardiac phenotype with an increase in fetal gene program mRNA expression (e.g., α-skeletal actin, β myosin heavy chain, atrial natriuretic peptide) and an increase in cardiac mass [40]. Moreover, hypertrophy as measured by the increase in myocyte size and the rate of protein synthesis is also induced by angiotensin II [41,42]. Angiotensin II stimulation of the AT1 receptor is known to activate signaling pathways directly involved in the induction of factors that control gene transcription (i.e., the immediate early genes: *c-fos*, *c-jun*, *jun-B*, *Egr-1* and *c-myc* [41]). Modulation of the immediate early gene expression is implicated in the initial hypertrophic response. In addition, angiotensin II (through the AT1 receptor) induces fibroblast hyperplasia and collagen synthesis, with specific increases in fibronectin, TGF-β and collagen type I and III expression [40,43]. These data indicate the likely role of angiotensin II in the extracellular matrix remodeling and myocardial fibrosis associated with myocardial failure. Moreover, cardiac tissue can produce endogenous angiotensin through cellular angiotensin-converting enzyme and thus modulate the above pathway through an autocrine–paracrine type mechanism, the functional significance of which is currently unknown. Increased cellular necrosis and programmed cell death was noted with angiotensin II infusion [44].

While the overall abundance of angiotensin II receptors is low in the heart, the angiotensin II type 2 (AT2) receptor is expressed at a higher level than AT1 receptor in the human heart [45,46]. Whereas AT1 receptor protein expression is downregulated in end stage failure, the relative abundance of the AT2 receptor is unchanged or even upregulated [46,47]. The functional effects of AT2 receptor stimulation is not as well known but appears to be protective. AT2 receptor overexpression in transgenic mice resulted in preserved left ventricular function after myocardial infarction [48]. Furthermore, the downstream functional effects of the AT2 receptor appear distinct and largely opposite to the effects of the AT1 receptor. Specifically, stimulation of the AT2 receptor causes a relative suppression of the hypertrophic response [45], likely through the activation of phosphatases that inactivate the MAPK growth cascade [49].

Aldosterone

In heart failure, aldosterone levels are increased as much as 30-fold and are correlated with increased mortality [50]. In addition, aldosterone is directly linked to the induction of cardiac fibrosis and hypertrophy. Aldosterone is produced primarily by the adrenal glands, stimulated by elevated angiotensin II levels. However, other factors can trigger aldosterone release such as hyperkalemia, hyponatremia and adrenocorticotrophic hormone. Aldosterone modulates vascular tone and volume through affecting epithelial sodium transport and possibly through modulating vascular response through the α-adrenergic receptor or by upregulation of angiotensin II receptors in the vasculature [51].

Aldosterone is known to cause cardiac fibrosis and hypertrophy which largely appear to be independent of its vascular effects and is likely the direct effect of stimulation of the myocardial mineralocorticoid receptor [52]. When aldosterone binds the mineralocorticoid receptor (located in the cytoplasm), the receptor–ligand complex dimerizes, is translocated to the nucleus and ultimately binds to hormone response elements activating the transcription of target genes. Aldosterone increases the cardiac expression of the AT1 receptor and angiotensin-converting enzyme [53,54], thus contributing to the hypertrophic response through angiotensin II stimulation. Moreover, aldosterone has been linked to the upregulation in collagen I, collagen III, matrix metalloproteinase (MMP) expression as demonstrated in animal models of heart failure, facilitating the observed fibrosis associated with aldosterone stimulation of the heart

[55,56]. However, while the induction of hypertrophy and fibrosis with aldosterone is well documented, the specific nuclear signaling mechanisms are not completely understood. Activation (direct or indirect) of the transcription factors: activating protein 1, nuclear factor κB (NF κB) and nuclear factor of activated T cells (NFAT) have been implicated [56,57].

In addition, the effects of aldosterone can be observed over a number of minutes, which are likely nongenomic effects, and appear to be mediated through inositol 1,4,5-triphosphate, PKC and Ca^{2+} signaling [58]. Finally, local production of aldosterone by the heart possibly has a significant role in human heart failure [59]. Given the profound effect of mineralocorticoid receptor blockade on ventricular remodeling and mortality in clinical trials [60], further research in this area will likely yield more insight into the detrimental effects of aldosterone in heart failure.

Arginine vasopressin

Arginine vasopressin, also known as antidiuretic hormone (ADH) is released by the posterior pituitary and is known to facilitate free water reabsorption through the collecting ducts in the kidney. In addition, it is a potent vasoconstrictor, and thus through these two mechanisms can increase circulating blood volume and blood pressure. Vasopressin is elevated in heart failure [61]. There are three vasopressin receptor isoforms. V_{1a} is located in the adrenal cortex, myocardium and the vasculature. V_2 is located predominantly in the kidney and is responsible for its antidiuretic effect. V_{1b} is involved in the regulation of adrenocorticotropic hormone release from the pituitary gland [60]. Vasopressin directly stimulates hypertrophy and hyperplasia in cardiac myocytes and fibroblasts, respectively, as is demonstrated by increased protein and DNA synthesis [63,64]. Studies in cultured myocytes suggests that the V_{1a} receptor is G protein coupled activating inositol 1,4,5-triphosphate, and diacylglycerol through phospholipase C [65] ultimately triggering PKC, MAPK and mediated calcium signaling [63,64]. The nuclear transcription factors, c-Fos and c-Jun, are associated with the hypertrophic response [64]. Vasopressin receptor blockade is an area of active basic and clinical research.

Endothelin

Endothelin was first discovered in 1988 as a very potent vasoconstrictor. Endothelin 1 is the predominant isoform of endothelin in the cardiovascular system and is the cleavage product of endothelin-converting enzyme. Endothelin 1 is primarily produced from endothelial cells but also can be produced by other cells including cardiac myocytes and vascular smooth muscle cells [66]. In the heart, endothelin works primarily by autocrine and paracrine mechanisms with the majority of the endothelin being produced locally [67]. There are two isoforms of the endothelin (ET) receptor (A and B). ET_A and ET_B receptors are located on cardiac myocytes and vascular smooth muscle cells, whereas only ET_B is located on endothelial cells. The endothelin receptors are $G\alpha q$.

The acute effects of endothelin are an increase in stroke volume and vascular resistance through enhanced contractility of the myocardium and vascular smooth muscle. Circulating endothelin levels are chronically elevated in human heart failure [68]. It is likely that chronic elevation of endothelin has deleterious effects on cardiovascular function as endothelin blockade has been demonstrated to improve myocardial contractility and survival in animal models of heart failure [69]. While the acute effects of endothelin are nongenomic, the longer and deleterious effects of endothelin receptor stimulation involve transcriptional changes that contribute to adverse ventricular remodeling and contractile dysfunction. Overexpression of endothelin in a transgenic mouse model induced a dilated cardiac phenotype with activation of cytokines and transcription factor, NF κB [70]. Interestingly, endothelin-converting enzyme was found to be differentially regulated in one gene array study of human ischemic and nonischemic dilated cardiomyopathy [71]. Despite mounting evidence to invoke endothelin in the pathophysiology of cardiac remodeling and failure, clinical trials using endothelin blockers both nonspecific (i.e., blockade of ET_A and ET_B such as with bosentan) or specific (e.g., ET_A as with darusentan) have so far yielded largely neutral or negative results.

Cytokines

While cytokines have traditionally been viewed as systemic inflammatory modulators released by the immune system, it is now understood that they can be generated by a variety of cell types including cardiac myocytes, fibroblast and endothelium, and can mediate inflammatory responses through an autocrine or paracrine mechanism. The effects of inflammatory modulators were confirmed in several gene array studies which found that several genes involved in inflammatory responses are upregulated in samples from failing human hearts [72–74]. The proinflammatory cytokines, tumor necrosis factor α (TNF-α), interleukin-1β (IL-1β), IL-2, IL-6 and IL-18, have all been implicated in the pathophysiology of heart failure [75,76]. Specifically, circulating cytokine levels correlate with the severity of myocardial dysfunction and overall mortality [77,78]. In addition, infusion of these cytokines in animal models is associated with immediate improvement and delayed (sustained) suppression of myocardial contractile function implicating nongenomic and genomic effects, respectively [79]. The immediate effect of cardiac contractility appears to be through alterations in calcium cycling through a variety of mechanisms [80,81].

Cardiac nitric oxide synthase (cNOS) activation resulting in increased nitric oxide production by cytokines has been implicated in the delayed suppression of contractile function. Nitric oxide is known to affect transcriptional gene regulation through the activation of guanylate cyclase and generation of cyclic GMP which in turn activates protein kinase G (PKG). Through this process the activities of several transcription factors, activating protein 1 (c-Fos, Jun B), NF κB and NFAT) are increased in cardiocytes [82]. Furthermore, activation of transcription factor GATA4 is reported with IL-18 stimulation [75]. Of these transcription factors, it appears that NF-κB may be playing a dominant part [83]. The net effect of chronic cytokine stimulation hypertrophy in isolated myocytes and in transgenic overexpression models is a re-expression of a fetal gene program (e.g., ANP, BNP, skeletal α actin and β myosin heavy chain), increased expression of the MMPs and induction of apoptosis. Ultimately, this leads to cellular hypertrophy, increased extracellular fibrosis, a dilated hypocontractile ventricular phenotype resulting in increased mortality [76,79,84].

Intracardiac factors in heart failure genomics

Sarcomeric proteins

There are two cardiac myosin heavy chain (MHC) isoforms, designated as α and β MHC. These two isoforms are functionally distinct as evidenced by the threefold greater rate of ATP hydrolysis by α MHC [85]. In response to hemodynamic overload, small mammals (containing predominantly α-MHC) rapidly increase the proportion of β-MHC present in the left ventricle [86,87]. While normal human myocardium contains predominantly β-MHC, recent studies have demonstrated small amounts of α-MHC in normal human myocardium, ranging from a few to as much as 14% of the total MHC [88–90]. An increase in β-MHC expression (and a corresponding decrease in α-MHC expression) has been reported in failing human myocardium using both standard methods and gene arrays [88–93]. To determine if this small myosin isoform shift is of any functional significance, myosin isolated from end stage failing human myocardium and nonfailing control tissue was directly compared in several studies [90,94–96]. These *in vitro* experiments failed to demonstrate differences in ATPase or mechanical performance, suggesting that a shift in myosin isoform is not of functional significance in human heart failure. However, much larger changes in MHC expression are seen at the mRNA level [97] and thus can be used as an index of fetal gene program activation in human hypertrophy and failure.

The MHC contains two associated light chains which provide structural support to the neck region of the myosin molecule. There is an up to 10% increase in expression of the atrial isoform of the essential light chain in the left ventricle of patients with heart failure [98]. While the functional implications of such shifts are not completely known, the magnitude of this isoform shift correlated with a change in the calcium sensitivity of the myofilament. However, a definitive role of changes in light chain isoform in heart failure has not been established to date.

Titin is a very large structural protein of the sarcomere which integrates the thick filament and the Z-disc. It is largely responsible for the passive tension of the myocyte. Two isoforms of titin are co-expressed in the human heart. A shift to the more compliant titin isoform has been described in end stage human heart failure and correlated with decreased passive stiffness [99]. The implication in an actively beating heart is not known. Perhaps more important is the fact that titin has recently been recognized as a biomechanical sensor, and is involved in increased expression of atrial natriuretic peptide (ANP) and B-type natriuretic peptide (BNP) with mechanical stress [100].

The human heart α-cardiac and α-skeletal actin isoforms. There is a large shift in actin isoform expression (α-cardiac to α-skeletal) in the developing human heart [101]. In addition, increased expression of α-skeletal actin mRNA has been reported in a rat myocardial infarction model [102,103]. An increase in α-skeletal actin expression is a common measure of fetal gene program activation in animal models of heart failure. However, no changes in actin isoform protein expression levels have been demonstrated in end stage human cardiomyopathy tissue compared with age-matched controls [101,104].

Thin filament-associated regulatory proteins of the sarcomere control cardiac contractility in a calcium dependent fashion. There is only one isoform of cardiac troponin C. While tropomyosin, troponin I and troponin T manifest developmental isoform variation, no isoform shifts of troponin I or tropomyosin at the protein level have been detected in human or experimental heart failure [105–107]. In contrast, there are now a number of reports demonstrating altered troponin T isoform expression in cardiac failure. In most cases there is increased expression of a fetal isoform in the disease state [108–110]. Anderson *et al.* [106] first reported a troponin T isoform shift in human heart failure with an 8% increase in expression of a fetal isoform compared with normal human myocardium. However, two recent studies which directly examined the effect of troponin T isoforms on thin filament function found only modest functional differences between the two isoforms [111,112]. Thus, it is unlikely that a troponin T isoform shift represents the primary cause of myofibrillar dysfunction in human heart failure.

In summary, while changes in actin, MHC, myosin light chain and troponin T isoform have been observed in animal models and human heart failure, these changes are likely the result of a re-expression of a fetal gene program and to date have not been shown to have any significant functional impact.

Myocardial calcium handling

In most forms of hypertrophy and heart failure, alterations of intracellular calcium handling contribute to impaired contraction and relaxation [113]. Regulation of intracellular calcium levels has a central role in the contraction–relaxation cycle of the mammalian heart. Calcium levels in cardiac myocytes vary by a factor of 10 between diastole and systole and are 10,000-fold lower than extracellular levels [114]. Changes in cytosolic calcium concentrations are mechanically translated as contraction and relaxation through the activation of sarcomeric proteins. This is a highly cooperative system, where small changes in calcium levels result in large changes in force.

To control these changes in intracellular calcium concentration, cardiac myocytes have two interacting systems:

1 one system is located in the outer cell (sarcolemmal) membrane and regulates calcium movements between the cytosol and the extracellular space; and
2 the other is a specialized intracellular calcium storing organelle, the sarcoplasmic reticulum (Fig. 7.2). Upon depolarization of the sarcolemmal membrane, voltage gated L-type calcium channels (dihydropyridine receptors) release a small amount of calcium into the cardiac myocyte. This in turn triggers the release of calcium (calcium-induced calcium release) from the sarcoplasmic reticulum. The rate of calcium flux from the sarcoplasmic reticulum is the major determinant of positive dP/dT and peak ventricular pressure.

In contrast, myocardial relaxation is determined by the kinetics of calcium removal from the myofilaments. The calcium pump of the sarcoplasmic reticulum (SERCA) is an ATP consuming pump that mediates the reuptake of calcium into the sarcoplasmic reticulum. The pump activity of SERCA is regulated by the small inhibitory protein phospholamban. The inotropic effects of beta-agonists are largely mediated by PKA dependent phosphorylation of phospholamban leading to a disinhibition of SERCA and therefore increased calcium uptake into the sarcoplasmic reticulum ultimately leading to stronger subsequent contractions. Because the capacity for calcium storage within the sarcoplasmic reticulum is limited, a sodium gradient driven calcium exchanger (Na/Ca exchanger) maintains a net balance between calcium influx and efflux at the sarcolemmal membrane.

Gene expression during cardiac hypertrophy and heart failure mimics in many aspects the fetal expression pattern, in which the sarcolemmal calcium transport proteins are abundantly expressed and the calcium handling proteins of the sarcoplasmic reticulum show reduced expression [115].

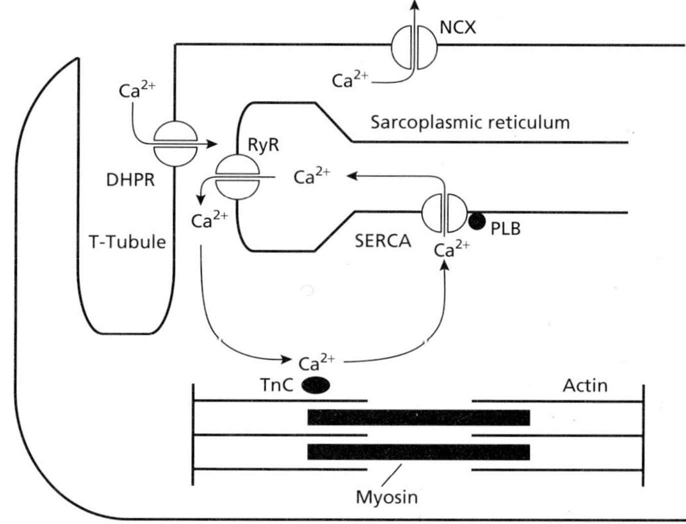

Figure 7.2 Calcium handling proteins and sarcomeric proteins. Calcium enters the cardiac cell via the dihydropyridine channels (DHPR) to trigger calcium release from the sarcoplasmic reticulum calcium channel (RyR, ryanodine receptor) into the cytoplasm. By binding to troponin C (TnC) calcium leads to actin–myosin interaction and contraction. In diastole calcium is sequestered into the sarcoplasmic reticulum by the calcium pump of the sarcoplasmic reticulum (SERCA) which is inhibited by phospholamban (PLB), and extruded to the extracellular space by the Na/Ca exchanger (NCX). Both phospholamban and troponin I are important phosphorylation targets of protein kinase A which results in increased contractility and relaxation.

It was first noted that calcium transients in heart failure were prolonged mostly because of a slowing of cytosolic calcium removal. This translates to slowed relaxation and consequently reduced contractility. Analysis of the candidate genes and proteins revealed that in most models of cardiac hypertrophy and heart failure, SERCA gene transcription and protein expression is markedly downregulated [116]. This has been confirmed in animal models of hypertrophy and heart failure and in end stage human heart failure [113,117]. Conversely, Na/Ca exchanger gene transcription and protein expression was found to be upregulated in heart failure [113,117,118]. Interestingly, many single gene defects that result in hypertrophy or heart failure lead to a similar change in expression pattern of calcium cycling proteins. The inciting factors that have been described to lead to these changes in calcium transients include:

1 Neurohormonal signaling, which is elevated in heart failure and acts in part to stimulate cardiac myocyte growth similar to the fetal situation; and
2 Mechanical tension and stretch, which activates intracellular signaling cascades resulting in a growth stimulus on cardiac myocytes.

SERCA is by far the most studied calcium cycling protein in hypertrophy and heart failure. Reported changes in the expression levels of the ryanodine receptor (calcium channel of the sarcoplasmic reticulum) and phospholamban are less consistent but suggest lower expression levels as well. While ryanodine receptor levels might be reduced, more importantly, the ryanodine receptor appears to be hyperphosphorylated in heart failure leading to a small calcium leak which potentially increases the probability for malignant ventricular tachyarrhythmias [119]. The changes in the expression pattern of calcium cycling protein expression have been demonstrated in failing hearts with nonischemic or ischemic dilated cardiomyopathy and in animal models of concentric hypertrophy, thus demonstrating a commonality across species and disease etiology. Data from gene array experiments and proteomics approaches confirmed a significant downregulation of SERCA in ventricular tissue from patients with hypertrophic and dilated cardiomyopathy [91,93,120–122]. The inability of the gene array to detect other changes is likely a result of the considerable variability within cardiomyopathic hearts and the small absolute changes in mRNA levels.

To summarize, the disequilibrium in calcium in heart failure stems from a shift of systolic and diastolic calcium cycling towards the sarcolemmal membrane and away from the sarcoplasmic reticulum which leads to decreased calcium release from the sarcoplasmic reticulum, impaired reuptake of calcium by the sarcoplasmic reticulum and possibly elevated diastolic calcium levels. This is seen functionally in the intact heart as a delay in relaxation, slowed systolic upstroke, reduced systolic pressures and potentially increased end-diastolic ventricular pressures (caused by incomplete sarcomeric calcium clearance resulting in residual cross-bridge cycling).

The interstitium

The cardiac interstitium is comprised of the extracellular matrix and fibroblasts. The extracellular matrix can be viewed as the scaffolding of the myocardium. Under normal physiologic conditions, extracellular matrix remains under the control of several signaling pathways that can affect extracellular matrix proteins synthesis and/or degradation. The primary proteases that are responsible for the degradation of the extracellular matrix are the MMPs which constitute a large family of proteases with isoform specific actions (e.g., fibrillar collagen or basement membrane proteins). Remodeling of the extracellular matrix has a key role in the development of cardiac hypertrophy and in the alteration of ventricular shape that is associated with the progression to end stage cardiac failure. Specifically, the deposition of collagen in the extracellular matrix leads to increased myocardial fibrosis and myocardial stiffness, where as activation of MMPs leads to digestion of the extracellular matrix. Increased MMP activity directly contributes to the remodeling (e.g., chamber dilation) associated with hypertrophy and myocardial failure in animal models and humans [123]. Many large-scale gene array studies have confirmed increased transcription of structural matrix proteins (e.g., collagen isoforms I and III) and the activation of genes involved in the turnover of extracellular components (e.g., MMPs) [72–74,93,122,124,125].

Cytokines likely have a major role in activation of the MMPs in myocardial hypertrophy and failure. The inflammatory cytokines IL-1β, TNF-α and IL-6 decrease collagen synthesis and procollagen mRNA levels in fibroblasts [126]. IL-1β and TNF-α cause transcriptional increase in the expression of MMPs. Thus, these inflammatory cytokines directly contribute to chamber dilatation in the cardiac remodeling process of myocardial failure. Furthermore, reactive oxygen species, as is present in end stage cardiac failure and ischemic heart disease, decrease transcriptional synthesis of collagen [127], and thus likely contribute to the degradation of the extracellular matrix by the inflammatory cytokines. TGF-β increases synthesis of collagen, decreases MMP activity and increases tissue inhibitors of MMPs [128], thus may limit chamber dilatation by counteracting some of the effects of the inflammatory cytokines.

Neurohormonal signaling by catecholamines, angiotensin II and endothelin all stimulate fibroblast synthesis of collagen. Angiotensin II stimulates fibroblast proliferation and the expression of collagen and other extracellular matrix proteins such as fibronectin and laminin [129]. The end result is increased myocardial fibrosis which is attenuated in animal models with angiotensin II blockade [130]. In contrast, endothelin appears to increase fibrosis through both activation of collagen synthesis and diminished MMP activity. Catecholamines likely stimulate myocardial fibrosis through a decrease in MMP activity as well as the increase in tissue inhibitors of MMPs [131]. Aldosterone has been shown to induce myocardial fibrosis, a process in which sodium appears to be a co-factor [132]. The mechanisms of aldosterone induced fibrosis involve the induction of fibroblast proliferation [133] as well as an increase in collagen synthesis [130] and possibly the induction of MMP activity [134].

Finally, the activation of collagen synthesis is likely through c-Jun and c-Fos which dimerize to form activating protein-1 which binds to specific gene promoter sites. Similarly, the induction MMP transcription is through the transcription factors, activating protein-1 and NF-κB to the promoter region of specific MMP genes [131]. Thus, the integrity of the interstitium lies in the balance between MMP activity and collagen synthesis, a process that is governed by similar signaling pathways and likely similar transcription factors.

Receptors and signal transduction

Several neurohormonal signals through receptor specific targeting directly affect the hypertrophy and remodeling associated with heart failure. Many of these receptors are members of the seven transmembrane family (the receptor spans the cell membrane seven times). Receptors in this class include β_1 and β_2-adrenergic, α-adrenergic, endothelin, AT1 and AT2 (angiotensin II receptor). Common to all these receptors is the close association with specific guanosine triphosphate (GTP) binding proteins (also known as G proteins comprised of three subunits α, β, γ) and an effector enzyme (Fig. 7.3). When an agonist binds to the receptor, GDP is released from the α subunit allowing GTP to bind and leading to the disassociation of α-effector enzyme, and the β–γ subunit from the agonist receptor. Typically, effector enzymes are associated with the α subunit but certain effector enzymes are associated with the β–γ subunit. G protein subunit isoform variation allows association with specific effector enzymes [135].

Gαs signal transduction

β-Adrenergic signaling employs the Gαs subunit which is associated with adenlylate cylase to generate cyclic AMP from ATP. Cyclic AMP functions as a second messenger to trigger the activation of PKA. The immediate primary function of PKA is to increase contractile function through enhanced calcium cycling by means of phosphorylating several calcium channels or regulatory proteins (i.e., the L-type calcium channel, sarcoplasmic reticulum calcium release channel – the ryanodine receptor – and phospholamban). However, with prolonged β-adrenergic stimulation as seen in heart failure, several likely genomically initiated mechanisms come into play. Specifically, there is a downregulation of the β_1-receptor and its mRNA in heart failure (as large as 50%) which is partially restored with beta-blockade, thus suggesting a direct β-adrenergic mediated mechanism [32]. β-Adrenergic receptor sensitivity is reduced in heart failure. This is largely because of the increase of expression of β-adrenergic receptor kinase (which through phosphorylation inhibits G protein binding to the β-adrenergic receptor), and the increased expression of an inhibitory G protein Gαi [136,137]. Chronic β-adrenergic stimulation in *in vitro* cell

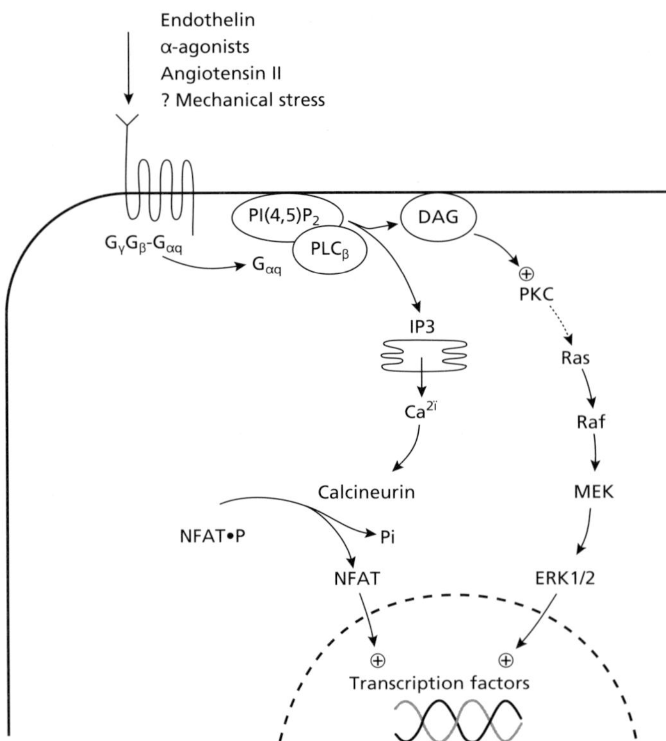

Figure 7.3 Gαq–Gα11 signal transduction. The seven transmembrane receptors for several agonists, including endothelin, catacholamines (α₁ adrenergic) and angiotensin II, are mediated through G protein subunits αq and α11. Phospholipase CβPLCβ) is coupled to the Gαq subunit. Upon receptor activation, PLCβ hydrolyses cell membrane bound phosphatidylinositol, PI(4,5)P₂ to yield diacylglycerol (DAG) and inositol 1,4,5-triphosphate (IP3). DAG in turn activates PKC which ultimately activates the extra cellular signal related kinases 1 and 2 (ERK 1/2) of the mitogen activated protein kinase (MAPK) family. ERK 1/2 is responsible for the induction of several transcription factors. On the other hand, IP3 triggers intracellular calcium release activating calcineurin, leading to the dephosphorylation of nuclear factor of activated T cells (NFAT), permitting its nuclear transport and activation of genes associated with hypertrophy. Finally, these pathways also appear to be activated by mechanical stress.

preparations and β₁-adrenergic receptor overexpression in a mouse model lead to initial myocyte hypertrophy followed by ventricular fibrosis and failure [138,139]. Furthermore, β-adrenergic stimulation has been implicated in the induction of apoptosis [32].

Gαq–Gα11 signal transduction

Signaling through the endothelin, α₁-adrenergic and angiotensin II receptors is largely mediated through G protein subunits αq and α11 (Fig. 7.3). Upon agonist activation of the receptor, Gα coupled phospholipase Cβ is activated, hydrolyzing the membrane bound phospholipid, phosphatidylinositol 4,5 bisphosphate, to yield diacylglycerol (DAG) and inositol 1,4,5-triphosphate. DAG is the primary activator of most, but not all, PKC isoforms. In the hypertrophic response mediated by Gαq/11 PKC isoforms α and ε appear to have a prominent role based on increased expression levels and changes in cellular distribution [140]. However, other isoforms likely have a role as is evidenced by the transgenic overexpression of PKC β1 which resulted in the induction of a hypertrophic

gene program [141]. Overall, the mechanisms of isoform specific PKC mediated hypertrophic response are not currently well understood, with likely several isoforms being involved.

It is now recognized that PKC activates the extracellular signal related kinases 1 and 2 (ERK 1/2) of the MAPK family. There are three MAPK cascades: ERK, the c-Jun N terminal kinases (JNKs) and the p38 MAPKs (the latter two are discussed below). MAPK are regulated upstream by MAPK kinases (MAPKK) which in turn are activated by MAPKK kinases (MAPKKK). Activation of the ERK 1/2 pathway has been demonstrated to be a principal mediator of the hypertrophic response in heart failure [142]. Current evidence indicates that the cascade is triggered by PKC mediated activation of the "small" GTP binding protein Ras, although the specific mechanisms of activation are not currently understood. Ras in turn activates Raf (a MAPKKK) which phosphorylates ERK 1/2, resulting in the activation of several transcription factors and inducing a hypertrophy gene program. Furthermore, the ERK 1/2 pathway has been shown to be activated by several growth factors as well as mechanical stress mediated signaling mechanisms [141]. Finally, ERK 1/2 signaling appears to have anti-apoptotic properties with several potential mechanisms being implicated [142].

Gαq/Gα11 protein activation result in the production of inositol 1,4,5-trisphosphate. This in turn promotes release of calcium from the endoplasmic reticulum. Elevation of calcium in turn activates the phosphatase, calcineurin. The details of how calcium activates calcineurin in the cardiocytes, where intracellular calcium levels increase at least 10-fold with each cardiac contraction, are not known. Through a series of elegant experiments conducted over the past several years, calcineurin has been firmly established as key mediator of the hypertrophic response in heart failure [144]. Calcineurin is known to dephosphorylate the transcription factor NFAT, thus promoting its nuclear transport.

Glycogen synthase 3β

Glycogen synthase 3β inhibits the activation of several transcription factors involved in the hypertrophic response including GATA-4, β-catenin, C-myc and c-Jun [140]. Moreover, glycogen synthase 3β phosphorylates NFAT blocking its nuclear translocation and thus counteracting calcineurin induced hypertrophy [145]. PKC, PKA and mechanical stress (by an unknown mechanism) phosphorylate glycogen synthase 3β and release its inhibition on gene transcription and promote hypertrophy by this mechanism.

Stress activated kinases

Both the MAPK p38 and JNK are activated by cellular stress by mechanisms that are not completely understood. The activation of p38 in transgenic models through the overexpression of upstream effectors resulted in a dilated ventricular phenotype with increased fibrosis and premature mortality [146]. p38 is known to activate the transcription factor GATA. It was later determined that p38 MAPK phosphorylates the transcription factor NFAT promoting nuclear export and counteracting calcineurin mediated hypertrophy [147]. Thus, it appears that p38 signaling results in the triggering of gene expression which promotes ventricular dilatation and fibrosis as is observed in end stage myocardial failure. Interestingly, p38 activity have

been reported to be elevated in end stage human heart failure, albeit not consistently. The JNK pathway is similar to p38 MAPK in that it likely blunts the hypertrophic response through phosphorylation (deactivation) of NFAT [143]. There is evidence to suggest that both p38 and JNK induce cellular apoptosis [148].

PI3K–Akt signal transduction

There are two PI3K isoforms directly involved in the hypertrophic response. The tyrosine kinase receptors (receptors for insulin, IGF-1 and other growth factors) and integrins (i.e., mechanoreceptors) are thought to mediate normal physiologic (adaptive) hypertrophy through the activation of the PI3K isoform p110α (Fig. 7.4). In contrast, activation of PI3K p110γ through G protein coupled receptors mediates a pathologic hypertrophic response. However, while the above discussed pathways are through the Gα subunit, signal transduction in this pathway is through the Gβγ subunit. Activation of either of the p110 isoforms results in phosphorylation of membrane bound phosphatidylinositols. This step leads to the downstream activation of Akt (also referred to as protein kinase B) by 3-phosphoinositide-dependent protein kinase D (PKD). Akt is known to activate mammalian target of rapamycin (mTOR), a transcription factor and a central mediator of protein synthesis. In addition, Akt phosphorylates GSK-3, thus blocking its inhibitory effects on transcription. The net effect is the development of hypertrophy through the activation of the Akt pathway. The PI3K–Akt pathways are dominant signaling pathways in the hypertrophic response as is demonstrated by the phenotypic response in several transgenic models. The mechanisms by which activation PI3K p110γ leads to pathologic hypertrophy and PI3K p100α results in adaptive hypertrophy are not currently known. It has been suggested that the development of pathologic hypertrophy requires the co-activation PI3K–Akt with other signaling pathways such as calcineurin–NFAT [145].

In summary, the signaling pathways involved in hypertrophy and failure are complex, and our understanding of these processes and their relative importance continues to evolve. Other pathways not discussed likely play a contributing part (e.g. janus kinases – signal transducers and activators of

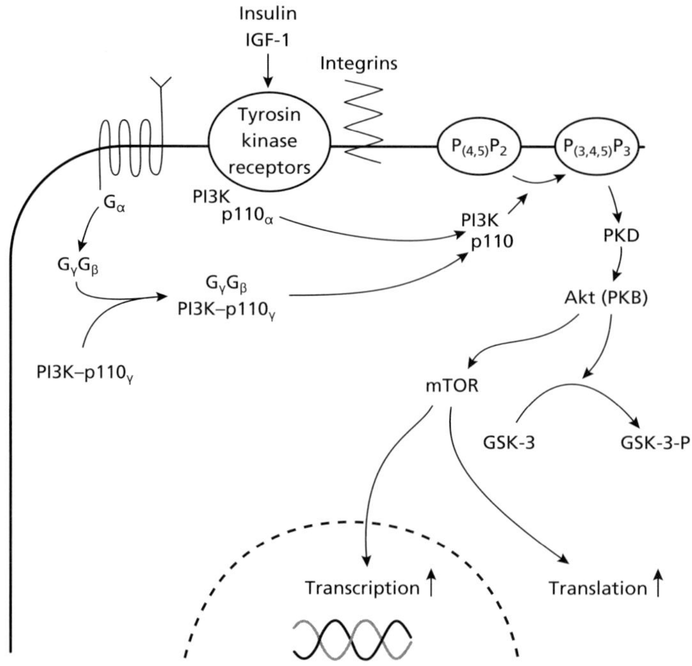

Figure 7.4 PI3K–Akt signal transduction. Phosphoinositide 3-OH kinase (PI3K) is activated by two mechanisms. The PI3K–p110α isoform is activated by the tyrosine kinase receptor via growth factors such as insulin and insulin like growth factor 1 (IGF-1). In addition integrins, trans-membrane proteins that function as mechanoreceptors, also activate PI3K–p110α. Alternatively; PI3K–p100γ isoform is activated through G protein coupled receptors by association with the G protein subunits βγ. Both activated PI3K–p110 isoforms phosphorylate the membrane bound phosphatidylinositol (PI(4,5)P$_2$ → PI(3,4,5)P$_3$). This permits the binding of both 3-phosphoinositide-dependent protein kinase (PKD) and Akt (as known as protein kinase B) to PIP3 (not shown). PKD then phosphorylates Akt which is turn activates mammalian target of rapamycin (mTOR). Furthermore, Akt phosphorylates glycogen synthase kinase 3β (GSK-3) thus blocking its inhibitory effects on transcription. mTOR facilitates the transcription of hypertrophic response genes as well as promoting ribosomal biosynthesis thus increasing transcription. Tyrosine kinase receptor activation is associated with adaptive hypertrophy whereas G protein coupled receptor activation predominates in cardiac failure. Thus both adaptive and pathologic hypertrophy can in part be mediated through this pathway.

transcription [JAK-STAT]; for review see [149]). The PIK3–Akt pathway is involved in both physiologic and pathologic hypertrophy. In contrast, ERK 1/2 and calcineurin–NFAT pathways have a dominant role in animal models of heart failure and are likely central mechanisms in the transition to failure in human cardiomyopathy. It should be emphasized that it is the interplay of several signaling pathways that, in concert through numerous transcription factors, ultimately guide gene transcription.

Energy metabolism

The contractile performance of a heart is very dependent on oxygen supply. The impairment of oxygen delivery has immediate contractile consequences which illustrates the very limited substrate storage capacity of the myocyte and the high dependency on mitochondrial oxidation [150]. Phosphocreatine is the most important energy store for ATP. To satisfy hemodynamic demands, a healthy adult heart depends mainly on two main energy substrate groups. Fatty acid oxidation supplies 60–90% of the energy and the oxidation of glucose and lactic acid supplies 10–40%. Only a small amount of ATP can be generated via anaerobic glycolysis. Most of the ATP is spent on cardiac contraction followed by calcium handling, maintenance of sarcolemmal ion gradients and other basic housekeeping functions. Acute changes in cardiac workload induce rapid adaptations of substrate uptake and activate the enzymatic cascades involved into ATP generation to satisfy myocardial energy demands. Only within the last two decades has it been recognized that chronic changes in cardiac workload as they occur in hypertrophy and heart failure have an effect on substrate preference, enzyme expression patterns and mitochondrial oxidative capacity. Hypertrophy and heart failure lead to a progressive decline in the activity of mitochondrial oxidative pathways, ultimately decreasing the capacity for ATP production [151]. Indeed, animal and human studies reveal a 20–30% reduction in ATP concentration in failing hearts and a reduced phosphocreatine : ATP ratio in hypertrophy and heart failure, indicating that myocardial energy reserves are disproportionately reduced. However, most ATP consuming cellular processes at normal cardiac workloads are not affected by these reductions in ATP levels and do not support the hypothesis that cardiac energy deficiency is a general cause of heart failure with the clinically important exception of heart failure from ischemia.

Interestingly, myocardial substrate preference has been shown to switch from predominantly fatty acids to glucose and lactic acid in heart failure and hypertrophy [152,153]. This mimics the metabolic situation in the fetal heart which was first described in animal models. Initially, this could be only indirectly demonstrated in cardiac tissue samples from patients with heart failure by analysis of the expression patterns of enzymes involved in the two metabolic pathways [152,154,155]. Recently, metabolic imaging studies with positron emission tomography (PET) in patients with hypertrophy and heart failure provided direct evidence for decreased fatty acid oxidation and increased glucose metabolism [156,157].

The molecular mechanism leading to this switch in substrate preference is the subject of intensive research. The role of key nuclear proteins involved in the transcriptional control of metabolic pathways and mitochondrial production has been recently delineated [158]. Peroxisome proliferator-activated receptors (PPARs) and the transcriptional co-activator (PCG-1), which are activated by an array of signal transduction pathways, have been shown to be involved in regulating metabolic pathways, mitochondrial metabolism and mitochondrial synthesis (Fig. 7.5). As nuclear receptors PPARs bind to DNA in the regulatory region of genes and activate gene transcription, whereas PCG-1 activates PPAR to augment gene transcription. Both proteins have been shown to promote

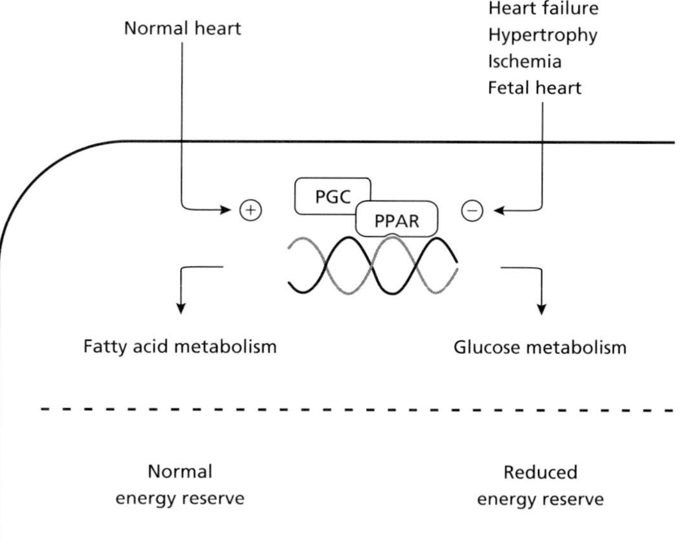

Figure 7.5 Energy metabolism in heart failure. In heart failure and hypertrophy glucose metabolism is activated and fatty acid metabolism is attenuated. This change is mediated by altered transcriptional control of the metabolic genes involved in both pathways. One key regulating mechanism is the activation or inhibition of the nuclear factor peroxisome proliferator-activated receptor (PPAR) and its co-activator (PCG-1). Numerous pathways which are activated in heart failure (e.g. beta-adrenergic pathway, nitric oxide, calcium dependent kinases) affect PCG-1 levels and activity. PPAR has additional regulatory effects on mitochondrial biogenesis. In heart failure, reduced adenosine triphosphate and phosphocreatine levels under baseline hemodynamic conditions do not affect cardiac function but become apparent with increased work loads.

the expression of genes involved in fatty acid oxidation and mitochondrial ATP-producing capacity. Several PCG-1 activating signaling pathways have been delineated that translate physiologic stimuli such as stress and exercise into changes in gene transcription. Alterations in expression levels of PCG-1 and PPARs represent another mechanism of how gene transcription can be affected. High levels of PCG-1 and PPARs promote fatty acid metabolism and low levels favor glucose oxidation. Indeed, in the fetal heart and animal models of hypertrophy and heart failure, PCG-1 and PPAR levels were found to be depressed, providing a possible explanation of why glucose and lactic acid oxidation is the dominant metabolic pathway in heart failure.

Gene array studies have confirmed changes in metabolic and mitochondrial gene expression in human hearts with ischemic and nonischemic cardiomyopathy [71,73,74,91–93,120,122,124,125]. Furthermore, alterations in mitochondrial and energy metabolism protein levels were also detected employing proteomic methodologies in a dog model of heart failure and in human tissue samples from patients with heart failure [121,159].

In summary, cardiac hypertrophy and heart failure result in a depressed fatty acid metabolism and increased glucose and lactic acid oxidation, with the later being the primary source for ATP. This change in primary substrate preference is orchestrated by nuclear transcription factors which lead to changes in cellular substrate uptake, expression levels of pathway-specific metabolic enzymes and changes in mitochondrial function. These changes are part of a final pathway in heart failure involving the activation of a fetal gene expression pattern which favors glucose and lactic acid metabolism.

Cell death and regeneration

Apoptosis is an evolutionary, highly conserved, cellular suicide process which has a critical role in embryogenesis and tissue remodeling. Conversely, dysregulation of apoptosis resulting in increased or diminished cell death can result in disease. Because programmed cell death is an ubiquitous process, the progressive loss of cardiomyocytes through apoptosis has also emerged as an important issue in heart failure research [160,161]. The two major modes of cardiac cell death are apoptosis, which occurs at a low rate in chronic heart failure, and necrosis as the major mode of cell death in myocardial infarction. An important feature of apoptosis compared to necrosis appears to be that dead cells are cleared without a significant inflammatory reaction, which leads to fibrosis [162].

The signaling pathways that lead to apoptosis are the intrinsic pathway which integrates a broad spectrum of cellular stresses, and the extrinsic pathway which is mediated by a cellular membrane receptor. The intrinsic pathway of apoptosis has been shown to be activated by many different stimuli such as oxidative stress, reoxygenation, metabolic substrate deprivation, radiation, metabolic inhibition, neuroendocrine stimulation and mechanical stress. Because many of these factors are present in heart failure it is not surprising to find increased rates of apoptosis. Whereas the normal apoptosis count in a healthy human heart is <2 cells per 100,000 cardiac myocytes, the apoptosis count in heart failure is increased to 8–250 cells [163–165]. In myocardial infarction, apoptosis was demonstrated not only in the border zones but also in necrotic areas [166].

The receptor mediated extrinsic pathway in apoptosis involves the binding of a ligand to a cell surface receptor (death receptor) which induces apoptosis (Fig. 7.6). Such ligands can be membrane proteins of other cells such as the CD95–Apo-1 (Fas) ligand or soluble factors such as TNF-α. Activation of the death receptor activates caspace-8 which in turn activates caspase-3. This last enzymatic step is shared between the extrinsic and the intrinsic pathway of apoptosis and leads to proteolysis of the cellular substrates. In the intrinsic pathway, cellular death signals such as oxidative stress, neurohormones, stretch or hypoxia, are transmitted to the mitochondria and stimulate the release of cytochrome c and other factors from the mitochondria. Cytosolic cytochrome c then assembles with other proteins to form the apoptosome which activates caspase-3, resulting in degradation of proteins and cell death. A large group of intracellular proteins, which have been originally found in B-cell lymphomas (Bcl-2 group) have been identified to either activate or inhibit the intrinsic pathway at the level of the mitochondria. The most prominent proapoptotic proteins are called Bcl-2-associated X protein (Bax) and Bcl-2-antagonist/killer (Bak).

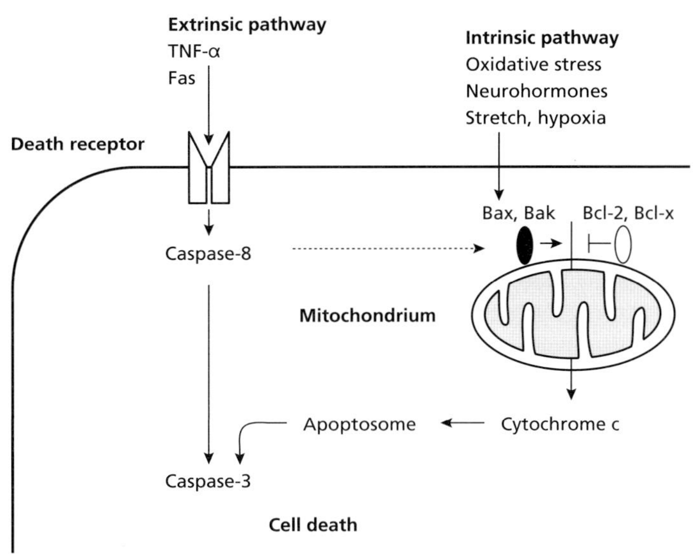

Figure 7.6 Apoptosis in heart failure. The extrinsic pathway is activated by tumor necrosis factor α (TNF-α) and by cell surface receptors (Fas) activating caspase-8 which in turn activates caspase-3. This enzyme leads to proteolysis of the intracellular proteins and cell death. The intrinsic pathway is activated by multiple stress induced signals (both intra- and extracellular) which are known to be activated in heart failure. This leads to Bax and Bak mediated mitochondrial release of cytochrome c. This process can be inhibited by the antiapoptotic Bcl-2 and Bcl-x. Cytochrome c released from the mitochondria forms multimeres with other proteins to activate caspase-3. The extrinsic pathway can modulate the intrinsic pathway at the mitochondrial level.

Either one of these proteins can trigger the release of cytochrome c from the mitochondria to initiate cell death. The proapoptotic subgroup of Bcl-2 proteins is comprised of many more proteins which integrate specific upstream signals to then activate Bax and Bak. The proapoptotic proteins are antagonized by the antiapoptotic proteins Bcl-2 and Bcl-x which prevent the release of cytochrome c, thus inhibiting apoptosis. In genetically engineered mouse models it was demonstrated that infarct size after occulation of the left anterior descending artery (LAD) is reduced by promoting antiapoptotic signaling, providing additional support to the role of apoptosis in myocardial infarction. Heart failure was observed in mice in which proapoptotic signaling was induced. Novel cellular pathways and factors that affect apoptosis continue to be described. This includes the control of apoptotic protein expression by transcription factors such as p53, or calcium mediated apoptosis, the role of which is still ill defined in cardiac myocytes where there are large cyclical changes in intracellular calcium [167]. Clearly, work is required to further delineate these factors, especially in human heart failure. So far it remains controversial whether apoptosis is cause or a consequence (or both) of heart failure.

The direct contribution of apoptosis towards heart failure is even more in question after it was convincingly demonstrated that there is a significant turnover of myocardial cells. Cardiac myocytes have been thought to be terminally differentiated and unable to divide. Moreover, it was thought that there was a lack of undifferentiated myocytes or stem cells that could develop into cardiac myocytes. However, there is now good evidence that there are undifferentiated cells that can develop into cardiac myocytes, even in the adult heart. Cardiac neomyogenesis in humans was first demonstrated in sex-mismatched transplanted hearts in which male patients with a female donor heart ended up with a small number of Y chromosome containing cardiac myocytes, a mechanism that only could be explained by the differentiation and homing of male cells within the female donor heart [168]. Some of the cell types that have been shown to be able to differentiate into cardiac myocytes are bone marrow stem cells, partly differentiated endothelial progenitor cells or a newly found pool of cardiac stem cells in the myocardium [169]. Since this steady turnover of cardiac myocytes was reported, basic scientists and clinicians have been developing strategies to take advantage of cardiac regeneration in both heart failure and myocardial infarctions [170,171]. However, to date, even in a nonfailing heart, we do not know the net effect of cell death and cardiac regeneration, and conflicting results complicate the issue. Gene array analysis has yielded differential expression patterns in some genes involved in apoptotic pathways in tissue samples from human hearts with heart failure when compared with normal hearts [91–93,122,124].

It is apparent that several genes that are implicated in apoptosis are activated in all stages of heart failure. In addition, a number of factors that characterize the hypertrophic and failing heart (mechanical stress, calcium overload and neuroendocrine stimulation) also trigger apoptotic pathways. Furthermore, reactive oxygen species induce apoptosis, suggesting a mechanism by which oxidative stress might contribute to heart failure. Regenerative, cardioprotective and apoptotic mechanisms continue to be characterized and further our understanding of their role in heart failure.

Oxidative injury, hypoxia and nitrous oxide

Although oxygen is a critical determinant of cardiac function, it has become clear that oxygen can also play a detrimental part because of the generation of reactive oxide species. The same pathways of cellular metabolism and mitochondrial oxidative phosphorylation that are necessary to provide cardiac myocytes with energy can lead to the generation of reactive oxide species. The oxygen atom contains two unpaired electrons in the outer shell, designating it as a free radical. In molecular oxygen (O_2), two oxygen atoms assemble to neutralize the unpaired electrons which greatly reduce its chemical reactivity. The reaction leading from molecular oxygen to carbon dioxide and water requires the donation of four electrons which lead to the generation of intermediate reactive oxide species. The donation of a single electron to molecular oxygen (O_2) results in the formation of a superoxide radical (O_2^-). The addition of a second electron yields peroxide which then reacts with hydrogen to become hydrogen peroxide (H_2O_2). Further donation of a third electron yields the highly reactive hydroxyl radical (OH) which yields water (H_2O) after a fourth electron is added. Reactive oxide species can interact with other molecules to lead to secondary radicals such as peroxynitrite ($ONOO^-$) which is formed from nitric oxide and superoxide. Reactive oxide species formation is induced by many of the neurohormones implicated in heart failure including angiotensin II, catecholamines, endothelin, cytokines and other growth factors [172–174].

Several cellular mechanisms counterbalance the production of reactive oxide species. Among the best characterized is superoxide dismutase and glutathione peroxidase which catalyze the reaction of O2$^-$ to H_2O_2 and H_2O_2 to water, respectively. In addition to nonenzymatic antioxidant agents such as vitamins C and E, cells have several enzymatic pathways that provide and regenerate antioxidative substrates such as ubiquinon (Q10) or glutathione (the reducing substrate for the glutathione peroxidase reaction). Whenever the cellular antioxidant defenses are overwhelmed, reactive oxide species can react with structural proteins, enzymes, lipids and DNA to cause cell damage, mutations and potentially cell death via necrosis or apotosis.

The contribution of reactive oxide species in the pathogenesis of vascular disease has been demonstrated in oxidized low density lipoprotein (LDL), plaque development and rupture, endothelial dysfunction and the stimulation of smooth muscle growth. These often subclinical processes are linked to an inflammatory response which can be followed by serum C-reactive protein levels which is now accepted as an independent predictor of vascular disease progression. However, the measurement of urinary or blood levels of free radical-catalyzed products of the arachidonic acid metabolism called isoprostanes (iPs) allows a more direct measure of oxidative stress [175]. At the level of the myocardium and more specifically in heart failure, reactive oxide species have been implicated to contribute to stunning, reperfusion injury and cardiac remodeling after myocardial infarction. In animal models of heart failure and cell culture experiments, several heart failure activated signaling pathways (e.g., angiotensin II, endothelin, catecholamines) have been found to increase the cellular generation of reactive oxygen species which can affect ion channel function and excitation–contraction coupling by interference with calcium cycling and contractile proteins. In human heart failure, alcohol-mediated and anthrocycline-induced cardiomyopathy have been proposed to result from oxidative damage [176,177]. In the recently published A-HEFT, which showed a marked decrease in mortality by adding the combination of hydralazine and nitrates to an optimal heart failure regimen, it was speculated that the antioxidative effects of hydralazine might contribute to the reduction in mortality [178]. Furthermore, antioxidant properties of statins and carvedilol have been suggested to have a role in their beneficial effects. However, other oral antioxidants (most of them vitamins) have not been found to be beneficial in protecting from cardiovascular disease, including heart failure [177,179–182].

Beyond the effects of oxygen on cellular function and on gene expression by chemically reacting with other molecules, it is now recognized that changes in oxygen levels have a direct effect on gene expression. This is best studied in hypoxia which induces hypoxia-inducible factor (HIF-1), a transcription factor involved in angiogenesis, glucose metabolism and apoptosis. In myocardium from patients with ischemic cardiomyopathy, this factor was found to be elevated, suggesting a role in neovascularization and the metabolic switch from fatty acid turnover to glucose metabolism [172,183].

The gas nitric oxide is an endothelium derived vasodilator which can add to oxidative injury by the formation of peroxynitrite and the formation of covalent chemical bonds to cysteine residues of proteins (S-nitrosylation). In addition, nitric oxide can modulate cardiac contractility by interfering with adrenergic signaling by changing cyclic AMP levels. In patients with heart failure it has been shown that the response to nitric oxide is attenuated, leading to increased vascular tone and increased afterload, but direct evidence for a causal pathologic role in heart failure is lacking [184] and only a few results from gene array studies suggest altered gene expression in antioxidative pathways [93].

To summarize from existing clinical and experimental data, it can be concluded that reactive oxide species can participate in the pathogenesis of heart failure by affecting basic cellular functions in cardiac and vascular cells, resulting in contractile dysfunction and cardiac remodeling. The level of contribution of reactive oxygen species towards the disease progression is difficult to assess and therefore poorly defined.

Transcription factors in cardiac hypertrophy and failure

Transcription factors are regulatory proteins that control gene transcription. Thus, they are the terminal effectors of most signaling pathways that affect gene expression. Regulation of gene expression is highly complex involving several hundred transcription factors through a variety of mechanisms. We discuss the major transcription factors that have been more commonly observed to be involved in gene transcriptional regulation in the cardiac remodeling of failure.

GATA4

GATA4 is a transcription factor in the activation of many cardiac specific genes under basal condition including α-MHC, essential myosin light chain, cardiac troponin I and troponin C. GATA4 is activated by several signaling pathways. The MAPK, ERK 1/2 and p38 both activate GATA4 [185]. Thus, neurohormones working through the Gqα coupled receptor will activate GATA4. β-Adrenergic stimulation, mechanical stress and IL-18 have also been shown to activate GATA4 [75,186]. Finally, GATA4 can also be activated through calcineurin–NFAT signaling. Glycogen synthase kinase 3β negatively regulates GATA4. Given this commonality of activation it is not surprising that GATA is a co-factor in the upregulation of gene specific to myocardial hypertrophy and failure including β-MHC, sodium calcium exchanger, angiotensin II type 1 receptor, brain natriuretic peptide and atrial natriuretic peptide [186].

Activating protein 1

This transcription factor is the result of dimerization of Jun and Fos protein family members. Thus, activating protein 1 (AP-1) is capable of activating a variety of genes depending on the dimerized combination of transcription factors. The increase in expression of c-Fos and c-Jun in hypertrophy and failure has been well documented [187]. c-Jun is the terminal effector of the c-Jun N terminal kinase (JNK), which is a MAPK pathway [188]. Both c-Jun and c-Fos are proto-oncogenes and part of an immediate early gene expression program. G protein coupled receptors, cellular stress and cytokines can all activate their transcription. Activation of other Jun and Fos members is demonstrated in hypertrophy models and thus are also likely involved in AP-1 gene regulation. In general, AP-1 is associated with the overall increase in protein synthesis and with the specific activation of several genes including atrial natriuretic factor, brain natriuretic peptide, MMP and collagen [131,188]. Interestingly, expression of a dominant negative c-Jun in cardiocytes blocked hypertrophy by Gαq coupled agonists suggesting a centralized role in the hypertrophic response [189].

Nuclear factor κB

NF κB can be activated by a variety of mechanisms. Perhaps the best established is the indirect activation by cytokines (most notably TNF-α and IL-1). However, other signaling pathways can activate NF κB including angiotensin II, endothelin, oxidative stress, aldosterone and catacholamines [188,189]. NF κB is directly involved in the inflammatory–apoptotic response of the cell with the upregulated expression of TNF-α, IL-1, caspase-8 and caspase-11 [192]. Interestingly, NF κB also increases the expression of antiapoptotic genes. NF κB likely in conjunction with AP-1 modulates the expression of MMPs, collagen and adhesion molecules, thus having a critical role in the adverse remodeling of myocardial failure [191,193]. Finally, Gαq receptor linked cellular hypertrophy appears to be mediated in part by NF κB [190].

Nuclear factor of activated T cells

The calcineurin–NFAT pathway is one of the better characterized mechanisms of the pathologic hypertrophic response. In contrast, calcineurin–NFAT signaling does not appear to be involved in physiologic hypertrophy [145]. Calcineurin is a phosphatase that dephosphorylates NFAT, permitting its transport into the nucleus. Glycogen synthase 3β kinase functions in a counter-regulatory manner through the phosphorylation of NFAT. Similarly, JNK, p38 MAPK and PKA all inhibit nuclear localization of NFAT [144,194]. Many studies in animal models of failure have demonstrated the attenuated hypertrophic response when the animals are treated with the calcineurin inhibitor, ciclosporin [144]. These studies demonstrated the integral role of calcineurin and NFAT in the pathologic hypertrophic response.

mTOR

Thus far we have focused on transcription factors that modulate gene expression. The mammalian target of rapamyosin (mTOR) affects transcription not only through increasing the transcription of hypertrophic response genes but also through the induction of the ribosomal biosynthesis, thus facilitating the translation of mRNA [195]. mTOR is activated in both physiologic and pathologic hypertrophy [140]. The transcription factor mTOR is activated primarily through the PI3K–Akt pathway but may also be activated through the ERK pathway [195].

Gene array studies in heart failure

Over the past few years several studies using large-scale gene arrays have been reported using both commercial and custom made gene chip systems. Most of these studies compare samples from non-failing and failing human myocardium obtained at the time of transplantation (see Table 7.1 for a list of the large-scale high-density gene arrays studies in heart failure). Some studies compared samples before and after left ventricular assist device (LVAD) implantation or compared right and left ventricles from failing hearts [72,196]. Other groups investigated the feasibility of performing gene array studies from percutaneously obtained right ventricular endomyocardial biopsies [71,124].

Despite the difficulties in comparing studies because of differences in approach and analysis (e.g., variable cutoff values for differential expression), common findings have been reported in all studies. For example, ANP and BNP have been confirmed to be largely universally upregulated in heart failure, further underscoring their value as biomarkers in this disease. Another common theme is the upregulation of extracellular matrix genes including different isoforms of collagen and several members of small leucin-rich repeat proteoglycans (SLRPs) which are involved in binding and regulating collagen assembly. The previously established changes in the expression of sarcomeric genes and genes involved in calcium cycling have been confirmed by many gene array studies and are often used to validate their results. Most gene array studies also report heart failure induced changes in genes involving energy metabolism, antioxidative and apoptotic pathways, signal transduction, cell-cycle regulators and transcription factors. However, because of large variations between samples from different patients and stringently applied statistical algorithms, the altered expression of other genes, which is commonly accepted to be changed

Table 7.1 List of large scale high density gene arrays in heart failure.

Year	Array design	Number of genes	Subject	Disease	Reference
2000	Affymetrix Hu6800 GeneChip	7085	Human	ICM, DCM	Yang *et al.* [123]
2001	Cardio Chip	10368	Human	HCM	Barrans *et al.* [118]
	Affymetrix Rat U34A Array		Rat	MI	Jin *et al.* [71]
2002	Cardio Chip	10848	Human	DCM	Barrans *et al.* [89]
	Custom-made Chip	10272	Human	DCM, HCM	Hwang *et al.* [120]
	Affymetrix Hu6800Fl GeneChip	6606	Human	DCM	Tan *et al.* [72]
2003	Affymetrix Hu GeneFL Chip	6800	Human	pre – post LVAD	Blaxall *et al.* [70]
	Incyte Human UniGem V	10176	Human	DCM	Boheler *et al.* [195]
	Agilent Human 1 Catalogue Array	12814	Human	pre – post LVAD	Chen *et al.* [194]
	Human Uni Gene RZPD1	30336	Human	DCM	Grzeskowiak *et al.* [122]
	Affymetrix HG-U95A	12626	Human	ICM, DCM	Steenman *et al.* [90]
2004	Affymetrix HG-U133A	22283	Human	DCM	Yung *et al.* [91]
	Incyte Rat Gem2/3 cDNA library	12336	Rat	Ren-2	Schroen *et al.* [206]
	Affymetrix U133A	22283	Human	DCM, ICM	Kittleson *et al.* [69]

DCM, dilated cardiomyopathy; HCM, hypertrophic cardiomyopathy; ICM, ischemic cardiomyopathy; LVAD, ; MI, myocardial infarction.

in heart failure, was not reported in some studies, indicating that subtle changes in gene expression may not be recognized by this approach, irrespective of the functional importance.

Studies that compared gene-expression in ischemic versus idiopathic dilated cardiomyopathy indicate that most genes are similarly regulated in both conditions, supporting the concept of a common final pathway in end stage heart failure. However, a subset of genes was demonstrated to be differentially regulated. [71,72,92], which may present an opportunity for diagnostic and/or therapeutic targets if confirmed in larger patient trials. However, other studies caution that this will not be easy to accomplish because age and sex appear to have a significant effect on gene transcription, thus further complicating the analysis of gene arrays [197]. The expression of genes in right and left ventricular samples from patients with dilated cardiomyopathy revealed similar expression patterns but a small number of genes were found to be differentially regulated between right and left ventricular samples [92]. There are limited array data patients with concentric hypertrophy and little is known on gene regulation as a function of the different stages of heart human failure [120,122,196].

In summary, the first results from large-scale gene array studies are very promising and confirm previous results using target gene approaches. This suggests that diagnostic and prognostic fingerprinting might become a reality in heart failure. However, so far the results are mostly hypothesis generating by hinting at novel pathways and gene regulation mechanisms involved in heart failure, because all studies to date have used small patient samples or analyzed pooled RNA from different patients. The available data suggest that larger scale patient trials will promote gene array technology to an accepted diagnostic and possibly prognostic tool in heart failure.

Gene expression as a function of the stages of heart failure

Information regarding the changes in gene expression in the various stages of human heart failure is limited at best. Studies are limited by biopsy sampling location (e.g., right versus left ventricle; endocardium versus epicardium; septal versus free wall), the inherent variability in patient populations, and accurately determining the stage of heart failure. Furthermore, small animal studies may not reflect the changes in gene expression inherent to human cardiomyopathy. Despite these concerns there are a small number of studies that indicate differential activation of signaling pathways and gene as a function of disease stage. Molkentin and colleagues demonstrated that calcineurin activity was increased in the human hypertrophied and failing heart [198]. Furthermore, all three MAPK pathways (ERK 1/2, JNK and p38) and the PI3K–Akt are increased in end stage failure but not in hypertrophied hearts. Recent evidence indicates that cytokine expression (TNF-α, IL-1β, IL-6) is greatest in compensated pressure overload hypertrophy (aortic stenosis) but decrease toward normal level in more advanced stages of the disease [199], suggesting that cardiac expression of cytokines may have an adaptive role in hypertrophy. In a recent study in aortic stenosis patients with preserved or impaired left ventricular function it was demonstrated that MMP activity was elevated in all stages of heart failure but that fibroblast hyperplasia and collagen synthesis was increased most markedly in patients with left ventricular ejection fractions of <30% [200], suggesting that the balance of MMP activity to collagen synthesis allows for degradation of the extracellular matrix in the earlier stages of the disease, facilitating myocyte slippage and chamber dilatation, whereas in the later stages of heart failure myocardial fibrosis is more prominent [201].

Effect of therapies on gene expression

Therapy with beta-blockers results in improved mortality in systolic heart failure patients and is associated with alterations of gene expression. Specifically, increased expression of SERCA and α-MHC, and decreased expression of the β-MHC has been reported, indicating a reversal of the fetal gene program that is activated in heart failure [202]. Interestingly, in this small study beta-blockers did not independently affect the expression levels of the β-adrenergic receptors over a 6-month period [202]. While to our knowledge there are no studies directly assessing the effect of ACE-I or angiotensin receptor blockers on gene expression in human cardiomyopathy, several studies have demonstrated that these therapies significantly affect gene expression. Most notably, treatment in animal models of cardiac failure have demonstrated decreased myocardial fibrosis, hypertrophy, calcineurin activity and decreased expression of collagen I, total collagen and TGF-β [203,204]. Mechanically unloading the heart and decreasing neurohormonal stimulation with the LVAD is associated with altered expression of numerous gene as elucidated by gene array including decreased expression BNP, TNF-α,

MMP, collagen, IL-8 and GATA4 [72,205]. It is of interest that the expression levels of many genes increased with LVAD support indicating suppressed expression in heart failure. Wall stress in either failing or dyssynchronous ventricles is known to directly contribute to the altered gene expression [206,207]. Attenuation of wall stress with a passive restraint device (ACORN) was associated with a favorable change in MHC expression, p38 MAPK, c-fos and attenuated myocyte hypertrophy [207].

Conclusions

This chapter presents an introduction into the changes in gene regulation and expression that occur through the development and progression of heart failure. We have not been able to cover all aspects of genomics in heart failure but have sought to cover major pathways and mechanisms. As is clear, many factors control gene regulation, thus making it a highly regulated and complex process. Furthermore, differences in the genetic composition of individuals can affect the pathophysiology of heart failure. Thus, more studies need to be conducted to better define the primary mechanisms of gene regulation and expression. It is through such an approach that tailored and target therapeutic approaches to heart failure can be implemented.

Acknowledgments

This work was funded by grants from the National Institutes of Health, HL077637 and HL65586 (PVB). We thank Drs. Martin LeWinter and David Maughan for their valuable comments on the manuscript.

References

1 American Heart Association. *Heart Disease and Stroke Statistics, 2005 Update.* American Heart Association, Dallas, TX, 2005.

2 Redfield MM, Jacobsen SJ, Burnett JC Jr, Mahoney DW, Bailey KR, Rodeheffer RJ. Burden of systolic and diastolic ventricular dysfunction in the community: appreciating the scope of the heart failure epidemic. *JAMA* 2003; 289: 194–202.

3 Levy D, Kenchaiah S, Larson MG *et al.* Long-term trends in the incidence of and survival with heart failure. *N Engl J Med* 2002; 347: 1397–1402.

4 Ho KK, Anderson KM, Kannel WB, Grossman W, Levy D. Survival after the onset of congestive heart failure in Framingham Heart Study subjects. *Circulation* 1993; 88: 107–115.

5 Krumholz HM, Larson M, Levy D. Sex differences in cardiac adaptation to isolated systolic hypertension. *Am J Cardiol* 1993; 72: 310–313.

6 Petrie MC, Dawson NF, Murdoch DR, Davie AP, McMurray JJ. Failure of women's hearts. *Circulation* 1999; 99: 2334–2341.

7 Ghali JK. Sex-related differences in heart failure and beta-blockers. *Heart Fail Rev* 2004; 9: 149–159.

8 Yancy CW. Special considerations for carvedilol use in heart failure. *Am J Cardiol* 2004; 93: 64B–68B.

9 Huang A, Kaley G. Gender-specific regulation of cardiovascular function: estrogen as key player. *Microcirculation* 2004; 11: 9–38.

10 Pelzer T, Loza PA, Hu K *et al.* Increased mortality and aggravation of heart failure in estrogen receptor-beta knockout mice after myocardial infarction. *Circulation* 2005; 111: 1492–1498.

11 Patten RD, Pourati I, Aronovitz MJ *et al.* 17beta-estradiol reduces cardiomyocyte apoptosis *in vivo* and *in vitro* via activation of phospho-inositide-3 kinase/Akt signaling. *Circ Res* 2004; 95: 692–699.

12 Schwartzbauer G, Robbins J. Matters of sex: sex matters. *Circulation* 2001; 104: 1333–1335.

13 Smith GL, Shlipak MG, Havranek EP *et al.* Race and renal impairment in heart failure: mortality in blacks versus whites. *Circulation* 2005; 111: 1270–1277.

14 Carson P, Ziesche S, Johnson G, Cohn JN. Racial differences in response to therapy for heart failure: analysis of the vasodilator-heart failure trials. Vasodilator-Heart Failure Trial Study Group. *J Card Fail* 1999; 5: 178–187.

15 Dries DL, Exner DV, Gersh BJ, Cooper HA, Carson PE, Domanski MJ. Racial differences in the outcome of left ventricular dysfunction. *N Engl J Med* 1999; 340: 609–616.

16 The Beta-Blocker Evaluation of Survival Trial Investigators. A trial of the beta-blocker bucindolol in patients with advanced chronic heart failure. *N Engl J Med* 2001; 344: 1659–1667.

17 Sackner-Bernstein JD, Skopicki HA. Racing away from bias. *J Am Coll Cardiol* 2004; 43: 785–786.

18 Taylor AL, Ziesche S, Yancy C *et al.* Combination of isosorbide dinitrate and hydralazine in blacks with heart failure. *N Engl J Med* 2004; 351: 2049–2057.

19 Kalinowski L, Dobrucki IT, Malinski T. Race-specific differences in endothelial function: predisposition of African Americans to vascular diseases. *Circulation* 2004; 109: 2511–2517.

20 Small KM, Wagoner LE, Levin AM, Kardia SL, Liggett SB. Synergistic polymorphisms of β_1- and α_{2C}-adrener-gic receptors and the risk of congestive heart failure. *N Engl J Med* 2002; 347: 1135–1142.

21 Yancy CW. Does race matter in heart failure? *Am Heart J* 2003; 146: 203–206.

22 Andersson B, Sylven C. The DD genotype of the angiotensin-converting enzyme gene is associated with increased mortality in idiopathic heart failure. *J Am Coll Cardiol* 1996; 28: 162–167.

23 Katz AM. *Heart Failure: Pathophysiology, Molecular Biology and Clinical Management.* Lippincott, Williams & Wilkins, 2000.

24 Starling EH. The Linacre Lecture on the Law of the Heart, London, UK: Longmans, Green and Co; 1918.

25 Bowditch HP. Über die Eigenthümlichkeiten der Reizbarkeit, welche die Musklefasern des Herzens zeigen. *Berichte der Kön Sächs Gesellschaft der Wissenschatten Mathematisch Physische Classe.* 1871; 23: 652–689.

26 Mann DL. Basic mechanisms of left ventricular remodeling: the contribution of wall stress. *J Card Fail* 2004; 10 (6 Suppl): S202–S206.

27 Wiese S, Breyer T, Dragu A *et al.* Gene expression of brain natriuretic peptide in isolated atrial and ventricular human myocardium: influence of angiotensin II and diastolic fiber length. *Circulation* 2000; 102: 3074–3079.

28 Houser SR, Margulies KB. Is depressed myocyte contractility centrally involved in heart failure? *Circ Res* 2003; 92: 350–358.

29 Franciosa JA, Park M, Levine TB. Lack of correlation between exercise capacity and indexes of resting left ventricular performance in heart failure. *Am J Cardiol* 1981; 47: 33–39.

30 Quinones MA, Greenberg BH, Kopelen HA *et al.* Echocardiographic predictors of clinical outcome in patients with left ventricular dysfunction enrolled in the SOLVD registry and trials: significance of left ventricular hypertrophy. Studies of Left Ventricular Dysfunction. *J Am Coll Cardiol* 2000; 35: 1237–1244.

31 Francis GS, Cohn JN, Johnson G, Rector TS, Goldman S, Simon A. Plasma norepinephrine, plasma renin activity, and congestive heart failure. Relations to survival and the effects of therapy in V-HeFT II. The V-HeFT VA Cooperative Studies Group. *Circulation* 1993; 87 (Supplement): VI40–VI48.

32 Lohse MJ, Engelhardt S, Eschenhagen T. What is the role of beta-adrenergic signaling in heart failure? *Circ Res* 2003; 93: 896–906.

33 Communal C, Colucci WS. The control of cardio-myocyte apoptosis via the beta-adrenergic signaling pathways. *Arch Mal Coeur Vaiss* 2005; 98: 236–241.

34 Port JD, Bristow MR. Altered beta-adrenergic receptor gene regulation and signaling in chronic heart failure. *J Mol Cell Cardiol* 2001; 33: 887–905.

35 Hata JA, Williams ML, Koch WJ. Genetic manipulation of myocardial beta-adrenergic receptor activation and desensitization. *J Mol Cell Cardiol* 2004; 37: 11–21.

36 Forleo C, Resta N, Sorrentino S *et al.* Association of beta-adrenergic receptor polymorphisms and progression to heart failure in patients with idiopathic dilated cardiomyopathy. *Am J Med* 2004; 117: 451–458.

37 Kaye DM, Smirk B, Finch S, Williams C, Esler MD. Interaction between cardiac sympathetic drive and heart rate in heart failure: modulation by adrenergic receptor genotype. *J Am Coll Cardiol* 2004; 44: 2008–2015.

38 Small KM, McGraw DW, Liggett SB. Pharmacology and physiology of human adrenergic receptor polymorphisms. *Annu Rev Pharmacol Toxicol* 2003; 43: 381–411.

39 Pasotti M, Repetto A, Tavazzi L, Arbustini E. Genetic predisposition to heart failure. *Med Clin North Am* 2004; 88: 1173–1192.

40 Kim S, Ohta K, Hamaguchi A, Yukimura T, Miura K, Iwao H. Angiotensin II induces cardiac phenotypic modulation and remodeling *in vivo* in rats. *Hypertension* 1995; 25: 1252–1259.

41 Sadoshima J, Izumo S. Molecular characterization of angiotensin II-induced hypertrophy of cardiac myocytes and hyperplasia of cardiac fibroblasts. Critical role of the AT1 receptor subtype. *Circ Res* 1993; 73: 413–423.

42 Schunkert H, Sadoshima J, Cornelius T *et al.* Angiotensin II-induced growth responses in isolated adult rat hearts. Evidence for load-independent induction of cardiac protein synthesis by angiotensin II. *Circ Res* 1995; 76: 489–497.

43 Manabe I, Shindo T, Nagai R. Gene expression in fibroblasts and fibrosis: involvement in cardiac hypertrophy. *Circ Res* 2002; 91: 1103–1113.

44 Kajstura J, Cigola E, Malhotra A *et al.* Angiotensin II induces apoptosis of adult ventricular myocytes *in vitro*. *J Mol Cell Cardiol* 1997; 29: 859–870.

45 Regitz-Zagrosek V, Friedel N, Heymann A *et al.* Regulation, chamber localization, and subtype distribution of angiotensin II receptors in human hearts. *Circulation* 1995; 91: 1461–1471.

46 Wharton J, Morgan K, Rutherford RA *et al.* Differential distribution of angiotensin AT2 receptors in the normal and failing human heart. *J Pharmacol Exp Ther* 1998; 284: 323–336.

47 Haywood GA, Gullestad L, Katsuya T *et al.* AT1 and AT2 angiotensin receptor gene expression in human heart failure. *Circulation* 1997; 95: 1201–1206.

48 Yang Z, Bove CM, French BA *et al.* Angiotensin II type 2 receptor overexpression preserves left ventricular function after myocardial infarction. *Circulation* 2002; 106: 106–111.

49 Fischer TA, Singh K, O'Hara DS, Kaye DM, Kelly RA. Role of AT1 and AT2 receptors in regulation of MAPKs and MKP-1 by ANG II in adult cardiac myocytes. *Am J Physiol* 1998; 275: H906–H916.

50 Swedberg K, Eneroth P, Kjekshus J, Wilhelmsen L. Hormones regulating cardiovascular function in patients with severe congestive heart failure and their relation to mortality. CONSENSUS Trial Study Group. *Circulation* 1990; 82: 1730–1736.

51 Connell JM, Davies E. The new biology of aldosterone. *J Endocrinol* 2005; 186: 1–20.

52 Young M, Head G, Funder J. Determinants of cardiac fibrosis in experimental hypermineralocorticoid states. *Am J Physiol* 1995; 269: E657–E662.

53 Harada E, Yoshimura M, Yasue H *et al.* Aldosterone induces angiotensin-converting-enzyme gene expression in cultured neonatal rat cardiocytes. *Circulation* 2001; 104: 137–139.

54 Robert V, Heymes C, Silvestre JS, Sabri A, Swynghedauw B, Delcayre C. Angiotensin AT1 receptor subtype as a cardiac target of aldosterone: role in aldosterone-salt-induced fibrosis. *Hypertension* 1999; 33: 981–986.

55 Qin W, Rudolph AE, Bond BR *et al.* Transgenic model of aldosterone-driven cardiac hypertrophy and heart failure 1. *Circ Res* 2003; 93: 69–76.

56 Matsumoto R, Yoshiyama M, Omura T *et al.* Effects of aldosterone receptor antagonist and angiotensin II type I receptor blocker on cardiac transcriptional factors and mRNA expression in rats with myocardial infarction. *Circ J* 2004; 68: 376–382.

57 Takeda Y, Yoneda T, Demura M, Usukura M, Mabuchi H. Calcineurin inhibition attenuates mineralocorticoid-induced cardiac hypertrophy. *Circulation* 2002; 105: 677–679.

58 Takeda Y. Pleiotropic actions of aldosterone and the effects of eplerenone, a selective mineralocorticoid receptor antagonist. *Hypertens Res* 2004; 27: 781–789.

59 Mizuno Y, Yoshimura M, Yasue H *et al.* Aldosterone production is activated in failing ventricle in humans. *Circulation* 2001; 103: 72–77.

60 Zannad F, Alla F, Dousset B, Perez A, Pitt B. Limitation of excessive extracellular matrix turnover may contribute to survival benefit of spironolactone therapy in patients with congestive heart failure: insights from the randomized aldactone evaluation study (RALES). Rales Investigators 4. *Circulation* 2000; 102: 2700–2706.

61 Goldsmith SR, Francis GS, Cowley AW Jr, Levine TB, Cohn JN. Increased plasma arginine vasopressin levels in patients with congestive heart failure. *J Am Coll Cardiol* 1983; 1: 1385–1390.

62 Chatterjee K. Neurohormonal activation in congestive heart failure and the role of vasopressin. *Am J Cardiol* 2005; 95: 8B–13B.

63 Fukuzawa J, Haneda T, Kikuchi K. Arginine vasopressin increases the rate of protein synthesis in isolated per-

fused adult rat heart via the V1 receptor. *Mol Cell Biochem* 1999; 195: 93–98.

64 Nakamura Y, Haneda T, Osaki J, Miyata S, Kikuchi K. Hypertrophic growth of cultured neonatal rat heart cells mediated by vasopressin V(1A) receptor. *Eur J Pharmacol* 2000; 391: 39–48.

65 Xu Y, Hopfner RL, McNeill JR, Gopalakrishnan V. Vasopressin accelerates protein synthesis in neonatal rat cardiomyocytes. *Mol Cell Biochem* 1999; 195: 183–190.

66 Rich S, McLaughlin VV. Endothelin receptor blockers in cardiovascular disease. *Circulation* 2003; 108: 2184–2190.

67 Boerrigter G, Burnett JC. Endothelin in neurohormonal activation in heart failure. *Coron Artery Dis* 2003; 14: 495–500.

68 Stewart DJ, Cernacek P, Costello KB, Rouleau JL. Elevated endothelin-1 in heart failure and loss of normal response to postural change. *Circulation* 1992; 85: 510–517.

69 Zolk O, Quattek J, Sitzler G *et al.* Expression of endothelin-1, endothelin-converting enzyme, and endothelin receptors in chronic heart failure. *Circulation* 1999; 99: 2118–2123.

70 Yang LL, Gros R, Kabir MG *et al.* Conditional cardiac overexpression of endothelin-1 induces inflammation and dilated cardiomyopathy in mice. *Circulation* 2004; 109: 255–261.

71 Kittleson MM, Ye SQ, Irizarry RA *et al.* Identification of a gene expression profile that differentiates between ischemic and nonischemic cardiomyopathy. *Circulation* 2004; 110: 3444–3451.

72 Blaxall BC, Tschannen-Moran BM, Milano CA, Koch WJ. Differential gene expression and genomic patient stratification following left ventricular assist device support. *J Am Coll Cardiol* 2003; 41: 1096–1106.

73 Jin H, Yang R, Awad TA *et al.* Effects of early angiotensin-converting enzyme inhibition on cardiac gene expression after acute myocardial infarction. *Circulation* 2001; 103: 736–742.

74 Tan FL, Moravec CS, Li J *et al.* The gene expression fingerprint of human heart failure. *Proc Natl Acad Sci USA* 2002; 99: 11387–1192.

75 Chandrasekar B, Mummidi S, Claycomb WC, Mestril R, Nemer M. Interleukin-18 is a pro-hypertrophic cytokine that acts through a phosphatidylinositol 3-kinase-phosphoinositide-dependent kinase-1-Akt-GATA4 signaling pathway in cardiomyocytes. *J Biol Chem* 2005; 280: 4553–4567.

76 Mann DL. Inflammatory mediators and the failing heart: past, present, and the foreseeable future. *Circ Res* 2002; 91: 988–998.

77 Deswal A, Petersen NJ, Feldman AM, Young JB, White BG, Mann DL. Cytokines and cytokine receptors in advanced heart failure: an analysis of the cytokine database from the Vesnarinone trial (VEST). *Circulation* 2001; 103: 2055–2059.

78 Torre-Amione G, Kapadia S, Benedict C, Oral H, Young JB, Mann DL. Proinflammatory cytokine levels in patients with depressed left ventricular ejection fraction: a report from the Studies of Left Ventricular Dysfunction (SOLVD). *J Am Coll Cardiol* 1996; 27: 1201–1206.

79 Mehra VC, Ramgolam VS, Bender JR. Cytokines and cardiovascular disease. *J Leukoc Biol* 2005; 78: 805–818.

80 Yokoyama T, Vaca L, Rossen RD, Durante W, Hazarika P, Mann DL. Cellular basis for the negative inotropic effects of tumor necrosis factor-alpha in the adult mammalian heart. *J Clin Invest* 1993; 92: 2303–2312.

81 Prabhu SD. Cytokine-induced modulation of cardiac function. *Circ Res* 2004; 95: 1140–53.

82 Pilz RB, Casteel DE. Regulation of gene expression by cyclic GMP. *Circ Res* 2003; 93: 1034–1046.

83 Kawamura N, Kubota T, Kawano S *et al.* Blockade of NF-kappaB improves cardiac function and survival without affecting inflammation in TNF-alpha-induced cardiomyopathy. *Cardiovasc Res* 2005; 66: 520–529.

84 Yokoyama T, Nakano M, Bednarczyk JL, McIntyre BW, Entman M, Mann DL. Tumor necrosis factor-alpha provokes a hypertrophic growth response in adult cardiac myocytes. *Circulation* 1997; 95: 1247–1252.

85 VanBuren P, Harris DE, Alpert NR, Warshaw DM. Cardiac V1 and V3 myosins differ in their hydrolytic and mechanical activities *in vitro*. *Circ Res* 1995; 77: 439–444.

86 Kameyama T, Chen Z, Bell SP, VanBuren P, Maughan D, LeWinter MM. Mechanoenergetic alterations during the transition from cardiac hypertrophy to failure in Dahl salt-sensitive rats. *Circulation* 1998; 98: 2919–2929.

87 Alpert NR, Mulieri LA. Increased myothermal economy of isometric force generation in compensated cardiac hypertrophy induced by pulmonary artery constriction in the rabbit. A characterization of heat liberation in normal and hypertrophied right ventricular papillary muscles. *Circ Res* 1982; 50: 491–500.

88 Miyata S, Minobe W, Bristow MR, Leinwand LA. Myosin heavy chain isoform expression in the failing and nonfailing human heart. *Circ Res* 2000; 86: 386–390.

89 Reiser PJ, Portman MA, Ning XH, Schomisch MC. Human cardiac myosin heavy chain isoforms in fetal and failing adult atria and ventricles. *Am J Physiol Heart Circ Physiol* 2001; 280: H1814–H1820.

90 Noguchi T, Camp P, Alix SL *et al.* Myosin from failing and non-failing human ventricles exhibit similar contractile properties. *J Mol Cell Cardiol* 2003; 35: 91–97.

91 Barrans JD, Allen PD, Stamatiou D, Dzau VJ, Liew CC. Global gene expression profiling of end-stage dilated cardiomyopathy using a human cardiovascular-based cDNA microarray. *Am J Pathol* 2002; 160: 2035–2043.

92 Steenman M, Chen YW, Le CM *et al.* Transcriptomal analysis of failing and nonfailing human hearts. *Physiol Genomics* 2003; 12: 97–112.

93 Yung CK, Halperin VL, Tomaselli GF, Winslow RL. Gene expression profiles in end-stage human idiopathic dilated cardiomyopathy: altered expression of apoptotic and cytoskeletal genes. *Genomics* 2004; 83: 281–297.

94 Nguyen TT, Hayes E, Mulieri LA *et al.* Maximal acto-myosin ATPase activity and *in vitro* myosin motility are unaltered in human mitral regurgitation heart failure. *Circ Res* 1996; 79: 222–226.

95 Pagani ED, Alousi AA, Grant AM, Older TM, Dziuban SWJ, Allen PD. Changes in myofibrillar content and Mg-ATPase activity in ventricular tissues from patients with heart failure caused by coronary artery disease, cardiomyopathy, or mitral valve insufficiency. *Circ Res* 1988; 63: 380–385.

96 Alousi AA, Grant AM, Etzler JR, Cofer BR, Van dB, Melvin D. Reduced cardiac myofibrillar Mg-ATPase activity without changes in myosin isozymes in patients with end-stage heart failure. *Mol Cell Biochem* 1990; 96: 79–88.

97 Nakao K, Minobe W, Roden R, Bristow MR, Leinwand LA. Myosin heavy chain gene expression in human heart failure. *J Clin Invest* 1997; 100: 2362–2370.

98 Morano I, Hadicke K, Haase H, Bohm M, Erdmann E, Schaub MC. Changes in essential myosin light chain iso-form expression provide a molecular basis for isometric force regulation in the failing human heart. *J Mol Cell Cardiol* 1997; 29: 1177–1187.

99 Makarenko I, Opitz CA, Leake MC *et al.* Passive stiffness changes caused by upregulation of compliant titin iso-forms in human dilated cardiomyopathy hearts. *Circ Res* 2004; 95: 708–716.

100 Granzier HL, Labeit S. The giant protein titin: a major player in myocardial mechanics, signaling, and disease. *Circ Res* 2004; 94: 284–295.

101 Boheler KR, Carrier L, de la Bastie D *et al.* Skeletal actin mRNA increases in the human heart during ontogenic development and is the major isoform of control and failing adult hearts. *J Clin Invest* 1991; 88: 323–330.

102 Meggs LG, Tillotson J, Huang H, Sonnenblick EH, Capasso JM, Anversa P. Noncoordinate regulation of alpha-1 adrenoreceptor coupling and reexpression of alpha skeletal actin in myocardial infarction-induced left ventricular failure in rats. *J Clin Invest* 1990; 86: 1451–1458.

103 Hanatani A, Yoshiyama M, Kim S *et al.* Inhibition by angiotensin II type 1 receptor antagonist of cardiac phe-notypic modulation after myocardial infarction. *J Mol Cell Cardiol* 1995; 27: 1905–1914.

104 Schwartz K, Carrier L, Lompre AM, Mercadier JJ, Boheler KR. Contractile proteins and sarcoplasmic reticulum calcium-ATPase gene expression in the hyper-

105 Gao L, Kennedy JM, Solaro RJ. Differential expression of TnI and TnT isoforms in rabbit heart during the peri-natal period and during cardiovascular stress. *J Mol Cell Cardiol* 1995; 27: 541–550.

106 Hunkeler NM, Kullman J, Murphy AM. Troponin I iso-form expression in human heart. *Circ Res* 1991; 69: 1409–1414.

107 Sasse S, Brand NJ, Kyprianou P *et al.* Troponin I gene expression during human cardiac development and in end-stage heart failure. *Circ Res* 1993; 72: 932–938.

108 Anderson PA, Malouf NN, Oakeley AE, Pagani ED, Allen PD. Troponin T isoform expression in the normal and failing human left ventricle: a correlation with myofibrillar ATPase activity. *Basic Res Cardiol* 1992; 87 (Supplement 1): 117–127.

109 Wolff MR, Buck SH, Stoker SW, Greaser ML, Mentzer RM. Myofibrillar calcium sensitivity of isometric ten-sion is increased in human dilated cardiomyopathies: role of altered beta-adrenergically mediated protein phosphorylation. *J Clin Invest* 1996; 98: 167–176.

110 Molina MI, Kropp KE, Gulick J, Robbins J. The sequence of an embryonic myosin heavy chain gene and isolation of its corresponding cDNA. *J Biol Chem* 1987; 262: 6478–6488.

111 VanBuren P, Alix SL, Gorga JA, Begin KJ, LeWinter MM, Alpert NR. Cardiac troponin T isoforms demon-strate similar effects on mechanical performance in a regulated contractile system. *Am J Physiol Heart Circ Physiol* 2002; 282: H1665–H1671.

112 Gomes AV, Guzman G, Zhao J, Potter JD. Cardiac tro-ponin T isoforms affect the Ca^{2+} sensitivity and inhibi-tion of force development. Insights into the role of troponin T isoforms in the heart. *J Biol Chem* 2002; 277: 35341–35349.

113 Bers DM. Cardiac excitation–contraction coupling. *Nature* 2002; 415: 198–205.

114 Bers DM. Calcium fluxes involved in control of cardiac myocyte contraction. *Circ Res* 2000; 87: 275–281.

115 Wankerl M, Schwartz K. Calcium transport proteins in the nonfailing and failing heart: gene expression and function. *J Mol Med* 1995; 73: 487–496.

116 Hasenfuss G, Meyer M, Schillinger W, Preuss M, Pieske B, Just H. Calcium handling proteins in the failing human heart. *Basic Res Cardiol* 1997; 92 (Supplement 1): 87–93.

117 Meyer M, Schillinger W, Pieske B *et al.* Alterations of sarcoplasmic reticulum proteins in failing human dilated cardiomyopathy. *Circulation* 1995; 92: 778–784.

118 Studer R, Reinecke H, Bilger J *et al.* Gene expression of the cardiac Na^+-Ca^{2+} exchanger in end-stage human heart failure. *Circ Res* 1994; 75: 443–453.

119 Wehrens XH, Lehnart SE, Huang F *et al.* FKBP12.6 deficiency and defective calcium release channel (ryan-

odine receptor) function linked to exercise-induced sudden cardiac death. *Cell* 2003; 113: 829–840.

120 Barrans JD, Stamatiou D, Liew C. Construction of a human cardiovascular cDNA microarray: portrait of the failing heart. *Biochem Biophys Res Commun* 2001; 280: 964–969.

121 Heinke MY, Wheeler CH, Chang D *et al.* Protein changes observed in pacing-induced heart failure using two-dimensional electrophoresis. *Electrophoresis* 1998; 19: 2021–2030.

122 Hwang JJ, Allen PD, Tseng GC *et al.* Microarray gene expression profiles in dilated and hypertrophic cardiomyopathic end-stage heart failure. *Physiol Genomics* 2002; 10: 31–44.

123 Chapman RE, Spinale FG. Extracellular protease activation and unraveling of the myocardial interstitium: critical steps toward clinical applications. *Am J Physiol Heart Circ Physiol* 2004; 286: H1–H10.

124 Grzeskowiak R, Witt H, Drungowski M *et al.* Expression profiling of human idiopathic dilated cardiomyopathy. *Cardiovasc Res* 2003; 59: 400–411.

125 Yang J, Moravec CS, Sussman MA *et al.* Decreased SLIM1 expression and increased gelsolin expression in failing human hearts measured by high-density oligonucleotide arrays. *Circulation* 2000; 102: 3046–3052.

126 Siwik DA, Chang DL, Colucci WS. Interleukin-1beta and tumor necrosis factor-alpha decrease collagen synthesis and increase matrix metalloproteinase activity in cardiac fibroblasts *in vitro*. *Circ Res* 2000; 86: 1259–1265.

127 Siwik DA, Pagano PJ, Colucci WS. Oxidative stress regulates collagen synthesis and matrix metalloproteinase activity in cardiac fibroblasts. *Am J Physiol Cell Physiol* 2001; 280: C53–C60.

128 Seeland U, Haeuseler C, Hinrichs R *et al.* Myocardial fibrosis in transforming growth factor-beta(1) (TGF-beta(1)) transgenic mice is associated with inhibition of interstitial collagenase. *Eur J Clin Invest* 2002; 32: 295–303.

129 Rosenkranz S. TGF-beta1 and angiotensin networking in cardiac remodeling. *Cardiovasc Res* 2004; 63: 423–432.

130 Weber KT, Brilla CG, Campbell SE, Guarda E, Zhou G, Sriram K. Myocardial fibrosis: role of angiotensin II and aldosterone. *Basic Res Cardiol* 1993; 88 (Supplement 1): 107–124.

131 Tsuruda T, Costello-Boerrigter LC, Burnett JC Jr. Matrix metalloproteinases: pathways of induction by bioactive molecules. *Heart Fail Rev* 2004; 9: 53–61.

132 Brilla CG, Weber KT. Mineralocorticoid excess, dietary sodium, and myocardial fibrosis. *J Lab Clin Med* 1992; 120: 893–901.

133 Stockand JD, Meszaros JG. Aldosterone stimulates proliferation of cardiac fibroblasts by activating Ki-RasA and MAPK1/2 signaling. *Am J Physiol Heart Circ Physiol* 2003; 284: H176–H184.

134 Rude MK, Duhaney TA, Kuster GM *et al.* Aldosterone stimulates matrix metalloproteinases and reactive oxygen species in adult rat ventricular cardiomyocytes. *Hypertension* 2005; 46: 555–561.

135 Rockman HA, Koch WJ, Lefkowitz RJ. Seven-transmembrane-spanning receptors and heart function. *Nature* 2002; 415: 206–212.

136 Qiu Z, Wang J, Perreault CL, Meuse AJ, Grossman W, Morgan JP. Effects of endothelin on intracellular Ca^{2+} and contractility in single ventricular myocytes from the ferret and human. *Eur J Pharmacol* 1992; 214: 293–296.

137 Feldman AM, Cates AE, Veazey WB *et al.* Increase of the 40,000-mol wt pertussis toxin substrate (G protein) in the failing human heart. *J Clin Invest* 1988; 82: 189–197.

138 Engelhardt S, Hein L, Wiesmann F, Lohse MJ. Progressive hypertrophy and heart failure in beta1-adrenergic receptor transgenic mice. *Proc Natl Acad Sci USA* 1999; 96: 7059–7064.

139 Simpson PC, Kariya K, Karns LR, Long CS, Karliner JS. Adrenergic hormones and control of cardiac myocyte growth. *Mol Cell Biochem* 1991; 104: 35–43.

140 Dorn GW, Force T. Protein kinase cascades in the regulation of cardiac hypertrophy. *J Clin Invest* 2005; 115: 527–537.

141 Wakasaki H, Koya D, Schoen FJ *et al.* Targeted overexpression of protein kinase C beta2 isoform in myocardium causes cardiomyopathy. *Proc Natl Acad Sci USA* 1997; 94: 9320–9325.

142 Bueno OF, Molkentin JD. Involvement of extracellular signal-regulated kinases 1/2 in cardiac hypertrophy and cell death. *Circ Res* 2002; 91: 776–781.

143 Liang Q, Molkentin JD. Redefining the roles of p38 and JNK signaling in cardiac hypertrophy: dichotomy between cultured myocytes and animal models. *J Mol Cell Cardiol* 2003; 35: 1385–1394.

144 Wilkins BJ, Molkentin JD. Calcium-calcineurin signaling in the regulation of cardiac hypertrophy. *Biochem Biophys Res Commun* 2004; 322: 1178–1191.

145 Wilkins BJ, Dai YS, Bueno OF *et al.* Calcineurin/NFAT coupling participates in pathological, but not physiological, cardiac hypertrophy. *Circ Res* 2004; 94: 110–118.

146 Liao P, Georgakopoulos D, Kovacs A *et al.* The *in vivo* role of p38 MAP kinases in cardiac remodeling and restrictive cardiomyopathy. *Proc Natl Acad Sci USA* 2001; 98: 12283–12288.

147 Gomez del AP, Martinez-Martinez S, Maldonado JL, Ortega-Perez I, Redondo JM. A role for the p38 MAP kinase pathway in the nuclear shuttling of NFATp. *J Biol Chem* 2000; 275: 13872–13878.

148 Baines CP, Molkentin JD. STRESS signaling pathways that modulate cardiac myocyte apoptosis. *J Mol Cell Cardiol* 2005; 38: 47–62.

149 Booz GW, Day JNE, Baker KM. Interplay between the cardiac renin angiotensin system and JAK-STAT signaling: Role in cardiac hypertrophy, ischemia/reperfusion dysfunction, and heart failure. *J Mol Cell Cardiol* 2002; 34: 1443–1453.

150 Taegtmeyer H. Energy metabolism of the heart: from basic concepts to clinical applications. *Curr Probl Cardiol* 1994; 19: 59–113.

151 Ingwall JS, Weiss RG. Is the failing heart energy starved? On using chemical energy to support cardiac function. *Circ Res* 2004; 95: 135–145.

152 Bishop SP, Altschuld RA. Increased glycolytic metabolism in cardiac hypertrophy and congestive failure. *Am J Physiol* 1970; 218: 153–159.

153 Taegtmeyer H, Overturf ML. Effects of moderate hypertension on cardiac function and metabolism in the rabbit. *Hypertension* 1988; 11: 416–426.

154 Razeghi P, Young ME, Alcorn JL, Moravec CS, Frazier OH, Taegtmeyer H. Metabolic gene expression in fetal and failing human heart. *Circulation* 2001; 104: 2923–2931.

155 Sack MN, Rader TA, Park S, Bastin J, McCune SA, Kelly DP. Fatty acid oxidation enzyme gene expression is downregulated in the failing heart. *Circulation* 1996; 94: 2837–2842.

156 De Las FL, Herrero P, Peterson LR, Kelly DP, Gropler RJ, vila-Roman VG. Myocardial fatty acid metabolism: independent predictor of left ventricular mass in hypertensive heart disease. *Hypertension* 2003; 41: 83–87.

157 vila-Roman VG, Vedala G, Herrero P et al. Altered myocardial fatty acid and glucose metabolism in idiopathic dilated cardiomyopathy. *J Am Coll Cardiol* 2002; 40: 271–277.

158 Huss JM, Kelly DP. Mitochondrial energy metabolism in heart failure: a question of balance. *J Clin Invest* 2005; 115: 547–555.

159 McGregor E, Dunn MJ. Proteomics of heart disease. *Hum Mol Genet* 2003; 12: R135–R144.

160 Kang PM, Izumo S. Apoptosis in heart: basic mechanisms and implications in cardiovascular diseases. *Trends Mol Med* 2003; 9: 177–182.

161 Crow MT, Mani K, Nam YJ, Kitsis RN. The mitochondrial death pathway and cardiac myocyte apoptosis. *Circ Res* 2004; 95: 957–970.

162 Olivetti G, Quaini F, Sala R et al. Acute myocardial infarction in humans is associated with activation of programmed myocyte cell death in the surviving portion of the heart. *J Mol Cell Cardiol* 1996; 28: 2005–2016.

163 Olivetti G, Abbi R, Quaini F et al. Apoptosis in the failing human heart. *N Engl J Med* 1997; 336: 1131–1141.

164 Saraste A, Pulkki K, Kallajoki M et al. Cardiomyocyte apoptosis and progression of heart failure to transplantation. *Eur J Clin Invest* 1999; 29: 380–386.

165 Guerra S, Leri A, Wang X et al. Myocyte death in the failing human heart is gender dependent. *Circ Res* 1999; 85: 856–866.

166 Fliss H, Gattinger D. Apoptosis in ischemic and reperfused rat myocardium. *Circ Res* 1996; 79: 949–956.

167 Marks AR. Calcium and the heart: a question of life and death. *J Clin Invest* 2003; 111: 597–600.

168 Quaini F, Urbanek K, Beltrami AP et al. Chimerism of the transplanted heart. *N Engl J Med* 2002; 346: 5–15.

169 Dimmeler S, Zeiher AM, Schneider MD. Unchain my heart: the scientific foundations of cardiac repair. *J Clin Invest* 2005; 115: 572–583.

170 Perin EC, Dohmann HF, Borojevic R et al. Transendocardial, autologous bone marrow cell transplantation for severe, chronic ischemic heart failure. *Circulation* 2003; 107: 2294–2302.

171 Strauer BE, Brehm M, Zeus T et al. Repair of infarcted myocardium by autologous intracoronary mononuclear bone marrow cell transplantation in humans. *Circulation* 2002; 106: 1913–1918.

172 Giordano FJ. Oxygen, oxidative stress, hypoxia, and heart failure. *J Clin Invest* 2005; 115: 500–508.

173 Griendling KK, FitzGerald GA. Oxidative stress and cardiovascular injury: Part I: basic mechanisms and *in vivo* monitoring of ROS. *Circulation* 2003; 108: 1912–1916.

174 Griendling KK, FitzGerald GA. Oxidative stress and cardiovascular injury: Part II: animal and human studies. *Circulation* 2003; 108: 2034–2040.

175 Roberts LJ, Morrow JD. Products of the isoprostane pathway: unique bioactive compounds and markers of lipid peroxidation. *Cell Mol Life Sci* 2002; 59: 808–820.

176 Jaatinen P, Saukko P, Hervonen A. Chronic ethanol exposure increases lipopigment accumulation in human heart. *Alcohol Alcohol* 1993; 28: 559–569.

177 Maack C, Kartes T, Kilter H et al. Oxygen free radical release in human failing myocardium is associated with increased activity of rac1-GTPase and represents a target for statin treatment. *Circulation* 2003; 108: 1567–1574.

178 Taylor AL, Ziesche S, Yancy C et al. Combination of isosorbide dinitrate and hydralazine in blacks with heart failure. *N Engl J Med* 2004; 351: 2049–2057.

179 Packer M, Coats AJ, Fowler MB et al. Effect of carvedilol on survival in severe chronic heart failure. *N Engl J Med* 2001; 344: 1651–1658.

180 Yusuf S, Dagenais G, Pogue J, Bosch J, Sleight P. Vitamin E supplementation and cardiovascular events in high-risk patients. The Heart Outcomes Prevention Evaluation Study Investigators. *N Engl J Med* 2000; 342: 154–160.

181 Hennekens CH, Buring JE, Manson JE et al. Lack of effect of long-term supplementation with beta carotene

on the incidence of malignant neoplasms and cardiovascular disease. *N Engl J Med* 1996; 334: 1145–1149.

182 De CR, Cipollone F, Filardo FP *et al.* Low-density lipoprotein level reduction by the 3-hydroxy-3-methylglutaryl coenzyme-A inhibitor simvastatin is accompanied by a related reduction of F2-isoprostane formation in hypercholesterolemic subjects: no further effect of vitamin E. *Circulation* 2002; 106: 2543–2549.

183 Lee SH, Wolf PL, Escudero R, Deutsch R, Jamieson SW, Thistlethwaite PA. Early expression of angiogenesis factors in acute myocardial ischemia and infarction. *N Engl J Med* 2000; 342: 626–633.

184 Kubo SH, Rector TS, Bank AJ, Williams RE, Heifetz SM. Endothelium-dependent vasodilation is attenuated in patients with heart failure. *Circulation* 1991; 84: 1589–1596.

185 Akazawa H, Komuro I. Roles of cardiac transcription factors in cardiac hypertrophy. *Circ Res* 2003; 92: 1079–1088.

186 Pikkarainen S, Tokola H, Kerkela R, Ruskoaho H. GATA transcription factors in the developing and adult heart. *Cardiovasc Res* 2004; 63: 196–207.

187 Iwaki K, Sukhatme VP, Shubeita HE, Chien KR. Alpha- and beta-adrenergic stimulation induces distinct patterns of immediate early gene expression in neonatal rat myocardial cells. fos/jun expression is associated with sarcomere assembly; Egr-1 induction is primarily an alpha 1-mediated response. *J Biol Chem* 1990; 265: 13809–13817.

188 Clerk A, Cullingford TE, Kemp TJ, Kennedy RA, Sugden PH. Regulation of gene and protein expression in cardiac myocyte hypertrophy and apoptosis. *Adv Enzyme Regul* 2005; 45: 94–111.

189 Omura T, Yoshiyama M, Yoshida K *et al.* Dominant negative mutant of c-Jun inhibits cardiomyocyte hypertrophy induced by endothelin 1 and phenylephrine. *Hypertension* 2002; 39: 81–86.

190 Purcell NH, Tang G, Yu C, Mercurio F, DiDonato JA, Lin A. Activation of NF-kappa B is required for hypertrophic growth of primary rat neonatal ventricular cardiomyocytes. *Proc Natl Acad Sci USA* 2001; 98: 6668–6673.

191 Valen G, Yan Zq, Hansson GK. Nuclear factor kappa-B and the heart. *J Am Coll Cardiol* 2001; 38: 307–314.

192 Purcell NH, Molkentin JD. Is nuclear factor κB an attractive therapeutic target for treating cardiac hypertrophy? *Circulation* 2003; 108: 638–640.

193 Takemoto Y, Yoshiyama M, Takeuchi K *et al.* Increased JNK, AP-1 and NF-kappa B DNA binding activities in isoproterenol-induced cardiac remodeling. *J Mol Cell Cardiol* 1999; 31: 2017–2030.

194 Fiedler B, Wollert KC. Interference of antihypertrophic molecules and signaling pathways with the Ca^{2+}-calcineurin-NFAT cascade in cardiac myocytes. *Cardiovasc Res* 2004; 63: 450–457.

195 Proud CG. Ras, PI3-kinase and mTOR signaling in cardiac hypertrophy. *Cardiovasc Res* 2004; 63: 403–413.

196 Chen MM, Ashley EA, Deng DX *et al.* Novel role for the potent endogenous inotrope apelin in human cardiac dysfunction. *Circulation* 2003; 108: 1432–1439.

197 Boheler KR, Volkova M, Morrell C *et al.* Sex- and age-dependent human transcriptome variability: implications for chronic heart failure. *Proc Natl Acad Sci USA* 2003; 100: 2754–2759.

198 Haq S, Choukroun G, Lim H *et al.* Differential activation of signal transduction pathways in human hearts with hypertrophy versus advanced heart failure. *Circulation* 2001; 103: 670–677.

199 Vanderheyden M, Paulus WJ, Voss M *et al.* Myocardial cytokine gene expression is higher in aortic stenosis than in idiopathic dilated cardiomyopathy. *Heart* 2005; 91: 926–931.

200 Polyakova V, Hein S, Kostin S, Ziegelhoeffer T, Schaper J. Matrix metalloproteinases and their tissue inhibitors in pressure-overloaded human myocardium during heart failure progression. *J Am Coll Cardiol* 2004; 44: 1609–1618.

201 Spinale FG. Matrix metalloproteinases: regulation and dysregulation in the failing heart. *Circ Res* 2002; 90: 520–530.

202 Lowes BD, Gilbert EM, Abraham WT *et al.* Myocardial gene expression in dilated cardiomyopathy treated with beta-blocking agents. *N Engl J Med* 2002; 346: 1357–1365.

203 Nagata K, Somura F, Obata K *et al.* AT1 receptor blockade reduces cardiac calcineurin activity in hypertensive rats. *Hypertension* 2002; 40: 168–174.

204 Yu CM, Tipoe GL, Wing-Hon Lai K, Lau CP. Effects of combination of angiotensin-converting enzyme inhibitor and angiotensin receptor antagonist on inflammatory cellular infiltration and myocardial interstitial fibrosis after acute myocardial infarction. *J Am Coll Cardiol* 2001; 38: 1207–1215.

205 Hall JL, Grindle S, Han X *et al.* Genomic profiling of the human heart before and after mechanical support with a ventricular assist device reveals alterations in vascular signaling networks. *Physiol Genomics* 2004; 17: 283–291.

206 Spragg DD, Leclercq C, Loghmani M *et al.* Regional alterations in protein expression in the dyssynchronous failing heart. *Circulation* 2003; 108: 929–932.

207 Sabbah HN, Sharov VG, Gupta RC *et al.* Reversal of chronic molecular and cellular abnormalities due to heart failure by passive mechanical ventricular containment. *Circ Res* 2003; 93: 1095–1101.

208 Schroen B, Heymans S, Sharma U *et al.* Thrombospondin-2 is essential for myocardial matrix integrity: increased expression identifies failure-prone cardiac hypertrophy. *Circ Res* 2004; 95: 515–522.

CHAPTER 8

The implications of genes on the pathogenesis, diagnosis and therapeutics of hypertension

Kiat Tsong Tan, MD, MRCP, FRCR *& Choong-Chin Liew,* PhD

Introduction

Hypertension is a major cause of mortality and morbidity in the developed world and affects up to 60–75% of Americans over the age of 60. Genetic factors account for 30–60% of an individual's risk for developing hypertension [1]. However, elucidation of the genes involved in hypertension has proven to be difficult. Blood pressure is inherited in a polygenic fashion, with multiple genes contributing to the final phenotype. Environmental factors also make a major contribution to blood pressure, further complicating the understanding and investigation of the molecular and genetic aspects of hypertension. For example, increased salt intake has been linked to hypertension. Other environmental factors that may influence the development of hypertension include smoking, excessive alcohol intake, psychologic stress, lack of aerobic exercise and central obesity. In addition, there is some evidence that developmental factors may also predispose to high blood pressure. A low birth weight is associated with the later development of hypertension. It has been hypothesized that this is related to a congenital reduction of nephrons, which may subsequently impair salt excretion [2]. Unfortunately, it is not always possible to dissect out the role of nature versus nurture in the etiology of hypertension. For example, genetic causes may predispose towards smoking and excessive alcohol intake, which may then contribute to the development of hypertension.

However, the role of genetics in hypertension is not limited to the determination of its etiology; advances in the understanding of the genetics and molecular biology of hypertension may also improve the diagnosis and treatment of the disorder. A large proportion of hypertensive patients do not have adequate blood pressure control on their current medication [3,4]. Although poor patient medication compliance is likely to account for a proportion of these, suboptimal drug efficacy may also contribute to this number. In addition, 10–20% of treatment failures are a result of drug side effects.

Thus, pharmacogenomic-based strategies for rationalizing treatment in hypertension would be a highly desirable goal. To some extent, clinicians are already matching patient profiles with drug treatment. For example, a hypertensive patient with left ventricular hypertrophy is likely to benefit from angiotensin-converting enzyme (ACE) inhibition, while patients with concurrent ischemic heart disease are better treated with a beta-blocker. However, much of the current treatment for hypertension is carried out on an empirical basis. A recent study suggests that there is great variation in patient response to antihypertensive drugs and selection of the best drug for a given patient is currently a matter of trial and error, a costly and time-consuming process [5]. In addition, the efficient control of hypertension can decrease the risk of blood pressure induced target organ damage, the prevalence of which increases with the length of time in which blood pressure remains uncontrolled. An improved understanding of the genes of hypertension may therefore enable the physician to better tailor medication to the patient with hypertension.

Pathophysiology of hypertension

The regulation of extracellular fluid volume by sodium excretion renal is responsible for the control of blood pressure. This mechanism is made possible by the ability of the kidneys to vary renal tubular sodium excretion in accordance with renal perfusion pressure [6]. In the healthy individual, any increase in mean arterial blood pressure would lead to a proportional rise in renal perfusion pressure, which would increase renal sodium excretion. The loss of sodium leads to shrinkage of extracellular fluid volume, causing a drop in cardiac output, thus normalizing the blood pressure. A drop in renal perfusion pressure has the opposite effect and would result in sodium retention. Therefore, in order to maintain blood pressure homeostasis in the healthy individual, the kidney has to match sodium excretion to sodium intake.

Hypertension results from the inability of the kidneys to excrete enough sodium to match intake or the inability of kidneys to excrete sufficient sodium in response to raised mean arterial pressure. Impairment of renal perfusion by renal artery stenosis, for example, results in a drop in renal perfusion pressure, leading to a rise in mean arterial pressure in order to maintain kidney perfusion. A similar mechanism accounts for the hypertensive change observed in coarctation of the aorta. Other causes of hypertension include renal failure, where the remaining nephrons are unable to excrete enough sodium to match blood pressure, Conn and Cushing syndromes, and the various monogenetic forms of hypertension discussed below.

However, up to 90% of patients with hypertension do not have a discernible cause for their high blood pressure. These patients are described as having "essential hypertension." There is often no histologic abnormality in the kidneys of patients in the early stages of essential hypertension. However, a proportion of patients with essential hypertension, often referred to as salt-sensitive hypertensives, have a blunted natriuretic response to a rise in blood pressure.

Molecular pathways and hypertension

Although the following discussion classifies the molecular pathways involved in the pathogenesis of hypertension into neat systems for the sake of clarity, it should be remembered that these systems are closely interrelated. Changes in one system may have profound effects on others.

The adrenergic system

The sympathetic nervous system is important in the pathogenesis of hypertension [7]. Qi *et al.* [7] have hypothesized that an increase in sodium intake increases sodium plasma concentration, which then triggers an increase in sympathetic activity. The increase in sympathetic activity causes a rise in blood pressure by increasing cardiac output and/or total peripheral resistance. Sympathetic activity also blunts the rise in natriuresis induced by increased mean arterial pressure, resulting in hypertension. The mechanism behind this blunting of pressure natriuresis may be the stimulation of renal arterial vasoconstriction by sympathetic activity, which could result in a decrease in renal perfusion. In addition, a reduction in renal perfusion would also result in the secretion of renin by the kidney, which would further increase systemic blood pressure. Indeed, there is evidence to suggest that there is an increase in renal sympathetic activity in early hypertension and that renal denervation may attenuate the rise in blood pressure in renal hypertension [8].

The adrenergic system is involved in the regulation of blood pressure both centrally and peripherally. The genomic revolution has allowed the identification of nine subtypes of adrenergic receptors [9]. Most of our knowledge of the adrenoceptors is centered upon the four major subgroups of the receptor: α_1, α_2, β_1 and β_2. However, the study of the physiology of the adrenergic system is complicated by interspecies and intertissue variation of the distribution of the receptors. For example, the exposure of the dog saphenous vein to epinephrine results in vasoconstriction while the effect of epinephrine on the rabbit facial vein is one of relaxation.

The α_1-adrenoceptor has an important role in regulating vascular tone. The receptor stimulates smooth muscle contraction and proliferation on exposure to adrenergic stimulation. Indeed, α_1-blockers, such as doxazosin, are in widespread clinical use in the treatment of hypertension.

α_2-Adrenoceptors are involved in both the central and peripheral regulation of blood pressure. Stimulation of central α_2-adrenoceptors causes a drop in blood pressure and in sedation [9]. α_2-Adrenoceptors are also found in vascular smooth muscle cells. Stimulation of peripheral α_2-adrenoceptors is associated with vasoconstriction. Therefore, the administration of the clinically used α_2-agonist, clonidine, elicits a transient hypertensive response before its central hypotensive effect predominates.

The development of knockout mice has allowed further elucidation of the function of the α_2-adrenoceptor subtypes. α_2-Adrenoceptors can be further subclassified into α_{2A}, α_{2B} and α_{2C} subtypes. The initial hypertensive response to α_2 stimulation is abolished in α_{2B} knockout mice. The deletion of the α_{2B}-adrenoceptor gene did not affect the hypotensive effect of clonidine. These findings suggest that the α_{2B}-adrenoceptor is responsible for the initial pressor effect of α-agonists. In addition to their lack of pressor response to the administration of clonidine, it has been shown that α_{2B}-adrenoceptor knockout mice are resistant to the development of salt-sensitive hypertension. Inhibition of the translation α_{2B}-adrenoceptor mRNA by the use of antisense oligonucleotides also has a hypotensive effect. It is believed that the hypotensive effect of α_{2B} antagonism is mediated centrally. Therefore, it is likely that selective inhibition of the α_{2B}-adrenoceptor is a promising target for antihypertensive agents.

In contrast to the α_{2B} knockout mice, mice with defective α_{2A}-adrenoceptors have a normal initial pressor response to clonidine but do not demonstrate a hypotensive response. Indeed, α_{2A}-adrenoceptor represents the most common receptor subtype in central cardiovascular regulatory centers.

The β-adrenoceptors are found in the myocardium, vascular smooth muscle and visceral smooth muscle. β_1-Adrenoceptors in the heart mediate positive chrono- and inotropic effects. β_2-Adrenoceptors in vascular and bronchial smooth muscle mediates relaxation. However, the hypertensive effects of β stimulation usually predominates, which accounts for the use of beta-blockers in the treatment of hypertension. It is also not surprising that peripheral vasoconstriction and bronchospasm can very occasionally complicate beta-blockade. β_3-Adrenoceptor mediates vascular smooth muscle relaxation in some species. It is also believed to have an important role in lipolysis. However, its exact role in human disease is uncertain.

Renin–angiotensin system

The renin–angiotensin system (RAS) has a vital role in the regulation of blood pressure [10]. The production of renin is the mechanism by which a fall in renal perfusion pressure causes sodium and water retention. Renin is secreted by the juxtaglomerular cells of the kidneys in response to a fall in the delivery of sodium ions to the distal renal tubules. In addition, a fall in intra-arteriolar pressure at the site of the juxtaglomerular cells results in increased renin production and vice versa. Increased sympathetic tone also acts to enhance renin secretion. Renin acts on angiotensinogen, secreted by the kidney, to produce the angiotensin I.

Angiotensin I is then converted to angiotensin II by ACE, which occurs predominantly in the lungs. ACE also inactivates bradykinin (see below). Angiotensin II has a half-life of approximately 2 minutes and has a potent hypertensive effect. Angiotensin II type 1 receptor (AGTR1) which is coupled via a G protein to phospholipase C. Therefore, stimulation of AGTR1 by angiotensin II stimulates an increase in intracellular calcium. AGTR1 mediates vasoconstriction; increased secretion of aldosterone, adrenocorticotrophic hormone, norepinephrine and vasopressin; decreased glomerular filtration rate and increased water intake. Angiotensin II and vasopressin provides negative feedback to inhibit renin secretion.

Angiotensin II is converted to the less vasoactive angiotensin III by enzymes termed angiotensinases. Angiotensin III is then broken down to biologically inactive metabolites.

The importance of the RAS in the regulation of blood pressure is clearly illustrated in many animal studies. Rodent models show that the number of copies of functional angiotensinogen genes can be related to the plasma concentration of angiotensinogen, which can then be connected to blood pressure [11]. Knockout of the angiotensinogen gene produces hypotension.

AGTR1 expression has also been shown to be important in blood pressure regulation. AGTR1 −/− mice have a lower blood pressure than AGTR1 −/+ mice, which are hypotensive when compared with their AGTR +/+ wild-type littermates [12]. Transplantation of a kidney derived from an AGTR −/− mouse into an AGTR1 +/+ recipient had a hypotensive effect, which illustrates the importance of renal AGTR1 in blood pressure regulation. Therefore, it is not surprising that AGTR1 inhibition is used clinically in the treatment of human hypertension.

Inhibition of ACE has also been shown to be effective in treating hypertension. It is therefore surprising that genetically influenced plasma level of ACE does not have an effect on blood pressure. In humans, the presence of an insertion/deletion (I/D) polymorphism of the ACE gene can have dramatic effects on the plasma level of ACE [13]. However, the I/D polymorphism in humans is not associated with blood pressure differences. In the mouse, ACE activity increases with the number of copies of the ACE gene [14]. However, blood pressure is not affected by the concentration of ACE in the blood. It is believed that a slight impairment of ACE activity, as seen in a mouse with one copy of the ACE gene, results in the build-up of angiotensin I, which then drives the production of angiotensin II. However, more complete inhibition of ACE, as occurs during ACE inhibitor therapy, will impair angiotensin II production and lower blood pressure.

Dopaminergic system

The dopaminergic system is an important regulator of blood pressure through renal, adrenal, central and gastrointestinal functions [15]. Proximal renal tubular cells can secrete dopamine, which then acts in an autocrine/paracrine fashion to reduce sodium uptake from the renal tubule. This mechanism may account for an increase in sodium excretion of more than 50% during periods of increased salt intake. In a similar manner, the jejunal cells can also produce dopamine to reduce intestinal uptake of sodium.

Dopaminergic receptors can be classified into two families, which are both coupled to G proteins.

The D1-like receptors comprise the D1 and D5 subtypes. These receptors increase intracellular cAMP via stimulatory G proteins $G\alpha_s$ and $G\alpha_{olf}$. The D2-like receptors consist of D2, D3 and D4. Stimulation of D2-like receptors results in the activation of the inhibitory G proteins $G\alpha_i$ and G_o, and leads to the inhibition of adenylyl cyclase.

Activation of the D1-like receptors in the renal tubular cells leads to the inhibition of sodium reabsorption by various transporters, including NHE1, NHE3, NA/HCO_3 and the Na^+/K^+ ATPase. The importance of the D1 receptor in the regulation of blood pressure is illustrated by many studies. Mice lacking at least one copy of the D1 receptor gene show raised blood pressure. In addition, there is an uncoupling of the D1 receptor from its effector G protein in some rat models of hypertension, leading to increased sodium reabsorption.

Abnormalities in the dopaminergic system may be responsible for abnormal high blood pressure in some patients. There may be reduced synthesis of renal dopamine in a subset of hypertensives. In others, the renal production of dopamine may actually be normal or increased. This suggests that there may be a defect in dopaminergic signal induction in some hypertensives. Experimental studies with the use of the D1 receptor agonist, fenoldopam, showed that renal tubular cells from some patients with hypertension have an attenuated cAMP response to D1 stimulation. The impaired coupling of D1 stimulation to intracellular responses can be brought about by excessive phosphorylation of the receptor by a G protein coupled receptor kinase. GRK4 is the most important member of the G protein coupled receptor kinase family in the regulation of D1 activity in the proximal renal tubule. GRK4 activity accounts for the desensitization of renal D1 in some hypertensives. Indeed, the use of antisense oligonucleotides directed against GRK4 can attenuate the development of hypertension in the spontaneously hypertensive rat (SHR).

Disruption of the D5 dopaminergic receptor also causes hypertension, possibly because of increased sympathetic outflow [16]. There is also a reduction in the expression of D5 receptors in the renal cortex of SHR. However, the significance of this is incompletely understood.

Stimulation of D2-like receptors can inhibit the secretion of renin by the juxtaglomerular cells. The

D3 receptor is found in rat juxtaglomerular cells and its selective inhibition results in an increase in renin secretion. In addition, −/− D3 mice have high renin salt-sensitive hypertension [17].

The D4 −/− mouse also has a hypertensive phenotype [18]. This observation is hypothesized to be a result of its influence on angiotensin II type 1 receptor expression in the brain and kidney.

Nitric oxide

Nitric oxide (NO) is produced by nitric oxide synthase (NOS) and is an important mediator of vasodilatation [19]. NOS exists in three isoforms: endothelial (eNOS), neuronal (nNOS) and inducible (iNOS). As the name implies, eNOS is the isoform that is primarily involved in the regulation of arterial tone. Inhibition of the production of arterial nitric oxide results in vasoconstriction, leading to an increase in systemic resistance, which then leads to a rise in blood pressure [20]. In addition, eNOS −/− mice have higher blood pressure than wild-type controls. There is also evidence to suggest that NO activity is impaired in hypertension [21].

eNOS may provide the link between arterial hypertension and its thrombotic complications, such as stroke and myocardial infarction. NO has important antiplatelet effects. Inhibition of NOS in humans can reduce bleeding time, which is a clinical measure of platelet function. In addition, mice that lack a functional copy of the eNOS gene have decreased bleeding times and platelets that are more easily activated.

Mice that are deficient in eNOS demonstrate the triad of hypertension, insulin resistance and dyslipidaemia: the "metabolic syndrome" [22]. eNOS may contribute to glucose uptake into striated muscle by promoting muscular blood flow. It may also have a direct effect in promoting the uptake of glucose by striated muscle. The mechanism by which eNOS disruption can result in dyslipidaemia is uncertain.

Natriuretic peptides

The natriuretic peptides are molecules that regulate sodium balance, vascular tone and have important effects on cellular proliferation [23]. The beneficial effects of natriuretic peptides have resulted in the introduction of recombinant natriuretic peptides in the treatment of heart failure.

Natriruretic peptides include:

1 Atrial natriuretic peptide (ANP);
2 Brain natriuretic peptide (BNP);
3 C-type natriuretic peptide (CNP).

More recently, denroaspis natriuretic peptide (DNP) has been described.

There are three main classes of natriuretic receptors: natriuretic peptide receptor-A (NPRA), which preferentially binds to ANP and BNP; natriuretic peptide receptor-B (NPRB), which has a predilection for CNP; and natriuretic peptide receptor-C (NPRC), which binds all natriuretic peptides [24]. NPRA and NPRB can be found in the vasculature, kidneys, lungs and adrenals. NPRB is the predominant natriuretic peptide receptor in the brain. NPRC can be found in most tissues and acts as a clearance receptor. Ligand binding to NPRA and NPRB results in the production of cGMP via the activation of guanylate cyclase.

ANP and BNP are produced by the atria and ventricles, respectively, in response to cardiac wall tension. CNP is produced by vascular endothelium and is believed to regulate vascular tone. Alternative processing of the ANP precursor yields urodilantin in the kidney. This molecule is secreted into the lumen of the distal nephron and promotes natriuresis.

The natriuretic peptides cause both venous and arterial dilatation. This effect leads to an immediate reduction in both preload (and hence cardiac output) and peripheral resistance, changes that can result in a fall in arterial blood pressure. In addition, the reflex increase in sympathetic output that usually accompanies a fall in blood pressure is suppressed.

Natriuretic peptides also promote the excretion of sodium. These molecules relax the afferent renal arterioles while causing the efferent arterioles to contract thus increasing glomerular filtration. They also block the reabsorption of sodium in the renal tubules. Natriuretic peptides antagonize the effects of antidiuretic hormone and angiotensin II. ANP −/− mice have exaggerated left ventricular hypertrophy and high renin salt-sensitive hypertension. In addition, the disruption of the NPRA gene leads to hypertension. Indeed, blood pressure is decreased by increasing the number of copies of the functional NPRA gene in mice [25].

In addition to their beneficial hemodynamic effects in hypertension, natriuretic peptides also have a positive influence on cardiac remodeling and may have a cardioprotective effect. Therefore, they may attenuate target organ damage by high blood pressure.

Endothelin

The endothelin (ET) family consists of three closely related peptides made up of 21 amino acid residues. The constituents of the endothelin family are referred to as ET-1, ET-2 and ET-3 [26]. ET-1 represents the most important subtype of the family. The molecule is a potent mediator of vasoconstriction, vascular inflammation and cell proliferation.

The precursor of ET-1, preproendothelin, is produced by the endothelial cells. Preproendothelin is cleaved to form proendothelin, which can subsequently be converted to bigET-1. bigET-1 is then further cleaved by endothelin-converting enzyme to ET-1. The production of ET-1 is believed to be stimulated by angiotensin II. It has been shown in rats that chronic angiotensin II induced hypertension is associated with increases in preproendothelin mRNA and ET-1 expression. In addition, the administration of endothelin antagonists attenuates the hypertension resulting from chronic angiotensin II infusion.

ET-1 mediates its effects through two G protein coupled receptors, ETA and ETB, which are encoded by different genes and have different biologic functions. ETA is found in cardiac myocytes and vascular smooth muscle cells. Activation of ETA leads to an increase in intracellular calcium via the phospholipase C pathway, leading to vasoconstriction.

ETB is found in endothelial cells, the inner medullary collecting duct and the medullary thick ascending limb [27]. This receptor also mediates its effects through the phospholipase C pathway. In addition, ETB may also activate inhibitory G protein. Renal ETB activation is associated with a hypotensive effect. Indeed, chronic ETB receptor blockade is associated with hypertension. In addition, ETB knockout mice become hypertensive on a high salt diet. Therefore, it is believed that ETB is a regulator of sodium excretion in the kidneys.

Kallikrein–kinin system

The Kallikrein–kinin system (KKS) mediates a wide range of physiologic functions, from inflammation through cell proliferation to salt balance [28]. The KKS is composed of kininogens, kallikreins and the active kinins.

Kinins, such as bradykinin and lysylbradykinin, are formed from kininogens by the action of kallikreins. Kininogens are coded for by a single gene and circulating kininogen is synthesized primarily by the liver. Tissue kininogens can be found in various tissues, such as the kidney, and are synthesized locally.

Kallikreins can be classified into plasma kallikrein, which circulates with the blood in an inactive form, and tissue kallikrein, circulates within the bloodstream found in cells involved in electrolyte transport. Inactive plasma kallikrein is activated by activators derived from fragments of clotting factor XII. Conversely, plasma kallikrein can also activate factor XII. Therefore, kallikreins provide a vital link between the coagulation pathway and a myriad of physiologic functions.

Kinins are inactivated by peptidases such as kininase I and ACE (also referred to as kininase II). Kinins exert their effects via two receptors, termed "B1" and "B2". B1 expression is usually only detectable in pathologic conditions such as inflammation. In contrast, B2 is expressed constitutively in most tissues. It regulates renal sodium transport and mediates the release of nitric oxide and prostacyclin by vascular endothelium. Administration of bradykinin elicits natriuresis.

The urinary excretion of kallikreins is found to be reduced in patients with essential hypertension and in rodent models of hypertension. The overexpression of human kallikrein in genetically engineered mice results in hypotension. In addition, the introduction of the kallikrein as gene therapy in rats attenuates the development of hypertension [29]. Rats that were inbred for low kallikrein activity also had raised blood pressure, providing further evidence for the role of kallikrein in hypertension.

B2 receptor −/− mice have an exaggerated pressor response to salt loading [30]. They also have an enhanced response to antidiuretic hormone (ADH), suggesting that kinins may antagonize the effects of ADH. Conversely, mice overexpressing the B2 receptor are hypotensive.

Oxidative stress

Vascular oxidative stress has been shown to be increased in experimental models of hypertension [31]. SHRs have increased NADPH driven production of superoxide in their arteries. Patients with severe hypertension also have increased levels of plasma thiobarbituric acid-reactive substances and 8-epi-isoprostanes, which represent markers of increased oxidative stress.

Oxidative stress may contribute to the development of hypertension by the destruction of nitric oxide. Increased reactive oxygen species also contributes to inflammation, vascular smooth muscle proliferation and the deposition of matrix proteins. These processes have an important role in the development of vascular damage associated with hypertension. Therefore, it is not surprising that the use of inhibitors of reactive oxygen species production or the use of free radical scavengers may attenuate the development of hypertension in animal models. However, the large trials on the use of antioxidant supplements in the management of human cardiovascular disease have mostly yielded negative results.

Inflammation and hypertension

There is an association between inflammation and hypertension. The plasma level of C-reactive protein (CRP), as determined by the high-sensitivity (hs) method, is higher in patients with hypertension when compared with healthy controls [32]. A raised hsCRP level is also associated with an increased future risk of developing hypertension [33]. Many risk factors that are associated with vascular inflammation, such as central obesity and smoking, are also linked to hypertension. Indeed, the presence of inflammation may also aggravate the increase in peripheral resistance observed in hypertension. For example, inflammation is associated with a decrease in vascular nitric oxide and an increase in isoprostanes, which are conditions that promote vascular smooth muscle constriction [21]. In addition, many of the complications of hypertension, such as stroke and myocardial infarction, often occur with the pro-thrombotic state associated with significant inflammation. A more detailed discussion on inflammation and the pro-thrombotic state is given in Chapter 7.

Approaches to the study of essential hypertension

Currently there is controversy as to the best method for investigating the genes involved in the etiology of hypertension. Although there are numerous studies in this area of research, virtually all have provided conflicting results. It is likely that in addition to the part played by the environment, epistatic interactions (i.e., interactions among various genes) may influence the final phenotype, large sample sizes are required if these interactions are to be studied in detail. In addition, the effect of age on the pathogenesis of hypertension should not be ignored. As hypertension is more common in older people, it is possible that the influence of genes on blood pressure may also vary with age, or that gene changes related to aging may be among the risk factors for hypertension.

Identification of the genes involved in hypertension

The search for the genes involved in the causation of hypertension can be classified as methods that utilize:

1 The candidate gene approach;
2 Genome-wide linkage screens;
3 The study of monogenic hypertension; and
4 Animal models of hypertension.

Candidate genes

This approach investigates particular genes that are believed to be involved in blood pressure regulation. Most of the early studies on the genes involved in the etiology of hypertension were based in a traditional candidate gene approach and were often dependent on prior knowledge of the gene product in question. However, genes whose products are unknown or believed to be "unimportant" in the pathogenesis of hypertension are likely to be ignored.

Genome-wide linkage screens

Genome-wide linkage screens utilize genetic markers spread throughout the genome. The use of densely spread markers in the genome allows the identification of quantitative trait loci (QTL) associated with hypertension. QTL refers to a position in the chromosome that may influence the trait of interest. This approach eliminates the need for prior knowledge of gene function and allows the identification of loci that may not previously have been suspected to be involved in blood pressure regulation.

Although genome-wide scans are useful in detecting loci that may be responsible for hypertension, it should be noted that the possible pathophysiologic role(s) of these sites remain to be determined. A potential source of error in the use of genome-wide linkage screens is the absence of linkage disequilibrium between the gene of interest and the marker, because of recombination during meiosis. As the risk of recombination increases with the distance between the two loci, it follows that genome-wide scans require the use of closely spaced markers in order to produce a meaningful result.

The simultaneous study of large numbers of loci increases the chance of obtaining a false positive result. Therefore, the conventional threshold of significance of $P <0.05$ may be inappropriate in many genetic studies. Indeed, the independent replication of a positive result is of particular importance in the study of the genes of hypertension.

Study of monogenic hypertension

The identification of monogenic (Mendelian) forms of hypertension has contributed greatly to our understanding of the pathophysiology of hypertension. These forms of hypertension are rare but their Mendelian pattern of inheritance has greatly facilitated their study. However, studies aimed at determining whether the genes identified by this modality are involved in essential hypertension have often yielded conflicting results.

Molecular tools in animal models of hypertension

Since the classic studies of Goldblatt *et al.* [34] in 1934, animal models of hypertension have been crucial in advancing our understanding of the pathogenesis of this disorder. The use of animal studies overcomes many of the problems associated with human studies. For example, environmental conditions (e.g., salt intake) and genetic homogeneity can be strictly controlled. The identification of QTL or candidate genes in the rodent models of hypertension can lead to the study of homologous loci or genes in humans. Indeed, the public availability of genomic sequences in rats, mice and humans has greatly facilitated this process in recent times.

The selective breeding of animals displaying elevated blood pressure has resulted in the development of useful rodent models of hypertension, such as the SHR, Dahl salt-sensitive and Lyon strains [35]. Indeed, Dahl was one of the first researchers to show that the pressor effect of salt could be related to genetic factors through his work on the rat strain that bears his name [36]. Crosses between inbred rodent strains can yield valuable information on the QTLs associated with blood pressure.

A congenic strain can be defined as the strain in which the chromosomal region featuring the QTL of interest in one strain, the recipient, has been replaced by that of another, the donor. A difference in blood pressure between the recipient and congenic strains would indicate that the QTL in question is associated in the regulation of blood pressure.

The use of genetically engineered animals in the study of the genetics of hypertension was introduced in the late 1980s and has become increasingly widespread. This approach, often referred to as physiologic genomics, usually involves the use of animals in which the gene of interest is overexpressed, underexpressed or deleted. The RAS has been studied extensively in genetically engineered rodent strains. For example, the overexpression of the rat angiotensinogen gene in the mouse results in hypertension [37]. In addition, angiotensinogen knockout mice are hypotensive [38]. The application of this approach is of particular use in the study of complex diseases such as hypertension, as it allows the detailed dissection of the various (patho-) physiologic pathways. However, the results of experiments with knockout mice should be interpreted with care, as some of the functions of the deleted gene may be taken over by other genes.

The introduction of techniques to "silence" a specific gene, without the manipulation of the host genome, are also useful in the study of hypertension. This can be achieved by one of two mechanisms:

1 The use of antisense molecules; or

2 By the introduction of RNA interference.

Antisense technology refers to the use of oligonucleotide molecules that possess a complementary sequence to the mRNA sequence being targeted [39]. This allows the antisense oligonucleotide to form a double-stranded complex with the target mRNA, which is then broken down by the host cell. Indeed, the admininistration of antisense oligonucleotides directed against the constituents of the RAS has resulted in hypotensive effects in rats.

RNA interference refers to an evolutionarily conserved mechanism by which gene expression is suppressed by the presence of homologous RNA [40]. In RNA interference, the presence of short segments of double-stranded mRNA of 21–23 nucleotide pairs has been shown to block the expression of homologous mRNA for a period of up to 3 weeks [41]. RNA interference has an advantage over the use of knockout mice in that it allows the blockade of the expression of one or more genes without the need for time-consuming crosses. Interfering RNA has been used successfully to demonstrate the role of WNK-1 gene, which has a major effect on blood pressure, in influencing the ERK-5 pathway (see section on "Monogenic hypertension" for a more detailed discussion on the WNK-1 gene) [42].

In addition, it is possible that RNA interference may be exploited as a therapeutic tool in the treatment of hypertension. For example, it has been shown that a single dose of interfering mRNA has beneficial effects on blood pressure and renal function in a rat model of hypertension [41].

Methods used to determine the involvement of the gene or QTL in human hypertension

Linkage analysis

Linkage analysis refers to the use of the link between the presence of the trait in question (e.g., hypertension) and the inheritance of a particular allele as studied either in sibling-pairs or in two to three generations of affected families. Linkage analysis has good specificity but poor sensitivity for the detection of genes involved in hypertension, and many of the linkage studies published over the past 10 years have been inconclusive because of the large sample size required [43].

Association studies

Association studies are based on a case–control design, which allows the recruitment of cases that are unrelated to the healthy controls. In contrast to linkage studies, association studies have good sens-

itivity but are prone to false positive results. Another problem with many association studies is that such studies concentrate on each candidate gene individually, disregarding other candidate genes. Indeed, most of the association studies published to date have been used to test the involvement of candidate genes. This approach may be inappropriate, as it ignores the complex interactions between the genes involved in hypertension. Williams *et al.* [44] showed that interactions between the various alleles of different genes may be more important than allelic variation at a single locus in the determination of blood pressure. Potential confounding factors (e.g., linkage disequilibrium both within a gene and between neighboring genes) are also frequently ignored in most association studies. These shortcomings may explain the inconsistent and often contradictory results generated by the various studies on the genes involved in the etiology of hypertension. Another criticism of association studies in polygenic disorders is that many of these are of a poor quality [45]. Finally, an association between a particular allele and high blood pressure does not imply "cause."

Epistatic interactions and haplotype analysis

Epistatic interactions refer to interactions between the allele of one particular gene and that of another. Studies on epistatic interactions may be of particular use in polygenic disorders such as essential hypertension and are often based on a modification of the association study. For example, a study by Staessen *et al.* [46] has shown a strong association between hypertension and the presence of a DD homozygosity of the ACE gene, a CC homozygosity of the aldosterone synthase gene and a Gly460Trp heterozygosity of the α-adducin gene.

Haplotype analysis refers to the simultaneous study of multiple allelic variations of a QTL or candidate gene [47]. In a modified association study, Brand-Herrmann *et al.* [47] studied four different polymorphisms of the angiotensinogen gene – C-532T, A-20C, C-18T and G-6A – to determine which of these represented the best gene for functional analysis. They found that C-532T and G-6A had the strongest associations with blood pressure. The complexity of epistatic studies and haplotype analysis means that large numbers of patients are required for a meaningful result to be obtained.

Combination of the above

The most convincing evidence for the involvement of a particular genes in the etiology of hypertension has come from combining the methods and models mentioned above. As the studies below illustrate, such combinations of methods using different models have proven to be of greatest effectiveness in developing an understanding of the genes underlying and influencing essential hypertension, and in developing therapies for hypertension.

Study of monogenic hypertension and genes related to essential hypertension

Some rare forms of hypertension are caused by the mutation of a single gene. Many single gene forms of hypertension have been extensively studied, and this research has greatly contributed to our understanding of the pathophysiology and pharmacogenomics of hypertension. To date, all known forms of monogenic hypertension are caused by mutations of the genes involved in renal sodium handling.

Mutations of genes encoding epithelial sodium channel

Liddle syndrome is an autosomal dominant disorder characterized by low renin hypertension with hypokalemic alkalosis [48,49]. The disease is caused by mutations affecting the epithelial sodium channel (ENac), which is made up of α-, β- and γ-subunits on chromosomes 12p13 (α-subunit) and 16p12–p13 (β- and γ-subunits) [50,51]. Mutations that either delete or cause a frameshift mutation in the last 45–76 amino acids of the C-terminal end of either the β- or γ-subunits of the ENac can cause Liddle syndrome. These mutations reduce clearance of ENac from the cell surface, leading to an increased reabsorption of sodium [48]. In addition, an Asn530Ser mutation of the γ-subunit has been described in a Finnish patient with Liddle syndrome [52]. The Asn530Ser mutation is believed to increase the probability of the sodium channel being open, with consequent increase in renal sodium reabsorption.

As Liddle syndrome is brought about by a defect in the sodium channel, the condition responds to sodium channel antagonists such as triamterene and amiloride but not to spironolactone, an aldoster-

one antagonist [49]. Salt restriction is also important in the management of patients with Liddle syndrome. Thiazide and loop diuretics may be useful therapies but, as these agents can aggravate hypokalaemia, serum potassium needs to be monitored closely.

A T594M polymorphism of the β-subunit of ENac has also been reported to be associated with (non-Liddle) essential hypertension [53]. Lymphocyte ENac coded by the 594M variant shows greater response to cAMP stimulation when compared with wild-type controls, an effect possibly brought about by loss of protein kinase C inhibition [54]. It is interesting to note that amiloride is capable of inhibiting the 594M variant [54]. Whether some subset of hypertensives might benefit especially from amiloride, however, remains to be demonstrated.

Hypertensives of African descent often present with increased sodium sensitivity, low plasma renin activity and respond better to diuretic treatment [55,56]. The urine aldosterone : potassium ratio – low in Liddle syndrome – tends also to be lower in those of African descent compared to other ethnic groups [56]. Aldosterone secretion is also lower in children of African descent [57]. These observations are consistent with a hypothesis of increased ENac activity. Of interest in this context, it has been proposed that increased ENac activity, and hence enhanced sodium retention, may be beneficial in ancestral habitats where sodium is relatively scarce. A G442V polymorphism has been shown to be associated with a lower aldosterone : potassium ratio in healthy young people of African descent. This polymorphism, however, has not been shown to be associated with hypertension [56].

ACE inhibitors are often ineffective when used to treat hypertensives of African descent. Lack of efficacy is attributed to the low plasma renin activity associated with this group of patients. Diuretics are often very useful in lowering the blood pressure of hypertensives of African descent. These observations are again consistent with increased ENac activity in this population.

Mutations of aldosterone synthase gene

Glucocorticoid suppressible hypertension (GSH) is a form of autosomal dominant hypertension characterized by decreased renin activity and normal or increased levels of aldosterone [48,58]. The disorder is associated with hypokalemia and metabolic alkalosis. Aldosterone levels in GSH can be suppressed by the administration of exogenous glucocorticoid. GSH is caused by a chimeric gene on chromosome 8q21–22 with the regulatory region of 11b-hydroxylase and the coding portion of aldosterone synthase, presumably the result of an unequal cross-over during meiosis. This gene is regulated by adrenocorticotrophic hormone (ACTH) but codes for aldosterone synthase. Therefore, adrenal production of aldosterone is stimulated by the presence of endogenous ACTH. Excess circulating aldosterone stimulates salt and water retention, which results in hypertension. The condition is treated with glucocorticoids, which suppresses ACTH production, with or without the addition of a conventional antihypertensive. Thiazide or loop diuretics should be used with care because of the risk of precipitating hypokalaemia.

The discovery of GSH has led to an interest in the role of the aldosterone synthase gene in the etiology of essential hypertension. A T-344C polymorphism in the promoter region of the gene may be associated with increased aldosterone levels [59–62]. However, studies on the role of the T-344C polymorphism in the etiology of hypertension have been conflicting [59–62]. Some researchers have found an association between the T allele and hypertension [60,62], while others believe the C allele is more important in the pathophysiology of the disease [59,61,63]. The fact that the T-344C single nucleotide polymorphism (SNP) is in complete linkage disequilibrium with another polymorphism, Lys173Arg, further complicates matters. It is uncertain which of these polymorphisms are of clinical importance [64]. Recently, a small Japanese study has shown the additive beneficial effect of spironolactone in the regression of left ventricular hypertrophy in hypertensive patients treated with ACE inhibition [65]. This observation provides supplementary evidence of the benefit of mineralocorticoid antagonism in cardiac hypertrophy/failure, a treatment that was recently shown to be beneficial in severe heart failure patients by the Randomized Aldactone Evaluation Study (RALES) [66]. Another study of Chinese and Japanese hypertensives detected through serendipity a significant

association between two SNPs in the aldosterone synthase gene (one in T-344C and one resulting in a lysine/argenine substitution at amino acid 173) and plasma glucose levels and patients' diabetes status, an intriguing finding suggesting an unexpected link between aldosterone and glucose homeostasis [64].

11-β Hydroxylase deficiency and 17α-hydroxylase deficiency

11-β Hydroxylase deficiency (11βHD) is a rare cause of congenital adrenal hyperplasia [58,67]. Classic 11βHD patients present with masculinization of the female external genitalia, precocious pseudopuberty in both sexes and hypertension. Nonclassic forms may produce menstrual abnormalities, hirsutism and acne. The prevalence of nonclassic 11βHD is unknown, but the condition may affect as many as 8.4% and 0.6–6.5% of women with polycystic ovary syndrome and hirsutism, respectively [68,69].

11-β Hydroxylase is responsible for the production of cortisol. Multiple nonsense, missense and insertion mutations have been described in patients with 11βHD [69]. Defects in this enzyme result in an accumulation of steroid precursors which are shunted into the androgenic pathway, resulting in masculinization. Hypertension results from the elevated levels of steroid precursors with aldosterone-like actions (e.g., deoxycorticosterone).

Milder forms of 11βHD may well represent an intermediate hypertensive phenotype [70,71]. Indeed, there is some evidence to suggest that 11-β hydroxylase activity may be impaired in some essential hypertensives [70–72].

Cytochrome P450c17 has both 17α-dehydroxylase and 17,20-lyase activity. Defects in cytochrome P450c17 are another cause of congenital adrenal hyperplasia. The pathophysiology of this condition is similar to that of 11βHD in that there is impairment of secretion of cortisol, with compensatory hypersecretion of ACTH, which stimulates the production of large quantities of deoxycorticosterone by the adrenal glands. In addition, as 17,20-lyase is required for the synthesis of gonadal sex hormones, affected males are usually born with female genitalia. Females affected by the condition have primary amenorrhoea and hypogonadism. Multiple mutations of the P450c17 gene have been described to cause the disease [73,74].

Apparent mineralocorticoid excess

Patients with the autosomal recessive condition, apparent mineralocorticoid excess (AME), usually present with hypertension, hypokalaemic alkalosis, an increased cortisol : cortisone ratio, and reduced plasma renin activity [48,75]. AME is caused by mutations affecting the gene coding for the enzyme 11β-hydroxysteroid dehydrogenase type 2 isozyme (11βHsD2). This enzyme is responsible for converting cortisol to cortisone in the kidney. This process protects the distal tubular mineralocorticoid receptor from activation by endogenous cortisol, which has the same affinity as aldosterone for the mineralocorticoid receptor. Mutations that deactivate or reduce the activity of 11βHsD2 will result in excessive activation of the mineralocorticoid receptor, leading to salt retention. The clinical picture is virtually identical to hypertension that is caused by the chronic ingestion of liquorice, an inhibitor of 11βHsD2.

AME is treated with the aldosterone antagonist, spironolactone [75]. Sodium channel blockade with amiloride and triamterene are effective alternatives.

Mild forms of AME resemble essential hypertension [75]. Li *et al.* [76] described the heterozygote father of a child with AME with mineralocorticoid hypertension. This has given rise to the hypothesis that some cases of essential hypertension may be caused by defects of the 11βHsD2 [77].

Studies of essential hypertension have shown that the half-life of cortisol is significantly prolonged in hypertensive patients [72]. An impairment in the conversion of cortisol to inactive metabolites has also been reported in young men with hypertension [75]. Lovati *et al.* [78] have found that salt-sensitivity and 11βHsD2 activity is associated with a polymorphic CA repeat in the 11βHsD2 gene. A G534A mutation of the 11βHsD2 may also be important in determining salt sensitivity, the G allele being associated with a greater increase in blood pressure with a salt load [78]. These findings may account for the observation that spironolactone may be a useful add-on therapy in the treatment of some hypertensives with low renin hypertension who are resistant to conventional antihypertensives [79]. The relationship between the polymorphic CA repeat and hypertension, however, remains controversial and a recent study failed to find an association between them [80].

Mutation in the mineralocorticoid receptor

A missense mutation that results in the substitution of leucine for serine at position 810 of the mineralocorticoid receptor has recently been described [81]. Individuals heterozygous for the mutant receptor develop hypertension at a young age. The mutant receptor shows activity in the absence of added steroid, but normal activation by aldosterone. This receptor is also activated by the mineralocorticoid receptor antagonists, progesterone and spironolactone. The former explains the pregnancy-induced exacerbation of hypertension in some of these patients. The mutation may also be related to the early onset of heart failure, providing more evidence in support of the role of mineralocorticoids in heart failure. As spironolactone activates the mutant receptor, unlike in the wild-type receptor where it acts as an antagonist, its use is likely to cause deterioration in these patients. The exact prevalence of this mutation is unknown, although it is likely to be extremely rare.

Mutations in the peroxisome proliferator-activated receptor γ

Peroxisome proliferator-activated receptor γ (PPARγ) is one of the three members of the PPAR family of nuclear receptors, and acts by modifying the transcription of various target genes [82]. PPARγ mRNA is abundant in adipose tissue and is downregulated in starvation and in insulin-deficient states (i.e., untreated type I diabetes). PPARγ activation is associated with increased insulin sensitivity and growth of adipose tissue. The newest class of oral hypoglycaemic agents, thiazolidinediones, are potent PPARγ agonists.

Barroso *et al.* [83] described a syndrome of type 2 diabetes and hypertension linked to dominant missense PPARγ gene mutations, which act to decrease the activity of the gene product. They postulate that PPARγ mutations may result in hypertension through effects on vascular tone [83]. PPARγ is expressed in endothelial cells [84]. Indeed, thiazolidinediones have been shown to have a blood pressure lowering effect in rats and appears to have a similar effect in hypertensive diabetics [85]. More studies are needed to study the role of thiazolinediones in hypertension.

Gordon syndrome

Gordon syndrome (pseudohypoaldosteronism type II) is characterized by low-renin hypertension, hyperkalaemia, hyperchloremic metabolic acidosis and normal or high plasma aldosterone levels, all of which respond well to thiazide diuretics [86–88]. Some forms of this syndrome have been ascribed to mutations in either WNK1 or WNK4 genes [89], which belong to the WNK family of serine/threonine kinases [90]. Wild-type WNK4 is an inhibitor of the tubular thiazide-sensitive sodium-chloride co-transporter, an activity that is absent in the mutant variant associated with Gordon syndrome [91].

WNK1 is believed to inhibit WNK4 activity [92]. Therefore, gain of function mutations in the WNK1 gene would result in inhibition of WNK4 activity, leading to increased sodium and chloride reabsorption.

An SNP near the promoter region of the WNK1 gene has been shown to be associated with blood pressure [93]. It has therefore been suggested that increased expression of WNK1 may have a role in the pathogenesis of essential hypertension [93]. Indeed, a recent study suggests that a common WNK1 polymorphism can predict response to treatment with thiazide diuretics [94]. However, further studies are required to confirm this finding before any definite conclusions can be made.

Hypertension with brachydactyly

An autosomal dominant form of salt-resistant hypertension associated with brachydactyly has been described in Turkish, US and Canadian kindreds [95]. The condition appears to be related to genetic changes in the short arm of chromosome 12 [95]. Indeed, a study has recently found a locus for essential hypertension on chromosome 12p in Chinese patients [96].

Familial hyperaldosteronism type II

This disorder has an autosomal dominant inheritance and is characterized by nonglucocorticoid remediable hyperaldosteronism with bilateral adrenal hyperplasia or aldosterone producing ade-

nomas [97]. The genetic cause of this condition is not known.

Although single-gene disorders have led to intriguing clues to hypertension, most of the research on essential hypertension has involved association and linkage studies focused on candidate genes. Although some gene variants appear to be associated with hypertension, results have not been consistently reproducible. Some of the candidate genes are listed in Table 8.1 [98–155]. A few of the genes that have been studied in detail are discussed in the following section.

Table 8.1 Candidate genes that may have a role in essential hypertension.

Genes involved in hypertension	Reference
Aldosterone synthase	[59–62]
Alpha-adducin	[98–102]
α_2-Adrenoceptor	[106–108]
Alpha subunit of Gs protein	[111]
Angiotensin II receptor type I	[139–141,205]
Angiotensin converting enzyme (ACE)	[142–145,205]
Angiotensinogen	[130–138]
Atrial natriuretic peptide	[119–121]
Beta-adducin	[206]
β_2-Adrenoceptor	[106,146–152]
β_3-Adrenoceptor	[109,110]
β_2-Bradykinin receptor gene	[124,125]
11β-Hydroxysteroid dehydrogenase	[207]
Chemokine receptor 2 gene	[208]
Dopamine D2 receptor	[209]
Endothelial nitric oxide synthase	[103–105]
Endothelin-1	[112]
Endothelin-2	[113]
Epithelial sodium channel beta-subunit	[53]
Glucagon receptor	[128,129]
Glucokinase	[127]
Human natriuretic peptide receptor type A	[210]
Human natriuretic peptide receptor type B	[211]
Insulin receptor	[116]
Lipoprotein lipase	[122,123]
Pertussis toxin sensitive G protein β-subunit	[153–155]
Renin	[114,115]
Transforming growth factor β1	[117,118]
Tyrosine hydroxylase	[126]
WNK1	[93]
WNK4	[212]

Candidate genes and essential hypertension

α-Adducin

Studies in the Milan hypertensive rat (MHR) suggested that alterations in the renal tubular absorption of salt may account for the raised blood pressure found in this strain of rat [156]. It was subsequently found that cytoskeletal proteins are involved in mediating the increased salt reabsorption. Cross-immunization techniques between the MHR and its normotensive control revealed differences in the cDNA of the α-subunit of the adducin molecule. α-Adducin is a cytoskeletal protein involved in the regulation of the assembly of the actin cytoskeleton in renal tubular cells [156]. As the actin cytoskeleton may be able to influence the surface expression of the Na-K-ATPase, it is possible that disordered regulation of actin assembly may lead to increased sodium reabsorption. In addition, α-adducin has been shown to have a direct effect on the Na-K-ATPase responsible for the tubular reabsorption of salt [156]. Indeed, in the MHR, mutations of the α-adducin gene have been linked to hypertension and to increased Na-K pump activity.

In humans, a Gly460Trp mutation in α-adducin has been associated with increased sodium retention, probably secondary to increased numbers of renal tubular Na-K pumps [157,158]. The 460Trp allele may be associated with lower plasma renin activity and with a better response to thiazide diuretics, consistent with the increased sodium retention [98,159]. This finding could be of particular relevance in antihypertensive therapy as the 460Trp allele is present in 20% of European Caucasians. Unfortunately, an increase in Na-K pump activity has not been associated with the 460Trp allele in some studies [99]. Similarly, the 460Trp allele has been related to hypertension in some populations but not others [98–102].

Angiotensinogen

Multiple SNPs have been described in the angiotensinogen gene [130–132,160]. The M235T variant has been studied extensively. The T235 allele is associated with raised plasma angiotensinogen [130], which in turn may be correlated with blood pressure [161]. T235 has been associated with hypertension in some studies, but not in others [131–135]. M235T is in linkage disequilibrium with the G-6A and A-20C polymorphisms in the promoter region of the angiotensinogen gene [130,131]. The G-6 allele is associated with a lower transcription rate for angiotensinogen, while the presence of a -20C allele can increase angiotensinogen transcription [130,162]. M235T is also in linkage disequilibrium with at least three other polymorphisms: T+68C, C–18T and T+31C [131,160,163].

Another polymorphism, M174T, may be involved in hypertension in some populations [136–138], but not others [134,135]. Other angiotensinogen gene polymorphisms include C-532T, which is in linkage disequilibrium with G-6A, and an Arg-30Pro substitution in the signal peptide of angiotensinogen [164,165]. These polymorphisms may also influence plasma angiotensinogen levels.

Hunt *et al.* [166] showed that individuals with the -6A allele are more salt sensitive and more likely to benefit from salt restriction. Patients with the -6A allele may demonstrate a larger fall in blood pressure following weight loss when compared with -6G homozygotes. As Hunt *et al.* [166] point out, there may be a particular genotype that would especially benefit from increased vigilance in salt intake and/or weight monitoring. However, Giner *et al.* [167] failed to find an association between salt sensitivity and the M235T allele.

Hypertensives with at least one T235 allele have been shown to require a higher dosage level of antihypertensive medication compared with M235 homozygotes [168]. The angiotensinogen gene locus may influence an individual's response to ACE inhibition [169]. However, more studies are required to confirm the pharmacogenomic role of the angiotensinogen locus.

Angiotensin II receptor type I

Multiple single nucleotide polymorphisms have been found in the angiotensin II receptor type I (AGTRI) gene [139,170]. An A1166C polymorphism has been associated with hypertension in some studies, but not in others [139–141]. The mechanism by which this polymorphism affects blood pressure remains to be elucidated [171].

Angiotensin-converting enzyme

Plasma ACE levels are linked to the D allele of an I/D polymorphism of the ACE gene [172]. Linkage and association data from the Framingham Heart Study supports a relationship between the D variant and hypertension in males [142]. This hypothesis is supported by Fornage *et al.* [143] and Higaki *et al.* [144]. However, it should be noted that many studies fail to demonstrate a relationship between the ACE I/D polymorphism and hypertension [145].

The effects of the ACE I/D polymorphism and the response to ACE inhibition have been studied extensively. Some studies have found an association between the blood pressure response to ACE inhibition and ACE genotype, whereas others have not [145,173–175]. Moreover, even the results of the positive association studies are inconsistent, with some studies showing a greater blood pressure response in the DD group and others in the II genotype [173–175].

Left ventricular hypertrophy is an independent risk factor for mortality and morbidity from hypertension [176]. In addition, left ventricular hypertrophy in normotensive individuals is a risk factor for the later development of hypertension [176]. Evidence for the role of angiotensin and aldosterone comes from the observation that the intravenous administration of these agents can cause left ventricular hypertrophy and fibrosis [177]. In addition, blockade of the RAS with ACE inhibitors can cause regression of left ventricular hypertrophy [178]. The role of the ACE I/D polymorphism in hypertensive left ventricular hypertrophy has been studied, with inconclusive results [179,180].

β_2-Adrenoceptor

The β_2-adrenoceptor locus has been linked to hypertension [106,146]. Two polymorphisms, Arg16Gly and Gln27Glu, have been studied in detail [147–152]. The Gly16 variant of the receptor demonstrates increased downregulation after stimulation by isoproterenol compared to the Arg16 isoform [146]. Such a response can lead to impaired vasodilatation to β_2-adrenergic stimulation. Kotanko *et al.* [148] found an association between the Gly16 allele and hypertension. However, Timmerman *et al.* [149] discovered that the Gly16 allele is associated with lower blood pressure. Other studies have also shown conflicting results [150–152].

Atenolol and other beta-blockers used to treat hypertension can cause or aggravate bronchocon-striction. Therefore, it is of great clinical interest that the β_2-adrenoceptor polymorphisms may have a role in determining airway reactivity. Patients homozygous for the Glu27 allele have decreased bronchial sensitivity to methacholine compared to those with at least one Gln27 allele [128]. The presence of the Gly16 allele may be associated with increased downregulation, resulting in agonist desensitization when the patient is treated with beta-agonists [129]. Indeed, asthmatic children with the Arg16 allele are more likely to respond to albuterol compared to Gly16 homozygotes [130]. A better understanding of the interaction between beta-blockers, β_2-adrenoceptor polymorphisms and bronchoconstriction could lead to the identification of a subset of patients who are likely to suffer from this adverse effect. However, it is premature to comment on the implications of the β_2-adrenoceptor polymorphisms on the treatment of hypertension.

A critical evaluation of genes in essential hypertension

Although many of the candidate genes have genuine pathophysiologic effects on blood pressure, no single gene has consistently been shown to influence the development of hypertension in humans in classic linkage and association studies. However, the lack of consistent results from association and linkage studies should not be viewed in isolation. For example, in spite of the poor reproducibility of the association studies on the role of allelic variation of α-adducin in human essential hypertension, there is strong evidence from animal as well as pharmacologic and physiologic studies on its role in the regulation of blood pressure. Therefore, even though the Gly460Trp polymorphism has not been shown conclusively to be associated with hypertension, much has been learnt on the physiologic role of adducin on the regulation of blood pressure using genomic technology. There are many reasons as to the poor reproducibility of genetic studies of essential hypertension in humans. Poor study design, laboratory errors and/or lack of statistical power could partially account for poor reproducibility between the various association and linkage studies. The complexity of gene–gene and gene–environment interactions should not be discounted.

In addition, studies on the genes involved in the regulation of blood pressure should also be viewed in the context of the target population. For example, results from a study performed in the Japanese may not be extrapolable to a European population. Similarly, age and gender differences should be taken into account when interpreting the results of these studies.

Another cause of confusion is the presence of multiple polymorphisms within the same gene that are in linkage disequilibrium with one another, an example typified by the multiple polymorphisms found in the angiotensinogen gene. Therefore, differences in blood pressure associated with a particular polymorphism may, in fact, be brought about by variation elsewhere in the gene. In addition, the effects (if any) of intragene interactions between various polymorphisms should be thoroughly investigated before any definite conclusions can be drawn.

Recently, many investigators have tried to get around many of these problems by performing meta-analyses of previous studies in well-defined ethnic groups [133,181]. Indeed, these studies appear to show that the 235T allele of the angiotensinogen gene is associated with hypertension. However, reporting and positive publication biases can result in erroneous results in meta-analyses. In addition, because of design variation, not all studies can be compared satisfactorily. For example, the authors of one of these meta-analyses concluded that many of the studies they used had considerable heterogeneity, and the evidence was borderline.

Many believe that genome screens involving large numbers of subjects will provide definitive evidence for the genes involved in hypertension. The results from the BRIGHT study of 2010 sibling-pairs in a white British population suggests that a few loci on chromosomes 2q, 5q, 6q and 9q may be involved in blood pressure regulation [182]. Chromosomes 6q and 9q may have particularly strong influences on blood pressure, with LOD scores of 3.21 and 2.24, respectively. Results from the HyperGen, GENOA and GenNet studies indicated some – extremely weak – influences on blood pressure from loci in chromosomes 2p, 1p and 1, respectively, providing further evidence on the polygenic nature of hypertension [183–185]. Data from HyperGen also suggests that genetic variation may be a more important determinant of early onset hypertension in African-Americans than in white population [186]. However, the results from the large genome screens remain to be replicated. Indeed, it is possible that the relative importance of the individual genetic polymorphism in essential hypertension will never be conclusively shown and that future efforts should concentrate on the elucidation of the role of these polymorphisms.

Genomics and risk stratification

The main objective for the identification and treatment of hypertension is to prevent the development or progression of target organ damage [187]. In addition, secondary hypertension should be identified in order for the underlying cause of the high blood pressure to be treated. In current clinical practice, a combination of clinical, biochemical and imaging data are obtained in order to achieve these aims. The growth of genomics has great potential to aid risk stratification. Some of the data from basic science studies are already being used for this purpose. For example, the use of ANP and hsCRP in risk stratification have arisen as a direct result of our knowledge of the disease process.

In addition, there is great interest in the use of gene expression profiling in risk stratification [188]. This process has been greatly aided by the development of microarray technology, where the simultaneous analysis of the expression profile of the entire genome can be carried out. The gene expression profile of an individual is the result of a complex interplay between genetic and environmental factors. Changes in the gene expression profile usually accompany phenotypic change. Moreover, changes in gene expression patterns may occur before the occurrence of any phenotypic change. Indeed, the gene expression profile of circulating monocytes from patients with carotid atherosclerosis is different from that of normal controls [189].

Genes and the treatment of hypertension

Advances in our understanding of genetics has contributed to the development of new strategies of treating hypertension, namely in the fields of pharmacogenetics and gene therapy.

Pharmacogenomics

Pharmacogenomics can be defined as the application of genomic technology, such as gene sequencing and microarray technology, to the development of a pharmacologic agent. The term thus encompasses the fields of "pharmacogenetics" and "gene therapy," which fields refer, respectively, to the study of the effects of genotype on the response to a particular drug and the use of genetic material as a treatment for a disorder.

Pharmacogenomic technology can therefore potentially be applied to any stage in drug development: from the initial development of the pharmacologic agent, through the preclinical tests, to the final clinical trials. In addition, it is hoped that the application of pharmacogenomics can result in the development of personalized medicine, in which the use of a particular drug or drug combination can be tailored to the genotype of a patient. This is particularly true for a condition such as hypertension, for which a bewildering array of medications are available and which shows considerable interindividual variation to treatment by any given drug.

Drug isolation

Ferrari *et al.* [190] have shown that a digitoxigenin derivative, PST 2238, can decrease the blood pressure and Na-K pump activity in the MHR but has no effect on normotensive controls. This agent also has more effect in the Na-K pump of cell lines expressing mutant α-adducin compared with wild-type controls. These observations are of great interest, as they suggest PST2238 (or related compounds) may potentially be used in the future in the subgroup of hypertensives with the 460Trp allele.

Interindividual response to treatment

Early pharmacogenetic studies mostly focused on the effects of genetic variation on the metabolism (i.e., pharmacokinetics) of antihypertensive drugs. Cytochrome P450 CYP2D6 (CYP2D6) metabolizes many of the commonly used agents (including the beta-blocker, metoprolol), and polymorphisms in this enzyme can affect drug plasma levels and half-life. A clinically important polymorphism of the CYP2D6 enzyme was first noticed during a trial involving the adrenergic ganglion-blocking antihypertensive drug debrisoquine [191]. It was found that patients who are poor metabolizers of debrisoquine have an exaggerated hypotensive effect to treatment by the drug, which is one of the reasons why debrisoquine is not widely used in clinical practice. However, although individuals who are poor metabolizers of debrisoquine also have prolonged elimination half-life of metoprolol, they do not appear to have an increased incidence of adverse effects at therapeutic doses of the beta-blocker [192,193].

Catechol-O-methyltransferase (COMT) is responsible for the methylation of catecholamines and drugs containing the catechol group (e.g., levodopa and methyldopa) [194]. Variation in COMT activity is most commonly because of two different alleles, one of which codes for an enzyme with high activity while the other has low activity. Individuals who have the high activity form of COMT exhibit enhanced breakdown of methyldopa [195]. Therefore, these individuals may need a higher dose of the drug for the desired pharmacologic effect. Conversely, individuals with the low activity form of the enzyme may be more susceptible to the toxic effects of the drug.

Another enzyme, acetyltransferase, is responsible for the acetylation of the drug hydralazine [195]. Patients who are slow acetylators of the drug are at higher risk of developing hydralazine-induced systemic lupus erythematosus.

More recently, efforts have been made to identify polymorphisms in the genes involved in blood pressure regulation that may be implicated in the interindividual variation to treatment by antihypertensive drugs. In other words, these studies aim to research the direct interaction (i.e., the pharmacodynamic relationship) between drugs and polymorphisms of these genes. Genes that have been evaluated for their pharmacodynamic effects include α-adducin, angiotensinogen, pertussis toxin sensitive G protein β-subunit, ACE I/D polymorphism, β_1-adrenergic receptor, WNK1 and AGTR1 [94,196,197]. However, at present, there is no consensus as to the effect (if any) of polymorphisms of these genes on the pharmacodynamics of essential hypertension.

Gene therapy

The recent advances in our understanding of hypertension have stimulated interest into gene therapy for hypertension, although research in this field is still restricted to animal studies. Research into gene therapy in hypertension can be classified into studies based on the overexpression of a gene known to have a hypotensive effect and those suppressing the effect of genes that have pressor effect on blood pressure. Both of these approaches require a thorough knowledge of the genetic sequence of the gene of interest.

An example of the latter approach is the use of antisense oligonucleotides against the mRNA of the gene of interest. Antisense oligonucleotides are short segments of DNA which are designed to bind to the corresponding parts of the target mRNA [198]. This interaction prevents the translation of the target mRNA by ribosomes and stimulates its breakdown by RNAse H. The use of antisense therapy has so far focused on the mRNA of genes of the RAS. Animal models have shown a marked reduction of blood pressure in rats after the injection of antisense DNA directed against AGTR1 [198]. In addition, the hypotensive effect was prolonged, with a reduction in blood pressure for up to 9 weeks. However, it should be borne in mind that the study was carried out by the injection of the DNA-containing viral vector into the cerebral ventricles or the hypothalamus. More recently, it was shown that the use of intravenous antisense oligonucleotides directed against AGTR1 can reduce the blood pressure effectively for 9 days after a single injection [199]. Table 8.2 summarizes some of the other genes that have been targeted successfully by the antisense approach.

The overexpression of genes known to have a hypotensive effect has also been shown to be successful in lowering blood pressure in animal models. The administration of the human kallikrein gene using an adenovirus vector resulted in a marked reduction in blood pressure in the deoxycorticosterone acetate salt-sensitive (DOCA) rat [200]. In addition, the use of kallikrein gene therapy may prevent end organ damage in this rat model of hypertension [201]. It has been hypothesized that the protective effect of kallikrein is mediated by the prevention of end organ fibrosis

Table 8.2 Some of the genes which have been targeted with antisense oligonucleotides in animal models of hypertension.

Target	Reference
Angiotensinogen	[213]
AGTR1	[198,199]
β_1-Adrenergic receptor	[214]
c-fos	[215]
CYP4A1	[216]
Thyrotropin releasing hormone	[217]
Thyrotropin releasing hormone receptor	[218]
Urinary kininase	[219]

and the suppression of free radical production [201]. Other genes whose overexpression has led to a decrease in blood pressure include the adrenomedullin gene [202], atrial natriuretic peptide [203] and parathyroid hormone-related protein [204].

Drug synthesis

For the past few hundred years, the medical treatment of disease has been by the use of chemical agents that are either extracted from natural sources or synthesized chemically. In the last decade, the use of recombinant biologic agents in the treatment of disease has become increasingly common. This is a development that would not have been possible without the use of genomic technology. An example of this class of agent is the use of the natriuretic peptides.

Future perspectives

The use of genomic technology has not led to the detection of a genetic polymorphism that may lead to essential hypertension in humans. However, there have been great advances in our understanding of the role of various genes in blood pressure regulation. Indeed, this has been marked by the increasing use of genetically engineered animals and inhibition of gene expression to investigate the effect of modulating the effect of the gene of interest.

It is likely that a better knowledge of the molecular pathophysiology of hypertension will provide novel targets for pharmacologic intervention, such as PST 2238. In addition, better understanding may facilitate the development of gene therapy for the treatment of hypertension. Gene therapy for hypertension provides a potential route for effective prolonged blood pressure control with a single course of treatment, thus overcoming the problem of treatment failure resulting from poor compliance [198]. Finally, a better understanding of the interaction between genes, the environment and drugs may facilitate the matching of a particular drug to the clinical requirements of the individual patient.

References

1 Kaplan NM. Systemic hypertension: mechanisms and diagnosis. In: Braunwald E, Zipes DP, Libby P, eds. *Heart Disease: A Textbook of Cardiovascular Medicine*, 6th edn. WB Saunders Company, Philadelphia, 2001: 941–971.

2 Brenner BM, Garcia DL, Anderson S *et al.* Glomeruli and blood pressure. Less of one, more the other? *Am J Hypertens* 1988; 1: 335–347.

3 Weber MA. Unsolved problems in treating hypertension: rationale for new approaches. *Am J Hypertens* 1998; 11: 145S–149S.

4 Berlowitz DR, Ash AS, Hickey EC *et al.* Inadequate management of blood pressure in a hypertensive population. *N Engl J Med* 1998; 339: 1957–1963.

5 Dickerson JE, Hingorani AD, Ashby MJ, Palmer CR, Brown MJ. Optimisation of antihypertensive treatment by crossover rotation of four major classes. *Lancet* 1999: 353: 2008–2013.

6 Guyton AC. The surprising kidney-fluid mechanism for pressure control – its infinite gain! *Hypertension* 1990; 16: 725–730.

7 Qi N, Rapp JP, Brand PH *et al.* Body fluid expansion is not essential for salt-induced hypertension in SS/Jr rats. *Am J Physiol Regul Integr Comp Physiol* 1999; 277: R1392–R1400.

8 DiBona GF, Kopp UC. The neural control of renal function. *Physiol Rev* 1997; 77: 75–197.

9 Guimaraes S, Moura D. Vascular adrenoceptors: an update. *Pharmacol Rev* 2001; 53: 319–356.

10 Paul M, Poyan Mehr A, Kreutz R. Physiology of local renin-angiotensin systems. *Physiol Rev* 2006; 86: 747–803.

11 Kim HS, Krege JH, Kluckman KD *et al.* Genetic control of blood pressure and the angiotensinogen locus. *Proc Natl Acad Sci USA* 1995; 92: 2735–2739.

12 Crowley SD, Gurley SB, Oliverio MI *et al.* Distinct roles for the kidney and systemic tissues in blood pressure regulation by the renin-angiotensin system. *J Clin Invest* 2005; 115: 1092–1099.

13 Rigat B, Hubert C, Alhenc-Gelas F *et al.* An insertion/deletion polymorphism in the angiotensin I-converting enzyme gene accounting for half the variance of serum enzyme levels. *J Clin Invest* 1990; 86: 1343–1346.

14 Smithies O, Kim HS, Takahashi N *et al.* Importance of quantitative genetic variations in the etiology of hypertension. *Kidney Int* 2000; 58: 2265–2280.

15 Zeng C, Sanada H, Watanabe H *et al.* Functional genomics of the dopaminergic system in hypertension. *Physiol Gen* 2004; 19: 233–246.

16 Hollon TR, Bek MJ, Lachowicz JE *et al.* Mice lacking D5 dopamine receptors have increased sympathetic tone and are hypertensive. *J Neurosci* 2002; 22: 10801–10810.

17 Asico LD, Ladines C, Fuchs S *et al.* Disruption of the dopamine D3 receptor gene produces renin-dependent hypertension. *J Clin Invest* 1998; 102: 493–498.

18 Bek MJ, Wang X, Asico LD *et al.* Angiotensin-II type 1 receptor-mediated hypertension in D4 dopamine receptor-deficient mice. *Hypertension* 2006; 47: 288–295.

19 Huang PL, Huang Z, Mashimo H *et al.* Hypertension in mice lacking the gene for endothelial nitric oxide synthase. *Nature* 1995; 377: 196–197.

20 Stamler JS, Loh E, Roddy MA *et al.* Nitric oxide regulates basal systemic and pulmonary vascular resistance in healthy humans. *Circulation* 1994; 89: 2035–2040.

21 Tan KT, Lip GY. Platelets, atherosclerosis and the endothelium: new therapeutic targets? *Expert Opin Investig Drugs* 2003; 12: 1765–1776.

22 Duplain H, Burcelin R, Sartori C *et al.* Insulin resistance, hyperlipidemia, and hypertension in mice lacking endothelial nitric oxide synthase. *Circulation* 2001; 104: 342–345.

23 Silver MA. The natriuretic peptide system: kidney and cardiovascular effects. *Curr Opin Nephrol Hypertens* 2006; 15: 14–21.

24 Matsukawa N, Grzesik WJ, Takahashi N *et al.* The natriuretic peptide clearance receptor locally modulates the physiological effects of the natriuretic peptide system. *Proc Natl Acad Sci USA* 1999; 96: 7403–7408.

25 Oliver PM, John SW, Purdy KE *et al.* Natriuretic peptide receptor 1 expression influences blood pressures of mice in a dose-dependent manner. *Proc Natl Acad Sci USA* 1998; 95: 2547–2551.

26 Levin ER. Endothelins. *N Engl J Med* 1995; 333: 356–363.

27 Pollock DM. Endothelin, angiotensin, and oxidative stress in hypertension. *Hypertension* 2005; 45: 477–480.

28 Emanueli C, Madeddu P. Role of the kallikrein-kinin system in the maturation of cardiovascular phenotype. *Am J Hypertens* 1999; 12: 988–999.

29 Zhao C, Wang P, Xiao X *et al.* Gene therapy with human tissue kallikrein reduces hypertension and hyperinsulinemia in fructose-induced hypertensive rats. *Hypertension* 2003; 42: 1026–1033.

30 Alfie ME, Sigmon DH, Pomposiello SI *et al.* Effect of high salt intake in mutant mice lacking bradykinin-B2 receptors. *Hypertension* 1997; 29: 483–487.

31 Touyz R. Reactive oxygen species, vascular oxidative stress, and redox signalling in hypertension. *Hypertension* 2004; 44: 248–252.

32 Schillaci G, Pirro M, Gemelli L *et al.* Increased C-reactive protein concentrations in never-treated hypertension: the role of systolic and pulse pressures. *J Hypertens* 2003; 21: 1841–1846.

33 Sesso HD, Buring JE, Rifai N *et al.* C-reactive protein and the risk of developing hypertension. *JAMA* 2003; 290: 2945–2051.

34 Goldblatt H, Lynch J, Hanzal RF, Summerville WW. Studies of experimental hypertension I: Production of persistent elevation of systolic blood pressure by means of renal ischaemia. *J Exp Med* 1934; 59: 347–379.

35 Okamoto K, Akoi K. Development of a strain of spontaneously hypertensive rat. *Jpn Circ J* 1963; 27: 282–293.

36 Dahl LK, Heine M, Thompson K. Genetic influence of the kidneys on blood pressure. *Circ Res* 1974; 34: 94–101.

37 Kimura S, Mullins JJ, Bunneman B, Metzger R, Hilgenfeldt U. High blood pressure in transgenic mice carrying the rat angiotensinogen gene. *EMBO J* 1992; 11: 821–827.

38 Kim HS, Krege JH, Kluckman KD *et al.* Genetic control of blood pressure and the angiotensinogen locus. *Proc Natl Acad Sci USA* 1995; 92: 2735–2739.

39 Raizada MK, DerSakissian S. Potential of gene therapy strategy for the treatment of hypertension. *Hypertension* 2006; 47: 6–9.

40 Zamore PD, Tuschl T, Sharp PA *et al.* RNAi: double-stranded RNA directs the ATP-dependent cleavage of mRNA at 21 to 23 nucleotide intervals. RNAi: double-stranded RNA directs the ATP-dependent cleavage of mRNA at 21 to 23 nucleotide intervals. *Cell* 2000; 101: 25–33.

41 Wang X, Skelley L, Cade R *et al.* AAV delivery of mineralocorticoid receptor shRNA prevents progression of cold-induced hypertension and attenuates renal damage. *Gene Ther* 2006; 13: 1097–1103.

42 Xu BE, Stippec S, Lenertz L *et al.* WNK1 activates ERK5 by an MEKK2/3-dependent mechanism. *J Biol Chem* 2004; 279: 7826–7831.

43 Jones HB. The relative power of linkage and association studies for the detection of genes involved in hypertension. *Kidney Int* 1998; 53: 1446–1448.

44 Williams SM, Addy JH, Phillips JA *et al.* Combinations of variations in multiple genes are associated with hypertension. *Hypertension* 2000; 36: 2–6.

45 Bonnici F, Keavney B, Collins R, Danesh J. Angiotensin converting enzyme insertion or deletion polymorphism and coronary restenosis: meta-analysis of 16 studies. *Br Med J* 2002; 325: 517–520.

46 Staessen JA, Wang JG, Brand E *et al.* Effects of three candidate genes on the prevalence and incidence of hypertension in a Caucasion population. *J Hypertens* 2001; 19: 1349–1358.

47 Brand-Herrmann SM, Kopke K, Reichenberger F *et al.* Angiotensinogen promoter haplotypes are associated with blood pressure in untreated hypertensives. *J Hypertens* 2004; 22: 1289–1297.

48 Lifton RP, Gharavi AG, Geller DS. Molecular mechanisms of hypertension. *Cell* 2001; 104: 545–556.

49 Warnock DG. Liddle syndrome: An autosomal dominant form of human hypertension. *Kidnet Int* 1998; 53: 18–24.

50 Voilley N, Lingueglia E, Champigny G *et al.* The lung amiloride-sensitive Na$^+$ channel: biophysical properties, pharmacology, ontogenesis, and molecular cloning. *Proc Natl Acad Sci USA* 1994; 91: 247–251.

51 Voilley N, Bassilana F, Mignon C *et al.* Cloning, chromosomal localization and physical linkage of the beta and gamma subunits of the human epithelial amiloride-sensitive sodium channel. *Genomics* 1994; 29: 560–565.

52 Hiltunen TP, Hannila-Handelberg T, Petajaniemi N *et al.* Liddle's syndrome associated with a point mutation in the extracellular domain of the epithelial sodium channel gamma subunit. *J Hypertens* 2002; 20: 2383–2390.

53 Baker EH, Dong YB, Sagnella GA *et al.* Association of hypertension with T594M mutation in beta subunit of epithelial sodium channels in black people resident in London. *Lancet* 1998; 351: 1388–1392.

54 Cui Y, Su YR, Rutkowski M, Reif M, Menon AG, Pun RY. Loss of protein kinase C inhibition in the beta-T594M variant of the amiloride-sensitive sodium channel. *Proc Natl Acad Sci USA* 1997; 94: 9962–9966.

55 Hall WD. A rational approach to the treatment of hypertension in special populations. *Am Fam Phys* 1999; 60: 156–162.

56 Ambrosius WT, Bloem LJ, Zhou L *et al.* Genetic variants in the epithelial sodium channel in relation to aldosterone and potassium excretion and risk for hypertension. *Hypertension* 1999; 34: 631–637.

57 Pratt JH, Jones JJ, Miller JZ, Wagner MA, Fineberg NS. Racial differences in aldosterone excretion and plasma aldosterone concentrations in children. *N Engl J Med* 1989; 321: 1152–1157.

58 White PC. Inherited forms of mineralocorticoid hypertension. *Hypertension* 1996; 28: 927–936.

59 Tamaki S, Iwai N, Tsujita Y, Kinoshita M. Genetic polymorphism of CYP11B2 gene and hypertension in Japanese. *Hypertension* 1999; 33: 266–270.

60 Brand E, Chatelain N, Mulatero P *et al.* Structural analysis and evaluation of the aldosterone synthase gene in hypertension. *Hypertension* 2005; 32: 198–204.

61 Komiya I, Yamada T, Takara M *et al.* Lys(173)Arg and -344T/C variants of CYP11B2 in Japanese patients with low-renin hypertension. *Hypertension* 2005; 35: 699–703.

62 Davies E, Holloway CD, Ingram MC *et al.* Aldosterone excretion rate and blood pressure in essential hypertension are related to polymorphic differences in the aldosterone synthase gene CYP11B2. *Hypertension* 1999; 33: 703–707.

63 Kupari M, Hautanen A, Lankinen L *et al.* Association between human aldosterone synthase (CYP11B2) gene polymorphisms and left ventricular size, mass and function. *Circulation* 1998; 97: 569–575.

64 Ranade K, Wu KD, Risch N *et al.* Genetic variation in aldosterone synthase predicts plasma glucose levels. *Proc Natl Acad Sci USA* 2001; 98: 13219–13224.

65 Sato A, Suzuki Y, Saruta T. Effects of spironolactone and angiotensin-converting enzyme inhibitor on left ventricular hypertrophy in patients with essential hypertension. *Hypertens Res* 1999; 22: 17–22.

66 Pitt B, Zannad F, Remme WJ *et al.* The effect of spironolactone on morbidity and mortality in patients with severe heart failure. Randomized Aldactone Evaluation Study Investigators. *N Engl J Med* 1999; 341: 709–717.

67 Deaton MA, Glorioso JE, McLean DB. Congenital adrenal hyperplasia: Not really a zebra. *Am Fam Physician* 1999; 59: 1190–1199.

68 Joehrer K, Geley S, Strasser-Wozak EM *et al.* CYP11B1 mutations causing non-classic adrenal hyperplasia due to 11 beta-hydroxylase deficiency. *Hum Mol Genet* 1997; 6: 1829–1834.

69 Sahin Y, Kalestimur F. The frequency of late-onset 21-hydroxylase and 11 beta-hydroxylase deficiency in women with polycystic ovarian syndrome. *Eur J Endocrinol* 1997; 137: 670–674.

70 Geley S, Kapelari K, Johrer K *et al.* CYP11B1 mutations causing congenital adrenal hyperplasia due to 11 beta-hydroxylase deficiency. *J Clin Endocrinol Metab* 1996; 81: 2896–2901.

71 de Simone G, Tommaselli AP, Rossi R *et al.* Partial deficiency of adrenal 11-hydroxylase. A possible cause of primary hypertension. *Hypertension* 1985; 7: 204–210.

72 Valentino R, Tommaselli AP, Savastano S *et al.* Dysregulation of adrenal 11 beta-hydroxylase activity in hypertensive subjects: usefulness of the ACTH 1–17 stimulation test. *Nutr Metab Cardiovasc Dis* 1999; 9: 192–195.

73 Lam CW, Arlt W, Chan CK *et al.* Mutation of Proline 409 to Arginine in the meander region of cytochrome P450c17 causes severe 17-alpha hydroxylase deficiency. *Mol Genet Metab* 2001; 72: 254–259.

74 Auchus RJ. The genetics, pathophysiology and management of human deficiencies of P450c17. *Endocrinol Metab Clin North Am* 2001; 30: 101–119.

75 Connell JM, Kenyon CJ, Ingram M *et al.* Corticosteroids in essential hypertension: multiple loci and phenotypic variation. *Clin Exp Pharmacol Physiol* 1996; 23: 369–374.

76 Li A, Li KX, Marui S *et al.* Apparent mineralocorticoid excess in a Brazilian kindred: Hypertension in the heterozygote state. *J Hypertens* 1997; 15: 1397–1402.

77 Ferrari P, Krozowski Z. Role of 11beta-hydroxysteroid dehydrogenase type 2 in blood pressure regulation. *Kidney Int* 2000; 57: 1374–1381.

78 Lovati E, Ferrari P, Dick B *et al.* Molecular basis of human salt sensitivity: the role of the 11beta-hydroxysteroid dehydrogenase type 2. *J Clin Endocrinol Metab* 1999; 84: 3745–3749.

79 Pratt JH. Low-renin hypertension: more common than we think. *Cardiol Rev* 2000; 8: 202–206.

80 Brand E, Kato N, Chatelain N *et al.* Structural analysis and evaluation of the 11beta-hydroxysteroid dehydrogenase type 2 (11beta-HSD2) gene in human essential hypertension. *J Hypertens* 2005; 16: 1627–1633.

81 Geller DS, Farhi A, Pinkerton N *et al.* Activating mineralocorticoid receptor mutation in hypertension exacerbated by pregnancy. *Science* 2000; 289: 119–123.

82 Auwerx J. PPARγ, the ultimate thrifty gene. *Diabetologia* 1999; 42: 1033–1049.

83 Barroso I, Gurnell M, Crowley VE *et al.* Dominant negative mutations in human PPARgamma associated with severe insulin resistance, diabetes mellitus and hypertension. *Nature* 1999; 402: 880–883.

84 Iijima K, Yoshizumi M, Ako J *et al.* Expression of peroxisome proliferator-activated receptor gamma (PPARγ) in rat aortic smooth muscle cells. *Biochem Biophys Res Commun* 1998; 247: 353–356.

85 Kotchen TA. Attenuation of hypertension by insulin-sensitizing agents. *Hypertension* 1996; 28: 219–223.

86 Disse-Nicodeme S, Achard JM, Desitter I *et al.* A new locus on chromosome 12p13.3 for pseudohypoaldosteronism type II, an autosomal form of hypertension. *Am J Hum Genet* 2000; 67: 302–310.

87 O'Shaughnessy KM, Fu B, Johnson A, Gordon RD. Linkage of Gordon's syndrome to the long arm of chromosome 17 in a region recently linked to familial essential hypertension. *J Hum Hypertens* 1998; 12: 675–678.

88 Mansfield TA, Simon DB, Farfel Z *et al.* Multilocus linkage of familial hyperkalaemia and hypertension, pseudohypoaldosteronism type II, to chromosomes 1q31–42 and 17p11–q21. *Nat Genet* 1997; 16: 205.

89 Wilson FH, Disse-Nicodeme S, Choate KA *et al.* Human hypertension caused by mutations in WNK kinases. *Science* 2001; 293: 1107–1112.

90 Xu B, English JM, Wilsbacher JL, Stippec S, Goldsmith EJ, Cobb MH. WNK1, a novel mammalian serine/threonine protein kinase lacking the catalytic lysine in subdomain II. *J Biol Chem* 2000; 275: 16795–16801.

91 Wilson FH, Kahle KT, Sabath E *et al.* Molecular pathogenesis of inherited hypertension with hyperkalaemia: the Na-Cl cotransporter is inhibited by wild-type but not mutant WNK4. *Proc Natl Acad Sci USA* 2003; 100: 680–684.

92 Yang CL, Zhu X, Wang Z, Subramanya AR, Ellison DH. Mechanisms of WNK1 and WNK4 interaction in the regulation of thiazide-sensitive NaCl cotransport. *J Clin Invest* 2005; 115: 1379–1387.

93 Newhouse SJ, Wallace C, Dobson R *et al.* Haplotypes of the WNK1 gene associate with blood pressure variation in severely hypertensive population from the British Genetics of Hypertension study. *Hum Mol Genet* 2005; 14: 1805–1814.

94 Turner ST, Schwartz GL, Chapman AB, Boerwinkle E. WNK1 kinase polymorphism and blood pressure response to a thiazide diuretic. *Hypertension* 2005; 46: 758–765.

95 Bahring S, Rauch A, Toka O *et al.* Autosomal dominant hypertension with type E brachydactyly is caused by rearrangement on the short arm of chromosome 12. *Hypertension* 2004; 43: 471–476.

96 Gong M, Zhang H, Schulz H *et al.* Genome-wide linkage reveals a locus for human essential hypertension on chromosome 12p. *Hum Mol Genet* 2003; 12: 1273–1277.

97 Stowasser M, Gordon RD. Primary aldosteronism: from genesis to genetics. *Trends Endocrinol Metab* 2003; 14: 310–317.

98 Cusi D, Barlassina C, Azzani T *et al.* Polymorphism of alpha-adducin and salt sensitivity in patients with essential hypertension. *Lancet* 1997; 349: 1353–1357.

99 Kamitani A, Wong ZY, Fraser R *et al.* Human alpha-adducin gene, blood pressure, and sodium metabolism. *Hypertension* 1998; 32: 138–143.

100 Province MA, Arnett DK, Hunt SC *et al.* Association between the alpha-adducin gene and hypertension in the HyperGEN Study. *Am J Hypertens* 2000; 13: 710–718.

101 Bray MS, Li L, Turner ST, Kardia SL, Boerwinkle E. Association and linkage analysis of the alpha-adducin gene and blood pressure. *Am J Hypertens* 2000; 13: 699–703.

102 Schork NJ, Chakravarti A, Thiel B *et al.* Lack of association between a biallelic polymorphism in the adducin gene and blood pressure in whites and African Americans. *Am J Hypertens* 2000; 13: 693–698.

103 Kato N, Sugiyama T, Morita H *et al.* Lack of evidence for association between the endothelial nitric oxide synthase gene and hypertension. *Hypertension* 1999; 33: 933–936.

104 Lacolley P, Gautier S, Poirier O, Pannier B, Cambien F, Benetos A. Nitric oxide synthase gene polymorphisms, blood pressure and aortic stiffness in normotensive and hypertensive subjects. *J Hypertens* 1998; 16: 31–35.

105 Miyamoto Y, Saito Y, Kajiyama N *et al.* Endothelial nitric oxide synthase gene is positively associated with essential hypertension. *Hypertension* 1998; 27: 3–8.

106 Svetkey LP, Timmons PZ, Emovon O, Anderson NB, Preis L, Chen YT. Association of hypertension with beta2- and alpha2c10-adrenergic receptor genotype. *Hypertension* 1996; 27: 1210–1215.

107 Lockette W, Ghosh S, Farrow S *et al.* Alpha 2-adrenergic receptor gene polymorphism and hypertension in blacks. *Am J Hypertens* 1995; 8: 390–394.

108 Umemura S, Hirawa N, Iwamoto T *et al.* Association analysis of restriction fragment length polymorphism for alpha 2-adrenergic receptor genes in essential hypertension in Japan. *Hypertension* 1994; 23: 203–206.

109 Fujisawa T, Ikegami H, Yamato E *et al.* Trp64Arg mutation of beta3-adrenergic receptor in essential hypertension: insulin resistance and the adrenergic system. *Am J Hypertens* 1997; 10: 101–105.

110 Bendlova B, Mazura I, Vcelak J *et al.* Is a mutation of the beta-3-adrenergic receptor gene related to NIDDM and juvenile hypertension in the Czech population. *Ann NY Acad Sci* 1997; 827: 135–143.

111 Jia H, Hingorani AD, Sharma P *et al.* Association of the G(s)alpha gene with essential hypertension and response to beta-blockade. *Hypertension* 1999; 4: 8–14.

112 Stevens PA, Brown MJ. Genetic variability of the ET-1 and the ETA receptor genes in essential hypertension. *J Cardiovasc Pharmacol* 1995; 26S: S9–S12.

113 Sharma P, Hingorani A, Jia H, Hopper R, Brown MJ. Quantitative association between a newly identified molecular variant in the endothelin-2 gene and human essential hypertension. *J Hypertens* 1999; 17: 1281–1287.

114 Frossard PM, Lestringant GG, Malloy MJ, Kane JP. Human renin gene BglI dimorphism associated with hypertension in two independent populations. *Clin Genet* 1999; 56: 428–433.

115 Chiang FT, Hsu KL, Tseng CD, Lo HM, Chern TH, Tseng YZ. Association of the renin gene polymorphism with essential hypertension in a Chinese population. *Clin Genet* 1997; 51: 370–374.

116 Morris BJ. Insulin receptor gene in hypertension. *Clin Exp Hypertens* 1997; 19: 551–565.

117 Cambien F, Ricard S, Troesch A *et al.* Polymorphisms of the transforming growth factor-beta 1 gene in relation to myocardial infarction and blood pressure. The Etude

Cas-Temoin de l'Infarctus du Myocarde (ECTIM) Study. *Hypertension* 1996; 28: 881–887.

118 Li B, Khanna A, Sharma V, Singh T, Suthanthiran M, August P. TGF-beta1 DNA polymorphisms, protein levels, and blood pressure. *Hypertension* 1999; 33: 271–275.

119 Rutledge DR, Sun Y, Ross EA. Polymorphisms within the atrial natriuretic peptide gene in essential hypertension. *J Hypertens* 1995; 13: 953–955.

120 Kato N, Sugiyama T, Morita H *et al.* Genetic analysis of the atrial natriuretic peptide gene in essential hypertension. *Clin Sci* 2000; 98: 251–258.

121 Daniel HI, Munroe PB, Kamdar SM *et al.* The atrial natriuretic peptide gene and essential hypertension in African-Caribbeans from St Vincent and the Grenadines. *J Hum Hypertens* 1997; 11: 113–117.

122 Wu DA, Bu X, Warden CH *et al.* Quantitative trait locus mapping of human blood pressure to a genetic region at or near the lipoprotein lipase gene locus on chromosome 8p22. *J Clin Invest* 1996; 97: 2111–2118.

123 Hunt SC, Province MA, Atwood LD *et al.* No linkage of the lipoprotein lipase locus to hypertension in Caucasians. *J Hypertens* 1997; 17: 39–43.

124 Gainer JV, Brown NJ, Bachvarova M *et al.* Altered frequency of a promoter polymorphism of the kinin B2 receptor gene in hypertensive African-Americans. *Am J Hypertens* 2000; 13: 1268–1273.

125 Mukae S, Aoki S, Itoh S *et al.* Promoter polymorphism of the beta2 bradykinin receptor gene is associated with essential hypertension. *Jpn Circ J* 1999; 63: 759–762.

126 Sharma P, Hingorani A, Jia H *et al.* Positive association of tyrosine hydroxylase microsatellite marker to essential hypertension. *Hypertension* 1998; 32: 676–682.

127 Chiang FT, Chiu KC, Tseng YZ, Lee KC, Chuang LM. Nucleotide(-258) G-to-A transition variant of the liver glucokinase gene is associated with essential hypertension. *Am J Hypertens* 1997; 10: 1049–1052.

128 Morris BJ, Chambers SM. Hypothesis: glucagon receptor glycine to serine missense mutation contributes to one in 20 cases of essential hypertension. *Clin Exp Pharmacol Physiol* 1996; 23: 1035–1037.

129 Brand E, Bankir L, Plouin PF, Soubrier F. Glucagon receptor gene mutation (Gly40Ser) in human essential hypertension: the PEGASE study. *Hypertension* 1999; 34: 15–17.

130 Inoue I, Nakajima T, Williams CS *et al.* A nucleotide substitution in the promoter of human angiotensinogen is associated with essential hypertension and affects basal transcription *in vitro. J Clin Invest* 1997; 99: 1786–1797.

131 Sato N, Katsuya T, Rakugi H *et al.* Association of variants in critical core promoter element of angiotensinogen gene with increased risk of essential hypertension in Japanese. *Hypertension* 1997; 30: 321–325.

132 Kunz R, Kreutz R, Beige J, Distler A, Sharma AM. Association between angiotensinogen 235T variant and essential hypertension in whites: a systematic review and methodological appraisal. *Hypertension* 1997; 30: 1331–1337.

133 Staessen JA, Kuznetsova T, Wang JG, Emelianov D, Vlietinck R, Fagard R. M235T angiotensinogen gene polymorphism and cardiovascular risk. *J Hypertens* 1999; 17: 9–17.

134 Niu T, Chen C, Yang J *et al.* Blood pressure and the T174M and M235T polymorphisms of the angiotensinogen gene. *Ann Epidemiol* 1999; 9: 245–253.

135 Rotimi C, Cooper R, Ogunbiyi O *et al.* Hypertension, serum angiotensinogen, and molecular variants of the angiotensinogen gene among Nigerians. *Circulation* 1997; 95: 2348–2350.

136 Tiret L, Ricard S, Poirier O *et al.* Genetic variation at the angiotensinogen locus in relation to high blood pressure and myocardial infarction: the ECTIM study. *J Hypertens* 1995; 13: 311–317.

137 Hegele RA, Brunt JH, Connelly PW. A polymorphism of the angiotensinogen gene associated with variation in blood pressure in a genetic isolate. *Circulation* 1994; 90: 2207–2212.

138 Jeunemaitre X, Soubrier F, Kotelevtsev YV *et al.* Molecular basis of human hypertension: role of angiotensinogen. *Cell* 1992; 71: 169–180.

139 Bonnardeaux A, Davies E, Jeunemaitre X *et al.* Angiotensin II type 1 receptor gene polymorphisms in human essential hypertension. *Hypertension* 1994; 24: 63–69.

140 Wang WY, Zee RY, Morris BJ. Association of angiotensin II type 1 receptor gene polymorphism with essential hypertension. *Clin Genet* 1997; 51: 31–34.

141 Schmidt S, Beige J, Walla-Friedel M, Michel MC, Sharma AM, Ritz E. A polymorphism in the gene for the angiotensin II type 1 receptor is not associated with hypertension. *J Hypertens* 1997; 15: 1385–1388.

142 O'Donnell CJ, Lindpaintner K, Larson MG *et al.* Evidence for association and genetic linkage of the angiotensin-converting enzyme locus and blood pressure in men but not women in the Framingham Heart Study. *Circulation* 1998; 97: 1766–1772.

143 Fornage M, Amos CI, Kardia S, Sing CF, Turner ST, Boerwinkle E. Variation in the region of the angiotensin-converting enzyme gene influences interindividual differences in blood pressure levels in young white males. *Circulation* 1998; 97: 1773–1779.

144 Higaki J, Baba S, Katsuya T *et al.* Deletion allele of angiotensin-converting enzyme gene increases risk of essential hypertension in Japanese men: the Suita Study. *Circulation* 2000; 101: 2060–2065.

145 Dudley C, Keavney B, Casadei B, Conway J, Bird R, Ratcliffe P. Prediction of patient responses to antihyper-

tensive drugs using genetic polymorphisms: investigation of renin-angiotensin system genes. *J Hypertens* 1996; 14: 259–262.

146 Bray MS, Krushkal J, Li L *et al.* Positional analysis identifies the beta-2-adrenergic receptor gene as a susceptibility locus for human hypertension. *Circulation* 2000; 101: 2877–2882.

147 Cockcroft JR, Gazis AG, Cross DJ *et al.* Beta(2)-adrenergic receptor polymorphism determines vascular reactivity in humans. *Hypertension* 2000; 36: 371–375.

148 Kotanko P, Binder A, Tasker J *et al.* Essential hypertension in African Caribbeans associates with a variant of the beta-2-adrenoceptor. *Hypertension* 1997; 30: 773–776.

149 Timmermann B, Mo R, Luft FC *et al.* Beta-2-adrenoceptor genetic variation is associated with genetic predisposition to essential hypertension: The Bergen Blood Pressure Study. *Kidney Int* 1998; 53: 1455–1478.

150 Jia H, Sharma P, Hopper R, Dickerson C, Lloyd DD, Brown MJ. Beta2-adrenoceptor gene polymorphisms and blood pressure variations in East Anglia caucasians. *J Hypertens* 2000; 18: 687–693.

151 Herrmann V, Buscher R, Go MM *et al.* Beta2-adrenergic receptor polymorphisms at codon 16, cardiovascular phenotypes and essential hypertension in whites and African Americans. *J Hypertens* 2000; 13: 1021–1026.

152 Kato N, Sugiyama T, Morita H *et al.* Association analysis of beta(2)-adrenergic receptor polymorphisms with hypertension in Japanese. *Hypertension* 2001; 37: 286–292.

153 Siffert W, Rosskopf D, Siffert G *et al.* Association of a human G-protein beta3 subunit variant with hypertension. *Nat Genet* 1998; 18: 45–48.

154 Beige J, Hohenbleicher H, Distler A, Sharma AM. G-Protein beta3 subunit C825T variant and ambulatory blood pressure in essential hypertension. *Hypertension* 1999; 33: 1049–1051.

155 Brand E, Herrmann SM, Nicaud V *et al.* The 825C/T polymorphism of the G-protein subunit beta3 is not related to hypertension. *Hypertension* 1999; 33: 1175–1178.

156 Ferrandi M, Salardi S, Tripodi G *et al.* Evidence for an interaction between adducin and Na-K-ATPase: relation to genetic hypertension. *Am J Physiol Hear Circ Physiol* 1999; 277: 338–349.

157 Manunta P, Cusi D, Barlassina C *et al.* Alpha-adducin polymorphisms and renal sodium handling in essential hypertension patients. *Kidney Int* 1998; 53: 1471–1478.

158 Barlassina C, Citterio L, Bernardi L *et al.* Genetics of renal mechanisms of primary hypertension: the role of adducin. *J Hypertens* 1997; 15: 1567–1571.

159 Glorioso N, Manunta P, Filigheddu F *et al.* The role of alpha-adducin polymorphism in blood pressure and sodium handling regulation may not be excluded by a negative association study. *Hypertension* 1999; 34: 649–654.

160 Ishikawa K, Baba S, Katsuya T *et al.* T+31C polymorphism of angiotensinogen gene and essential hypertension. *Hypertension* 2001; 37: 281–285.

161 Walker WG, Whelton PK, Saito H, Russell RP, Hermann J. Relation between blood pressure and renin, renin substrate, angiotensin II, aldosterone and urinary sodium and potassium in 574 ambulatory subjects. *Hypertension* 1979; 1: 287–291.

162 Zhao YY, Zhon J, Narayanan CS, Cui Y, Kumar A. Role of C/A polymorphism at -20 on the expression of human angiotensinogen gene. *Hypertension* 1999; 33: 108–115.

163 Ishigami T, Tamura K, Fujita T *et al.* Angiotensinogen gene polymorphism near transcription start site and blood pressure: role of a T-to-C transition at intron I. *Hypertension* 1999; 34: 430–434.

164 Paillard F, Chansel D, Brand E *et al.* Genotype–phenotype relationships for the renin-angiotensin-aldosterone system in a normal population. *Hypertension* 1999; 34: 423–429.

165 Nakajima T, Cheng T, Rohrwasser A *et al.* Functional analysis of a mutation occurring between the two in-frame AUG codons of human angiotensinogen. *J Biol Chem* 1999; 274: 35749–35755.

166 Hunt SC, Cook NR, Oberman A *et al.* Angiotensinogen genotype, sodium reduction, weight loss, and prevention of hypertension: trials of hypertension prevention, phase II. *Hypertension* 1998; 32: 393–401.

167 Giner V, Poch E, Bragulat E. Renin-angiotensin genetic polymorphism and salt sensitivity in essential hypertension. *Hypertension* 2000; 35: 512.

168 Schunkert H, Hense HW, Gimenez-Roqueplo AP. The angiotensinogen T235 variant and the use of antihypertensive drugs in a population based cohort. *Hypertension* 1997; 29: 628–633.

169 Hingorani AD, Jia H, Stevens PA, Hopper R, Dickerson JE, Brown MJ. Renin-angiotensin system gene polymorphisms influence blood pressure and the response to angiotensin converting enzyme inhibition. *J Hypertens* 1995; 13: 1602–1609.

170 Rolfs A, Weber-Rolfs I, Regitz-Zagrosek V, Kallisch H, Riedel K, Fleck E. Genetic polymorphisms of the angiotensin II type 1 (AT1) receptor gene. *Eur Heart J* 1994; 15D: 108–112.

171 Hilgers KF, Langenfeld MR, Schlaich M, Veelken R, Schmieder RE. 1166 A/C polymorphism of the angiotensin II type 1 receptor gene and the response to short-term infusion of angiotensin II. *Circulation* 1999; 100: 1394–1399.

172 Tiret L, Rigat B, Visvikis S *et al.* Evidence, from

combined segregation and linkage analysis, that a variant of the angiotensin I-converting enzyme (ACE) gene controls plasma ACE levels. *Am J Hum Genet* 1992; 51: 197–205.

173 Stavroulakis GA, Makris TK, Krespi PG *et al.* Predicting response to chronic antihypertensive treatment with fosinopril: the role of angiotensin-converting enzyme gene polymorphism. *Cardiovasc Drugs Ther* 2000; 14: 427–432.

174 Ueda S, Meredith PA, Morton JJ, Connell JM, Elliott HL. ACE (I/D) genotype as a predictor of the magnitude and duration of the response to an ACE inhibitor drug (enalaprilat) in humans. *Circulation* 1998; 98: 2148–2153.

175 Ohmichi N, Iwai N, Uchida Y, Shichiri G, Nakamura Y, Kinoshita M. Relationship between the response to the angiotensin converting enzyme inhibitor imidapril and the angiotensin converting enzyme genotype. *Am J Hypertens* 1997; 10: 951–955.

176 Lip GY, Felmeden DC, Li-Saw-Hee FL, Beevers DG. Hypertensive heart disease. A complex syndrome or a hypertensive "cardiomyopathy"? *Eur Heart J* 2000; 21: 1653–1665.

177 Sun Y, Weber KT. Angiotensin II and aldosterone receptor binding: response to chronic angiotensin II or aldosterone adminstration. *J Lab Clin Med* 1993; 122: 404–411.

178 Pitt B. Regression of left ventricular hypertrophy in patients with hypertension: Blockade of the renin-angiotensin-aldosterone system. *Circulation* 1998; 98: 1987–1999.

179 Schunkert H, Hense HW, Holmer SR *et al.* Association between a deletion polymorphism of the angiotensin-converting-enzyme gene and left ventricular hypertrophy. *N Engl J Med* 1994; 330: 1634–1638.

180 Lindpainter K, Lee M, Larson MG *et al.* Absence of association or genetic linkage between the angiotensin-converting enzyme gene and left ventricular mass. *N Engl J Med* 1996; 334: 1023–1028.

181 Kato N, Sugiyama T, Morita H, Kurihara H, Yamori Y, Yazaki Y. Angiotensinogen gene and essential hypertension in the Japanese: extensive association study and meta-analysis on six reported studies. *J Hypertens* 1999; 17: 757–763.

182 MRC British Genetics of Hypertension Study. Genome-wide mapping of human loci for essential hypertension. *Lancet* 2003; 361: 2118–2123.

183 Thiel BA, Chakravarti A, Cooper R *et al.* A genome-wide linkage analysis investigating the determinants of blood pressure in whites and African Americans. *Am J Hypertens* 2003; 16: 151–153.

184 Kardia SL, Rozek LS, Krushkal J *et al.* Genome-wide linkage analyses for hypertension genes in two ethnically and geographically diverse population. *Am J Hypertens* 2003; 16: 154–157.

185 HyperGen Network. A genome-wide affected sibpair linkage analysis of hypertension: the HyperGen network. *Am J Hypertens* 2003; 16: 148–150.

186 Wilk JB, Djousse L, Arnett DK *et al.* Genome-wide linkage analyses for age at diagnosis of hypertension and early onset hypertension in the HyperGen study. *Am J Hypertens* 2004; 17: 839–844.

187 Erdine S, Ari O. ESH-ESC Guidelines for the management of hypertension. *Herz* 2006; 31: 331–338.

188 Liew CC. Expressed genome molecular signatures of heart failure. *Clin Chem Lab Med* 2005; 43: 462–469.

189 Patino WD, Mian OY, Kang JG *et al.* Circulating transcriptome reveals markers of atherosclerosis. *Proc Natl Acad Sci USA* 2005; 102: 3423–3428.

190 Ferrari P, Ferrandi M, Torielli L, Tripodi G, Bianchi G. PST 2238: A new antihypertensive compound that modulates Na,K-ATPase in genetic hypertension. *J Pharmacol Exp Ther* 1999; 288: 1074–1083.

191 Gonzalez FJ, Skoda RC, Kimura S *et al.* Characterisation of the common genetic defect in humans deficient in debrisoquine metabolism. *Nature* 1988; 331: 442–446.

192 Zineh I, Beitelshees AL, Gaedigk A, Walker JR, Pauly DF. Pharmacokinetics and CYP2D6 genotypes do not predict metoprolol adverse events or efficacy in hypertension. *Clin Pharm Ther* 2004; 76: 536–544.

193 Kirchheiner J, Heesch C, Bauer S *et al.* Impact of the ultrarapid metabolizer genotype of cytochrome P450 2D6 on metoprolol pharmacokinetics and pharmacodynamics. *Clin Pharm Ther* 2004; 76: 302–312.

194 Weinshilboum RM. Human pharmacogenetics of methyl comjugation. *Fed Proc* 1984; 43: 2303–2307.

195 West WL, Knight EM, Pradhan S, Hinds TS. Interpatient variability: Genetic predisposition and other genetic factors. *J Clin Pharm* 1997; 37: 635–648.

196 Mellen PB, Herrington DM. Pharmacogenomics of blood pressure response to antihypertensive treatment. *J Hypertens* 2005; 23: 1311–1325.

197 Koopmans RP, Insel PA, Michel MC. Pharmacogenetics of hypertension treatment: a structured review. *Pharmacogenetics* 2003; 13: 705–713.

198 Phillips MI, Mohuczy-Dominiak D, Coffey M *et al.* Prolonged reduction of high blood pressure with an in vivo, nonpathogenic, adeno-associated viral vector delivery of AT1-mRNA antisense. *Hypertension* 1997; 29: 374–380.

199 Galli SM, Phillips MI. Angiotensin II AT receptor antisense lowers blood pressure in acute 2 kidney, 1-clip hypertension. *Hypertension* 2001; 38: 674–678.

200 Dobrzynski E, Yoshida H, Chao J, Chao L. Adenovirus-mediated kallikrein gene delivery attenuates hyperten-

sion and protects against renal injury in deoxycorticosterone-salt rats. *Immunopharmacology* 1999; 44: 57–65.

201 Xia CF, Beldsoe G, Chao L, Chao J. Kallikrein gene transfer reduces renal fibrosis, hypertrophy and proliferation in DOCA-salt hypertensive rats. *Am J Physiol Renal Physiol* 2005; 289: F622–F631.

202 Dobrzynski E, Wang C, Chao J, Chao L. Adrenomedullin. gene delivery attenuates hypertension, cardiac remodelling, and renal injury in deoxycorticosterone acetate-salt hypertensive rats. *Hypertension* 2000; 36: 995–1001.

203 Schillinger KJ, Tsai SY, Taffet GE *et al.* Regulatable atrial natriuretic peptide gene therapy for hypertension. *Proc Natl Acad Sci USA* 2005; 102: 13789–13794.

204 Landa MS, Garcia SI, Liberjen L *et al.* Parathyroid hormone-related protein overexpression decreases blood pressure in spontaneously hypertensive rats. *Clin Exp Hypertens* 2005; 27: 343–354.

205 Tiret L, Blanc H, Ruidavets JB *et al.* Gene polymorphisms of the renin-angiotensin system in relation to hypertension and parental history of myocardial infarction and stroke: the PEGASE study. Projet d'Etude des Genes de l'Hypertension Arterielle Severe a moderee Essentielle. *J Hypertens* 1998; 16: 37–44.

206 Wang JG, Staessen JA, Barlassina C *et al.* Association between hypertension and variation in the alpha- and beta-adducin genes in a white population. *Kidney Int* 2002; 62: 2152–2159.

207 Agarwal AK, Giachetti G, Lavery G *et al.* CA-repeat polymorphism in intron 1 of HSD11B2: effects on gene expression and salt sensitivity. *Hypertension* 2000; 36: 187–194.

208 Izawa H, Yamada Y, Okada T *et al.* Prediction of genetic risk for hypertension. *Hypertension* 2003; 41: 1035–1040.

209 Thomas GN, Tomlinson B, Critchley JA. Modulation of blood pressure and obesity with the dopamine D2 receptor gene TaqI polymorphism. *Hypertension* 2000; 36: 177–182.

210 Nakayama T, Soma M, Mizutani Y *et al.* A novel missense mutation of exon 3 in the type A human natri-

uretic peptide receptor gene: possible association with essential hypertension. *Hypertens Res* 2002; 25: 395–401.

211 Rehemudula D, Nakayama T, Soma M *et al.* Structure of the type B human natriuretic peptide receptor gene and association of a novel microsatellite polymorphism with essential hypertension. *Circ Res* 1999; 84: 605–610.

212 Kokubo Y, Kamide K, Inamoto N *et al.* Identification of 108 SNPs in TSC, WNK1 and WNK4 and their association with hypertension in a japanese general population. *J Hum Genet* 2004; 49: 507–515.

213 Makino N, Sugano M, Ohtsuka S, Sawada S. Intravenous injection with antisense oligonucleotides against angiotensinogen decreases blood pressure in spontaneously hypertensive rats. *Hypertension* 1998; 31: 1166–1170.

214 Zhang YC, Bui JD, Shen L, Phillips MI. Antisense inhibition of beta-1-adrenergic receptor mRNA in a single dose produces a profound and prolonged reduction in high blood pressure in spontaneously hypertensive rats. *Circulation* 2000; 101: 682–688.

215 Suzuki S, Pilowsky P, Minson J *et al.* c-fos antisense in rostral ventral medulla reduces arterial blood pressure. *Am J Physiol* 1994; 266: R1418–R1422.

216 Wang MH, Zhang F, Marji J *et al.* CYP4A1 antisense oligonucleotide reduces mesenteric vascular reactivity and blood pressure in SHR. *Am J Physiol Regul Integr Comp Physiol* 2001; 280: R255–R261.

217 Garcia SI, Alvarez AL, Porto PI, Garfunkel VM, Finkielman S, Pirola CJ. Antisense inhibition of thyrotropin-releasing hormone reduces arterial blood pressure in spontaneously hypertensive rats. *Hypertension* 2001; 37: 365–70.

218 Suzuki S, Pilowsky P, Minson J, Arnolda L, Llewellyn-Smith I, Chalmers J. Antisense to thyrotropin releasing hormone receptor reduces arterial blood pressure in spontaneously hypertensive rats. *Circ Res* 1995; 77: 679–683.

219 Hayashi I, Majima M, Fujita T *et al. In vivo* transfer of antisense oligonucleotide against urinary kininase blunts deoxycortisterone active-salt hypertension in rats. *Br J Pharmacol* 2000; 131: 820–826.

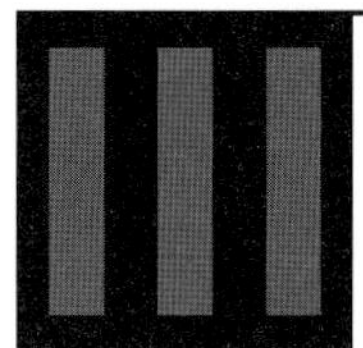

PART III

Therapies and applications

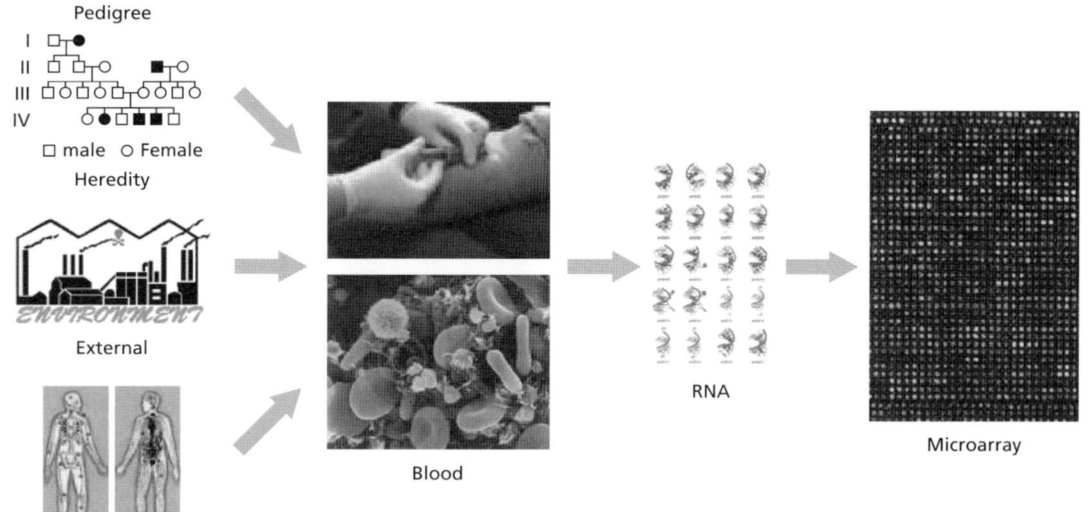

CHAPTER 9

Gene therapy for cardiovascular disease: inserting new genes, regulating the expression of native genes, and correcting genetic defects

Ion S. Jovin, MD *& Frank J. Giordano,* MD

Introduction

Genes are the blueprints that tell cells how to build specific proteins. Thus, the ultimate goal of all gene therapy is to either replace a defective or missing protein, or to alter the amount of one or more specific proteins produced by a cell. Despite this convergence of purpose, the manner in which genes accomplish this can be quite diverse and complex. When "gene therapy" was first conceived it was envisioned primarily as a treatment for hereditary diseases caused by a missing or defective gene [1–3]. The missing gene would be "replaced" with exogenous DNA encoding a copy of the normal gene. Since then, the definition has expanded considerably and now encompasses a formidable array of new technology, including novel approaches to turn on or off a patient's own genes, and methods to repair an individual's own defective DNA within a chromosome [4–6]. This chapter tours the various aspects of gene therapy and discuss how they may be applied to cardiovascular diseases. We begin, however, by laying a foundation of fundamental concepts relevant to gene therapy, including discussions of:

1 How therapeutic genes can be delivered into cells and tissues;

2 Factors that determine how long a therapeutic gene is expressed; and

3 Potential dangers of gene therapy.

Gene delivery

Methods to deliver genes to cells, tissues, and organs can be broadly categorized as either viral or nonviral [7–9]. Viral methods of gene delivery depend upon the use of viral *vectors*. These are naturally occurring viruses that have been altered such that they are "defective" and can no longer either cause disease or replicate themselves. This generally involves the deletion of genetic material from the virus that is required by the virus for its normal replication cycle. Deletion of this genetic material also makes space available within the *capsid* of the virus for the inclusion of therapeutic genes. The success of this approach requires that the virus retain the ability to enter cells and carry the genetic information it contains to the nucleus. Several viruses have been altered in this manner (see Fig. 9.1 for depiction of general scheme). The most widely used of these viral vectors include adenovirus, retrovirus, lentivirus and adeno-associated virus vectors [10–16]. Each of these has unique properties, advantages and limitations that are discussed below and summarized in Table 9.1.

Adenovirus vectors

Adenovirus vectors have been the most commonly used viral vectors for clinical trials of gene therapy for cardiovascular disease [17–19]. Adenoviruses

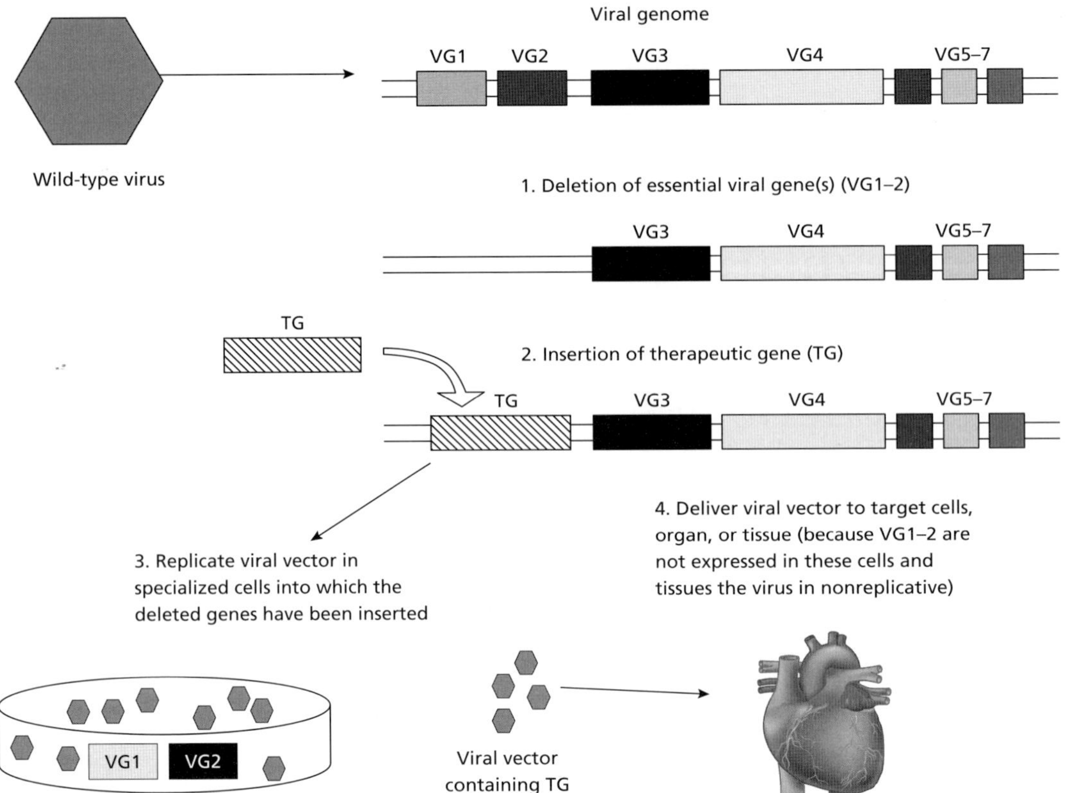

Figure 9.1 General scheme for the construction and production of viral gene therapy vectors. Although many different naturally occurring (wild-type) viruses have been used to make gene therapy vectors to carry therapeutic genes into cells and tissues, the general scheme for engineering these vectors is based on a common principle. Essential viral genes (VG) are deleted from the wild-type virus genome, rendering the virus nonreplicating and incapable of causing human disease (1). This also creates space within the viral genome to insert foreign genetic material, such as a therapeutic gene (TG; 2). Once the therapeutic gene is inserted the viral vector is replicated in specialized cells that express the essential wild-type viral genes that are required by the virus (3). After purification steps the viral vector is introduced into targeted cells, organs or tissues into which it carries the TG (4). These normal human cells do not contain the essential viral genes that are missing from the vector, thus the vector cannot replicate in these cells. The TG, however, is expressed in these cells.

are double-stranded DNA viruses that do not insert into the host cell genome and that can infect nondividing cells, such as cardiac myocytes. The family of adenoviridae contains over 40 serotypes, including adenoviruses 2 and 5 which are most commonly used to produce gene therapy vectors. In humans, adenoviral infection is associated with conjunctivitis and upper respiratory track infections. Wild-type adenovirus has also been associated with myocarditis. The "first generation" adenovirus vectors were based on deletion of the E1 gene from the virus [20]. This gene encodes proteins that are normally expressed early after the virus enters a cell, and without which normal replication cannot occur. In normal human cells these E1 deficient vectors cannot replicate. To produce adenovirus vectors for research and clinical use, they must be replicated in specialized cell lines that provide the missing E1 gene product.

The advantages of adenovirus vectors include that they can carry genes into most known cell types efficiently, they do not insert into the host DNA, they allow the insertion of relatively large therapeutic genes and are relatively easy to produce at titers high enough to allow clinical use [21,22]. Their disadvantages include that they can induce a signific-

Table 9.1 Viral vectors used in gene delivery.

Vector	Description	Associated human disease (wild-type virus)	Advantages	Disadvantages
Adenovirus	dsDNA virus; ~90 nm diameter	Upper respiratory infection; conjunctivitis	Relatively easy to produce in high titers; able to transduce most cell types; do not integrate into host genome	Inflammatory/immunogenic; early generation vectors in common use mediate only short term gene expression
Adeno-associated virus	ssDNA virus; ~25 nm diameter	No known human diseases for AAV-2	Relatively noninflammatory/immunogenic; small size facilitates egress from blood stream into tissues; mediates long-term gene expression without requirement for genomic insertion; current vector of choice for long-term gene expression in skeletal muscle and heart	More difficult to produce in high titer; require synthesis of second DNA strand to express transgene (delay in expression); small capsid with limited space for genetic material
Retrovirus	Enveloped dsRNA virus		Relatively noninflammatory/immunogenic; mediate long-term and heritable transgene expression because of genomic insertion	
Lentivirus	Enveloped dsRNA virus	Acquired immune deficiency syndrome (HIV)	Relatively noninflammatory/immunogenic; able to transduce nondividing cells; mediate long-term and heritable transgene expression because of genomic insertion	

ant inflammatory response and generally only allow for short-term expression of therapeutic genes (e.g., days to a few weeks) [23]. Some of these disadvantages may be overcome by the use of advanced generation adenovirus vectors, but these vectors have not generally been validated clinically. These include vectors with larger deletions of the viral DNA, the most extreme example being the "gutless" adenovirus vectors that retain only minimal viral genetic sequence [22,24]. Adenovirus vectors have been used clinically to carry angiogenic genes to heart muscle and skeletal muscle to stimulate the growth of new blood vessels [17,25–27]. For this application their ability to transduce nondividing cardiac muscle cells, and the relatively short-term gene expression they mediate has been advantageous.

Adeno-associated viruses

Adeno-associated viruses (AAV) belong to a subset of Parvoviruses called the Dependoviruses. They contain single-stranded DNA and are much smaller than adenoviruses (~95 nm vs. ~25 nm diameter for adenovirus and AAV, respectively). AAV-2, the serotype of AAV upon which the first AAV vectors were based, causes no known human disease. The life cycle of naturally occurring (wild-type) AAV depends on "helper" viruses to allow AAV to replicate. Adenovirus or herpes viruses can each provide this helper function. In the absence of helper viruses, AAV does not replicate and can integrate into the host cell genome, although the efficiency of integration is apparently low. Interestingly, wild-type AAV has a proclivity to integrate within a specific region of human chromosome 19. AAV-based vectors generally do not share this property, and it is generally held that they do not mediate efficient integration of therapeutic genes into the host cell DNA [28], although they can integrate more efficiently at sites of DNA strand breaks [29].

Because AAV is a single-stranded DNA virus it must form a second DNA strand in the cell in order to replicate or, in the case of gene therapy, to express a therapeutic gene. This process can be inefficient in some cells, and thus the full extent of AAV-mediated gene therapy often takes several weeks to manifest [30]. This issue may have been solved, however, with the recent development of AAV vectors that contain complementary DNA strands that can assemble into a double-stranded complex without the requirement for synthesis of a second strand [31]. The advantages of AAV vectors are that they transduce nondividing cells (e.g., cardiac muscle), are relatively noninflammatory, can mediate long-term expression of a therapeutic gene and are the smallest of the most widely used vectors (this may provide advantages for gene delivery) [15,32–34]. That AAV vectors are not efficient at integrating into the host cell genome is in some ways an additional advantage. The reason is because this decreases the possibility that during insertion the new DNA will cause a mutation in the host genome, a phenomenon known as mutational mutagenesis [28]. This is discussed in more detail in the description of retrovirus and lentivirus vectors. The disadvantages of AAV vectors include that their small viral capsid limits the size of the therapeutic gene that can be inserted into the vector, and that AAV is more difficult to make in high titer than adenovirus, although this latter limitation has been markedly improved [35]. Initially, AAV vectors were produced using adenovirus to provide the required adenoviral "helper" function, requiring removal of the contaminating adenovirus from the AAV vectors after production. This is no longer necessary as the essential adenovirus helper genes required for AAV replication are now known and can be supplied during AAV vector production without the need for helper virus. AAV vectors have been shown capable of long-term high efficiency gene delivery to the heart, and so have been the focus of efforts to develop gene-based therapies to treat heart failure [36,37].

Retrovirus vectors

Retrovirus vectors were one of the earliest vectors to be used for gene therapy [38]. Retroviruses are enveloped double-stranded RNA viruses that require a special enzyme, reverse transcriptase, to make complementary double-stranded DNA from the viral RNA template [39]. This double-stranded DNA is flanked by palindromic regions called long terminal repeats (LTRs) that, in conjunction with an integrase enzyme and other proteins that form a "pre-integration complex," facilitates insertion of this complementary double-strand DNA into the host cell genome [40]. Insertion is required for the expression of the therapeutic gene.

The advantages of retroviral vectors include that they are capable of carrying therapeutic genes into stem cells and that they mediate sustained gene transfer. Once inserted into the host cell genome, the therapeutic gene will be transferred to the daughter cells resulting from replication of the host cell. Thus, after retrovirous-mediated gene delivery to a hematopoietic stem cell, for example, all the progeny cells will contain the therapeutic gene over successive generations. This property of retrovirus vectors led to their use in treating children with X-linked severe combined immune deficiency syndrome (SCIDS-X). The corrective gene was inserted into bone marrow-derived stem cells with a retrovirus vector *ex-vivo*, and the stem cells were then given back to the children. The results of this therapy were initially one of the biggest successes in the field of gene therapy and these children were able to live normal lives outside of the sterile environments they required prior to gene therapy [41]. Tragically, several children latter developed an uncommon T-cell leukemia caused by the activation of a proto-oncogene by the insertion of the retrovirus [42,43]. This phenomenon of *insertional mutagenesis* remains a significant concern of all gene therapy vectors that integrate randomly into the host genome (Table 9.1). Thus, the ability to direct insertion of a therapeutic gene into the host genome is both an advantageous and disadvantageous property of retroviral vectors.

Other drawbacks of retroviral vectors include that they cannot transduce nondividing cells, and their ability to transduce many cell types is limited unless the vectors are manipulated to change the character of their envelope proteins (*pseudotyping*) [44,45]. Their use in cardiovascular medicine has been limited, although they have been used in clinical trials to replace defective low density lipoprotein (LDL) receptor genes in the liver of patients with familial hypercholesterolemia [46], and experimentally for delivering genes to the vessel wall to alter vascular remodeling [47].

Lentiviruses

Lentiviruses are enveloped double-stranded RNA viruses that, like retroviruses, require reverse transcription of the RNA to DNA, and then insertion of the double-stranded DNA into the host cell genome. Unlike retroviruses, however, lentivirus vectors can transduce nondividing cells and are thus capable of gene delivery to heart muscle [11,12,45,48]. Given that these vectors are based primarily on human, simian or feline immunodeficiency viruses, there has been concern that recombination could produce wild-type HIV. There is also concern that interaction of an HIV-based lentiviral vector and wild-type HIV in a cell could result in a hybrid virus. Finally, lentivirus vectors, like retrovirus vectors, mediate insertion into the host genome and thus engender the same concerns for insertional mutagenesis. These cumulative concerns have led to the development of newer generations of retrovirus and lentivirus vector systems with modifications to make them safer, including self-inactivating vectors, and vectors that do not integrate into the host genome [49,50]. In fact, a lentivirus vector carrying an antisense gene against HIV is now in clinical trial in the USA [51].

Nonviral gene delivery

Nonviral gene delivery includes a number of approaches, most frequently centered on plasmid DNA [9,52–55]. Plasmids are circularized double-stranded DNA that are best known for carrying antibiotic resistance genes into bacteria. For gene therapy an *expression cassette* encoding the therapeutic gene is inserted into the plasmid, and a large number of copies of this plasmid are made within bacteria. After isolation from the bacteria and purification to remove endotoxin and other contaminating bacterial products the plasmid is ready for delivery to human cells. Some cells, such as cardiac and skeletal myocytes, are able to efficiently take up plasmid DNA that is simply injected, in solution, into the heart or skeletal muscle bed. This approach has been used frequently in clinical trials to induce therapeutic blood vessel growth in the heart or ischemic limbs [55,56]. Other cell types require additional measures to facilitate plasmid

entry into the cell. These include combining the plasmid DNA with lipid formulations (*lipofection*), polymers (e.g., *hydrogels*) or precipitates (e.g., calcium phosphate-DNA co-precipitation) to facilitate entry through the cell membrane [9,57,58]. This type of DNA delivery to cells is designated *transfection*, as opposed to viral *transduction* of cells.

Other methods of nonviral gene delivery include receptor-mediated endocytosis and a variety of "mechanical" methods. *Receptor-mediated* gene delivery is based upon linking a natural ligand for a specific cell membrane receptor to DNA encoding a therapeutic gene [59,60]. Binding of the ligand to the receptor results in endocytosis of the receptor–ligand complex, along with the DNA that is attached to the ligand. An example is the use of asialoglycoproteins to target gene delivery to the liver [61]. Mechanical methods include electroporation, ultrasound-mediated gene delivery and methods based upon altered pressure gradients.

Electroporation is the application of a brief electrical impulse to cells (or tissues) to disrupt the cell membrane long enough to allow the entry of new genetic material. It can be performed on cells *in vitro* (e.g., stem cells) and on tissues *in vivo* [62]. It is a standard method used to deliver genetic material to stem cells, and has been considered for the delivery of plasmid DNA to skeletal muscle to facilitate therapeutic angiogenesis [63].

Ultrasound-mediated gene delivery is generally performed in conjunction with lipid "microbubbles" that can be loaded with, or attached to, the genetic material [64,65]. These microbubbles are injected into the general circulation and then destroyed selectively by ultrasound energy applied directly at the desired site, thus depositing the genetic material there.

Methods based on altered pressure gradients are exemplified by *retrograde coronary venous delivery* (RCVD), and by a specific method developed to deliver oligonucleotides to segments of veins used for coronary artery bypass surgery [66]. RCVD is predicated on the principle that injecting genes or drugs retrograde under pressure into the coronary venous system facilitates the bulk movement of these agents out of the circulation and into the myocardium via the postcapillary venules [67]. It is easier to achieve an effective pressure gradient across the venules by retrograde perfusion because in this manner the resistance vessels (i.e., arterioles) are bypassed.

Gene delivery considerations specific to the cardiovascular system

There are a wide variety of potential cardiovascular indications for gene therapy. These include treatment of heart failure, ischemic heart and peripheral vascular disease, dyslipidemias, inherited or acquired genetic diseases such as glycogen or lipid storage disorders, atherosclerosis, pulmonary hypertension and many others. For each of these there are unique considerations for how best to deliver the therapeutic gene, how many cells need to receive the gene and how long it needs to be expressed. A fundamental consideration is whether the protein encoded by the therapeutic gene is secreted or retained within the cell in which it was produced. Secreted proteins produced in a relatively small number of cells can have significant biologic effects by diffusing into surrounding tissues or by entering the systemic circulation. Gene therapy involving a gene that encodes a secreted protein can thus be effective even if a relatively small number of cells receive the therapeutic gene. An example is angiogenic gene therapy in which the therapeutic gene encodes a secreted protein that promotes the growth of new blood vessels. Conversely, for genes that encode nonsecreted proteins that function within the cell in which they are produced, a relatively large percentage of cells must receive the therapeutic gene in order to mediate a meaningful clinical effect (e.g., genes encoding calcium regulatory proteins such as SERCA2 which influence cardiac contractility). This section addresses the various routes by which genes can be delivered to the heart, skeletal muscle and the vasculature.

Gene delivery to the heart

Gene delivery to the heart can be accomplished either by the direct intramyocardial injection of plasmid DNA or viral vectors, or by an intravascular route [19,21,68]. The intramyocardial (IM) route is efficient and generally reproducible, but gene delivery is limited to the area into which the DNA or viral vector is administered. Injection volumes vary, but generally are in the range of 100–300 μL per site, with a few milliliters maximum injected into the entire heart. The area of gene delivery achieved by this method is limited largely by the relatively small amount of myocardium that can be infiltrated by this volume. This method, therefore, delivers genes to a small percentage of the total cells in the heart. This limited number of cells may be sufficient, however, to produce a therapeutic effect when the gene product is a secreted factor that can diffuse to adjacent areas of the myocardium. This has been the reasoning behind using IM gene delivery to stimulate angiogenesis in the heart.

IM gene delivery can be accomplished directly with a needle and syringe during open chest surgery with direct visualization of the myocardium, or via a thoroscopic approach [69–71]. IM injection can be accomplished percutaneously by specialized catheters that cross the aortic valve retrograde from the aorta and allow endocardial/myocardial injection of plasmid DNA, or viral vectors, via a needle at the end of the catheter [69]. Examples include the Stiletto catheter (no specialized guidance system; Boston Scientific, Inc.) [72], and the NOGA system (guided by electromagnetic mapping; Biosense Webster of J&J, Inc.) [56,73]. Further enhancements of the IM approach include the use of polymers or specific carrier peptides to augment delivery, and specialized needles to promote infiltration of a larger myocardial area.

The *intravascular route* for gene delivery to the heart is *theoretically* the most effective route to introduce a therapeutic gene to the largest number of cardiac muscle cells. This is because there is generally at least one capillary per cardiac myocyte. The caveat to this is that the endothelium of the coronary microvasculature is a continuous endothelium that rivals the cerebral microcirculation in its barrier function. Macromolecules, like viral vectors, have limited ability to traverse coronary capillaries and venules to gain access to the myocardium [74,75]. This issue became particularly relevant in recent clinical efforts to stimulate therapeutic blood vessel growth in the hearts of patients with severe ischemic heart disease. These clinical trials were based on an experimental study of intracoronary (IC) infusion of adenovirus in a pig model of myocardial ischemia. This study demonstrated that

IC delivery of adenovirus encoding the angiogenic gene FGF-5 relieved ischemia by stimulating new blood vessel growth (therapeutic angiogenesis) [21]. On the basis of these data and confirmatory animal studies Phase I and II clinical trials were performed in patients with advanced ischemic heart disease [17,26]. Although the approach appeared safe and there was evidence of a biologic signal, the therapeutic results in patients were significantly less robust than hoped. In addition, there was an indication of systemic distribution of adenovirus after IC injection, as evidenced by episodes of elevated liver transaminases after viral vector delivery to the coronary circulation. While it is not clear that inefficient gene delivery to the human heart was responsible for the disappointing clinical results, it is now widely accepted that simple IC administration of viral vectors is an inefficient method of gene delivery to the myocardium.

Nonetheless, IC administration remains the most promising approach to achieve widespread therapeutic gene delivery to the heart. Realization of the full potential of intravascular gene delivery to the heart and other target organs, however, requires the development of clinically feasible methods to overcome the endothelial barrier function of the microvasculature. Several diverse approaches have been pursued to accomplish this, including mechanical, pharmacologic and molecular methods.

Mechanical methods include RCVD in which a catheter is inserted into the coronary sinus of the heart, a balloon inflated to block venous outflow from the coronary sinus, and then a solution containing either viral vector or plasmid DNA is injected under pressure retrograde into the coronary venous system. The theory behind this approach, as discussed briefly above, is that retrograde delivery in this manner bypasses the coronary resistance vessels (i.e., the arterioles) and thus efficiently increases the driving pressure across the coronary venules and capillaries, forcing viral vector or DNA out of the vasculature and into the myocardium by bulk flow [67,76].

Other mechanical methods include the slowing or cessation of coronary flow to increase the dwell-time of the viral vector in the coronary circulation, or recirculating the vector through the coronary circulation to increase the time of exposure of the vasculature to the vector [75,77]. When combined with pharmacologic agents that increase vascular permeability (e.g., histamine) these methods can be very effective, achieving gene delivery to >80% of the cardiac myocytes in the heart [75,78]. Interestingly, in the setting of acute myocardial infarction (MI) the coronary microvasculature serving the peri-infarction zone becomes quite hyperpermeable to macromolecules [79]. Viral vectors can thus be delivered efficiently to the peri-infarction myocardium by IC infusion into the infarct-related coronary artery in this setting. This fact may facilitate post-MI gene therapy by IC administration, and also illustrates how important altering vascular permeability can be in achieving efficient gene delivery to the heart by the IC route.

Molecular approaches to increase gene delivery to the heart are also under development. Many of these approaches are predicated on the fact that certain macromolecules can efficiently traverse the microvascular endothelium, in some cases by using specific transport pathways. It has recently been discovered that certain serotypes of AAV (e.g., AAV-6, AAV-8 and AAV-9) have a greatly enhanced ability to cross endothelial barriers and mediate gene delivery to the heart and skeletal muscle [36,80,81]. It is assumed that this property is a result of unique configurations of the capsid proteins of these AAV serotypes that direct them to cross biologic barriers via specific molecular transport pathways. This principle is also driving efforts to engineer capsid mutants in AAV, adenovirus and other vectors that will facilitate transendothelial passage of these vectors and allow them to carry genes efficiently from the blood stream into targeted organs and tissues [16]. Similarly, this principle is being applied in attempts to develop nonviral vectors that have the capacity to transport out of the blood stream and target specific tissues. The ultimate application of this approach would be the ability to inject viral or nonviral vectors into the general circulation after which they would "home" to specific organs and tissues and deliver their genetic cargoes there exclusively.

Gene delivery to the vasculature

Gene delivery to the vasculature can be directed to many clinical indications, including the regression or prevention of atherosclerosis, the inhibition of vein graft disease, prevention of post intervention restenosis, prevention of transplant vasculopathy and inhibition of vascular inflammation. A variety of catheter-based methods for delivering genes to blood vessels have been devised. Only a few representative catheter systems are discussed here.

The simplest catheter system uses two occlusive balloons, between which are delivery ports through which viral vectors or DNA–lipid complexes can be administered [82,83]. The space between the two balloons is sequestered from blood flow, thus increasing the duration of contact between the gene delivery vectors and the vessel wall. As one can imagine, branching vessels create problems for this catheter, and the ability of genes to penetrate the vessel wall and reach the media by this method is limited.

In an attempt to further increase the delivery of therapeutic genes into the vessel wall, alternative catheters were developed that used either porous balloons or polymer (e.g., hydrogel) coated balloons that decreased the diffusion area from the catheter to the vessel wall and prolonged contact. The hydrogel balloon catheter (Boston Scientific, Inc.) is one example, and was used for a short period of time to deliver plasmid DNA encoding VEGF to the wall of arteries to stimulate therapeutic angiogenesis [84,85].

Another type of catheter uses either high-pressure fluid jets or fine needles to inject genes deeper into the vessel wall. There have been many iterations of this type of device [86–89], although the safety and efficacy of this approach in atherosclerotic and/or calcified vessel segments remains unclear.

One of the newest approaches is the use of transvenous needle catheters that obviate the issue of atherosclerosis and calcification. These catheters are directed into the coronary venous system via the coronary sinus, and then a needle is deployed through a thin-walled coronary vein, thus allowing peri-adventitial gene delivery to coronary arteries [90]. None of these catheter systems has yet achieved significant clinical use.

A relatively straightforward gene delivery system predicated on inducing a pressure gradient within the vessel lumen was used successfully to delivery decoy oligonucleotides to human vein segments, *ex vivo*, which were then used for coronary by pass surgery [91]. This clinical trial, and the decoy oligonucleotide therapeutic approach, is discussed later in this chapter.

Gene delivery to skeletal muscle

Gene delivery to skeletal muscle has been used in cardiovascular medicine primarily to stimulate therapeutic angiogenesis in patients with severe peripheral vascular disease. This has been performed primarily by direct intramuscular injection of plasmid DNA, although adenoviral vectors have also been used [27,92–94]. Variations of this approach include the use of specific polymers, electroporation, ultrasound energy, lipid formulations and specialized needle designs to augment gene delivery [93,95–98]. More global gene delivery to skeletal muscle has been achieved with specific serotypes of AAV vectors, and the use of these vectors is being pursued to treat muscular dystrophy [36,81,99]. Recently, there has also been experimentation in the use of artificial vascular matrices generated by culturing vascular cells or progenitor cells in collagen gels. The cells orient themselves into a vascular network, and when implanted under the skin these engineered vascular networks connect to the native circulation [100,101]. By genetically engineering these cells to secrete specific factors into the blood stream they can be used as finite molecular factories. There is ongoing research investigating the ability of these gels to increase skeletal muscle vascularization in ischemic limbs.

Ex vivo gene delivery

Finally, mention should be made of *ex vivo* gene delivery to stem cells and other cell types, with subsequent reintroduction of these cells into the circulation or into a specific organ or tissue. This has been largely an experimental approach, and has yet to make it to the clinic in the realm of cardiovascular disease, but it does have significant appeal for specific conditions. For example, endothelial progenitor cells (EPC) are bone marrow-derived cells found in the peripheral blood which are thought to home to areas of angiogenesis and participate in the formation of new blood vessels [102]. Introducing an angiogenic gene into EPC and then reinfusing them might lead to the delivery of the angiogenic gene to the area of active angiogenesis, augmenting the process [103]. There are innumerable settings such as this in which gene delivery to stem cells or other cell types could be part of a viable therapeutic strategy [104], including strategies for gene correction [6], discussed below. Current concerns and limitations to this approach include that some "stem cells" are more difficult to deliver genes to than other cell types, and that genetic manipulation of stem cells carries significant safety concerns, given the pluripotency of these cells.

Determinants of the duration of therapeutic gene expression

For some gene therapy applications it is desirable to maintain the expression of a therapeutic gene indefinitely. Examples include the replacement of a defective LDL receptor gene to ensure an enduring effect on lipid metabolism, or replacement gene therapy for patients with hemophilia. For other gene therapy applications, such as the stimulation of blood vessel growth (*therapeutic angiogenesis*), it may be desirable to express the therapeutic gene for a shorter period of time to avoid deleterious effects (e.g., formation of hemangiomas).

Several factors contribute to how long a therapeutic gene will be expressed in a host cell:

1 Whether or not the transgene is inserted into the host cell genome;

2 The susceptibility of the transgene DNA to degradation by cellular enzymes;

3 The propensity of the gene delivery vector to elicit an immune response;

4 The propensity for the transgene promoter (activating regulatory sequences) to undergo "silencing" (e.g., via "chemical alterations" of DNA, such as DNA methylation); and

5 The immunogenicity of the transgene product.

For long-term "stable" gene expression that will be transferred to all the progeny of the host cell that received the therapeutic gene, the best approach is to use a gene delivery method that results in insertion into the host genome. Retroviral and lentiviral vectors are well-established gene delivery vehicles for achieving genomic insertion, albeit with the potential risk of "insertional mutagenesis," discussed above and in the section below on potential dangers associated with gene therapy.

Mammalian cells, and indeed prokaryotic cells, have evolved mechanisms to eradicate foreign DNA, including cellular enzymes that degrade DNA not inserted into the host genome. These represent a significant barrier against the sustained presence of therapeutic genes that do not undergo genomic insertion [105]. AAV vectors overcome this limitation, in part, via the effects of their inverted terminal repeats (ITRs). ITRs are palindromic sequences located at each end of the linear DNA of AAV viruses. They are retained in AAV gene therapy vectors, and flank the therapeutic genes that have been inserted into these AAV vectors. Although not definitively understood, it is thought that the ITRs protect the therapeutic DNA from degradation in the host cell, possibly via the formation of circular episomal DNA [105,106]. Whatever the mechanism, AAV-mediated gene transfer into heart and skeletal muscle can mediate long-term expression of therapeutic genes, and thus AAV vectors have generated significant interest as vehicles for gene therapy of heart failure [107,108]. It should be noted that without integration, AAV-mediated gene therapy does not ensure that the therapeutic gene will be transferred to progeny cells produced by cell division. Although this limits the usefulness of AAV gene therapy for stem cells, it is not a problem for nondividing (or infrequently dividing) cells such as heart and skeletal muscle.

AAV vectors also have the advantage of being small (~25 nm in diameter) and relatively non-immunogenic. Larger nonenveloped viruses, such as adenoviruses (~90 nm diameter), do elicit an immune response, and this has led to issues with inflammation, limited duration of therapeutic gene expression and potentially problems with redosing (i.e., repeat administration of these vectors) [109]. The safety issues engendered by this issue are discussed further below. This dichotomy in immune responses to adenovirus and AAV correlates with the duration of therapeutic gene expression that they mediate, and thus the applications to which they are best suited. Adenovirus vectors mediate limited duration of gene expression (for first generation adenovirus vectors this is on the order of 1–3 weeks), and thus are favored for applications such as gene therapy to stimulate angiogenesis. A separate, but related issue is the immune response, if any, to the protein product of the therapeutic gene. This is a significant consideration when gene therapy is being used to express a protein that is not present at all at baseline, or is significantly mutated from its normal form. In this setting, tolerance to the normal protein has not been established during development, and thus the body sees the normal protein as foreign. It is not yet known how significant a hurdle this issue will be to successful gene therapy, and it is likely that its importance will vary for different genes.

The expression of therapeutic genes is dependent in part on the regulatory sequences (e.g., promoter elements) that activate them. The most common promoter used for gene therapy applications has been the cytomegalovirus (CMV) promoter. There are reports that this promoter can become "silenced" by DNA methylation [110,111]. If this happens, gene expression will cease irrespective of the presence of the therapeutic gene. It is possible that similar events, perhaps directed to the therapeutic gene itself, or other regulatory sequences, have a significant role in defining the longevity of therapeutic gene expression in certain situations. All of these issues are considerations when deciding which gene delivery method, or vector, to use. Table 9.1 delineates the general longevity of gene expression, and the inflammation-inducing properties of the most widely used gene delivery vectors.

Gene therapy by controlling a patient's own genes

As gene therapy has moved into areas such as therapeutic angiogenesis in which the biology is complex and the optimal gene or genes to use is not clear, the possibility of modeling complex biology by regulating the expression of a patient's own genes has generated significant interest. Efforts to develop this approach have focused primarily on regulating the transcription of genes from DNA to mRNA. We discuss two related methods that are used to achieve this transcriptional control:

1 The use of known naturally occurring transcription factors; and
2 The *de novo* engineering of specifically targeted transcription factors.

To understand these approaches most clearly, it helps to consider what transcription factors actually are, and how they regulate gene expression. Although the instruction set for making all of the proteins required by a particular type of cell is con-

tained within that cell's DNA, how those instructions are "interpreted" is a crucial determinant of what a cell will be. Bear in mind that all cells have the same instruction set, yet some become rod cells in the retina, some become neurons and others become heart muscle. In each of these cell types some genes are active and some are silent. The manner in which this occurs is complex and remains incompletely understood. However, it is known that the configuration of the protein coat of chromosomal DNA, the *chromatin* structure, is a critical control point [112–115]. If the protein coat around a particular gene is not tightly packed (the "open" chromatin configuration), the gene is accessible to the "transcription machinery" and more easily expressed. If tightly packed in a "closed" chromatin configuration, gene expression is conversely repressed. The chromatin configuration of a neuron is distinct from that of a cardiac myocyte, and this contributes significantly in determining the differences in gene expression that distinguish these two cell types.

Transcription factors work, in part, by binding to DNA in the regulatory region of a particular gene and directing the reconfiguration of the chromatin at that location [116–118]. "Activating" transcription factors promote a looser chromatin structure, thus allowing access of RNA polymerase and the rest of the transcriptional machinery to the gene. They may also participate in the recruitment of other co-factors that support transcription. "Repressor" transcription factors, and the co-factors they recruit, induce a more tightly packed chromatin and thus decrease the transcription of a particular gene. Transcription factors generally contain:

1 A DNA binding domain that binds to a specific DNA sequence; and
2 A functional domain that directs chromatin reconfiguration, recruits transcription co-factors, or both.

The biology of transcription factors can be considerably more complex than this, but for purposes of understanding how they can be used in gene therapy this modular concept is helpful. It is the DNA sequence to which it is targeted that determines what gene or genes a particular transcription factor regulates, and the character of its functional domain that determines the type of regulation. This

(a)

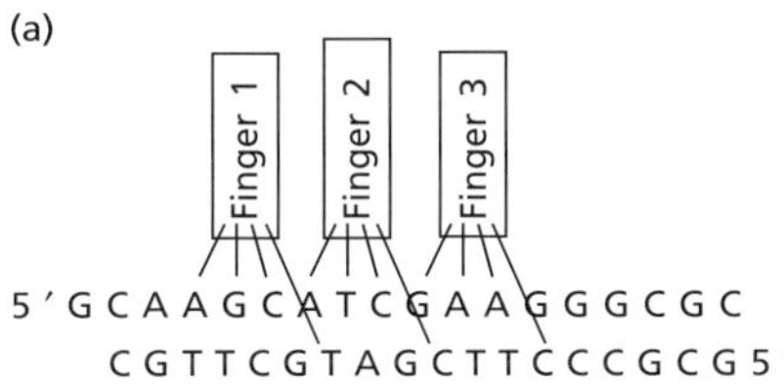

Each zinc finger binds three bases on the sense strand and interacts with a fourth base on the antisense strand. By combining multiple fingers longer DNA sequences can be targeted, potentially increasing specificity

(b)

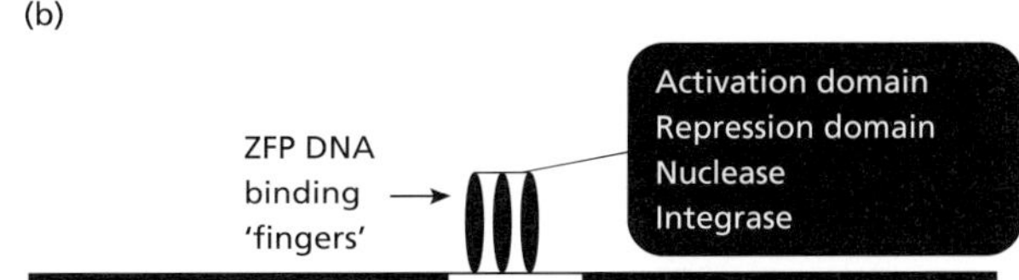

By fusing the DNA-binding zinc fingers to a functional domain, ZFPs can be engineered to activate or repress the expression of specific genes, direct integration into specific genomic sites, and to create specifically targeted cuts in the DNA to facilitate the repair of defective genes

1. Activating the transcription of a specific endogenous gene results in the expression of all the natural splice variants of that particular gene

2. ZFPs can be engineered to repress the transcription of a chosen gene and turn off its expression with single-gene specificity

3. When fused to an endonuclease domain ZFPs can be designed to cut DNA in high specific sites, faciliating the repair of defective genes

4. The ability to engineer ZFPs to bind specific genomic DNA sites with highly-specificity allows the potential use of engineered ZFPs to direct gene insertion into safe sites in the genome

Figure 9.2 The modular structure of zinc finger proteins can be exploited to engineer highly specific transcription factors or endonucleases. Transcription factors generally contain a DNA-binding domain that defines what gene(s) the factor will regulate, and a "functional" domain that defines what the nature of that regulation will be. Each "finger" of a zinc finger protein (ZFP) transcription factor binds a specific 3 base pair (bp) sequence on the sense strand of the targeted gene (a). By combining multiple fingers together longer and longer DNA sequences can be specifically targeted (e.g., a six finger ZFP will target an 18-bp sequence). By genetically fusing this DNA-binding domain to a specific functional domain (b), the ZFP fingers can be used to either turn on (activate) gene expression, or turn off (repress) gene expression. This same DNA-binding domain can be used to direct other functions such as DNA cutting (endonuclease function that can be used to repair defective genes) or site-specific insertion of genetic material (integrase function).

is illustrated in Fig. 9.2 in the context of zinc finger proteins.

One reason transcriptional control of endogenous genes has generated significant interest for gene therapy is that often a particular transcription factor controls a number of distinct genes that are all involved in one way or another in the same biologic process. An example of this is the hypoxia-inducible basic helix-loop-helix transcription factor HIF-1α. HIF-1α is an interesting transcription factor which regulates the expression of a broad repertoire of genes involved in angiogenesis, metabolism, hematopoiesis, apoptosis, inflammation and several other processes [119–122]. Levels of HIF-1α rise as tissue oxygen tension decreases, thus HIF-1α levels are elevated during ischemia. The fact that it is an "upstream" regulator of several known angiogenic genes and that it may control the expression of still other unknown angiogenesis-associated genes, has led to clinical testing of HIF-1α in patients with ischemic heart and peripheral vascular disease [123]. In this approach, the gene encoding HIF-1α is contained within an adenovirus vector, and constitutive expression of HIF-1α, independent of tissue oxygen levels, leads to the expression of multiple genes that are normally activated during hypoxia. It is hoped that this approach will recapitulate natural biology more effectively than standard gene therapy that leads to the expression of a single biologically active protein. Whether this "upstream" approach will prove superior to standard cDNA-based gene therapy remains unclear at this point.

The HIF-1α approach is an example of using a naturally occurring transcription factor to control a native gene or genes. A potential drawback of using endogenous transcription factors in this manner is that the number of genes they control can be quite high, thus raising the possibility of undesired biologic effects. HIF-1α, for example, can alter cellular metabolism and promote erythropoiesis, in addition to stimulating blood vessel growth [4,119–121, 124]. These effects of HIF-1α may actually be beneficial, but they are nonetheless somewhat "collateral" to the targeted biologic process of angiogenesis. One reason natural transcription factors regulate a wide repertoire of genes is because they target relatively short DNA sequences that are shared by multiple genes. This may be the result of coordinated co-evolution of these genes and their regulatory factors, and the ability of a single transcription factor to regulate the expression of multiple genes is biologically more efficient than having separate factors for each of tens of thousands individual genes. Although biologically more efficient, this non-specificity of native transcription factors can be a significant limitation for their use as therapeutics.

Zinc finger proteins

A novel way around this limitation is to engineer unique transcription factors that bind with high affinity and specificity to a single chosen gene, or a few biologically related genes [4,125–131]. The most straightforward way to accomplish this is to use a naturally occurring DNA binding protein as a starting point. However, this requires an intricate knowledge of how the structure of that protein directs sequence-specific DNA binding. Of all the known transcription factor classes that could be used for this purpose, it is the zinc finger protein backbone that has proven the most useful for retargeting DNA binding and creating novel engineered transcription factors [126,129,132–135]. Zinc finger proteins (ZFPs) are among the most highly represented protein motifs encoded in the entire mammalian genome. Further, the structure of ZFPs has been solved, and the relationship between ZFP protein configuration and DNA binding specificity is now understood well enough to allow the *de novo* design and engineering of ZFPs that will bind virtually any chosen DNA sequence with high affinity and specificity [4,129,135].

Each "finger" of a ZFP binds a 3 base pair (bp) DNA sequence in the major groove of DNA. This property of ZFPs, with certain limitations, allows for a modular design such that multiple fingers can be combined to bind specific DNA sequences of variable lengths (i.e., a three-finger ZFP binds a 9 bp DNA sequence, a six-finger ZFP an 18 bp sequence). Further, these ZFP fingers can then be fused to specific "functional" protein domains to direct a variety of biologic processes. To create, for example, a novel transcription factor that activates expression of the native erythropoietin gene, one would engineer a ZFP DNA-binding domain that targeted the regulatory region of the erythropoietin gene, and then link this to a transcription activation domain (TAD), such as the viral TAD VP16. To turn off erythropoietin expression one would use a repression domain, such as the KRAB domain of the naturally occurring transcription repressor KOX1, as the functional domain. In fact, the ability to target ZFP DNA binding very specifically allows the creation of multiple types of new therapeutic DNA-binding proteins. The ZFP motif has thus been exploited successfully to create ZFPs that activate specific native genes, repress specific native genes and to target specific DNA loci for the correction of deleterious gene mutations [4,6,136,137]. This is shown schematically in Figs 9.2 and 9.3, and discussed in further detail in the sections below on splicing, gene correction, angiogenesis, heart failure and targeted transcriptional repression.

The importance of splicing

The principle of one gene – one protein is not accurate. A single gene can encode many different proteins, either by using alternative transcriptional start-sites or, more commonly, by alternative splicing. As discussed elsewhere in this book, most genes contain several exons (the segment of the gene that encodes protein) and introns (noncoding). Pre-messenger RNA, the unprocessed RNA produced initially by transcription, undergoes a process of splicing in which introns are looped out and the exons are arranged next to one another. Depending upon what exons are including in this splicing process, a significant number of highly related but biologically distinct proteins can be produced by a single gene. The crucial biologic importance of these diverse splice variants is only recently becoming fully appreciated [4,138], and it is now recognized

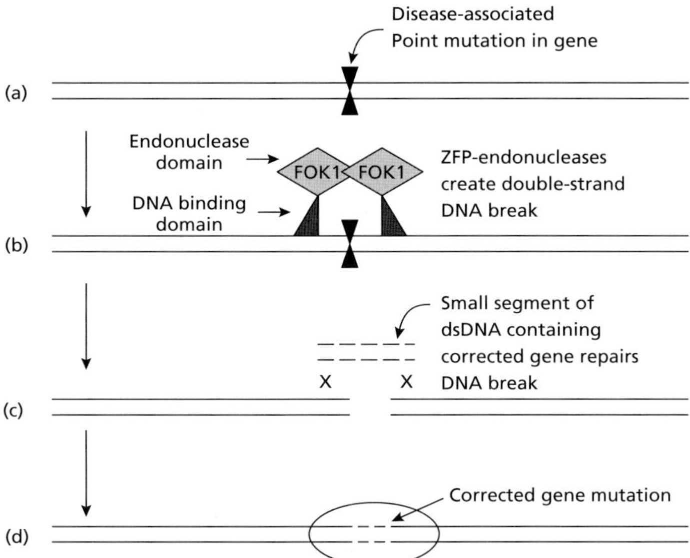

Figure 9.3 Correcting gene defects with engineered endonucleases. Many cardiovascular diseases are caused by or associated with gene mutations. The ability to engineer DNA-binding proteins that can bind with high specificity to targeted DNA sequences allows the creation of enzymes that will cut DNA at only the point specified. Two separate DNA-binding proteins are targeted to separate specific DNA sequences around a point mutation in a defective gene (a and b). These DNA-binding proteins are fused to the FOK1 endonuclease domain to create engineered site-specific DNA-cutting enzymes. FOK1 requires dimerization with a second FOK1 domain to become functional. Thus, DNA cutting will not occur unless both engineered endonucleases bind to their closely separated and unique targeted sites (b). Creation of a site-specific double-strand break in the DNA facilitates recruitment of the cellular DNA repair machinery (b,c). A small double-strand DNA segment encoding the normal (corrected) gene sequence is provided during DNA repair, and the repair process results in incorporation of this corrected gene sequence (c,d).

that defective gene splicing causes several specific human diseases [139,140].

From a gene therapy perspective, splice variants have taken on increasing importance. Gene therapy, until recently, primarily involved the delivery of a single cDNA construct encoding a single splice variant of a particular gene. For example, the gene encoding vascular endothelial growth factor A (VEGF) can produce at least four major VEGF proteins via alterative splicing [138]. Two of these, VEGF121 and VEGF165, were used separately in distinct clinical trials to produce therapeutic angiogenesis in the heart [18,69]. The reasoning for using only one of the major four splice variants included the assumption that each of these splice variants could produce near equivalent results. Several lines of evidence challenge this assumption and strongly suggest that the angiogenic process mediated by the combination of all the naturally occurring VEGF splice variants is distinct and potentially superior to

that produced by a single splice variant. To produce all the naturally occurring splice variants in their optimal stoichiometry, however, requires the delivery of a relatively large sequence of DNA that contains the genetic information for all the splice variants, or the transcriptional activation of the endogenous gene in the target cell genome.

The two most feasible current methods for activating the expression of a patient's own genes, and thus inducing the expression of all that gene's splice variants, are the use of naturally occurring transcription factors, or the engineering of gene specific transcription factors *de novo* using the ZFP backbone as discussed above. Each of these approaches can effectively activate expression of a targeted endogenous gene but, as also discussed, using naturally occurring transcription factors can lead to the activation of a significant number of collateral genes, raising concerns of off-target effects. In the case of VEGF-A, both the ZFP and the natural transcription factor (HIF-1α in this case) approaches have been used experimentally and clinically to activate the endogenous gene and induce expression of all the natural VEGF-A splice variants [4,123,125,141,241,242]. Most interestingly, the experimental studies indicated that activation of the endogenous VEGF-A gene led to a biologically distinct angiogenic response in which the new blood vessels that formed appeared to be more physiologically mature [4]. Whether this approach will prove superior in ongoing clinical studies awaits the completion of these clinical assessments.

Turning genes 'off'

In some circumstances reducing the expression of a particular gene, or its protein product, can be therapeutic. Examples include genes involved in deleterious inflammation, apoptosis, heart failure and in malignancies genes that promote angiogenesis. There are several methods available to accomplish this, including the use of repressor transcription factors, siRNA, antisense, ribozymes, transcription decoys and targeted genomic disruption. Each of these approaches are described briefly here.

siRNA, antisense and ribozymes

The term siRNA refers to small inferring RNA molecules; double-stranded RNA sequences that are ~20–25 bp in length. There is a natural mechanism in mammalian cells wherein double-stranded RNA is recognized as foreign and is targeted for destruction by the cell. This is termed RNA interference (RNAi) and also results in degradation of single-stranded RNA that has the same sequence as the double-stranded RNA. siRNA, although predicated on much smaller double-stranded RNA sequences, induces the same process. Using the siRNA approach therapeutically, siRNA specific for the mRNA of a targeted gene is introduced into a cell and induces the cell to degrade the single-stranded mRNA of that particular gene, effectively reducing the expression of that gene [142–145]. Antisense is a similar approach that generally uses longer single-stranded nucleic acid sequences designed to hybridize with specific mRNAs in the cell and prevent the translation of these mRNAs into protein. This can be accomplished with either single-stranded DNA or RNA. The DNA is introduced into cells as single-stranded DNA oligonucleotides (relatively short DNA sequences), and antisense RNA is introduced as the transcriptional product of an antisense gene (this can be simply the reversed cDNA for a particular gene that then encodes a full-length antisense RNA strand) [146–148]. Ribozymes are catalytic RNA-based molecules that have the capacity to hybridize to specific mRNAs and cut them, thus preventing the translational expression of the gene from which these mRNAs were transcribed [149].

The net effect of all the above methodologies is to reduce the amount of a specific protein a cell makes, without turning off the transcription of the gene encoding that protein. The newest and potentially most powerful of these approaches is currently thought to be the siRNA approach. However, questions remain regarding the specificity of these approaches and safety concerns regarding potential effects on off-target genes. Of note, all of the above modalities can be delivered to cells by viral vectors, by plasmid DNA or in some cases as oligonucleotides.

Targeted transcriptional repression and targeted genomic disruption

One of the primary mechanisms used by mammalian cells to limit the expression of specific genes is transcriptional repression mediated by specific DNA-binding factors [5,150–152]. Most naturally occurring repressor transcription factors have effects on multiple genes, and therefore have limited use as therapeutic agents. The ability to engineer highly specific ZFPs has led to the ability to make, in turn, highly specific repressor transcription factors that work with essentially single gene specificity [5,150,151]. Currently, this approach is being used to develop a repressor ZFP that turns off a major calcium regulatory gene in cardiac myocytes and thus increases their contractility. The specificity of this approach makes it applicable to a wide variety of cardiovascular targets, and this approach represents a strong rival alternative to siRNA technology.

Using a similar DNA-targeting approach, ZFPs can be engineered to bind a specific gene, create a double-strand break at that point and induce a frameshift that effectively silences the gene. This type of approach is heritable and would thus be passed on to all the progeny cells of the cell harboring the altered gene. If applied to stem cells, for instance, the silencing of a particular disease-associated gene would be maintained in each subsequent generation of cells produced. This type of approach can also be used for gene correction, and is discussed further below in that context.

Decoys

Another novel method to decrease gene expression is by using decoy oligonucleotides [153–156]. An example of this approach is the use of oligonucleotide decoys that contain DNA-binding sites for specific essential transcription factors. Naturally occurring transcription factors recognize and bind specific DNA sequences that occur in the regulatory regions of particular genes. When they bind genes at these specific sites they act in concert with co-factors and the general cellular transcription machinery to turn on (or repress) the expression of these specific genes. Thus, the levels of specific transcription factors in a cell are a major determinant of which genes are expressed, and at what level. Decoy oligonucleotides that contain copies of the specific binding site for a particular transcription factor can compete for that transcription factor and thus reduce its availability to the gene or genes it usually binds, thus altering their expression. This strategy was used in clinical trial to prevent vein graft failure after bypass surgery, and is discussed further in the section on clinical applications of gene therapy [157–159].

Gene correction

Many human diseases that directly involve or indirectly affect the cardiovascular system are caused by single gene mutations. Examples include Marfan's syndrome, heart failure caused by mutations in the gene for phospholamban, long Q-T syndrome, hemochromatosis, specific dyslipidemias and sickle cell anemia, among many others. An emerging approach to treat these types of genetic diseases is to correct, *in situ* within the chromosome, the causative gene defect. For diseases such as sickle cell anemia, correcting the defective gene in stem cells could be curative after a single intervention. Several approaches have been proposed to achieve *in situ* gene correction, including the use of triple-helix

forming oligonucleotides (TFOs) [160,161], and most recently the use of engineered ZFP endonucleases [6].

The triple-helix approach is based on the ability of specifically targeted TFOs to bind DNA in the major groove, thus forming triplex DNA. Areas of triplex formation promote nucleotide excision repair and can be used to insert corrective DNA sequences into targeted genes. Several hurdles remain for the effective use of TFOs in gene correction, however, and to date the efficiency of gene correction achieved with this approach has remained low. Conversely, over the past several years advances in ZFP-based technology has greatly increased the achievable efficiency of *in situ* gene correction. Porteus [162] and Porteus and Baltimore [163] showed that by designing pairs of ZFPs that could bind with high affinity to DNA sequences flanking a specific site in a target gene, it was possible to specifically direct the cutting of double-stranded DNA at this site. This was accomplished by fusing the ZFP DNA-binding domains to the FOK1 endonuclease domain, which effectively created engineered DNA-cutting enzymes. The FOK1 endonuclease domain must form a homodimer with another FOK1 domain to be able to cut DNA. Taking advantage of this, the ZFP gene correction approach requires that two FOK1 ZFP endonucleases bind close to one another on the target gene in order to facilitate dimerization and DNA cutting. This markedly increases the specificity of the approach in that both ZFP constructs must bind with high affinity to their target sites in order for DNA cutting to occur. Once a double-strand break has been induced, the DNA repair machinery is attracted to the site and attempts to repair the break. If at the same time a complementary corrective double-stranded donor DNA oligonucleotide is provided to the cell, this oligo can recombine with the native gene at the site of the double-strand break and replace the defective gene segment. Currently, the efficiency of this approach *in vitro* has reached 15–30%, which is orders of magnitude greater than other approaches and is high enough to be in the range of required to "cure" genetic diseases. While there are still several issues to be addressed, this advancement has raised hopes that clinically feasible gene correction is within reach [6]. This approach can also be used to "silence" targeted genes by inducing mutations in them. Figure 9.3 illustrates the use of the ZFP approach to achieve gene correction and gene silencing.

Potential dangers of gene therapy

Unlike drugs, which have a short and finite duration of action in the body, gene therapy can have much longer term consequences, especially when using gene therapy approaches that mediate genomic insertion. Nonetheless, even nonintegrating gene therapies can cause serious deleterious effects. Two of the most infamous and tragic outcomes from gene therapy include the death of a young man in Pennsylvania who received adenovirus-mediated gene therapy to the liver [164,165], and the development of T-cell leukemia in several young children who were administered stem cells from their own bone marrow after these cells had been genetically altered using retroviral vectors [43,166]. In the first case, the delivery of a high titer of adenovirus to the liver resulted in severe liver inflammation, consequent multiorgan system failure and death. Although there were additional contributing factors in this particular case, the potential for adenovirus vectors to cause serious inflammation was confirmed by this tragic event. In the case of the children who developed T-cell leukemia, this was the first clinical manifestation of a gene therapy complication that had been a theoretical concern until that time; *insertional mutagenesis*. As discussed above briefly in the sections on retroviral and lentiviral vectors, insertional mutagenesis is the process in which a therapeutic gene is inserted into the host cell genome in a place that inadvertently alters the expression of an important gene or genes that are near that site. This includes the possible activation of proto-oncogenes and the induction of malignancy. Insertional mutagenesis is a particular concern when transferring genes to stem cells. Currently, efforts are underway to develop vectors that will integrate specifically in areas of the human genome that do not harbor sensitive genes.

Other concerns include that the therapeutic gene will cause undesirable effects. This has been a concern in the use of gene therapy to stimulate angiogenesis. Stimulation of angiogenesis can be deleterious in several settings, including when the recipient patient has a vascular proliferative retinopathy or an occult malignancy. Thankfully, to date no clear evidence of a deleterious angiogenic effect has been observed in any of the clinical trials of therapeutic angiogenesis that have been performed. Still, the possibility that a therapeutic gene could have unanticipated and deleterious effects in patients remains a serious concern that must be addressed and assessed carefully for each individual gene therapy application.

Clinical application of gene therapy to cardiovascular disease

Probably the first attempted cardiovascular application of gene therapy was the use of retroviral vectors to treat hyperlipidemia brought about by LDL receptor deficiency [167,168]. Today gene therapy is well established in cardiac and cardiovascular disease research and several studies testing the efficacy and safety of various gene therapy modalities in humans have been reported or are underway (Table 9.2) [169,170]. The scope of cardiovascular gene therapy clinical targets is quite broad, and includes the stimulation of blood vessel growth in ischemic heart and peripheral vascular disease (therapeutic angiogenesis), treatment of heart failure, post-MI cardiac repair, atherosclerosis, restenosis, vein graft disease, myocardial protection, cardiac rhythm disturbances, amongst others.

Angiogenesis

Gene therapy to stimulate blood vessel growth (angiogenesis and/or arteriogenesis) in the heart and in the periphery is the cardiovascular application that has been the most visible and intensely studied, and a relatively large amount of data regarding the efficacy and safety of this approach has accumulated [19,171]. Although no significant safety issues have emerged, the clinical efficacy of angiogenic gene therapy has been modest at best, despite the high expectations that marked the initial preclinical and clinical experiences. There has been much speculation as to why the clinical results have been less robust than hoped. One of the most compelling explanations is that the biology of what controls blood vessel growth is not sufficiently understood to facilitate an optimal gene therapy approach. For example, until fairly recently the

Table 9.2 Clinical gene therapy trials in patients with cardiovascular diseases.

Trial/clinical target	Gene	Vector/delivery route	Phase	Patients enrolled	Results (controlled trials only)
Therapeutic angiogenesis					
Losordo *et al.* [182]	VEGF165	Naked plasmid/direct myocardial injection	I	5	CAD
Rosengart *et al.* [70]	VEGF121	Adenovirus/direct myocardial injection	I	21	CAD
Vale *et al.* [56]	VEGF-2	Naked plasmid/direct myocardial injection	I/II	6	CAD
Rajagopalan *et al.* [239]	VEGF121	Adenovirus/direct injection into lower extremity muscle	I	5	PVD
Rajagopalan *et al.*	Del1	Plasmid-poloxamer/injection into lower extremity muscle	II		PVD Negative (? Effect of poloxamer)
Morishita *et al.* [92]	HGF	Naked plasmid/injection into lower extremity muscle			PVD
Sarkar *et al.* [240]	VEGF165	Naked plasmid/direct myocardial injection	I	7	CAD
Grines *et al.* [17] (AGENT)	FGF4	Adenovirus/intracoronary infusion	I/II	79	CAD Equivocal
Losordo *et al.* [69]	VEGF-2	Naked plasmid/percutaneous intramyocardial injection	I/II	16	CAD Positive
Grines *et al.* [26] (AGENT-2)	FGF4	Adenovirus/intracoronary infusion	II	52	CAD Equivocal
Hedman *et al.* [184] (KAT)	VEGF165	Adenovirus or plasmid liposomes/to coronary vessel wall after PTCA	I/II	103	CAD +/– for AdVEGF, negative for plasmid liposomes
Kastrup *et al.* [68] (Euroinject One)	VEGF165	Naked plasmid/percutaneous intramyocardial injection	II	80	CAD Equivocal
Prevention of restenosis					
Laitinen *et al.* [82] Restenosis	VEGF	Plasmid liposome/to coronary vessel wall after PTCA	I/II	15	Equivocal
Prevention of coronary artery bypass graft failure					
Alexander *et al.* [229] (PREVENT IV) Coronary bypass vein grafts	E2F Decoy	Oligonucleotide E2F decoy delivered to vein graft by pressure at coronary bypass surgery	III	3014	Negative

CAD, coronary artery disease; FGF, fibroblast growth factor; hVEGF, human vascular endothelial growth factor; mVEGF, mouse vascular endothelial growth factor; PVD, peripheral vascular disease.

term "angiogenesis" was used to describe the growth of capillaries and larger vessels. It is now accepted that "arteriogenesis," the development of larger blood vessels of the type that can be seen on an angiogram "bypassing" occluded vascular segments, is distinct from angiogenesis [172–174]. In this review the term "angiogenesis" refers to both types of vascular growth.

Currently, there are hundreds of proteins purported to have a role in regulating blood vessel growth, and likely many more will be defined as research continues. The earliest angiogenic gene therapy approaches were based on the expression of a single angiogenic growth factor alone, somewhat ignoring the complexity of the biology [17,18,69,175–177]. Nonetheless, these approaches did yield significant biologic results in preclinical animal models, and there have been indications of biologic activity and therapeutic effects in patients. Several representative preclinical and clinical studies, using different genes and different delivery approaches, are addressed below and in Table 9.2.

The first clinical experience with angiogenic gene therapy was initiated by Isner *et al.* [175] who incorporated plasmid DNA encoding the gene for VEGF into the hydrogel matrix of a coated balloon catheter. The balloon was inflated in the femoral arteries of patients with severe peripheral vascular disease, resulting in the transfer of the plasmid into the artery wall where it was taken up and VEGF expressed. Although safe, Isner *et al.* quickly determined that more efficient gene delivery, and consequently higher levels of VEGF, could be obtained by injecting the plasmid DNA directly into the skeletal muscle of the ischemic limb [178,179]. These initial clinical trials, based on a large body of preclinical studies in hindlimb ischemia models, were all Phase I trials without control groups, but they were critically important in launching the field, demonstrating safety and provided compelling early evidence that a biologic effect could be achieved. Since then several clinical trials of gene therapy for peripheral vascular disease have been performed, with a number of different genes, including VEGF, fibroblast growth factor 1 (aFGF), hypoxia-inducible factor 1α (HIF-1α), hepatocyte growth factor, Del-1 (developmentally regulated endothelial locus-1), and an engineered transcription factor designed to turn on the endogenous VEGF gene. To date, no published double-blind study has shown definitive efficacy in peripheral vascular disease, but there has been evidence of a biologic effect in Phase I studies, and the results from several clinical studies have not yet been publically reported. One exception is the Del-1 trial that did not reach the pre-determined end-points for clinical efficacy [93,180].

Angiogenic gene therapy for coronary artery disease has also been the subject of a great deal of preclinical and clinical research. Preclinical studies with plasmid DNA and adenoviral vectors demonstrated efficacy in porcine models of coronary ischemia, and paved the way for clinical studies [18,21,70,181]. In one of the first clinical applications of gene therapy to the heart, Losordo *et al.*

[182] studied five patients with coronary disease who did not respond to conventional anti-anginal therapy and reported that direct intramyocardial delivery of naked plasmid encoding VEGF165 directly into the myocardium led to a reduction of anginal symptoms and an improvement in left ventricular function concomitant with reduced ischemia. In the first clinical trial of adenoviral gene therapy in the heart, Rosengart *et al.* [70] reported significant improvement in regional ventricular function and wall motion in the region of vector administration after intramyocardial delivery of adenovirus encoding VEGF121 in patients undergoing conventional coronary artery bypass grafting compared to patients receiving placebo. Using catheter-based delivery of naked VEGF165 assisted by electromechanical NOGA mapping of the left ventricle in patients with chronic myocardial ischemia, Vale *et al.* [56] reported significant reductions in the frequency of anginal attacks after gene delivery in the treated patients compared to patients receiving placebo.

Despite these encouraging early clinical experiences, attempts to validate angiogenic gene therapy in controlled clinical trials have yielded mixed results. In a placebo-controlled, double blind clinical trial headed by the same group that pioneered plasmid-based cardiac and peripheral vascular gene therapy, catheter-mediated intramyocardial injection of plasmid DNA encoding VEGF-2 resulted in a significant improvement in anginal symptom class. However, although there were strong trends towards improvement in exercise capacity and myocardial perfusion, the results for these end-points did not reach statistical significance [69]. Despite robust results in the initial proof-of-concept study [21], three separate clinical trials of intracoronary infusion of adenovirus encoding FGF-4 yielded only mixed results [17]. Despite promising trends toward improvement in exercise performance in the two smaller FGF-4 studies [17,183], a large controlled trial of intracoronary Ad-FGF-4 was stopped early when interim analysis indicated efficacy end-points would not be reached. In a Phase II trial designed to test the effects of VEGF gene therapy on restenosis, neither plasmid nor adenovirus-based VEGF165 gene transfer into the wall of the coronary artery at the time of angioplasty had any effect on restenosis, but patients receiving

Ad-VEGF165 demonstrated smaller perfusion defects on nuclear scans [184].

The most recently reported trial is the EUROINJECT II trial which compared the effects of intramyocardial plasmid VEGF165 gene transfer with placebo on myocardial perfusion, left ventricular function and clinical symptoms. Whereas VEGF gene transfer did not significantly improve stress-induced myocardial perfusion abnormalities compared to placebo, it did result in improved regional wall motion, as assessed both by electromechanical mapping (NOGA) and ventriculography, thus suggesting a favorable anti-ischemic effect [68]. Other controlled trials are currently in progress [93].

Although the failure, thus far, to achieve consistent robust results in clinical trials is disappointing, there have been encouraging indications of biologic and therapeutic effects, and the concept of angiogenic gene therapy remains highly promising. It is hoped that as the biology of blood vessel growth is better understood, more advanced angiogenic therapies will be devised. The trend in this field has been toward a greater appreciation of the differences between angiogenesis yielding capillarization and small vessels, and arteriogenesis yielding larger vessels such as the collaterals noted on coronary angiograms. Heil and Schaper [172,173] and Scholz *et al.* [174] were the first to investigate the differential biology of these processes, and much work is currently underway to further delineate this biology. Other advancements in the field have also been driven by a deeper appreciation of the complex biology controlling blood vessel growth. Examples include combining multiple "complementary" genes, and the use of engineered or endogenous transcription factors that activate the expression of multiple angiogenic genes or splice variants [125,185]. The marriage of angiogenic gene therapy with cell therapy using endothelial progenitor cells and other cell types is also very promising. The field remains young, and it is likely that many iterations of this therapeutic approach will be tested prior to achieving optimal clinical results.

Heart failure

Heart failure is another biologically complex and diverse condition that is nonetheless an extremely attractive gene therapy target. The prevalence of heart failure is high and growing. The morbidity and mortality associated with heart failure rival malignancies, the costs of treatment are high and current treatment options have limited efficacy. Many mechanistic links to heart failure have been established, including alterations in calcium handling [186–188], desensitization of the contractile apparatus to calcium [189], adrenergic receptor downregulation and desensitization [190–192], increased oxidative stress [193], inflammation, alterations in the extracellular matrix, and several others [194–200]. Accordingly, in contrast to the angiogenesis gene therapy trials where the choice of target genes was relatively limited, for heart failure the gamut of potential target genes is virtually unlimited and ranges from β-adrenergic receptors [201] to various metabolically active proteins [197], to proteins regulating calcium handling [202–206]. Despite this plethora of potential target genes, and experimental studies demonstrating the ability to augment contractility by gene therapy, there is to date no published clinical trial on gene therapy for heart failure. The approaches that are probably nearest to clinical testing are those directed to either cardiomyocyte calcium handling, or adrenergic signaling.

To augment contractility by altering calcium handling, the main targets have been the sarcoplasmic reticulum calcium pump SERCA2, and phospholamban (PLN), an inhibitor of SERCA2 function. SERCA2 is responsible for uptake of calcium into the sarcoplasmic reticulum (SR), and thus is responsible for lowering cytosolic calcium levels during myocyte relaxation, and loading the SR with calcium for release during the next excitation–contraction coupling event [202,203,206]. As such, SERCA2 has crucial effects on both systole and diastole. Overexpression of SERCA2 in transgenic mice, or as a consequence of gene delivery, alters cardiac calcium handling and augments contractility [204,206,207]. Abrogating or diminishing PLN expression by gene deletion, antisense or other approaches also increases cardiac contractility, as does interference of PLN interactions with SERCA2 by using PLN dominant negative mutants [78,208–210]. Currently, clinical trials of gene therapy using SERCA2 overexpression, or an engineered transcription factor that represses PLN expression, are in the planning stages.

Strategies targeted to adrenergic signaling include gene therapy to increase adrenergic receptor number or activity [211–213] and a strategy to increase cAMP levels via overexpression of adenylate cyclase [214,215]. Adenoviral transduction of a peptide inhibitor of β-adrenergic receptor kinase 1 (ARK1) was performed in an infarct model of heart failure in rabbits [216]. Despite unchanged β-adrenergic receptor (AR) density or βARK1 levels in the treated or control groups, a significantly higher βAR-stimulated adenyl cyclase activity was found in the treated group compared to the control group. Adenovirus-mediated intracoronary delivery of the gene encoding β2-AR led to improvements in basal and isoproterenol-stimulated left ventricle contractility and hemodynamic function in rabbits [211]. Moreover, this approach was shown to restore β-adrenergic signaling in cardiac myocytes from failing hearts [217]. Bypassing the adrenergic receptor and targeting camp levels directly, overexpression of adenylate cyclase was also shown capable of augmenting contractility *in vivo* [214,215]. Plans are also underway to clinically evaluate this strategy. Ultimately, irrespective of the target gene, one of the most important determinants of the success of any of these approaches is the ability to achieve gene transfer to a sufficient number of cardiac muscle cells to alter contractility.

Cardiac rhythm disturbances

Management of cardiac rhythm disturbances is now very heavily device oriented. Whereas genetic manipulation of complex re-entrant arrhythmias presents challenges that make gene therapy approaches unlikely to become clinically useful in the near future, there are instances in which gene therapy could very conceivably be used as either an adjunct to or replacement for device-based therapies [218]. One of the most intriguing is the development of biologic pacemakers. In a pioneering study by Miake *et al.* [219], cardiac myocytes were transduced with a gene encoding the K inward rectifier channel (Kir) 2.1AAA. This induced the electrophysiologic properties of cardiac pacemaker cells into ventricular myocytes *in vivo*, increasing their automaticity and resulting in spontaneous rhythmic electrical activity in the heart. Another group demonstrated that focal gene therapy-based

expression of β-adrenergic receptors in the atria increased the heart rates of pigs [220]. Still another group has shown the feasibility of using hyperpolarization activated cyclic nucleotide-gated potassium channel 2 (HCN2) gene transfer to create biologic pacemakers [221]. This group has recently shown that stem cells can be altered in this manner to produce cells with pacemaker activity [222].

Another arrhythmia related application of gene therapy in which proof-of-concept has been demonstrated is rate control for supraventricular arrhythmias; in particular, atrial fibrillation. In a porcine model of atrial fibrillation, investigators used adenovirus vectors to deliver to the AV node a gene encoding the inhibitory G protein $G\alpha_{i2}$. Gene delivery to the AV node was very efficient and resulted in increased refractoriness of the node accompanied by a reduction in ventricular heart rate [223]. Thus, gene therapy for cardiac rhythm disturbances is both feasible and attainable using currently available gene delivery technology. The fact that a therapeutic effect can be obtained with *focal* gene transfer makes this target particularly appealing. If effective, biologic pacemakers could have a marked impact on the cost of cardiac care by reducing the numbers of device implantations.

Myocardial protection

Myocardial protection as a target for gene therapy has been an appealing concept to reduce necrosis and apoptosis associated with myocardial infarction, reduce the myocyte loss that is purported to contribute to the development and progression of heart failure, and to provide protection to the heart during cardiac and noncardiac surgery. To date, the field has concentrated primarily on heart shock protein genes and antiapoptotic genes. The ability of heart shock protein expression to prevent myocardial necrosis was definitively demonstrated in transgenic mice constitutively expressing the inducible heat shock protein 70 (HSP-70i). In these mice, expression of HSP-70i reduced experimentally induced infarct size by 40% and preserved ventricular function [224]. That similar cardioprotective effects could be achieved by a gene therapy approach was demonstrated by adenovirus-mediated gene transfer [225], and by a unique gene delivery system that efficiently delivered the HSP70 gene to the heart by intracoronary infusion of a virus–liposome complex (hemagglutinating virus of Japan; HVJ) [226].

Chatterjee *et al.* [227] studied the effect of the antiapoptotic factor Bcl-2 in a rabbit model of ischemia followed by reperfusion. The experimental group treated with adeno-Bcl-2 was compared with a control group receiving empty vector adeno-null. Animals that received Bcl-2 maintained higher ejection fractions at 2, 4 and 6 weeks compared to controls. The Bcl-2 group had superior preservation of LV geometry with less ventricular dilatation and wall thinning. There was also reduced apoptosis compared to the controls [227]. The serine-threonine kinase Akt is activated by several ligand-receptor systems previously shown to be cardioprotective. Matsui *et al.* [228] examined the effects of a constitutively active mutant of Akt (myr-Akt) in a rat model of cardiac ischemia-reperfusion injury. *In vivo* gene transfer of myr-Akt reduced infarct and the number of apoptotic cells. Ischemia-reperfusion injury decreased regional cardiac wall thickening as well as the maximal rate of left ventricular pressure rise and fall (+dP/dt and −dP/dt). Akt activation restored regional wall thickening and +dP/dt and −dP/dt to levels seen in sham-operated rats.

These representative studies demonstrate that cardiac protection can indeed be achieved by gene transfer. Several major issues remain, however, prior to bringing this concept to clinical fruition. One major issue, shared with many other cardiovascular gene therapy targets, is the requirement for a safe and efficient method of delivering cardioprotective genes to a large enough percentage of cardiac muscle cells to be clinically meaningful. Another major issue is that of timing. Conceptually, one would have to know beforehand when a patient was going to have a major ischemic event so that gene transfer could be performed pre-emptively; or one would have to express these cardioprotective genes chronically in the heart so that they would be there prior to insult. Neither of these approaches is feasible, and if one knew when precisely a patient would have a major ischemic event there are numerous established clinical approaches that could be used to prevent this event. However, there are instances in which pre-emptive gene therapy-based cardioprotection would be both feasible and appealing, including cardiac surgery in high-risk patients (possibly including children with complex congenital heart disease) and noncardiac surgery in patients with advanced pre-existing heart disease.

Gene therapy directed to atherosclerosis and vascular remodeling

Gene therapy directed to vascular remodeling has focused on post intervention restenosis and on prevention of vein graft failure. In an elegant approach directed to saphenous vein bypass grafts, a oligonucleotide decoy targeting the E2F cell-cycle regulatory pathway was developed to prevent intimal hyperplasia and accelerated atherosclerosis in these grafts. In a small Phase I study, the E2F decoy decreased the cumulative number of graft occlusions, revisions or critical stenoses [91], demonstrating the feasibility and potential efficacy of the decoy approach. Disappointingly, the results of the PREVENT IV trial, a larger multicenter randomized trial testing the efficacy of E2F decoy in the prevention of coronary bypass saphenous vein disease, were negative, showing no advantage of the decoy over placebo in patients undergoing coronary artery bypass grafting [229].

Post intervention restenosis has been the type of vascular remodeling most intensely targeted by gene therapy. The repertoire of genes, vectors and approaches investigated has been quite diverse, and many of these approaches have been successful in animal models. These include gene therapy approaches directed to the cell cycle [230,231], extracellular matrix [232,233], and re-endothelialization [234]. To date, there have been relatively few clinical trials of gene therapy for restenosis, and none of these has shown a significant clinical improvement. On the other hand, drug-eluting stents have been enormously successful. It would seem then that restenosis is not a high priority target for gene therapy. It is possible that stents could be designed that deliver genes to the site of intraarterial deployment and prevent restenosis, and concomitantly thrombosis, and thus have an advantage over the current iterations of drug eluting stents.

The development of gene therapy for the treatment of atherosclerosis and plaque rupture is complicated by the fact that atherosclerosis is a multifactorial multistep process and that existing animal models do not faithfully reproduce human atherosclerosis, including coronary plaque rupture and local thrombosis. Nonetheless, several potential strategies have been put forth and tested in animal models [202,235–238]. With the tremendous therapeutic success of statin therapy, and the potential for various pharmacologic agents now in the pipeline to have an equal or greater impact, the impetus for the development of a gene-based therapy for atherosclerosis has probably diminished somewhat. However, the advent of new gene therapy technologies could revive enthusiasm. For example, gene-based expression of the Milano apolipoprotein [237] could be an important strategy for those patients who do not respond to pharmacologic therapies, or could be used as an adjunct for patients with advanced atherosclerotic disease to promote stabilization and regression. Also, specific genetic mutations that have been identified as associated with an increased risk of developing atherosclerosis or plaque rupture could potentially be targets for gene correction. As with all other disease processes that have been considered as possible targets for gene therapy, as the biology of the atherosclerotic process is better understood, more and more potential gene therapy targets will likely be identified.

The future

Gene therapy is only just beyond its infancy. The field has changed dramatically in just the past decade, and now includes such advanced technologies as targeted correction of genomic DNA mutations, engineered transcription factors to regulate endogenous gene expression, a diverse array of technologies to repress or silence the expression of pathology related genes, and a tremendous advancement in vectors. Although the clinical studies of gene therapy for cardiovascular disease conducted thus far have generally not yielded robust therapeutic results, there have been encouraging evidences of beneficial biologic activity. These can be taken as encouraging early steps along the way to the realization of the full potential of gene-based therapies, for cardiovascular disease and other indications.

References

1 Grosshans H. Gene therapy: when a simple concept meets a complex reality. Review on gene therapy. *Funct Integr Genomics* 2000; 1: 142–145.

2 Friedmann T. A brief history of gene therapy. *Nat Genet* 1992; 2: 93–98.

3 Friedmann T, Roblin R. Gene therapy for human genetic disease? *Science* 1972; 175: 949–955.

4 Rebar EJ, Huang Y, Hickey R *et al.* Induction of angiogenesis in a mouse model using engineered transcription factors. *Nat Med* 2002; 8: 1427–1432.

5 Snowden AW, Zhang L, Urnov F *et al.* Repression of vascular endothelial growth factor A in glioblastoma cells using engineered zinc finger transcription factors. *Cancer Res* 2003; 63: 8968–8976.

6 Urnov FD, Miller JC, Lee YL *et al.* Highly efficient endogenous human gene correction using designed zinc-finger nucleases. *Nature* 2005; 435: 646–651.

7 Thomas CE, Ehrhardt A, Kay MA. Progress and problems with the use of viral vectors for gene therapy. *Nat Rev Genet* 2003; 4: 346–358.

8 De Laporte L, Cruz Rea J, Shea LD. Design of modular non-viral gene therapy vectors. *Biomaterials* 2006; 27: 947–954.

9 Glover DJ, Lipps HJ, Jans DA. Towards safe, non-viral therapeutic gene expression in humans. *Nat Rev Genet* 2005; 6: 299–310.

10 Anderson JL, Hope TJ. Intracellular trafficking of retroviral vectors: obstacles and advances. *Gene Ther* 2005; 12: 1667–1678.

11 Delenda C. Lentiviral vectors: optimization of packaging, transduction and gene expression. *J Gene Med* 2004; 6 (Supplement 1): S125–S138.

12 Lever AM, Strappe PM, Zhao J. Lentiviral vectors. *J Biomed Sci* 2004; 11: 439–449.

13 Walther W, Stein U. Viral vectors for gene transfer: a review of their use in the treatment of human diseases. *Drugs* 2000; 60: 249–271.

14 Sadeghi H, Hitt MM. Transcriptionally targeted adenovirus vectors. *Curr Gene Ther* 2005; 5: 411–427.

15 Snyder RO, Francis J. Adeno-associated viral vectors for clinical gene transfer studies. *Curr Gene Ther* 2005; 5: 311–321.

16 Choi VW, McCarty DM, Samulski RJ. AAV hybrid serotypes: improved vectors for gene delivery. *Curr Gene Ther* 2005; 5: 299–310.

17 Grines CL, Watkins MW, Helmer G *et al.* Angiogenic Gene Therapy (AGENT) trial in patients with stable angina pectoris. *Circulation* 2002; 105: 1291–1297.

18 Rosengart TK, Lee LY, Patel SR *et al.* Six-month assessment of a phase I trial of angiogenic gene therapy for the treatment of coronary artery disease using direct

intramyocardial administration of an adenovirus vector expressing the VEGF121 cDNA. *Ann Surg* 1999; 230: 466–470; discussion 470–472.

19 Yeh JL, Giordano FJ. Gene-based therapeutic angiogenesis. *Semin Thorac Cardiovasc Surg* 2003; 15: 236–249.

20 Weitzman MD. Functions of the adenovirus E4 proteins and their impact on viral vectors. *Front Biosci* 2005; 10: 1106–1117.

21 Giordano FJ, Ping P, McKirnan MD *et al.* Intracoronary gene transfer of fibroblast growth factor-5 increases blood flow and contractile function in an ischemic region of the heart. *Nat Med* 1996; 2: 534–539.

22 Imperiale MJ, Kochanek S. Adenovirus vectors: biology, design, and production. *Curr Top Microbiol Immunol* 2004; 273: 335–357.

23 Cotter MJ, Muruve DA. The induction of inflammation by adenovirus vectors used for gene therapy. *Front Biosci* 2005; 10: 1098–1105.

24 Alba R, Bosch A, Chillon M. Gutless adenovirus: last-generation adenovirus for gene therapy. *Gene Ther* 2005; 12 (Supplement 1): S18–S27.

25 Freedman SB. Clinical trials of gene therapy for atherosclerotic cardiovascular disease. *Curr Opin Lipidol* 2002; 13: 653–661.

26 Grines CL, Watkins MW, Mahmarian JJ *et al.* A randomized, double-blind, placebo-controlled trial of Ad5FGF-4 gene therapy and its effect on myocardial perfusion in patients with stable angina. *J Am Coll Cardiol* 2003; 42: 1339–1347.

27 Mohler ER 3rd, Rajagopalan S, Olin JW *et al.* Adenoviral-mediated gene transfer of vascular endothelial growth factor in critical limb ischemia: safety results from a phase I trial. *Vasc Med* 2003; 8: 9–13.

28 McCarty DM, Young SM Jr, Samulski RJ. Integration of adeno-associated virus (AAV) and recombinant AAV vectors. *Annu Rev Genet* 2004; 38: 819–845.

29 Miller DG, Petek LM, Russell DW. Adeno-associated virus vectors integrate at chromosome breakage sites. *Nat Genet* 2004; 36: 767–773.

30 Ferrari FK, Samulski T, Shenk T, Samulski RJ. Second-strand synthesis is a rate-limiting step for efficient transduction by recombinant adeno-associated virus vectors. *J Virol* 1996; 70: 3227–3234.

31 Ren C, Kumar S, Shaw DR, Ponnazhagan S. Genomic stability of self-complementary adeno-associated virus 2 during early stages of transduction in mouse muscle *in vivo. Hum Gene Ther* 2005; 16: 1047–1057.

32 Blankinship MJ, Gregorevic P, Chamberlain JS. Gene therapy strategies for duchenne muscular dystrophy utilizing recombinant adeno-associated virus vectors. *Mol Ther* 2006; 13: 241–249.

33 Zaiss AK, Muruve DA. Immune responses to adeno-associated virus vectors. *Curr Gene Ther* 2005; 5: 323–331.

34 Flotte TR. Adeno-associated virus-based gene therapy for inherited disorders. *Pediatr Res* 2005; 58: 1143–1147.

35 Merten OW, Geny-Fiamma C, Douar AM. Current issues in adeno-associated viral vector production. *Gene Ther* 2005; 12 (Supplement 1): S51–S61.

36 Zhu T, Zhou L, Mori S *et al.* Sustained whole-body functional rescue in congestive heart failure and muscular dystrophy hamsters by systemic gene transfer. *Circulation* 2005; 112: 2650–2659.

37 Melo LG, Agrawal R, Zhang L *et al.* Gene therapy strategy for long-term myocardial protection using adeno-associated virus-mediated delivery of heme oxygenase gene. *Circulation* 2002; 105: 602–607.

38 Barquinero J, Eixarch H, Perez-Melgosa M. Retroviral vectors: new applications for an old tool. *Gene Ther* 2004; 11 (Supplement 1): S3–S9.

39 Katz RA, Skalka AM. The retroviral enzymes. *Annu Rev Biochem* 1994; 63: 133–173.

40 Varmus H. Retroviruses. *Science* 1988; 240: 1427–1435.

41 Hacein-Bey-Abina S, Le Deist F, Carlier F *et al.* Sustained correction of X-linked severe combined immunodeficiency by *ex vivo* gene therapy. *N Engl J Med* 2002; 346: 1185–1193.

42 Engel BC, Kohn DB, Podsakoff GM. Update on gene therapy of inherited immune deficiencies. *Curr Opin Mol Ther* 2003; 5: 503–507.

43 Hacein-Bey-Abina S, Von Kalle C, Schmidt M *et al.* LMO2-associated clonal T cell proliferation in two patients after gene therapy for SCID-X1. *Science* 2003; 302: 415–419.

44 Emi N, Friedmann T, Yee JK. Pseudotype formation of murine leukemia virus with the G protein of vesicular stomatitis virus. *J Virol* 1991; 65: 1202–1207.

45 Kafri T. Gene delivery by lentivirus vectors an overview. *Methods Mol Biol* 2004; 246: 367–390.

46 Grossman M, Rader DJ, Muller DW *et al.* A pilot study of *ex vivo* gene therapy for homozygous familial hypercholesterolaemia. *Nat Med* 1995; 1: 1148–1154.

47 Zhu NL, Wu L, Liu PX *et al.* Downregulation of cyclin G1 expression by retrovirus-mediated antisense gene transfer inhibits vascular smooth muscle cell proliferation and neointima formation. *Circulation* 1997; 96: 628–635.

48 Naldini L. Lentiviruses as gene transfer agents for delivery to non-dividing cells. *Curr Opin Biotechnol* 1998; 9: 457–463.

49 Terskikh AV, Ershler MA, Drize NJ, Nifontova IN, Chertkov JL. Long-term persistence of a nonintegrated lentiviral vector in mouse hematopoietic stem cells. *Exp Hematol* 2005; 33: 873–882.

50 Cefai D, Simeoni E, Ludunge KM *et al.* Multiply attenuated, self-inactivating lentiviral vectors efficiently transduce human coronary artery cells *in vitro* and rat arteries *in vivo. J Mol Cell Cardiol* 2005; 38: 333–344.

51 Morris KV, Rossi JJ. Anti-HIV-1 gene expressing lentiviral vectors as an adjunctive therapy for HIV-1 infection. *Curr HIV Res* 2004; 2: 185–191.

52 Tiera MJ, Winnik FO, Fernandes JC. Synthetic and natural polycations for gene therapy: state of the art and new perspectives. *Curr Gene Ther* 2006; 6: 59–71.

53 van Drunen S, den Hurk LV. Novel methods for the non-invasive administration of DNA therapeutics and vaccines. *Curr Drug Deliv* 2006; 3: 3–15.

54 Labhasetwar V. Nanotechnology for drug and gene therapy: the importance of understanding molecular mechanisms of delivery. *Curr Opin Biotechnol* 2005; 16: 674–680.

55 Shah PB, Losordo DW. Non-viral vectors for gene therapy: clinical trials in cardiovascular disease. *Adv Genet* 2005; 54: 339–361.

56 Vale PR, Losordo DW, Milliken CE *et al.* Randomized, single-blind, placebo-controlled pilot study of catheter-based myocardial gene transfer for therapeutic angiogenesis using left ventricular electromechanical mapping in patients with chronic myocardial ischemia. *Circulation* 2001; 103: 2138–2143.

57 Dass CR. Lipoplex-mediated delivery of nucleic acids: factors affecting *in vivo* transfection. *J Mol Med* 2004; 82: 579–591.

58 Kushibiki T, Tabata Y. A new gene delivery system based on controlled release technology. *Curr Drug Deliv* 2004; 1: 153–163.

59 Stefanidakis M, Koivunen E. Peptide-mediated delivery of therapeutic and imaging agents into mammalian cells. *Curr Pharm Des* 2004; 10: 3033–3044.

60 Cristiano RJ, Curiel DT. Strategies to accomplish gene delivery via the receptor-mediated endocytosis pathway. *Cancer Gene Ther* 1996; 3: 49–57.

61 Kim EM, Jeong HJ, Park IK *et al.* Asialoglycoprotein receptor targeted gene delivery using galactosylated polyethylenimine-graft-poly(ethylene glycol): *in vitro* and *in vivo* studies. *J Control Release* 2005; 108: 557–567.

62 Jaichandran S, Yap ST, Khoo AB, Ho LP, Tien SL, Kon OL. *In vivo* liver electroporation: optimization and demonstration of therapeutic efficacy. *Hum Gene Ther* 2006; 17: 362–375.

63 Li S. Electroporation gene therapy: new developments *in vivo* and *in vitro*. *Curr Gene Ther* 2004; 4: 309–316.

64 Oupicky D, Bisht HS, Manickam DS, Zhou QH. Stimulus-controlled delivery of drugs and genes. *Expert Opin Drug Deliv* 2005; 2: 653–665.

65 Unger EC, Hersh E, Vannan M, McCreery T. Gene delivery using ultrasound contrast agents. *Echocardiography* 2001; 18: 355–361.

66 Dillmann WH. Calcium regulatory proteins and their alteration by transgenic approaches. *Am J Cardiol* 1999; 83: 89H–91H.

67 Giordano FJ. Retrograde coronary perfusion: a superior route to deliver therapeutics to the heart? *J Am Coll Cardiol* 2003; 42: 1129–1131.

68 Kastrup J, Jorgensen E, Ruck A *et al.* Direct intramyocardial plasmid vascular endothelial growth factor-A165 gene therapy in patients with stable severe angina pectoris A randomized double-blind placebo-controlled study: the Euroinject One trial. *J Am Coll Cardiol* 2005; 45: 982–988.

69 Losordo DW, Vale PR, Hendel RC *et al.* Phase 1/2 placebo-controlled, double-blind, dose-escalating trial of myocardial vascular endothelial growth factor 2 gene transfer by catheter delivery in patients with chronic myocardial ischemia. *Circulation* 2002; 105: 2012–2018.

70 Rosengart TK, Lee LY, Patel SR *et al.* Angiogenesis gene therapy: phase I assessment of direct intramyocardial administration of an adenovirus vector expressing VEGF121 cDNA to individuals with clinically significant severe coronary artery disease. *Circulation* 1999; 100: 468–474.

71 Reilly JP, Grise MA, Fortuin FD *et al.* Long-term (2-year) clinical events following transthoracic intramyocardial gene transfer of VEGF-2 in no-option patients. *J Interv Cardiol* 2005; 18: 27–31.

72 Naimark WA, Leporc JJ, Klugherz BD *et al.* Adenovirus-catheter compatibility increases gene expression after delivery to porcine myocardium. *Hum Gene Ther* 2003; 14: 161–166.

73 Rutanen J, Rissanen TT, Markkanen JE *et al.* Adenoviral catheter-mediated intramyocardial gene transfer using the mature form of vascular endothelial growth factor-D induces transmural angiogenesis in porcine heart. *Circulation* 2004; 109: 1029–1035.

74 Wright MJ, Wightman LM, Latchman DS, Marber MS. *In vivo* myocardial gene transfer: optimization, evaluation and direct comparison of gene transfer vectors. *Basic Res Cardiol* 2001; 96: 227–236.

75 Ikeda Y, Gu Y, Iwanaga Y *et al.* Restoration of deficient membrane proteins in the cardiomyopathic hamster by *in vivo* cardiac gene transfer. *Circulation* 2002; 105: 502–508.

76 Hou D, Maclaughlin F, Thiesse M *et al.* Widespread regional myocardial transfection by plasmid encoding Del-1 following retrograde coronary venous delivery. *Catheter Cardiovasc Interv* 2003; 58: 207–211.

77 Asfour B, Byrne BJ, Baba HA *et al.* Effective gene transfer in the rat myocardium via adenovirus vectors using a coronary recirculation model. *Thorac Cardiovasc Surg* 1999; 47: 311–316.

78 Hoshijima M, Ikeda Y, Iwanaga Y *et al.* Chronic suppression of heart-failure progression by a pseudophosphorylated mutant of phospholamban via *in vivo* cardiac rAAV gene delivery. *Nat Med* 2002; 8: 864–871.

79 Saeed M, van Dijke CF, Mann JS *et al.* Histologic confirmation of microvascular hyperpermeability to macromolecular MR contrast medium in reperfused myocardial infarction. *J Magn Reson Imaging* 1998; 8: 561–567.

80 Kawamoto S, Shi Q, Nitta Y, Miyazaki J, Allen MD. Widespread and early myocardial gene expression by adeno-associated virus vector type 6 with a beta-actin hybrid promoter. *Mol Ther* 2005; 11: 980–985.

81 Wang Z *et al.* Adeno-associated virus serotype 8 efficiently delivers genes to muscle and heart. *Nat Biotechnol* 2005; 23: 321–328.

82 Laitinen M, Hartikainen J, Hiltunen MO *et al.* Catheter-mediated vascular endothelial growth factor gene transfer to human coronary arteries after angioplasty. *Hum Gene Ther* 2000; 11: 263–270.

83 Rome JJ, Shayani V, Newman KD *et al.* Adenoviral vector-mediated gene transfer into sheep arteries using a double-balloon catheter. *Hum Gene Ther* 1994; 5: 1249–1258.

84 Isner JM, Walsh K, Symes J *et al.* Arterial gene transfer for therapeutic angiogenesis in patients with peripheral artery disease. *Hum Gene Ther* 1996; 7: 959–988.

85 Riessen R, Rahimizadah H, Blessing E *et al.* Arterial gene transfer using pure DNA applied directly to a hydrogel-coated angioplasty balloon. *Hum Gene Ther* 1993; 4: 749–758.

86 Willard JE, Landau C, Glamann DB *et al.* Genetic modification of the vessel wall. Comparison of surgical and catheter-based techniques for delivery of recombinant adenovirus. *Circulation* 1994; 89: 2190–2197.

87 Enger C, Wolinsky H. Porous balloon catheters. *Semin Interv Cardiol* 1996; 1: 28–29.

88 Wolinsky H. Historical perspective. *Semin Interv Cardiol* 1996; 1: 3–7.

89 Deiner C, Schwimmbeck PL, Koehler IS *et al.* Adventitial VEGF(165) gene transfer prevents lumen loss through induction of positive arterial remodeling after PTCA in porcine coronary arteries. *Atherosclerosis* 2006; 189: 123–132.

90 Thompson CA *et al.* Percutaneous transvenous cellular cardiomyoplasty. A novel nonsurgical approach for myocardial cell transplantation. *J Am Coll Cardiol* 2003; 41: 1964–1971.

91 Mann MJ, Whittemore AD, Donaldson MC *et al. Ex vivo* gene therapy of human vascular bypass grafts with E2F decoy: the PREVENT single-centre, randomised, controlled trial. *Lancet* 1999; 354: 1493–1498.

92 Morishita R, Aoki M, Hashiya N *et al.* Safety evaluation of clinical gene therapy using hepatocyte growth factor to treat peripheral arterial disease. *Hypertension* 2004; 44: 203–209.

93 Rajagopalan S, Olin JW, Young S *et al.* Design of the Del-1 for therapeutic angiogenesis trial (DELTA-1), a phase II multicenter, double-blind, placebo-controlled trial of VLTS-589 in subjects with intermittent claudication secondary to peripheral arterial disease. *Hum Gene Ther* 2004; 15: 619–624.

94 Baumgartner I, Pieczek A, Manor O *et al.* Constitutive expression of phVEGF165 after intramuscular gene transfer promotes collateral vessel development in patients with critical limb ischemia. *Circulation* 1998; 97: 1114–1123.

95 Mir LM, Moller PH, Andre F, Gehl J. Electric pulse-mediated gene delivery to various animal tissues. *Adv Genet* 2005; 54: 83–114.

96 Tjelle TE, Salte R, Mathiesen I, Kjeken R. A novel electroporation device for gene delivery in large animals and humans. *Vaccine* 2006; 22: 4667–4670.

97 Dev SB, Dhar D, Krassowska W. Electric field of a six-needle array electrode used in drug and DNA delivery *in vivo*: analytical versus numerical solution. *IEEE Trans Biomed Eng* 2003; 50: 1296–1300.

98 Zhang Q, Wang Z, Ran HX *et al.* Enhanced gene delivery into skeletal muscles with ultrasound and microbubble techniques. *Acad Radiol* 2006; 13: 363–367.

99 Cordier L, Hack AA, Scott MO *et al.* Rescue of skeletal muscles of gamma-sarcoglycan-deficient mice with adeno-associated virus-mediated gene transfer. *Mol Ther* 2000; 1: 119–129.

100 Schechner JS, Nath AK, Zheng L *et al. In vivo* formation of complex microvessels lined by human endothelial cells in an immunodeficient mouse. *Proc Natl Acad Sci USA* 2000; 97: 9191–9196.

101 Enis DR, Shepherd BR, Wang Y *et al.* Induction, differentiation, and remodeling of blood vessels after transplantation of Bcl-2-transduced endothelial cells. *Proc Natl Acad Sci USA* 2005; 102: 425–430.

102 Liew A, Barry F, O'Brien T. Endothelial progenitor cells: diagnostic and therapeutic considerations. *Bioessays* 2006; 28: 261–270.

103 Iwaguro H, Asahara T. Endothelial progenitor cell culture and gene transfer. *Methods Mol Med* 2005; 112: 239–247.

104 Neff T, Beard BC, Kiem HP. Survival of the fittest: *In vivo* selection and stem cell gene therapy. *Blood* 2006; 107: 1751–1760.

105 Musatov SA, Scully TA, Dudus L, Fisher KJ. Induction of circular episomes during rescue and replication of adeno-associated virus in experimental models of virus latency. *Virology* 2000; 275: 411–432.

106 Duan D, Sharma P, Yang J *et al.* Circular intermediates of recombinant adeno-associated virus have defined structural characteristics responsible for long-term episomal persistence in muscle tissue. *J Virol* 1998; 72; 8568–8577.

107 Flierl A, Chen Y, Coskun PE, Samulski RJ, Wallace DC. Adeno-associated virus-mediated gene transfer of the heart/muscle adenine nucleotide translocator (ANT) in mouse. *Gene Ther* 2005; 12: 570–578.

108 Bouchard S, MacKenzie TC, Radu AP *et al.* Long-term transgene expression in cardiac and skeletal muscle following fetal administration of adenoviral or adeno-associated viral vectors in mice. *J Gene Med* 2003; 5: 941–950.

109 Schaack J. Induction and inhibition of innate inflammatory responses by adenovirus early region proteins. *Viral Immunol* 2005; 18: 79–88.

110 Loser P, Jennings GS, Strauss M, Sandig V. Reactivation of the previously silenced cytomegalovirus major immediate-early promoter in the mouse liver: involvement of NF κB. *J Virol* 1998; 72: 180–190.

111 Krishnan M, Park JM, Cao F *et al.* Effects of epigenetic modulation on reporter gene expression: implications for stem cell imaging. *FASEB J* 2006; 20: 106–108.

112 Tchurikov NA. Molecular mechanisms of epigenetics. *Biochemistry (Mosc)* 2005; 70: 406–423.

113 Roloff TC, Nuber UA. Chromatin, epigenetics and stem cells. *Eur J Cell Biol* 2005; 84: 123–135.

114 Cho KS, Elizondo LI, Boerkoel CF. Advances in chromatin remodeling and human disease. *Curr Opin Genet Dev* 2004; 14: 308–315.

115 Huang C, Sloan EA, Boerkoel CF. Chromatin remodeling and human disease. *Curr Opin Genet Dev* 2003; 13: 246–252.

116 Suzuki T, Matsumura T, Nagai R. Transcriptional regulation at the chromatin level in the cardiovasculature through protein – protein interactions and chemical modifications. *Trends Cardiovasc Med* 2005; 15: 125–129.

117 Jackson DA. The amazing complexity of transcription factories. *Brief Funct Genomic Proteomic* 2005; 4: 143–157.

118 Gilbert N, Ramsahoye B. The relationship between chromatin structure and transcriptional activity in mammalian genomes. *Brief Funct Genomic Proteomic* 2005; 4: 129–142.

119 Huang Y, Hickey RP, Yeh JL *et al.* Cardiac myocyte-specific HIF-1alpha deletion alters vascularization, energy availability, calcium flux, and contractility in the normoxic heart. *Faseb J* 2004; 18: 1138–1140.

120 Semenza GL. HIF-1, O_2, and the 3 PHDs: how animal cells signal hypoxia to the nucleus. *Cell* 2001; 107: 1–3.

121 Semenza GL. HIF-1 and mechanisms of hypoxia sensing. *Curr Opin Cell Biol* 2001; 13: 167–171.

122 Ryan HE, Lo J, Johnson RS. HIF-1 alpha is required for solid tumor formation and embryonic vascularization. *Embo J* 1998; 17: 3005–3015.

123 Genezyme. Genzyme Phase 2 Clinical Trial of HIF-1a for peripheral vascular disease; Phase 1 clinical trial of HIF-1a for ischemic heart disease. 2005.

124 Tang N *et al.* Loss of HIF-1alpha in endothelial cells disrupts a hypoxia-driven VEGF autocrine loop necessary for tumorigenesis. *Cancer Cell* 2004; 6: 485–495.

125 Yu J, Lei L, Liang Y *et al.* An engineered Vegf-activating zinc finger protein transcription factor improves blood flow and limb salvage in advanced-age mice. *FASEB J* 2006; 20: 479–481.

126 Beerli RR, Barbas CF 3rd. Engineering polydactyl zinc-finger transcription factors. *Nat Biotechnol* 2002; 20: 135–141.

127 Jantz D, Amann BT, Gatto GJ Jr, Berg JM. The design of functional DNA-binding proteins based on zinc finger domains. *Chem Rev* 2004; 104: 789–799.

128 Klug A. Zinc finger peptides for the regulation of gene expression. *J Mol Biol* 1999; 293: 215–218.

129 Pabo CO, Peisach E, Grant RA. Design and selection of novel Cys2His2 zinc finger proteins. *Annu Rev Biochem* 2001; 70: 313–340.

130 Segal DJ, Barbas CF 3rd. Custom DNA-binding proteins come of age: polydactyl zinc-finger proteins. *Curr Opin Biotechnol* 2001; 12: 632–637.

131 Urnov FD, Rebar EJ. Designed transcription factors as tools for therapeutics and functional genomics. *Biochem Pharmacol* 2002; 64: 919–923.

132 Klug A, Rhodes D. Zinc fingers: a novel protein fold for nucleic acid recognition. *Cold Spring Harb Symp Quant Biol* 1987; 52: 473–482.

133 Leon O, Roth M. Zinc fingers: DNA binding and protein–protein interactions. *Biol Res* 2000; 33: 21–30.

134 Liu Q, Segal DJ, Ghiara JB, Barbas CF 3rd. Design of polydactyl zinc-finger proteins for unique addressing within complex genomes. *Proc Natl Acad Sci USA* 1997; 94: 5525–5530.

135 Pavletich NP, Pabo CO. Zinc finger-DNA recognition: crystal structure of a Zif268-DNA complex at 2.1 A. *Science* 1991; 252: 809–817.

136 Snowden AW, Gregory PD, Case CC, Pabo CO. Gene-specific targeting of H3K9 methylation is sufficient for initiating repression *in vivo. Curr Biol* 2002; 12: 2159–2166.

137 Wolfe SA, Nekludova L, Pabo CO. DNA recognition by Cys2His2 zinc finger proteins. *Annu Rev Biophys Biomol Struct* 2000; 29: 183–212.

138 Grunstein J, Masbad JJ, Hickey R, Giordano F, Johnson RS. Isoforms of vascular endothelial growth factor act in a coordinate fashion To recruit and expand tumor vasculature. *Mol Cell Biol* 2000; 20: 7282–7291.

139 Nissim-Rafinia M, Kerem B. The splicing machinery is a genetic modifier of disease severity. *Trends Genet* 2005; 21: 480–483.

140 Bonne G, Carrier L, Bercovici J *et al.* Cardiac myosin binding protein-C gene splice acceptor site mutation is associated with familial hypertrophic cardiomyopathy. *Nat Genet* 1995; 11: 438–440.

141 Elson DA, Thurston G, Huang LE *et al.* Induction of hypervascularity without leakage or inflammation in transgenic mice overexpressing hypoxia-inducible factor-1alpha. *Genes Dev* 2001; 15: 2520–2532.

142 Geanacopoulos, M. An introduction to RNA-mediated gene silencing. *Sci Prog* 2005; 88: 49–69.

143 Morris KV. siRNA-mediated transcriptional gene silencing: the potential mechanism and a possible role in the histone code. *Cell Mol Life Sci* 2005; 62: 3057–3066.

144 Mahy BW. Therapeutic RNA? *Rev Med Virol* 2005; 15: 349–350.

145 Hammond SM. Dicing and slicing: the core machinery of the RNA interference pathway. *FEBS Lett* 2005; 579: 5822–5829.

146 Costa FF. Non-coding RNAs: new players in eukaryotic biology. *Gene* 2005; 357: 83–94.

147 Tamm I. Antisense therapy in malignant diseases: status quo and quo vadis? *Clin Sci (Lond)* 2006; 110: 427–442.

148 Lu Y. Recent advances in the stereocontrolled synthesis of antisense phosphorothioates. *Mini Rev Med Chem* 2006; 6: 319–330.

149 Sioud M, Iversen PO. Ribozymes, DNAzymes and small interfering RNAs as therapeutics. *Curr Drug Targets* 2005; 6: 647–653.

150 Reynolds L, Ullman C, Moore M *et al.* Repression of the HIV-1 5′ LTR promoter and inhibition of HIV-1 replication by using engineered zinc-finger transcription factors. *Proc Natl Acad Sci USA* 2003; 100: 1615–1620.

151 Tan S, Guschin D, Davalos A *et al.* Zinc-finger protein-targeted gene regulation: genomewide single-gene specificity. *Proc Natl Acad Sci USA* 2003; 100: 11997–12002.

152 Backs J, Olson EN. Control of cardiac growth by histone acetylation/deacetylation. *Circ Res* 2006; 98: 15–24.

153 Cozzani M, Giovannini I, Naccari R *et al.* Transcription factor decoy (TFD) as a novel approach for the control of osteoclastic resorption. *Prog Orthod* 2005; 6: 238–247.

154 Da Ros T, Spalluto G, Prato M *et al.* Oligonucleotides and oligonucleotide conjugates: a new approach for cancer treatment. *Curr Med Chem* 2005; 12: 71–88.

155 Dzau VJ. Transcription factor decoy. *Circ Res* 2002; 90: 1234–1236.

156 Tomita N, Kim JY, Gibbons GH *et al.* Gene therapy with an E2F transcription factor decoy inhibits cell cycle progression in rat anti-Thy 1 glomerulonephritis. *Int J Mol Med* 2004; 13: 629–636.

157 Ehsan A, Mann MJ, Dell'Acqua G, Dzau VJ. Long-term stabilization of vein graft wall architecture and prolonged resistance to experimental atherosclerosis after E2F decoy oligonucleotide gene therapy. *J Thorac Cardiovasc Surg* 2001; 121: 714–722.

158 Ehsan A, Mann MJ, Dell'Acqua G *et al.* Endothelial healing in vein grafts: proliferative burst unimpaired by genetic therapy of neointimal disease. *Circulation* 2002; 105: 1686–1692.

159 Mangi AA, Dzau VJ. Gene therapy for human bypass grafts. *Ann Med* 2001; 33: 153–155.

160 Casey BP, Glazer PM. Gene targeting via triple-helix formation. *Prog Nucleic Acid Res Mol Biol* 2001; 67: 163–192.

161 Seidman MM, Glazer PM. The potential for gene repair via triple helix formation. *J Clin Invest* 2003; 112: 487–494.

162 Porteus MH. Mammalian gene targeting with designed zinc finger nucleases. *Mol Ther* 2006; 13: 438–446.

163 Porteus MH, Baltimore D. Chimeric nucleases stimulate gene targeting in human cells. *Science* 2003; 300: 763.

164 Raper SE, Chirmule N, Lee FS *et al.* Fatal systemic inflammatory response syndrome in a ornithine transcarbamylase deficient patient following adenoviral gene transfer. *Mol Genet Metab* 2003; 80: 148–158.

165 Carmen IH. A death in the laboratory: the politics of the Gelsinger aftermath. *Mol Ther* 2001; 3: 425–428.

166 McCormack MP, Forster A, Drynan L, Pannell R, Rabbitts TH. The LMO2 T-cell oncogene is activated via chromosomal translocations or retroviral insertion during gene therapy but has no mandatory role in normal T-cell development. *Mol Cell Biol* 2003; 23: 9003–9013.

167 Miyanohara A, Sharkey MF, Witztum JL, Steinberg D, Friedmann T. Efficient expression of retroviral vector-transduced human low density lipoprotein (LDL) receptor in LDL receptor-deficient rabbit fibroblasts *in vitro*. *Proc Natl Acad Sci USA* 1988; 85: 6538–6542.

168 Chowdhury JR, Grossman M, Gupta S *et al.* Long-term improvement of hypercholesterolemia after *ex vivo* gene therapy in LDLR-deficient rabbits. *Science* 1991; 254: 1802–1805.

169 Melo LG, Pachori AS, Kong D *et al.* Gene and cell-based therapies for heart disease. *Faseb J* 2004; 18: 648–663.

170 Quarck R, Holvoet P. Gene therapy approaches for cardiovascular diseases. *Curr Gene Ther* 2004; 4: 207–223.

171 Simons M. Angiogenesis: where do we stand now? *Circulation* 2005; 111: 1556–1566.

172 Heil M, Schaper W. Influence of mechanical, cellular, and molecular factors on collateral artery growth (arteriogenesis). *Circ Res* 2004; 95: 449–458.

173 Heil M, Schaper W. Cellular mechanisms of arteriogenesis. *Exs* 2005; 181–191.

174 Scholz D, Cai WJ, Schaper W. Arteriogenesis, a new concept of vascular adaptation in occlusive disease. *Angiogenesis* 2001; 4: 247–257.

175 Isner JM, Pieczek A, Schainfeld R *et al.* Clinical evidence of angiogenesis after arterial gene transfer of phVEGF165 in patient with ischaemic limb. *Lancet* 1996; 348: 370–374.

176 Kalka C, Masada H, Takahashi T *et al.* Vascular endothelial growth factor (165) gene transfer augments circulating endothelial progenitor cells in human subjects. *Circ Res* 2000; 86: 1198–1202.

177 Simons M *et al.* Clinical trials in coronary angiogenesis: issues, problems, consensus: An expert panel summary. *Circulation* 2000; 102: E73–E86.

178 Rauh G, Pieczek A, Irwin W, Schainfeld R, Isner JM. *In vivo* analysis of intramuscular gene transfer in human subjects studied by on-line ultrasound imaging. *Hum Gene Ther* 2001; 12: 1543–1549.

179 Tsurumi Y, Takeshita S, Chen D *et al.* Direct intramuscular gene transfer of naked DNA encoding vascular endothelial growth factor augments collateral development and tissue perfusion. *Circulation* 1996; 94: 3281–3290.

180 News Release from Valentis, Inc. Valentis Reports Results in Phase II Clinical Trial of Deltavasc (TM) in Peripheral Arterial Disease. September 29, 2004. http:// salesandmarketingnetwork.com/news_release.php?ID= 2000806 (last accessed April 2007).

181 Lazarous DF, Shou M, Stiber JA *et al.* Adenoviral-mediated gene transfer induces sustained pericardial VEGF expression in dogs: effect on myocardial angiogenesis. *Cardiovasc Res* 1999; 44: 294–302.

182 Losordo DW, Vale PR, Symes JF *et al.* Gene therapy for myocardial angiogenesis: initial clinical results with direct myocardial injection of phVEGF165 as sole therapy for myocardial ischemia. *Circulation* 1998; 98: 2800–2804.

183 Grines C, Rubanyi GM, Kleiman NS, Marrott P, Watkins MW. Angiogenic gene therapy with adenovirus 5 fibroblast growth factor-4 (Ad5FGF-4): a new option for the treatment of coronary artery disease. *Am J Cardiol* 2003; 92: 24N–31N.

184 Hedman M, Hartikainen J, Syvanne M *et al.* Safety and feasibility of catheter-based local intracoronary vascular endothelial growth factor gene transfer in the prevention of postangioplasty and in-stent restenosis and in the treatment of chronic myocardial ischemia: phase II results of the Kuopio Angiogenesis Trial (KAT). *Circulation* 2003; 107: 2677–2683.

185 Dai Q, Huang J, Klitzman B *et al.* Engineered zinc finger-activating vascular endothelial growth factor transcription factor plasmid DNA induces therapeutic angiogenesis in rabbits with hindlimb ischemia. *Circulation* 2004; 110: 2467–2475.

186 Babu GJ, Bhupathy P, Petrashevskaya NN *et al.* Targeted overexpression of sarcolipin in the mouse heart decreases sarcoplasmic reticulum calcium transport and cardiac contractility. *J Biol Chem* 2006; 281: 3972–3979.

187 Marks AR. Cardiac intracellular calcium release channels: role in heart failure. *Circ Res* 2000; 87: 8–11.

188 Zhao W, Frank KF, Chu G *et al.* Combined phospholamban ablation and SERCA1a overexpression result in a new hyperdynamic cardiac state. *Cardiovasc Res* 2003; 57: 71–81.

189 MacGowan GA. The myofilament force–calcium relationship as a target for positive inotropic therapy in congestive heart failure. *Cardiovasc Drugs Ther* 2005; 19: 203–210.

190 Perrino C, Naga Prasad SV, Patel M, Wolf MJ, Rockman HA. Targeted inhibition of beta-adrenergic receptor kinase-1-associated phosphoinositide-3 kinase activity preserves beta-adrenergic receptor signaling and prolongs survival in heart failure induced by calsequestrin overexpression. *J Am Coll Cardiol* 2005; 45: 1862–1870.

191 Blaxall BC, Spang R, Rockman HA, Koch WJ. Differential myocardial gene expression in the development and rescue of murine heart failure. *Physiol Genomics* 2003; 15: 105–114.

192 Koch WJ, Lefkowitz RJ, Rockman HA. Functional consequences of altering myocardial adrenergic receptor signaling. *Annu Rev Physiol* 2000; 62; 237–260.

193 Giordano FJ. Oxygen, oxidative stress, hypoxia, and heart failure. *J Clin Invest* 2005; 115: 500–508.

194 Kawada T, Masui F, Kumagai H *et al.* A novel paradigm for therapeutic basis of advanced heart failure: assessment by gene therapy. *Pharmacol Ther* 2005; 107: 31–43.

195 Yu X, Burgess SC, Ge H *et al.* Inhibition of cardiac lipoprotein utilization by transgenic overexpression of Angptl4 in the heart. *Proc Natl Acad Sci USA* 2005; 102: 1767–1772.

196 Chiu HC, Kovacs A, Blanton RM *et al.* Transgenic expression of fatty acid transport protein 1 in the heart causes lipotoxic cardiomyopathy. *Circ Res* 2005; 96: 225–233.

197 Park SY, Cho YR, Finck BN *et al.* Cardiac-specific overexpression of peroxisome proliferator-activated receptor-alpha causes insulin resistance in heart and liver. *Diabetes* 2005; 54: 2514–2524.

198 Leotta E, Patejunas G, Murphy G *et al.* Gene therapy with adenovirus-mediated myocardial transfer of vascular endothelial growth factor 121 improves cardiac performance in a pacing model of congestive heart failure. *J Thorac Cardiovasc Surg* 2002; 123: 1101–1113.

199 Nayak L, Rosengart TK. Gene therapy for heart failure. *Semin Thorac Cardiovasc Surg* 2005; 17: 343–347.

200 Rothermel BA, Vega RB, Williams RS. The role of modulatory calcineurin-interacting proteins in calcineurin signaling. *Trends Cardiovasc Med* 2003; 13: 15–21.

201 Melo LG, Pachori AS, Gnecchi M, Dzau VJ. Genetic therapies for cardiovascular diseases. *Trends Mol Med* 2005; 11: 240–250.

202 Giordano FJ, He H, McDonough P *et al.* Adenovirus-mediated gene transfer reconstitutes depressed sarcoplasmic reticulum Ca^{2+}-ATPase levels and shortens

prolonged cardiac myocyte Ca^{2+} transients. *Circulation* 1997; 96: 400–403.

203 He H, Giordano F, Hilal-Dandan R *et al.* Overexpression of the rat sarcoplasmic reticulum Ca^{2+} ATPase gene in the heart of transgenic mice accelerates calcium transients and cardiac relaxation. *J Clin Invest* 1997; 100: 380–389.

204 He H, Meyer M, Martin JL *et al.* Effects of mutant and antisense RNA of phospholamban on SR Ca^{2+}-ATPase activity and cardiac myocyte contractility. *Circulation* 1999; 100: 974–980.

205 Minamisawa S, Hoshijama M, Chu G *et al.* Chronic phospholamban–sarcoplasmic reticulum calcium ATPase interaction is the critical calcium cycling defect in dilated cardiomyopathy. *Cell* 1999; 99: 313–322.

206 Hajjar RJ, Schmidt U, Matsui T *et al.* Modulation of ventricular function through gene transfer *in vivo*. *Proc Natl Acad Sci USA* 1998; 95: 5251–5256.

207 del Monte F, Williams E, Lebeche D *et al.* Improvement in survival and cardiac metabolism after gene transfer of sarcoplasmic reticulum Ca^{2+}-ATPase in a rat model of heart failure. *Circulation* 2001; 104: 1424–1429.

208 Dieterle T, Meyer M, Gu Y *et al.* Gene transfer of a phospholamban-targeted antibody improves calcium handling and cardiac function in heart failure. *Cardiovasc Res* 2005; 67: 678–688.

209 Meyer M, Belke DD, Trost SU *et al.* A recombinant antibody increases cardiac contractility by mimicking phospholamban phosphorylation. *Faseb J* 2004; 18: 1312–1314.

210 Meyer M, Bluhm WF, He H *et al.* Phospholamban-to-SERCA2 ratio controls the force–frequency relationship. *Am J Physiol* 1999; 276: H779–H785.

211 Maurice JP, Hata JA, Shah AS *et al.* Enhancement of cardiac function after adenoviral-mediated *in vivo* intracoronary β2-adrenergic receptor gene delivery. *J Clin Invest* 1999; 104: 21–29.

212 Shah AS, Lilly RE, Kypson AP *et al.* Intracoronary adenovirus-mediated delivery and overexpression of the β(2)-adrenergic receptor in the heart: prospects for molecular ventricular assistance. *Circulation* 2000; 101: 408–414.

213 Jones JM, Petrofskia JA, Wilson KH *et al.* β2 Adrenoceptor gene therapy ameliorates left ventricular dysfunction following cardiac surgery. *Eur J Cardiothorac Surg* 2004; 26: 1161–1168.

214 Lai NC, Roth DM, Gao MH *et al.* Intracoronary delivery of adenovirus encoding adenylyl cyclase VI increases left ventricular function and cAMP-generating capacity. *Circulation* 2000; 102; 2396–2401.

215 Lai NC, Roth DM, Gao MH *et al.* Intracoronary adenovirus encoding adenylyl cyclase VI increases left ventricular function in heart failure. *Circulation* 2004; 110: 330–336.

216 White DC, Hata JA, Shah AS *et al.* Preservation of myocardial beta-adrenergic receptor signaling delays the development of heart failure after myocardial infarction. *Proc Natl Acad Sci USA* 2000; 97: 5428–5433.

217 Akhter SA, Skaer CA, Kypson AP *et al.* Restoration of beta-adrenergic signaling in failing cardiac ventricular myocytes via adenoviral-mediated gene transfer. *Proc Natl Acad Sci USA* 1997; 94: 12100–12105.

218 Donahue JK. Gene therapy for cardiac arrhythmias. *Ann N Y Acad Sci* 2004; 1015: 332–337.

219 Miake J, Marban E, Nuss HB. Biological pacemaker created by gene transfer. *Nature* 2002; 419; 132–133.

220 Edelberg JM, Huang DT, Josephson ME, Rosenberg RD. Molecular enhancement of porcine cardiac chronotropy. *Heart* 2001; 86; 559–562.

221 Qu J, Barbuti A, Protas L *et al.* Expression and function of a biological pacemaker in canine heart. *Circulation* 2003; 107: 1106–1109.

222 Potapova I, Plotnikov IA, Lu Z *et al.* Human mesenchymal stem cells as a gene delivery system to create cardiac pacemakers. *Circ Res* 2004; 94: 952–959.

223 Donahue JK, Heldman AW, Fraser H *et al.* Focal modification of electrical conduction in the heart by viral gene transfer. *Nat Med* 2000; 6: 1395–1398.

224 Marber MS, Mestri R, Chi SH *et al.* Overexpression of the rat inducible 70-kD heat stress protein in a transgenic mouse increases the resistance of the heart to ischemic injury. *J Clin Invest* 1995; 95: 1446–1456.

225 Mestril R, Giordano FJ, Conde AG, Dillmann WH. Adenovirus-mediated gene transfer of a heat shock protein 70 (hsp 70i) protects against simulated ischemia. *J Mol Cell Cardiol* 1996; 28: 2351–2358.

226 Suzuki K, Sawa Y, Kaneda Y *et al.* In vivo gene transfection with heat shock protein 70 enhances myocardial tolerance to ischemia-reperfusion injury in rat. *J Clin Invest* 1997; 99: 1645–1650.

227 Chatterjee S, Stewart AS, Bish LT *et al.* Viral gene transfer of the antiapoptotic factor Bcl-2 protects against chronic postischemic heart failure. *Circulation* 2002; 106: I212–I217.

228 Matsui T, Tao J, del Monte F *et al.* Akt activation preserves cardiac function and prevents injury after transient cardiac ischemia *in vivo*. *Circulation* 2001; 104: 330–335.

229 Alexander JH, Hafley G, Harrington RA *et al.* Efficacy and safety of edifoligide, an E2F transcription factor decoy, for prevention of vein graft failure following coronary artery bypass graft surgery: PREVENT IV: a randomized controlled trial. *JAMA* 2005; 294: 2446–2454.

230 Chang MW, Barr E, Lu MM, Barton K, Leiden JM. Adenovirus-mediated over-expression of the cyclin/cyclin-dependent kinase inhibitor, p21 inhibits vascular smooth muscle cell proliferation and neointima

formation in the rat carotid artery model of balloon angioplasty. *J Clin Invest* 1995; 96: 2260–2268.

231 Chang MW, Barr E, Seltzer J *et al.* Cytostatic gene therapy for vascular proliferative disorders with a constitutively active form of the retinoblastoma gene product. *Science* 1995; 267: 518–522.

232 Kingston PA, Sinha S, Appleby CE *et al.* Adenovirus-mediated gene transfer of transforming growth factor-β3, but not transforming growth factor-β1, inhibits constrictive remodeling and reduces luminal loss after coronary angioplasty. *Circulation* 2003; 108: 2819–2825.

233 Nili N, Cheema AN, Giordano FJ *et al.* Decorin inhibition of PDGF-stimulated vascular smooth muscle cell function: potential mechanism for inhibition of intimal hyperplasia after balloon angioplasty. *Am J Pathol* 2003; 163: 869–878.

234 Rutanen J, Turunen P, Rutanen J *et al.* Gene transfer using the mature form of VEGF-D reduces neointimal thickening through nitric oxide-dependent mechanism. *Gene Ther* 2005; 12: 980–987.

235 Ishisaki A, Matsuno H. Novel ideas of gene therapy for atherosclerosis: modulation of cellular signal transduction of TGF-β family. *Curr Pharm Des* 2006; 12: 877–886.

236 Liu Y, Li D, Chen J *et al.* Inhibition of atherogenesis in LDLR knockout mice by systemic delivery of adeno-associated virus type 2-hIL-10. *Atherosclerosis* 2005; Nov 18 [Epub ahead of print].

237 Sharifi BG, Wu K, Wang L *et al.* AAV serotype-dependent apolipoprotein A-I Milano gene expression. *Atherosclerosis* 2005; 181: 261–269.

238 Lee KU, Lee IK, Han J *et al.* Effects of recombinant adenovirus-mediated uncoupling protein 2 overexpression on endothelial function and apoptosis. *Circ Res* 2005; 96: 1200–1207.

239 Rajagopalan S, Shah M, Luciano A, Crystal R, Nabel EG. Adenovirus-mediated gene transfer of VEGF(121) improves lower-extremity endothelial function and flow reserve. *Circulation* 2001; 104: 753–755.

240 Sarkar N, Ruck A, Kallner G *et al.* Effects of intramyocardial injection of phVEGF-A165 as sole therapy in patients with refractory coronary artery disease: 12-month follow-up: angiogenic gene therapy. *J Intern Med* 2001; 250; 373–381.

241 Giordano FJ. Oxygen, oxidative stress, hypoxia, and heart failure. *J Clin Invest.* 2005; 115(3): 500–508.

242 Giordano FJ. Therapeutic gene regulation: targeting transcription. *Circulation* 2007; 115(10); 1180–1183.

CHAPTER 10

Stem cell therapy for cardiovascular disease

Emerson C. Perin, MD, PhD, *& Guilherme V. Silva,* MD

Introduction

Coronary artery disease (CAD) remains a leading cause of death and disability in the USA. Research efforts have concentrated on treating acute myocardial infarction (AMI) so as to reduce mortality rates. AMI can cause variable degrees of damage to myocardial tissue depending on the amount of necrosis and patency of the arterial bed. Myocardial scarring may compromise myocardial performance, leading to remodeling of the left ventricle in response to increased mechanical wall stress, inadequate perfusion that compromises remaining viable myocardial segments, and finally ischemic heart failure.

Despite efforts to halt the progression to ischemic heart failure by means of revascularization, ventricular remodeling remains a danger. Efforts to treat severely compromised hearts refractory to medical treatment have focused on heart transplantation and, more recently, mechanical ventricular assistance [1]. Until recently, cardiologists believed that beyond revascularization and medical therapy this process was irreversible because the heart did not have the capacity to renew itself.

However, new insights into the mechanisms of cardiac repair have provided evidence that the adult heart may repair itself and that vasculogenesis may not occur solely during embryonic development, which has in turn sparked strong interest in stem cell therapy [2,3]. Prompted by evidence that adult bone marrow harbors a reservoir of enormously plastic cells [4], animal experiments have generated a wide array of evidence supporting the use of stem cells to repair cardiac tissue in diverse clinical scenarios [5].

Stem cell therapy for cardiac diseases is a reality. Several strategies have been tested with promising initial results. This chapter, which is aimed at the clinical cardiologist who will ultimately deliver this investigational therapy to patients, outlines the basic concepts of such therapy and reviews the clinical data available from Phase I and II trials.

The basics of stem cells

Definition

Stem cells are self-replicating cells capable of generating, sustaining and replacing terminally differentiated cells [6,7]. Stem cells can be subdivided into two main groups: embryonic and adult. Embryonic stem cells are present in the earliest stage of embryonic development – the blastocyst. Embryonic stem cells are pluripotent. This means they are capable of generating any terminally differentiated cell in the human body that is derived from any one of the three embryonic germ layers: ectoderm, mesoderm or endoderm [8]. Through a series of divisions and differentiations, all the organs of the human body arise from the original embryonic stem cells that form the blastocyst [3].

Adult stem cells are intrinsic to specific tissues of the postnatal organism and are committed to differentiate into those tissues [9]. Hematologists have studied them for four decades, ever since the successful clinical introduction of bone marrow transplantation. Theoretically, adult stem cells self-renew forever, yielding mature differentiated cells that are: (i) integrated into a particular tissue; and (ii) capable of performing the specialized function, or functions, of that tissue. Each type of differentiated cell has its own phenotype (i.e., observable characteristics), including shape or morphology, interactions with surrounding cells and extracellular matrix, expression of particular cell surface proteins (receptors) and behavior [8]. Adult tissue-specific stem cells are present in self-renewable organs including the liver, pancreas, skeletal muscle and skin.

Identification

Each adult stem cell subtype can be identified by cell surface *receptors* that selectively bind to particular *signaling* molecules. Differences in structure and binding affinity allow for a remarkable multiplicity of receptors. Normally, cells utilize these receptors and the molecules that bind to them to communicate with other cells and perform the proper function of the tissue to which they belong (e.g., contraction, secretion or synaptic transmission). Each type of adult stem cell has a certain receptor or combination of receptors (i.e., *marker*) that distinguishes it from other types of stem cells (Table 10.1).

Stem cell markers are often given letter and number codes based on the molecules that bind to them (Table 10.1). A cell presenting the stem cell antigen-1 receptor is identified as Sca-1$^+$. Cells exhibiting Sca-1 but not CD34 antigen or lineage-specific antigen (Lin) are identified as CD34$^-$Sca-1$^+$ Lin$^-$. This particular combination of surface receptors identifies mesenchymal stem cells (MSCs).

Unfortunately, the nomenclature can be confusing. Different researchers have given the same bone marrow cells different names. In some cases, surface marker designations within cell subtypes overlap. Most surface markers do not adequately identify stem cells because they may also be found on nonstem cells. Moreover, some markers may be expressed only under certain culture conditions or at a certain stage of cell development. Bone marrow stem cells are also highly plastic and may give rise to several subtypes.

Role in cardiovascular repair

The field of stem cell therapy has benefited greatly from the work of numerous basic and clinical researchers whose studies have greatly improved

Table 10.1 Development of differentiated tissues from embryonic germ layers.

Embryonic germ layer	Differentiated tissue
Endoderm	Gastrointestinal tract lining
	Larynx
	Liver
	Lung
	Pancreas
	Parathyroid gland
	Respiratory tract lining
	Thymus
	Thyroid gland
	Trachea
	Urethra
	Urinary bladder
	Vagina
Mesoderm	Adrenal cortex
	Bone marrow (blood)
	Cardiac muscle
	Connective tissues (cartilage, bone)
	Heart and blood vessels (vascular system)
	Lymphatic tissue
	Skeletal muscle
	Smooth muscle
Ectoderm	Adrenal medulla
	Connective tissue of head and face
	Ears
	Eyes
	Neural tissue (neuroectoderm)
	Pituitary gland
	Skin

our understanding of the processes involved in cardiac repair and neovascularization. The creation of new blood vessels (neovascularization) requires the formation of new mature endothelial cells. In this process, the new endothelial cells migrate or proliferate from existing vessels (angiogenesis) or arise from bone marrow-derived progenitor cells (vasculogenesis) [10]. Asahara *et al.* [2] were the first to describe a unique population of bone marrow-derived endothelial progenitor cells (EPCs) found in the peripheral circulation. These EPCs share similarities with bone marrow hematopoietic progenitor cells. Before EPCs were discovered, vasculogenesis was thought to occur only in the human embryonic phase. In animal models of ischemia, however, EPCs participate in new vessel development, thus establishing a new paradigm of postnatal vasculogenesis [11–30].

The importance of postnatal vasculogenesis to stem cell therapy has been highlighted by several recent studies. Bone marrow-derived EPCs have been shown to contribute functionally to vasculogenesis after AMI [1,13,14], during wound healing [31] and in limb ischemia [12–14,19,20,27,30]. They have also been implicated in the endothelialization of vascular grafts [13,21,28]. The number of circulating EPCs and their migratory capacity correlates inversely with risk factors for CAD, such as smoking and hypercholesterolemia [31]. EPCs have also been implicated in the pathogenesis of allograft transplant vasculopathy and coronary restenosis after stent implantation, after being recruited by appropriate cytokines, growth factors and hormones via autocrine, paracrine and endocrine mechanisms [32]. Endothelial progenitor cell mobilization is a natural response to vascular trauma, as seen in patients who undergo coronary artery bypass graft surgery or suffer burns [16] or an AMI [1].

Endothelial progenitor cells also appear to be crucial to vascular homeostasis. Werner *et al.* [33] followed a series of 519 patients with CAD for up to 1 year after coronary angiography and found that preprocedural EPC levels were prognostically valuable in predicting major cardiovascular events (MACE) and death from cardiovascular causes. Patients with higher EPC levels were less likely to suffer these untoward events even after adjustment for traditional prognostic and risk factors.

Adult tissue-specific stem cells are present in other self-renewable organs such as the liver, pancreas, skeletal muscle and skin. The heart, which until very recently was considered a terminally differentiated, post mitotic organ with a finite store of myocytes established at birth, might now be added to this list. It has recently been observed that hematopoietic stem cells (HSCs) can transdifferentiate into cardiomyocytes [34,35] and that stem cells may reside in the heart [36]. Such resident cardiac stem cells are thought to occupy niches in the atria and apex and have been observed in the border zones of myocardial infarcts [37,38]. These observations have in turn drastically changed our understanding of the cardiac repair process. It now appears that resident cardiac stem cells and possibly bone marrow-derived stem cells may be able to repair the damaged heart. Once adequate signaling is established with cytokines and growth factors, bone marrow cells are mobilized [32]. This concept is strengthened by evidence from animal studies showing that AMI repair involves bone marrow cells [39] and by evidence of chimerism in transplanted hearts [40].

Further evidence for a dynamic cardiac renewal process in the adult heart comes from the recent identification of a novel population of early tissue-committed stem cells that may be part of a group of circulating progenitor cells involved in cardiac repair [41]. The particularities and interactions of resident and circulating stem cells in this setting continue to be delineated.

Together, these strands of evidence suggest that stem cells have key roles in the adult heart's ability to dynamically repair itself and its vessels and in the body's ability to maintain homeostasis.

Stem cell types and characteristics

At present, most basic and clinical research concerns bone marrow-derived progenitor cells and skeletal myoblasts.

Adult bone marrow-derived stem cells

Adult bone marrow-derived stem cells are the cell type most widely utilized in cardiac stem cell therapy. A very heterogeneous subset, termed autologous bone-marrow-derived mononuclear cells (ABMMNCs), is composed of small amounts of stromal or mesenchymal stem cells (MSCs), hematopoietic progenitor cells (HPCs), endothelial progenitor cells (EPCs) and more committed cell lineages such as natural killer lymphocytes, T and B lymphocytes (Plate 10.1) [2].

Bone marrow stem cells are aspirated from the patient's iliac crest under local anesthesia. The mononuclear subfraction of the aspirate is isolated by means of Ficoll density centrifugation, filtered through a 100-μm nylon mesh to remove cell aggregates or bone spicules, and washed several times in phospate-buffered saline solution before being used immediately for therapy or expanded in an endothelial cell-specific culture medium. Endothelial progenitor cells can be harvested from peripheral blood.

So far, the most important bone marrow subtypes utilized for cardiac repair have been MSCs, EPCs or, alternatively, the whole ABMMNC fraction. Newly described bone marrow cell subtypes with therapeutic potential are discussed below.

Mesenchymal stem cells

An adult MSC is a cell from any adult tissue that can be expanded in culture and can renew itself and differentiate into several specific mesenchymal cell lineages. MSCs are present in different niches throughout the body such as bone marrow and adipose tissue [42]. MSCs are extremely plastic, with the potential to terminally differentiate *in vitro* and *in vivo* into mesenchymal phenotypes such as bone [43,44], cartilage [45], tendon [46,47], muscle [34,48], adipose tissue [49,50] and hematopoiesis-supporting stroma [50]. MSCs differentiate not only into mesenchymal tissues but also into cells derived from other embryonic layers, including neurons [51] and epithelia in the skin, lung, liver, intestine, kidney and spleen [52–54]. Their plasticity has increased interest in using them for cardiac regeneration.

In a proof-of-concept study in which Saito *et al.* [55] intravenously injected LacZ reporter gene-transfected MSCs into healthy rats, the MSCs preferentially engrafted in the bone marrow. When injected into rats subjected to several cycles of ischemia-reperfusion, however, the MSCs engrafted in the infarcted regions of the heart, where they participated in angiogenesis and expressed cardiomyocyte-specific proteins. When injected into rats 10 days after myocardial injury, MSCs preferentially homed in instead on the bone marrow, suggesting that in the first days after myocardial injury a specific cell homing signal causes the MSCs to home in on the damaged myocardium.

MSCs are CD45⁻CD34⁻ bone marrow cells that can be readily grown in culture. They are evidently rare in the bone marrow (<0.01% of nucleated cells, by some estimates) and thus 10 times less abundant than HSCs. MSCs need to be cultured for at least 20 days to obtain the numbers necessary for therapy, which would directly affect any clinical strategy for treating AMI that involves autologous MSCs.

Adult bone marrow MSCs, which are easy to manipulate genetically and weakly immunogenic,

represent a potential source of allogeneic stem cells [56]. Allogeneic MSCs actually inhibit T cells in culture [57], and several *in vivo* studies have achieved good engraftment of allogeneic MSCs without rejection [56].

Our group at the Texas Heart Institute was the first to study mesenchymal cell injections in a large-animal model of chronic myocardial ischemia [58]. In brief, we used ameroid constrictors to induce ischemia in 12 dogs and a month later directly injected the myocardium of each with 100 million MSCs or a saline control. Subsequent two-dimensional echocardiography showed improved systolic function both at rest and during stress in treated dogs (Fig. 10.1), and histopathologic studies showed that the MSCs had transdifferentiated into endothelial and smooth muscle cells (Fig. 10.2; Plate 10.2) and improved vascularization (Plate 10.3).

In light of the current evidence, MSCs should have strong clinical potential, especially if human safety studies confirm the lack of rejection seen so far in preclinical studies.

Endothelial progenitor cells

Endothelial progenitor cells can be isolated from the mononuclear fraction of the bone marrow or peripheral blood, as well as from fetal liver or umbilical cord blood [12,23,59,60]. In widely ranging animal models of ischemia, heterologous, homologous and autologous EPCs have been shown to engraft at sites of active neovascularization [60].

EPCs can differentiate into endothelial cells, smooth muscle cells, or cardiomyocytes both *in vitro* and *in vivo*. EPCs have been identified by different research groups using different methodologies [15,17,22,25]. The classic methods involve culture of total peripheral blood mononuclear cells or isolation via magnetic microbeads coated with anti-CD133 or anti-CD34 antibodies. After isolation, the cells are cultured in medium containing specific growth factors such as vascular endothelial growth factor (VEGF) and fibroblast growth factor to facilitate the growth of endothelial-like cells.

"Immature" or "primitive" EPCs have a profile similar to that of HSCs; both cell types are thought to result from a common precursor, the hemangioblast. Within the bone marrow, immature EPCs

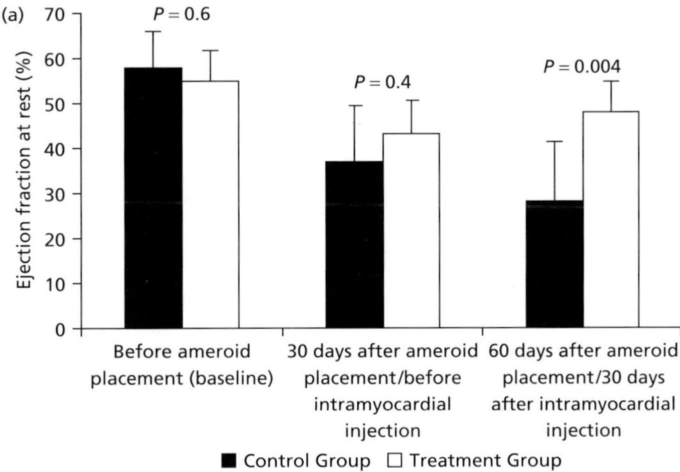

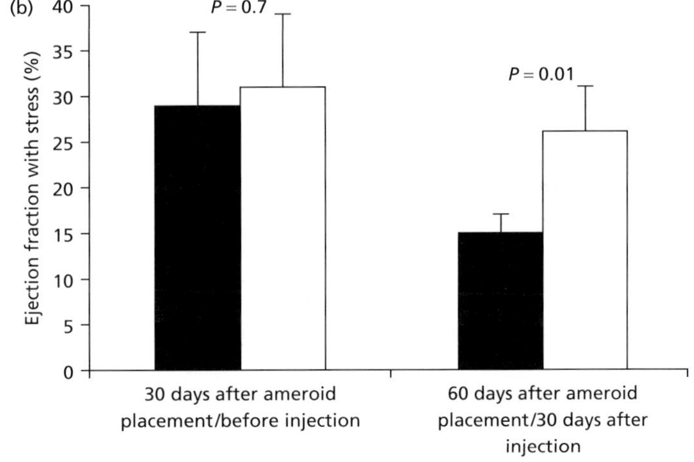

Figure 10.1 (a), LVEF at rest. Assessments were made at baseline before ameroid placement (left), 30 days later at time of cell or saline injection (middle), and 60 days after ameroid placement (right). (b), LVEF with stress. Assessments were made before and 30 days after intramyocardial injection.

and HSCs share common cell-surface markers: CD34, CD133 and VEGF receptor 2 (VEGFR-2, also known as KDR/FLK-1). Similarly, in the peripheral circulation, the more primitive cell population with the capacity of differentiating into EPCs expresses CD34, VEGFR-2 and CD133. In the peripheral circulation, the more committed EPCs lose CD133 but retain CD34 and VEGFR-2 expression. Some circulating EPCs and, to a greater extent, more differentiated EPCs start expressing the endothelial lineage-specific marker vascular endothelial (VE) cadherin or E selectin. However, when immature EPCs follow the hematopoietic path, the surface markers of CD133 and VEGFR-2 are extinguished because stem/progenitor cell markers are not expressed on differentiated hematopoietic cells.

Recent studies have challenged the more traditional view that primitive bone marrow-derived EPCs that lose CD133 become relatively committed EPCs that may subsequently differentiate into mature endothelial cells. EPCs (CD34/VEGF-2$^+$ cells) appear to originate from peripheral blood mononuclear cells that express the monocyte–macrophage markers CD14, MAC-1 and CD11-c, suggesting a possible monocyte–macrophage origin [61]. Harraz *et al.* [62] have observed, within populations of mononuclear peripheral cells, CD34$^-$ cells that are not only CD14$^+$ but also differentiate into endothelial cells. Adding to the debate is the recent description of "late" EPCs or outgrowth endothelial cells (OECs) that originate from a CD14$^-$ monocyte population [63]. EPCs

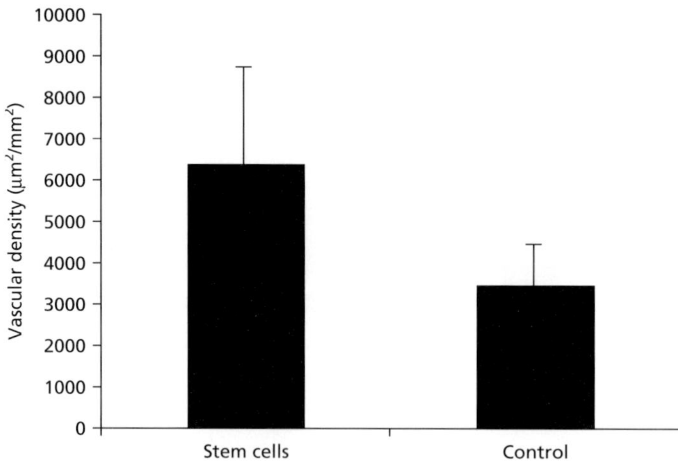

Figure 10.2 (a), Fibrosis (as evaluated by trichrome staining) in anterolateral wall of animals treated with MSCs and controls. There was a trend toward less fibrosis in treated dogs that did not reach statistical significance. (b), Vascular density was statistically greater in anterolateral walls of animals that received stem cells.

originating from the CD14$^+$ monocyte population (so-called "early" EPCs) were found to secrete angiogenic peptides such as VEGF, IL-8 and other enzymes such as matrix metalloproteinase 9 (MMP9). However, these "early" EPCs did not proliferate well. On the other hand, OECs did proliferate well but secreted only MMP2. Taken together, these findings argue for the plasticity of EPCs (CD34$^+$, VEGFR-2$^+$ and CD133$^+$), the different developmental stages of a common precursor progenitor cell, and the existence of distinct cell subtypes that might be further differentiated by surface markers yet to be discovered.

EPC numbers appear to decrease in the presence of risk factors for CAD and to correlate negatively with Framingham cardiovascular risk factors [31]. Therefore, stem cell therapy with EPCs may prove very useful in the clinical setting of cardiovascular disease. The kinetic and biologic properties of EPCs may be especially appropriate for autologous transplantation. EPCs may also be safe to use in elderly and diabetic patients, populations in which they do not tend to migrate as much or induce neovascularization [61].

Our ability to characterize EPCs and identify those subtypes most useful for cardiac cell therapy has advanced rapidly, but is still incomplete. Nonetheless, as the positive results of initial preclinical and clinical studies have shown, EPCs show great therapeutic promise.

Other bone marrow stem cells

Clearly, the bone marrow is a reservoir of cells whose regenerative capacity extends beyond the hematopoietic lineage. Identifying stem cells on the basis of cell-surface markers is a limited method that may delay the discovery of additional tissue-specific stem cell subtypes. Nevertheless, the stem cell field is advancing quickly. Recently, Kucia *et al.* [64] published the first evidence that postnatal bone marrow harbors a nonhematopoietic cell population that expresses markers for cardiac differentiation. This finding corroborates the early work of Deb *et al.* [65], who isolated Y chromosome-positive cardiac myocytes from female recipients of male bone marrow. The percentage of cardiomyocytes that harbored the Y chromosome was quite small (only 0.23%), but there was no evidence of either pseudonuclei or cell fusion. Two new bone marrow cardiac precursors have been identified: ABMMNCs expressing cardiac markers within a population of nonhematopoietic CXCR4$^+$/Sca-1$^+$/Lin$^-$/CD45$^-$ ABMMNCs in mice and within a population of nonhematopoietic CXCR4$^+$/CD34$^+$/AC133$^+$/CD45$^-$ ABMMNCs in humans. These nonhematopoietic ABMMNCs expressing cardiac precursors are mobilized into the peripheral blood after a myocardial infarction and home in on the infarcted myocardium in an SDF-1-CXCR4, HGF-c-Met and LIF-LIF-R dependent manner [64].

The identification of a direct cardiac precursor within the bone marrow cell population opens up a vast number of possibilities for the field of cardiac regeneration. In theory, *in vitro* expansion of this type of cell would be therapeutically attractive.

Skeletal myoblasts

Skeletal myoblasts are adult tissue-specific stem cells [66] located between the basal lamina and the sarcolemma on the periphery of the mature skeletal-muscle fiber [67]. Also known as muscle satellite cells, these small, mononuclear cells are activated by biochemical signals to divide and differentiate into fusion-competent cells after muscle injury.

The use of skeletal myoblasts for cardiac repair originated in earlier attempts to use fetal cardiomyocytes. When injected into the border zone of an AMI, these cells are able to engraft and survive [68]. Despite initial encouraging results in animal models, clinical use of fetal cardiomyocytes has not been pursued because of ethical issues and the limited availability of these cells.

Skeletal myoblasts have emerged as an attractive alternative [69]. The first therapeutic trials used skeletal myoblasts obtained under sterile conditions and local anesthesia 0.5–5.0 g muscle biopsy specimens. Individual cells were isolated by digestion with trypsin and collagenase, washed to remove red blood cells and debris, plated and cultured to obtain the numbers necessary for therapeutic use.

Skeletal myoblasts can survive prolonged periods of hypoxia [70]. They can also survive and engraft when injected into infarcted areas.

Embryonic stem cells

As gradually revealed over the past two decades, embryonic stem cells (ESCs) are derived from the cell mass of blastocysts in mice and humans. In the presence of leukemia inhibitor factor (LIF) or atop a layer of mitotically inactivated mouse embryonic fibroblasts, ESCs can proliferate indefinitely. Once removed from these conditions and transferred into a suspension culture, ESCs spontaneously form multicellular aggregates that turn into endoderm, mesoderm and ectoderm [70]. Murine ESC lines have been shown *in vitro* to differentiate into cells associated with each of these three layers: hematopoietic progenitors, adipocytes, hepatocytes, smooth muscle cells, endothelial cells, neurons and others [71,72]. More importantly, they have been shown to differentiate into cardiomyocytes [73] in response to appropriate stimuli and specific signaling factors. Hepatocyte growth factor (HGF), epidermal growth factor (EGF), basic fibroblast growth factor (bFGF), platelet-derived growth factor (PDGF), retinoic acid, vitamin C and overexpression of GAT4 all enhance this differentiation process *in vitro* [74]. However, the ideal combination of these factors for enhancing ESC differentiation into cardiomyocytes remains unknown.

Because they are pluripotent and can proliferate indefinitely, ESCs may have an important potential role in cardiac regeneration. Although ethical issues involving the use of human ESCs has slowed research in several countries including the USA, enthusiasm about their future clinical utilization remains high.

Resident cardiac stem cells

Myocyte replication is the failing heart's attempt to compensate for a limited capacity for hypertrophy. When Urbanek *et al.* [38] used Ki-67 (a nuclear protein expressed during cell division) to assess the mitotic activity of myocytes, they observed significantly greater mitotic activity at infarct border zones than in distant myocardium or healthy control hearts. The evidence that cardiac myocytes divide shortly after a myocardial infarction led investigators to search for the origin of the dividing myocytes [75]. This culminated in the description of resident cardiac stem cells (CSCs) [36–38].

Resident CSCs were first isolated in murine hearts. Characterization of these cells was based on the expression of the stem cell-related surface antigens c-Kit and Sca-1. In the first study, freshly isolated c-Kit$^+$/Lin$^-$ cells were shown to be clonogenic and to differentiate into myocytes, smooth muscle cells and endothelial lineage cells [36]. Those cells generated functional myocardium when injected into ischemic areas of the heart. The second study characterized CSCs as Sca-1/c-Kit$^-$. When treated in culture with 5-azacytidine, those cells differentiated into a myogenic lineage. Subsequently, intravenous injection of the cells in an ischemia-reperfusion model resulted in infarct healing with cardiomyocyte transdifferentiation [75]. In studies involving atrial and ventricular biopsies in sheep and humans, Messina *et al.* [76] isolated a cardiac progenitor cell that was c-Kit⁺ and capable of self-proliferating into a large number of cells. The authors also showed that human CSCs could participate in infarct repair in the murine model.

A detailed and uniform characterization of CSCs is still lacking, as are preclinical data from large-animal models. As further studies are performed and yield promising results, CSCs may be considered for utilization in clinical trials.

Alternative sources of stem cells

Despite successful preclinical and clinical utilization of bone marrow cells and skeletal myoblasts, the search continues for an ethical, easily accessible, high-yield source of stem cells. Mesenchymal stem cells have been isolated from adipose tissue, placental tissue and umbilical cord blood. A number of studies have shown adipose-derived mesenchymal stem cells (AMSCs) to be pluripotent and capable of differentiating into multiple cell lineages along the myogenic, osteogenic, neurogenic and hematopoietic pathways [77–79]. Additionally, AMSCs secrete VEGF, HGF, bFGF and transforming growth factor β (TGF-β), which have a potential angiogenic effect on ischemic myocardium [80]. These cells also express the cell-surface marker CD34, but it is uncertain whether their pluripotency is limited to the subgroup of cells that express this marker [81]. Research to better characterize AMSCs and evaluate the safety and efficacy of this stem cell type in preclinical studies is ongoing.

By means of dissection and proteinase digestion, large numbers of viable mononuclear cells can be harvested from the human placenta at term, and a mesenchymal cell population with characteristic expression of CD9, CD29 and CD73 can be obtained in culture. The *in vitro* growth behavior of such placenta-derived mesenchymal cells is similar to that of human bone marrow mesenchymal progenitor cells [82]. Transdifferentiation experiments have shown a potential for differentiation along osteogenic, chondrogenic, adipogenic and myogenic lines [82]. The human placenta at term might be an easily accessible, ample source of multipotent mesenchymal progenitor cells and is also under preclinical investigation.

Cord blood has long been used as a source of MSCs for bone marrow transplantation. The stem cell compartment is more abundant and less mature in cord blood than in bone marrow. Moreover, MSCs in cord blood have a higher proliferative potential because of their extended lifespan and longer telomeres [83–86]. Not only can they be harvested without morbidity to the donor, but they also display a robust *in vitro* capacity for directed or spontaneous differentiation into mesodermal, endodermal and ectodermal cell fates. Cord blood MSCs are CD45⁻ and HLA-II⁻ and can be expanded without losing their pluripotency. Therefore, cord blood is also undergoing preclinical evaluation as a possible easily accessible source of multipotent cells.

Stem cell delivery methods

The current understanding of stem cell biology and kinetics gives us important clues as to how we should deliver them. The efficacy of therapeutic stem cells will obviously depend largely on successful delivery. Stem cells have been delivered indirectly through peripheral and coronary veins and coronary arteries. Alternatively, they have been delivered directly by intramyocardial injections via surgical, transendocardial or transvenous approaches. Another potential delivery strategy is the mobilization of stem cells from the bone marrow by means of cytokine therapy with or without peripheral harvesting.

The main objective of any cell delivery method is to achieve the concentration of stem cells necessary for repairing the damaged region being targeted. To this end, the ideal modality should be safe; easy to use; cost-effective; clinically useful in a wide range of clinical disease settings and scenarios; easily, adequately and effectively targeted; and able to exert a long-lasting therapeutic effect.

Stem cell mobilization

In humans, progenitor cells from the bone marrow mobilize after an AMI. This suggests a "natural" attempt at cardiac repair [3]. In theory, therapeutic mobilization of bone marrow progenitor cells after an AMI would amplify the existing healing response. Because of its simplicity, mobilization of stem cells is therefore an attractive delivery strategy [87,88]. It would not only obviate the need for invasive harvesting or delivery procedures, but also take advantage of the clinical procedures already established for the use of progenitor cell-mobilizing granulocyte colony-stimulating factor (G-CSF) in treating hematologic disorders. However, because of the possibility of adverse events in a different patient population and the theoretical possibility of tumorigenesis, the safety of such "off-label" applications of G-CSF have been questioned.

Transvascular delivery
Peripheral (intravenous) infusion

Peripheral (intravenous) infusion of stem cells as performed in bone marrow transplantation would be a very convenient – not to mention simple, widely available and inexpensive – way of delivering therapeutic stem cells to myocardial targets. A study in a mouse model has confirmed that bone marrow cells infused into the peripheral circulation do indeed home in on peri-infarct areas [87]. However, the number of cells that reach the affected area is very small, and the technique would be most applicable only after an AMI, as it would rely on physiologic homing signals alone. Moreover, because peripherally infused stem cells home in on infarcted areas only when injected within a few days after an AMI, this delivery strategy would be much less suitable for treating chronic myocardial ischemia.

The major drawback to using an intravenous route of cell delivery is the possibility that the therapeutic cells would become trapped in the microvasculature of the lungs, liver and lymphoid tissues. This theoretical limitation of systemic transvenous delivery of stem cells has been confirmed experimentally. In a study by Toma *et al.* [89], human MSCs were injected into the left ventricular cavity of experimental mice; 4 days later, an estimated 0.44% of the injected cells remained in the myocardium, and the rest had localized to the spleen, liver and lungs. Other studies using the systemic delivery approach have produced similar results, with very low local cell retention rates of less than 5% [90,91]. Thus, the transvenous delivery route appears unlikely to achieve the local cell concentration needed to produce a significant therapeutic benefit.

Retrograde coronary venous delivery

Two methodologies have been described for delivering therapeutic agents via the coronary venous system. Low-pressure delivery aims to increase the time that the agent is in contact with the vessels without disrupting the venous endothelium [92–94]. High-pressure delivery aims to create a biologic reservoir of product by disrupting the tight endothelial junctions of the venocapillary vasculature and mechanically driving cells across them into the myocardial interstitium [95–97]. Another, newer technique involves a new catheter that has proximal and distal balloons that occlude coronary flow and therefore theoretically allow greater contact between therapeutic cells and the coronary venous system.

The clinical experience with retrograde venous infusion is limited. Several key issues (i.e., optimal delivery pressure, volume and infusion time) remain to be resolved.

Intracoronary infusion

Intracoronary infusion has been the most popular method of delivering stem cells in the clinical setting, especially after AMI. Intracoronary stem cell delivery 4–9 days after AMI is relatively safe [98–104]. The technique is similar to that for coronary angioplasty, which involves over-the-wire positioning of an angioplasty balloon in a coronary artery (Plate 10.4). Coronary blood flow is transiently stopped for 2–4 minutes while stem cells are infused under pressure. This maximizes their contact with the microcirculation of the infarct-related artery, thereby optimizing their homing time. Again, this delivery technique would be suitable only in the setting of acute ischemia when adhesion molecules and cytokine signaling are temporarily upregulated.

Results of recent studies have challenged the safety and effectiveness of intracoronary delivery. There is growing evidence of very low retention of stem cells in target regions and of increased restenosis rates associated with this delivery method.

Intramyocardial injection

Intramyocardial injection has been performed in the clinical setting of chronic myocardial ischemia. It is the preferred delivery route in patients with chronic total occlusion of coronary arteries and in patients with chronic conditions (e.g., congestive heart failure) that involve weaker homing signals. In theory, intramyocardial injection should be the most suitable route for delivering larger cells such as skeletal myoblasts and MSCs, which are prone to microvascular "plugging." Intramyocardial injection can be performed via transepicardial, transendocardial or transcoronary venous routes.

Transepicardial injection

Transepicardial injection of stem cells has been performed during open surgical revascularization procedures to deliver the cells to infarct border zones or areas of infarcted or scarred myocardium. Because a sternotomy is required, this approach is highly invasive and associated with surgical complications. However, in the setting of a planned open heart procedure, the ancillary delivery of cell therapy in this fashion can be easily justified. Interestingly, not all areas of the myocardium (e.g., the interventricular septum) can be reached via a direct external approach.

The main advantages of direct surgical injection are its proven safety in several preclinical and human trials [39,58,105–110] and ease of use. However, it is also costly and offers a very unsophisticated targeting opportunity. The surgeon chooses to inject the border of the infarcted area or scar tissue on the basis of visual assessment only. In addition, the safety of direct surgical injection in patients with recent AMI has not been tested in clinical trials. Nevertheless, direct surgical injection certainly might have a role in the future of stem cell therapy. One can easily envision the cardiac surgeon, during coronary artery bypass grafting surgery, bypassing all areas in which it is technically feasible to do so and then concomitantly injecting stem cells into those areas containing totally occluded epicardial coronary arteries.

Transendocardial injection

Transendocardial injection is performed via a percutaneous femoral approach. An injection-needle catheter is advanced in retrograde fashion across the aortic valve and positioned against the endocardial surface. Stem cells are then injected directly into targeted areas of the left ventricular wall. Three catheter systems are currently available for transendocardial cell delivery: the Stilleto™ (Boston Scientific, Natick, MA), the BioCardia™ (BioCardia South San Francisco, CA) and the Myostar™ (Biosense Webster, Diamond Bar, CA).

The Stilleto is used under fluoroscopic (usually biplanar) guidance. Drawbacks inherent in this approach are the bidimensional orientation and lack of precision associated with fluoroscopy. Another drawback is the inability to characterize the underlying or target myocardium. Nevertheless, this technology may be promising when used in association with imaging technologies such as magnetic resonance imaging (MRI) or when targeting of myocardial therapy is not necessary. To this end, in preclinical experiments, the Stilleto catheter has been coupled with real-time cardiac MRI, which permits online assessment of full-thickness myocardium and perfusion. Although still investigational and not currently practical in terms of clinical application, the simultaneous use of MRI offers three-dimensional spatial orientation. Few preclinical studies have been performed, and no safety data from human studies have been assessed [111]. Theoretically, the use of MRI also provides a unique opportunity to track the intramyocardial retention of therapeutic cells after direct injection. However, this will require the labeling of cells (specifically mesenchymal stem cells) with fluorescent iron particles that can be detected in the beating heart.

The BioCardia delivery system uses a catheter whose deflectable tip includes a helical needle for infusion. Initial preclinical and clinical experience with this system has provided preliminary evidence of its safety and feasibility [112,113]. Unlike the other two catheter delivery systems discussed here, the BioCardia catheter does not offer any additional navigational or targeting tool. More extensive preclinical experience with this catheter is needed before human trials can begin.

The Myostar injection catheter takes advantage of nonfluoroscopic magnetic guidance [114]. Injections are targeted with the help of a three-dimensional left ventricular "shell," or NOGA electromechanical map (EMM), representing the endocardial surface of the left ventricle. The shell is constructed by acquiring a series of electrocardiogram-gated points at multiple locations on the endocardial surface. Ultralow magnetic fields (10^- to 10^{-6} tesla) generated by a triangular magnetic pad positioned beneath the patient intersect with a sensor just proximal to the deflectable tip of a 7F mapping catheter, which helps determine the real-time location and orientation of the catheter tip inside the left ventricle. The NOGA system algorithmically calculates and analyzes the movement of the catheter tip or the location of an endocardial point throughout systole and diastole. That movement is then compared with the movement of neighboring points in an area of interest. The resulting value, called linear local shortening (LLS), is expressed as a percentage that represents the degree of mechanical function of the left ventricular region at that endocardial point. Data are obtained only when the catheter tip is in stable contact with the endocardium. This contact is determined automatically.

The mapping catheter also incorporates electrodes that measure endocardial electrical signals (unipolar or bipolar voltage) [114]. Voltage values are assigned to each point acquired during left ventricular mapping, and an electrical map is constructed concurrently with the mechanical map. Each data point has an LLS value and a voltage value. When the map is complete, all the data points are integrated by the NOGA workstation into a three-dimensional color-coded map of the endocardial surface, as well as 9- and 12-segment bull's-eye views that show average LLS and voltage values in each myocardial segment. These maps can be spatially manipulated in real time on a Silicon Graphics workstation (Mountain View, CA). The three-dimensional representations acquired during the cardiac cycle can also be used to calculate left ventricular volume and ejection fraction.

The three-dimensional EMM serves both therapeutic and diagnostic purposes. On the one hand, it allows the catheter to be maneuvered through the left ventricle and oriented for transendocardial injections. On the other hand, it allows ischemic areas (i.e., those with low LLS and preserved unipolar voltage [UniV]) to be distinguished from infarcted areas (i.e., those with low LLS and low UniV) [115]. Moreover, the Myostar catheter allows myocardial viability to be assessed at each specific injection site where the catheter touches the endocardial surface. The operator is thus able to target therapy to viable tissue (where neoangiogenesis may be possible) or nonviable tissue (where the target of cell therapy may be a scarred area). Because of the patchy nature of human ischemic heart disease, the ability to characterize the underlying myocardial tissue is important when delivering stem cells.

The EMM technology has been widely tested in both animals and humans and has an excellent safety profile [26,116–123]. Kornowski *et al.* [121] have studied the dynamics of transendocardial delivery using different needle extensions to inject 0.1 mL methylene blue dye as a tracer. A total of 152 injections were performed with needle extensions varying from 3 to 4 mm in length. Two myocardial regions were injected per animal, and injection sites were located after the animals were sacrificed acutely. Staining extended to a depth of 7.1 ± 2.1 mm (range, 2–11 mm) and to a width of 2.3 ± 1.8 mm (range, 1–9 mm). In 2.6% of cases (4 of 152), the injected dye stained the epicardial surface, suggesting pericardial extravasation; more importantly, three of those four injections were made in the apical area. There were no animal deaths, no instances of pericardial effusion or tamponade, and no episodes of sustained ventricular arrhythmia associated with the transendocardial injections.

Despite the limitations of the animal model, this preclinical experience has translated well into clinical trials. However, it is very important to note that the clinical safety profile of transendocardial delivery so far has entailed precise preinjection measurements of needle extension with the injection catheter tip deflected (to 90°) and not deflected, arbitrary insistence on a maximal needle : wall ratio of 0.6, and a conscious decision not to inject stem cells into cardiac walls that are less than 8 mm thick or into the true apical segment.

Transcoronary venous injection

Transcoronary venous injection is performed with a catheter system threaded percutaneously into the coronary sinus. Initial studies in swine have confirmed the feasibility and safety of this approach [112]. This delivery method has also been used to deliver skeletal myoblasts to scarred myocardium in cardiomyopathy patients [113]. With intravascular ultrasound guidance, this approach allows the operator to extend a catheter and needle away from the pericardial space and coronary artery into the adjacent myocardium. To date, human feasibility studies have had a good safety profile. This technique is limited, however, by coronary venous tortuosity, lack of site specific targeting and its own technically challenging nature. Unlike the transendocardial approach, in which cells are injected perpendicularly into the left ventricular wall, the transcoronary venous approach allows parallel cell injection, which may result in greater cell retention.

Comparisons of delivery methods

The biodistribution of intravenously injected allogeneic MSCs has been recently described [124]. Oxine-labeled MSCs were injected intravenously 72 hours after occlusion/reperfusion in seven dogs. Initially, cells were trapped in the lungs; within 24 hours after injection, they had been redistributed into the liver and spleen. Focal uptake and persistence of the stem cells was observed in a mid anterior wall location corresponding to the infarcted target area.

Few studies have compared the different modes of cell delivery. Hou *et al.* [125] have described the fate of peripheral blood mononuclear cells (PBMNCs) 1 hour after direct surgical injection, intracoronary infusion and retrograde venous infusion in an acute swine ischemia-reperfusion model. Overall, PBMNCs concentrated significantly more in the pulmonary vasculature and parenchyma than in the myocardium. Direct surgical injection resulted in significantly less pulmonary retention (26%) than did either intracoronary infusion (47%) or retrograde venous infusion (43%). Cells were scarcely present in the liver and spleen. Myocardial homing, even in a setting of intense homing signaling, was limited in all three approaches, although direct intramyocardial injection (11.3%) achieved better homing and engraftment than did either intracoronary infusion (2.6%) or retrograde venous infusion (3.2%).

Together, these data suggest that none of these three delivery strategies are more than modestly efficient at delivering cells to targeted regions. This is of special concern in the case of intracoronary delivery, which is the stem cell delivery method most widely used after AMI. This has several important clinical implications for the future of cardiac stem cell therapy: higher doses might be needed to achieve desired therapeutic effects, new (e.g., combined) delivery strategies need to be considered, myocardial homing and signaling must be better understood, and recipients of systemically delivered cells must be followed up carefully and closely.

Clinical trials of cardiac stem cell therapy

Clinical research with bone marrow-derived stem cells has focused on the period immediately after an AMI and on the chronic phase of ischemic heart disease. In these clinical scenarios, therapy has been targeted to viable myocardium with or without systolic heart failure. On the other hand, skeletal myoblast therapy has been used to treat ischemic heart failure involving nonviable myocardium or scar tissue and compromised systolic left ventricular function. In simple terms, skeletal myoblasts offer "myocyte replacement therapy" for scarred myocardial segments, and bone marrow stem cells offer "neoangiogenic and regenerative therapy" for acute and chronic ischemic heart disease involving viable myocardial tissue.

Most of the clinical experience gained with stem cells has involved therapy for AMI, particularly intracoronary infusion of bone marrow cells because skeletal myoblasts are too large for this purpose [126]. Table 10.2 summarizes the experience to date. In all of these trials, revascularization was performed promptly after the index myocardial infarction, and left ventricular systolic compromise was minor (in the BOOST trial, the baseline left ventricular ejection fraction [LVEF] was 50%).

In the Transplantation of Progenitor Cells and Regeneration Enhancement in Acute Myocardial Infarction (TOPCARE-AMI) trial, patients were randomized to receive either bone marrow-derived mononuclear cells or EPCs via intracoronary infusion [98]. Compared with the nonrandomized control patients, treated patients had a significantly improved global LVEF, as assessed by left ventricular angiography, regardless of cell type used. More recently, in a subgroup of this study population, LVEF was significantly increased on cardiac MRI and infarct size was reduced on late enhancement MRI [99]. Interestingly, the ability of infused cells to migrate was the most important predictor of infarct remodeling. Cell therapy also increased coronary flow reserve, possibly suggesting neovascularization.

The 1-year results of TOPCARE-AMI reinforce the notion that stem cells protect against ventricular

Table 10.2 Trials of intracoronary cell therapy in patients with acute myocardial infarction.

Study	[n]	Cell type	Dose	Time after AMI	Therapeutic effects	
					Improved	No change
Nonrandomized						
Strauer *et al.* [103]	10 treated 10 controls*	ABMMNC	$2.8 \pm 2.2 \times 10^7$	5–9 days	Regional wall motion↑ Perfusion↑ ↓ Infarct size	Global LVEF, LVEDV↑
TOPCARE-AMI [98,99,102]	29 ABMMNC 30 CPC 11 controls*	ABMMNC, CPC	$2.1 \pm 0.8 \times 10^8$ $1.6 \pm 1.2 \times 10^7$	5 ± 2 days	Regional wall motion↑ Global LVEF↑ ↓ Infarct size↑ Coronary flow↑	LVEDV↑
Fernandez-Aviles *et al.* [101]	20 treated 13 controls*	ABMMNC	$7.8 \pm 4.1 \times 10^7$	14 ± 6 days	Regional wall motion↑ Global LVEF↑	LVEDV↑
Randomized						
BOOST [104]	30 treated, 30 controls	NC	$2.5 \pm 0.9 \times 10^{10}$	6 ± 1 days	Regional wall motion Global LVEF	LVEDV Infarct size
Chen *et al.* [100]	34 treated 35 controls	MSC	$4.8 \pm 6.0 \times 10^{10}$	18 days	Regional wall motion Global LVEF ↓ Infarct size ↓ LVEDV	

ABMMNC, autologous bone-marrow-derived mononuclear cells; AMI, acute myocardial infarction; BOOST, Bone Marrow Transfer to Enhance ST Elevation Infarct Regeneration; CPC, circulating blood-derived progenitor cells; LVEDV, left ventricular end-diastolic volume; LVEF, left ventricular ejection fraction; MSC, mesenchymal stem cells; NC, bone-marrow-derived nucleated cells; TOPCARE, Transplantation of Progenitor Cells and Regeneration Enhancement.
* Nonrandomized control groups.
† Effects reported only within cell therapy groups. Values are mean ± standard deviation.

remodeling. Despite the limited number of patients, contrast-enhanced MRI revealed a significantly increased LVEF (P <0.001), significantly reduced infarct size (P <0.001) and the absence of reactive hypertrophy, suggesting that the infarcted ventricles had been functionally regenerated. Scientific criticism of this trial has focused on the cell delivery method, which included transient coronary occlusion and flow cessation, and its potential for ischemic preconditioning. Such preconditioning has been shown to improve outcomes during AMI and may have contributed to the functional improvement noted in this trial. Moreover, the occurrence of in-stent thrombosis in one patient 3 days after undergoing cell therapy has raised safety concerns.

The results of the randomized BOOST trial [104] – the most important clinical trial of intracoronary infusion to date – have recently been published. Patients received either bone marrow-derived ABMMNCs or no treatment at all (no placebo). Stem cell therapy resulted in an increased LVEF and a reduced end-systolic volume, as assessed by MRI. This improvement was attributed principally to increased contractility of the peri-infarct zones. Unlike earlier nonrandomized trials, the BOOST trial did not show a significant reduction in infarct size.

In a study by Bartunek *et al.* [127], 35 patients were infused with AC133$^+$ bone marrow cells after AMI. The mean dose was 12.6 million cells, and the

Table 10.3 Cell therapy trials in patients with ischemic cardiomyopathy.

Study	[n]	LVEF	Cell type	Dose	Time after MI	Delivery	Outcomes in treated groups
Menasche et al. [107]	10 treated	$24 \pm 4\%$	Myoblasts	$8.7 \pm 1.9 \times 10^8$	3–228 months	Transepicardial (during CABG)*	↑ Regional wall motion ↑ Global LVEF
Herreros et al. [106]	11 treated	$36 \pm 8\%$	Myoblasts	$1.9 \pm 1.2 \times 10^8$	3–168 months	Transepicardial (during CABG)†	↑ Regional wall motion ↑ Global LVEF ↑ Viability in infarct area
Siminiak et al. [108]	10 treated	25–40%	Myoblasts	$0.04–5.0 \times 10^7$	4–108 months	Transepicardial (during CABG)†	↑ Regional wall motion ↑ Global LVEF
Chachques et al. [142]	20 treated	$28 \pm 3\%$	Myoblasts	$3.0 \pm 0.2 \times 10^8$	Not reported	Transepicardial (during CABG)*	↑ Regional wall motion ↑ Global LVEF ↑ Viability in infarct area
Smits et al. [143]	5 treated	$36 \pm 11\%$	Myoblasts	$2.0 \pm 1.1 \times 10^8$	24–132 months	Transendocardial (guided by EMM)	↑ Regional wall motion ↑ Global LVEF
Stamm et al. [109,110]	12 treated	$36 \pm 11\%$	CD133+	$1.0–2.8 \times 10^6$	3–12 weeks	Transepicardial (during CABG)*	↑ Global LVEF ↑ Perfusion ↓ LVEDV
Assmus et al. [144]	51 ABMMNC 35 CPC 16 controls	$40 \pm 11\%$	ABMMNC CPC	$1.7 \pm 0.8 \times 10^8$ $2.3 \pm 1.2 \times 10^7$	3–144 months	IC	↑ Global LVEF (only in ABMMNC group)

* CABG of noninjected territories only.

† CABG of injected and noninjected territories.

Values are mean ± standard deviation.

ABMMNC, autologous bone marrow mononuclear cells; AMI, acute myocardial infarction; BM, bone marrow; CABG, coronary artery bypass grafting; CD133+, bone-marrow–derived CD133+ cells; CPC, circulating blood-derived progenitor cells; EMM, electromechanical mapping; IC, intracoronary; LVEDV, left ventricular end-diastolic volume; LVEF, left ventricular ejection fraction; MI, myocardial infarction; NYHA, New York Heart Association.

mean infusion was 11.4 days after the index event. At 4-month follow-up, treated patients had an improved mean LVEF but higher rates of stent restenosis, stent reocclusion and *de novo* coronary artery lesions than did the controls.

The intracoronary route has also been used to deliver autologous MSCs. Chen *et al.* [100] recently reported the first randomized clinical trial of these cells in 69 patients who underwent a primary percutaneous coronary intervention within 12 hours after an AMI. Either MSCs or saline was injected into the target coronary artery. At 3-month follow-up, left ventricular perfusion and the LVEF had significantly improved in the treatment group.

The feasibility and efficacy of G-CSF therapy and subsequent intracoronary infusion of collected peripheral blood stem cells were prospectively investigated in the MAGIC randomized clinical trial [128,129], which showed improved cardiac function and promotion of angiogenesis in myocardial infarction patients. However, the trial raised important safety questions. Intracoronary infusion of G-CSF-stimulated peripheral-blood stem cells apparently aggravated restenosis after coronary stenting, leading to early termination of the trial. Meanwhile, no temporal association between increased restenosis rate and stenting near the time of intracoronary cell administration has been noted in studies that did not use G-CSF stimulation.

A different therapeutic strategy using G-CSF involved mobilization of CD34$^+$ cells from the bone marrow to the peripheral blood [32]. Thirty patients in the superacute phase of MI underwent primary percutaneous revascularization. Eighty-five minutes after revascularization, 15 patients were randomized to begin receiving G-CSF stimulation for up to 6 days. At 1-year follow-up, the G-CSF treated patients had significantly improved LVEF and stable end-diastolic diameters.

Outside the AMI setting, stem cells have been used to treat patients with ischemic heart disease with or without systolic functional compromise and patients unsuitable for myocardial revascularization (Tables 10.3 and 10.4). Autologous bone marrow stem cells have been used to treat patients with chronic myocardial ischemia, including ischemic heart failure with or without systolic functional compromise, and patients ineligible for myocardial revascularization (Table 10.5). The preliminary

clinical evidence supports the efficacy of this new therapy and, at this point, all the evidence appears to substantiate its safety.

Tse *et al.* [123] have reported that transendocardial injection of ABMMNCs in eight patients with severe ischemic heart disease led to preserved left ventricular function. At 3-month follow-up, heart failure symptoms and myocardial perfusion had improved, especially in the ischemic region as shown by cardiac MRI (Plate 10.2).

Fuchs *et al.* [117] studied the clinical feasibility of transendocardial delivery of filtered unfractionated autologous bone marrow-derived (not mononuclear) cells in 10 patients with severe symptomatic chronic myocardial ischemia not amenable to conventional revascularization. Twelve targeted injections (0.2 mL each) were administered into ischemic noninfarcted myocardium identified previously by single-photon emission computed tomography (SPECT) perfusion imaging. No serious adverse effects (i.e., arrhythmia, infection, myocardial inflammation or increased scar formation) were noted. Moreover, even though treadmill exercise duration results did not change significantly (391 ± 155 vs. 485 ± 198 s; $P = 0.11$), there was improvement in Canadian Cardiovascular Society angina scores (3.1 ± 0.3 vs. 2.0 ± 0.94; $P = 0.001$) and in stress scores in segments within the injected regions (2.1 ± 0.8 vs. 1.6 ± 0.8; $P < 0.001$).

Our group performed the first clinical trial of transendocardial injection of ABMMCs to treat heart failure patients [26]. This study, performed in collaboration with physicians and scientists at the Hospital Pro-Cardiaco in Rio de Janeiro, Brazil, used EMM-guided transendocardial delivery of stem cells. The results of 2- and 4-month noninvasive and invasive follow-up evaluations [26] and of 6- and 12-month follow-up evaluation [122] have already been published.

A total of 21 patients were enrolled. The first 14 comprised the treatment group, and the last 7 patients the control group. Baseline evaluations included complete clinical and laboratory tests, exercise stress (ramp treadmill) studies, two-dimensional Doppler echocardiography, SPECT perfusion scanning and 24-hour Holter monitoring. ABMMNCs were harvested, isolated, washed and resuspended in saline for injection via NOGA catheter (15 injections of 0.2 mL, totalling 30×10^6

Table 10.4 Cell therapy trials in patients with myocardial ischemia and no revascularization option.

Study	[n]	LVEF	Cell Type	Dose	Delivery	Outcomes	
						Subjective	Objective
Hamano et al. [105]	5 treated	–	ABMMNC	0.3–2.2 × 10⁹	Transepicardial (during CABG)	–	↑ Perfusion†
Tse et al. [123]	8 treated	58 ± 11%	ABMMNC	From 40 mL BM	Transendocardial (guided by EMM)	↓ Angina†	↑ Perfusion† ↑ Regional wall motion†
Fuchs et al. [117]	10 treated	47 ± 10%	NC	7.8 ± 6.6 × 10⁷	Transendocardial (guided by EMM)	↓ Angina†	↑ Perfusion†
Perin et al. [26,122]	14 treated 7 controls*	30 ± 6%	ABMMNC	3.0 ± 0.4 × 10⁷	Transendocardial (guided by EMM)	↓ Angina ↓ NYHA class	↑ Perfusion ↑ Regional wall motion† ↑ Global LVEF

ABMMNC, autologous bone-marrow-derived mononuclear cells; BM, bone marrow; CABG, coronary artery bypass grafting; EMM, electromechanical mapping; LVEF, left ventricular ejection fraction; NC, bone-marrow-derived nucleated cells; NYHA, New York Heart Association.

* Nonrandomized control group.

† Effects reported only within cell therapy groups. Values are mean ± standard deviation.

cells per patient) in viable myocardium (unipolar voltage ≥6.9 mV). All patients underwent non-invasive follow-up tests at 2 months, and the treatment group also underwent invasive studies at 4 months, using standard protocols and the same procedures as at baseline. The demographic and exercise test variables did not differ significantly between the treatment and control groups. There were no procedural complications. At 2 months, there was a significant reduction in the total reversible defect in the treatment group and between the treatment and control groups ($P = 0.02$) on quantitative SPECT analysis. At 4 months, the LVEF improved from a baseline of 20–29% ($P = 0.003$) and the end-systolic volume decreased ($P = 0.03$) in the treated patients. Electromechanical mapping revealed significant mechanical improvement in the injected segments ($P < 0.0005$). In our opinion, this established the safety of transendocardial injection of ABMMNCs and warranted further investigation of this therapy's efficacy end-points. This trial was important because for the first time myocardial perfusion and cardiac function were observed to improve in a group of severely impaired patients treated solely with stem cells. The

significant improvement seen at 2 and 4 months was maintained at 6 and 12 months, even as exercise capacity improved slightly (Table 10.3). Monocyte, B cell, hematopoietic progenitor cell and early hematopoietic progenitor cell subpopulations correlated with improvement in reversible perfusion defects at 6 months (Table 10.4).

Clinical trials of skeletal myoblasts have focused on the treatment of patients with ischemic cardiomyopathy and systolic dysfunction. Overall, these trials have resulted in improved segmental contractility and global LVEF. The preferred delivery route has been surgical intramyocardial injection, and one feasibility trial of transendocardial injection has been reported in the literature so far.

Stem cell-induced functional improvement in chronic myocardial ischemia

Preclinical studies

Preclinical experiments have provided solid evidence supporting the efficacy of cardiac ABMMNC therapy; however, further investigation at the molecular level is needed to elucidate the mechanistic

Table 10.5 Comparison of clinical values for the treatment and control groups at baseline, 2 months, 6 months and 12 months.

Variable	Baseline		2 Months		6 Months		12 Months		P value*
	Rx	Control	Rx	Control	Rx	Control	Rx	Control	
SPECT									
Total reversible defect, %	14.8 ± 14.5	20 ± 25.4	4.45 ± 11.5	37 ± 38.4	8.8 ± 9	32.7 ± 37	11.3 ± 12.8	34.3 ± 30.8	0.01
Total fixed defect (50%), %	42.6 ± 10.3	38 ± 12	39.8 ± 6.9	39.1 ± 11.2	38 ± 6.7	36.4 ± 12	38.2 ± 8.5	35.2 ± 9.3	0.3
Ramp treadmill test									
VO_2 max, mL/kg/min	17.3 ± 8	17.5 ± 6.7	23.2 ± 8	18.3 ± 9.6	24.2 ± 7	17.3 ± 6	25.1 ± 8.7	18.2 ± 6.7	0.03
METS	5.0 ± 2.3	5.0 ± 1.91	6.6 ± 2.3	5.2 ± 2.7	7.2 ± 2.4	4.9 ± 1.7	7.2 ± 2.5	5.1 ± 1.9	0.02
LVEF	30 ± 6	37 ± 14	37 ± 6	27 ± 6	30 ± 10	28 ± 4	35.1 ± 6.9	34 ± 3	0.9
Functional class									
NYHA	2.2 ± 0.9	2.7 ± 0.8	1.5 ± 0.5	2.4 ± 1.0	1.3 ± 0.6	2.4 ± 0.5	1.4 ± 0.7	2.7 ± 0.5	0.01
CCSAS	2.6 ± 0.8	2.9 ± 1.0	1.8 ± 0.6	2.5 ± 0.8	1.4 ± 0.5	2 ± 0.1	1.2 ± 0.4	2.7 ± 0.5	0.002
PVCs, n	2507 ± 6243	672 ± 1085	901 ± 1236	2034 ± 4528	3902 ± 8267	1041 ± 1971	–	–	0.4
dQRS, ms	136 ± 15	145 ± 61	145.9 ± 25	130 ± 27	144.8 ± 25	140 ± 61	–	–	0.62
LAS 40, ms	50 ± 24	70 ± 76	54 ± 33	48 ± 20	25 ± 25	66 ± 79	–	–	0.47
RMS 40, μV	22.2 ± 22	23.3 ± 23	23.3 ± 19	24.6 ± 28	25 ± 25	30 ± 27	–	–	0.7

* *P* value for comparisons between the treatment and control groups, as assessed by ANOVA, relating to treatment over time.

CCSAS, Canadian Cardiovascular Society Angina Score; dQRS, filtered QRS duration; LAS 40, duration of terminal low-amplitude signal less than 40 mV; LEVF, left ventricular ejection fraction; METS, metabolic equivalents; NYHA, New York Heart Association; PVC, premature ventricular contraction; RMS 40, root mean square voltage in the terminal 40 ms of the QRS complex; Rx, treatment; SPECT, single-photon emission computer tomography; VO_2 max, maximal rate of oxygen consumption.

aspects of stem cell therapy – an area where more questions than answers remain.

Numerous research groups, using various detection methods in diverse experimental settings, have proposed different mechanisms for the apparent transformation of stem cells into cells of a variety of tissues [5]. Some investigators attribute the transformation to the transdifferentiation potential of stem cells [58,130,131], while others have it to be a result of cell fusion [132].

Initial evidence indicates that ABMMNCs transdifferentiate into endothelial cells and cardiac myocytes. Recent studies in mice, however, have controversially challenged this notion. In a recent study, Murry *et al.* [132] could detect no ABMMNC transdifferentiation into a cardiomyocyte phenotype, despite the use of sophisticated genetic techniques for following cell fate and engraftment. In experimental models, ABMMNCs have been shown to depend on external signals that trigger secretory properties and differentiation [133]. The local environment of viable myocardial cells may provide the milieu necessary for inducing ABMMNC myocyte differentiation [134]. In recent studies of occlusion-induced myocardial infarction in rats, few (if any) ABMMNCs might be expected to differentiate and express specific cardiac myocyte proteins, depending on the injection site. To further clarify the issue of transdifferentiation versus fusion, Zhang *et al.* [135] elegantly used flow cytometry analysis to study heart cell isolates from mice that had received human $CD34^+$ cells. HLA-ABC and cardiac troponin T or Nkx2.5 were used to identify cardiomyocytes derived from human $CD34^+$ cells, and HLA-ABC and VE-cadherin were used to identify the transformed endothelial cells. The double-positive cells were tested for the expression of human and mouse X chromosomes. As a result, 73.3% of nuclei derived from HLA^+ and troponin T^+ or Nkx2.5$^+$ cardiomyocytes contained both human and mouse X chromosomes, and 23.7% contained only human X chromosomes. In contrast, the nuclei of HLA^-, troponin T^+ cells contained only mouse X chromosomes. Furthermore, 97.3% of endothelial cells derived from $CD34^+$ cells contained human X chromosomes only. Thus, human $CD34^+$ cells both fused with and transdifferentiated into cardiomyocytes in this mouse model. In addition, human $CD34^+$ cells also transdifferentiated into endothelial cells.

The transdifferentiation of HSCs into a mature hematopoietic fate (e.g., endothelium) in the heart is less controversial [136]. In animal models of stem cell therapy in ischemic heart disease, the evidence points toward increased neovascularization (with reduced myocardial ischemia) and consequent improvement in cardiac function [137–139]. Bone marrow stem cells may directly contribute to an increase in contractility or, more likely, may passively limit infarct expansion and remodeling. Unfortunately, the limitations of the present animal models leave this question unanswered.

According to the current understanding of bone marrow stem cell engraftment, most cells die within the first days after delivery. Arteriogenesis and vasculogenesis have long been known to be highly dependent on vascular growth factors. In light of the notion, recently proposed by Kinnaird *et al.* [126,129], that MSCs contribute to angiogenesis by means of paracrine mechanisms, it may be that therapeutic bone marrow stem cells recruit circulating progenitor cells, activate resident cardiac stem cells, or both, via such paracrine means, thus triggering a cascade of events resulting in cardiac repair. The important role of resident cardiac stem cells in the process of cardiac repair should also be considered [75]. Urbanek *et al.* [38] were the first to describe evidence of myocyte formation from cardiac stem cells in human cardiac hypertrophy.

Clinical trials

We recently described the postmortem study of one of our patients who received ABMMNCs [140]. Eleven months after performing the treatment, we observed no abnormal or disorganized tissue growth, no abnormal vascular growth and no enhanced inflammatory reactions. Histologic and immunohistochemical findings from infarcted areas of the anterolateral ventricular wall (areas that had received bone marrow cell injections) were reported. The histologic findings from the anterolateral wall region were subsequently compared with findings from within the interventricular septum (which had normal perfusion in the central region and no cell therapy) and findings from the previously infarcted inferoposterior ventricular wall (which had extensive scarring and no cell therapy).

The observed effects of cell therapy were quite intriguing: first, the cell-treated infarcted areas of

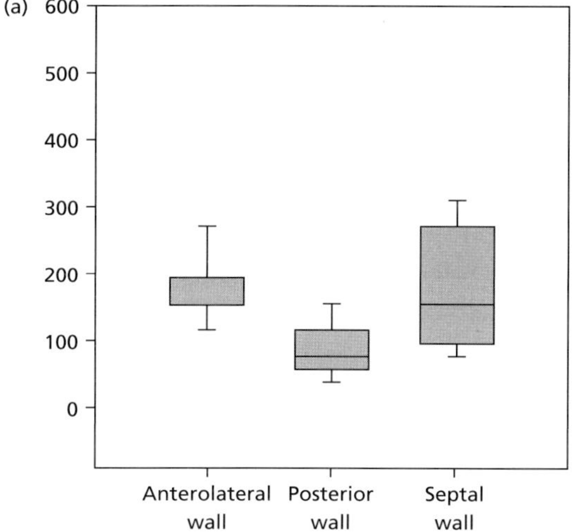

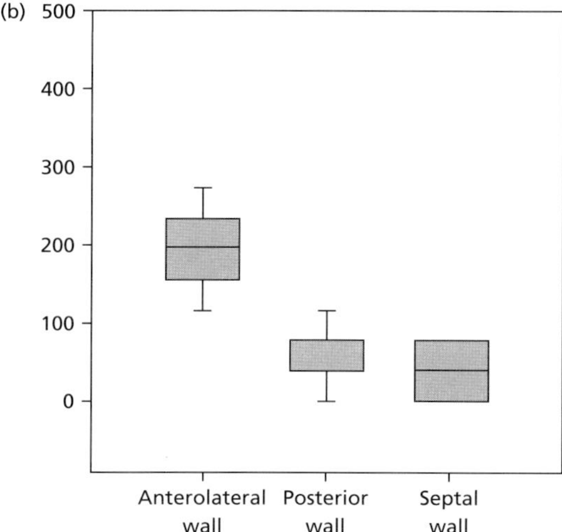

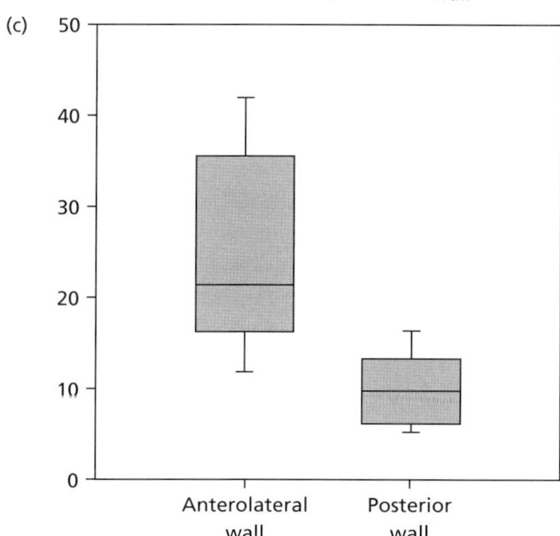

this patient's heart had a higher capillary density than did the nontreated infarcted areas (Fig. 10.3). Second, smooth muscle α-actin-positive pericytes and mural cells proliferated exclusively in the cell-treated area. Third, these pericytes and mural cells expressed specific cardiomyocyte proteins.

The angiogenesis literature makes clear that pericytes are essential for long-lasting physiologic angiogenesis. In our postmortem study, the cell-injected wall had marked areas of pericyte and mural cells hyperplasia. The observed hypertrophic pericytes, although still located in the vascular wall, expressed specific myocardial proteins and were found in locations distant from the vessel walls, suggesting detachment. Migratory pericytes and mural cells were found in adjacent tissue (in the vicinity of cardiomyocytes) either isolated or in small clumps. Closer to cardiomyocytes, the expression of myocardial proteins was enhanced, yielding brighter immunostaining throughout the whole cytoplasm. Within the posterior wall, none of this was seen and small blood vessels were rare. Although it would be premature to arrive at any definitive conclusions about ABMMNC efficacy on the basis of one postmortem study, the above findings in the cell-treated wall are consistent with neoangiogenesis. If confirmed in future human studies, these findings would corroborate most of the preclinical studies in chronic myocardial ischemia models.

Safety of stem cell therapy

With regard to left ventricular function, cardiac stem cell therapy is well tolerated overall. No proarrhythmic effects have been observed to date with ABMMNC therapy, although other deleterious

Figure 10.3 Number of capillaries per mm² in anterolateral, posterior, and septal walls of studied heart. (a), Anti-factor VIII-associated antigen counterstained with hematoxylin. (b), Anti-smooth muscle-actin antigen counterstained with hematoxylin. (c), Capillaries reacted with anti-factor VIII-associated antigen inside fibrotic areas only in anterolateral and posterior walls. (n = 108 microscope fields for (a); 96 microscope fields for (b); and 40 microscopic fields for (c).) Differences were statistically significant among all groups in pairwise comparisons (*P* <0.05, Newman-Keuls method) for (a) and (b). Differences were significantly different (*P* <0.05) between anterolateral and posterior walls in Mann–Whitney rank-sum test for (c).

effects are possible. Early concerns about abnormal transdifferentiation and tumorigenesis have subsided, but the potential for accelerated atherogenesis remains, given the limited clinical experience and the small number of patients treated. Because atherosclerosis is an inflammatory disease triggered and sustained by cytokines, adhesion molecules and cellular components such as monocytes and macrophages, intracoronary delivery is potentially risky. In addition, as already mentioned, post myocardial infarction intracoronary infusion has been associated with increased rates of restenosis and stent thrombosis. Given the small number of patients treated in all Phase I and II trials so far, this is a particular point of concern.

Another potential deleterious effect of bone marrow stem cell therapy is myocardial calcification. In a recent study, Yoon *et al.* [141] noted that direct transplantation of unselected bone marrow cells into acutely infarcted myocardium could induce significant intramyocardial calcification. In the same study, however, ABMMNCs did not.

Myoblast therapy raises the possibility of arrhythmogenic effects. Consequently, many clinical studies require the placement of cardiac defibrillators in patients receiving myoblasts.

Conclusions

Despite many unresolved issues related to treatment dose, timing and delivery, the clinical potential of stem cell therapy for cardiovascular disease is enormous. The expectations of both patients and clinicians for this new therapeutic modality, however, are high and will require continued cooperation and close collaboration between basic and clinical researchers.

References

1 Radovancevic, B, Vrtovec B, Frazier OH. Left ventricular assist devices: an alternative to medical therapy for end-stage heart failure. *Curr Opin Cardiol* 2003; 18: 210–214.

2 Asahara T, Murohara T, Sullivan A *et al.* Isolation of putative progenitor endothelial cells for angiogenesis. *Science* 1997; 275: 964–967.

3 Shintani S, Murohara T, Ikeda H *et al.* Mobilization of endothelial progenitor cells in patients with acute myocardial infarction. *Circulation* 2001; 103: 2776–2779.

4 Krause DS. Plasticity of marrow-derived stem cells. *Gene Ther* 2002; 9: 754–758.

5 Perin EC, Geng YJ, Willerson JT. Adult stem cell therapy in perspective. *Circulation* 2003; 107: 935–938.

6 Blau HM, Brazelton TR, Weimann JM. The evolving concept of a stem cell: entity or function? Cell 2001; 105: 829–841.

7 Weissman IL. Stem cells: units of development, units of regeneration, and units in evolution. Cell 2000; 100: 157–168.

8 Available from: www.nih.org

9 Korbling M, Estrov Z. Adult stem cells for tissue repair – a new therapeutic concept? *N Engl J Med* 2003; 349: 570–582.

10 Rumpold H, Wolf D, Koeck R *et al.* Endothelial progenitor cells: a source for therapeutic vasculogenesis? *J Cell Mol Med* 2004; 8: 509–518.

11 Asahara T, Masuda H, Takahashi T *et al.* Bone marrow origin of endothelial progenitor cells responsible for postnatal vasculogenesis in physiological and pathological neovascularization. *Circ Res* 1999; 85: 221–228.

12 Asahara T, Takahashi T, Masuda H *et al.* VEGF contributes to postnatal neovascularization by mobilizing bone marrow-derived endothelial progenitor cells. *EMBO J* 1999; 18: 3964–3972.

13 Bhattacharya V, McSweeney PA, Shi Q *et al.* Enhanced endothelialization and microvessel formation in polyester grafts seeded with CD34$^+$ bone marrow cells. *Blood* 2000; 95: 581–585.

14 Edelberg JM, Tang L, Hattori K *et al.* Young adult bone marrow-derived endothelial precursor cells restore aging-impaired cardiac angiogenic function. *Circ Res* 2002; 90: E89–E93.

15 Gehling UM, Ergun S, Schumacher U *et al.* In vitro differentiation of endothelial cells from AC133-positive progenitor cells. *Blood* 2000; 95: 3106–3112.

16 Gill M, Dias S, Hattori K *et al.* Vascular trauma induces rapid but transient mobilization of VEGFR2$^+$AC133$^+$ endothelial precursor cells. *Circ Res* 2001; 88: 167–174.

17 Gunsilius E, Petzer AL, Duba HC *et al.* Circulating endothelial cells after transplantation. *Lancet* 2001; 357: 1449–1450.

18 Hatzopoulos AK, Fokman J, Vasile E *et al.* Isolation and characterization of endothelial progenitor cells from mouse embryos. *Development* 1998; 125: 1457–1468.

19 Iwaguro H, Yamaguchi J, Kalka C *et al.* Endothelial progenitor cell vascular endothelial growth factor gene transfer for vascular regeneration. *Circulation* 2002; 105: 732–738.

20 Kalka C, Masuda H, Takahashi T *et al.* Transplantation of *ex vivo* expanded endothelial progenitor cells for therapeutic neovascularization. *Proc Natl Acad Sci USA* 2000; 97: 3422–3427.

21 Kaushal S, Amiel GE, Guleserian KJ *et al.* Functional small-diameter neovessels created using endothelial progenitor cells expanded *ex vivo*. *Nat Med* 2001; 7: 1035–1040.

22 Lin Y, Weisdorf DJ, Solovey A *et al.* Origins of circulating endothelial cells and endothelial outgrowth from blood. *J Clin Invest* 2000; 105: 71–77.

23 Murohara T, Ikeda H, Duan J *et al.* Transplanted cord blood-derived endothelial precursor cells augment postnatal neovascularization. *J Clin Invest* 2000; 105: 1527–1536.

24 Nieda M, Nicol A, Denning-Kendall P *et al.* Endothelial cell precursors are normal components of human umbilical cord blood. *Br J Haematol* 1997; 98: 775–757.

25 Peichev M, Neiyer AJ, Pereira D *et al.* Expression of VEGFR-2 and AC133 by circulating human CD34$^+$ cells identifies a population of functional endothelial precursors. *Blood* 2000; 95: 952–958.

26 Perin EC, Dohmann HF, Borojevic R *et al.* Transendocardial, autologous bone marrow cell transplantation for severe, chronic ischemic heart failure. *Circulation* 2003; 107: 2294–2302.

27 Schatteman GC, Hanlon HD, Jiao C *et al.* Blood-derived angioblasts accelerate blood-flow restoration in diabetic mice. *J Clin Invest* 2000; 106: 571–578.

28 Shi Q, Rafii S, Wu MH *et al.* Evidence for circulating bone marrow-derived endothelial cells. *Blood* 1998; 92: 362–367.

29 Springer ML, Chen AS, Kraft PE *et al.* VEGF gene delivery to muscle: potential role for vasculogenesis in adults. *Mol Cell* 1998; 2: 549–558.

30 Takahashi T, Kalka C, Masuda H *et al.* Ischemia- and cytokine-induced mobilization of bone marrow-derived endothelial progenitor cells for neovascularization. *Nat Med* 1999; 5: 434–438.

31 Vasa M, Fichtlscherer S, Aicher A *et al.* Number and migratory activity of circulating endothelial progenitor cells inversely correlate with risk factors for coronary artery disease. *Circ Res* 2001; 89: E1–E7.

32 Iwami Y, Masuda H, Asahara T. Endothelial progenitor cells: past, state of the art, and future. *J Cell Mol Med* 2004; 8: 488–497.

33 Werner N, Kosiol S, Schiegl T *et al.* Circulating endothelial progenitor cells and cardiovascular outcomes. *N Engl J Med* 2005; 353: 999–1007.

34 Ferrari G, Cusella-De Angelis G, Coletta M *et al.* Muscle regeneration by bone marrow derived myogenic progenitors. *Science* 1998; 279: 1528–1530.

35 Graf T. Differentiation plasticity of hematopoietic cells. *Blood* 2002; 99: 3089–3101.

36 Beltrami AP, Barlucchi L, Torella D *et al.* Adult cardiac stem cells are multipotent and support myocardial regeneration. *Cell* 2003; 114: 763–776.

37 Oh H, Bradfute SB, Gallardo TD *et al.* Cardiac progenitor cells from adult myocardium: homing, differentiation, and fusion after infarction. *Proc Natl Acad Sci USA* 2003; 100: 12313–12318.

38 Urbanek K, Quaini F, Tasca G *et al.* Intense myocyte formation from cardiac stem cells in human cardiac hypertrophy. *Proc Natl Acad Sci USA* 2003; 100: 10440–10445.

39 Orlic D, Kajstura J, Chimenti S *et al.* Bone marrow cells regenerate infarcted myocardium. *Nature* 2001; 410: 701–705.

40 Quaini F, Urbanek K, Beltrami AP *et al.* Chimerism of the transplanted heart. *N Engl J Med* 2002; 346: 5–15.

41 Kucia M, Ratajczak J, Ratajczak MZ. Bone marrow as a source of circulating CXCR4$^+$ tissue-committed stem cells. *Biol Cell* 2005; 97: 133–146.

42 Baksh D, Song L, Tuan RS. Adult mesenchymal stem cells: characterization, differentiation, and application in cell and gene therapy. *J Cell Mol Med* 2004; 8: 301–16.

43 Bruder SP, Jaiswal N, Haynesworth SE. Growth kinetics, self-renewal, and the osteogenic potential of purified human mesenchymal stem cells during extensive sub-cultivation and following cryopreservation. *J Cell Biochem* 1997; 64: 278–294.

44 Bruder SP, Kurth AA, Shea M *et al.* Bone regeneration by implantation of purified, culture-expanded human mesenchymal stem cells. *J Orthop Res* 1998; 16: 155–162.

45 Kadiyala S, Young RG, Thiede MA *et al.* Culture expanded canine mesenchymal stem cells possess osteo-chondrogenic potential in vivo and in vitro. *Cell Transplant* 1997; 6: 125–134.

46 Awad HA, Butler DL, Boivin GP *et al.* Autologous mesenchymal stem cell-mediated repair of tendon. *Tissue Eng* 1999; 5: 267–277.

47 Young RG, Butler DL, Weber W *et al.* Use of mesenchymal stem cells in a collagen matrix for Achilles tendon repair. *J Orthop Res* 1998; 16: 406–413.

48 Galmiche MC, Koteliansky VE, Briere J *et al.* Stromal cells from human long-term marrow cultures are mesenchymal cells that differentiate following a vascular smooth muscle differentiation pathway. *Blood* 1993; 82: 66–76.

49 Dennis JE, Merriam A, Awadallah A *et al.* A quadripotential mesenchymal progenitor cell isolated from the marrow of an adult mouse. *J Bone Miner Res* 1999; 14: 700–709.

50 Prockop DJ. Marrow stromal cells as stem cells for non-hematopoietic tissues. *Science* 1997; 276: 71–74.

51 Barry FP. Mesenchymal stem cell therapy in joint disease. *Novartis Found Symp* 2003; 249: 86–96; discussion 96–102, 170–174, 239–241.

52 Chapel A, Bertho JM, Bensidhoum M *et al.* Mesenchymal stem cells home to injured tissues when co-infused with hematopoietic cells to treat a radiation-induced multi-organ failure syndrome. *J Gene Med* 2003; 5: 1028–1038.

53 Deng Y, Guo X, Yuan Q *et al.* Efficiency of adenoviral vector mediated CTLA4Ig gene delivery into mesenchymal stem cells. *Chin Med J (Engl)* 2003; 116: 1649–1654.

54 Ortiz LA, Gambelli F, McBride C *et al.* Mesenchymal stem cell engraftment in lung is enhanced in response to bleomycin exposure and ameliorates its fibrotic effects. *Proc Natl Acad Sci USA* 2003; 100: 8407–8411.

55 Saito T, Kuang JQ, Bittira B *et al.* Xenotransplant cardiac chimera: immune tolerance of adult stem cells. *Ann Thorac Surg* 2002; 74: 19–24; discussion 24.

56 Pittenger MF, Martin BJ. Mesenchymal stem cells and their potential as cardiac therapeutics. *Circ Res* 2004; 95: 9–20.

57 Tse WT, Pendleton JD, Beyer WM *et al.* Suppression of allogeneic T-cell proliferation by human marrow stromal cells: implications in transplantation. *Transplantation* 2003; 75: 389–397.

58 Silva GV, Litovsky S, Assad JA, Sousa AL, Martin BJ, Vela D, Coulter SC, Lin J, Ober J, Vaughn WK, Branco RV, Oliverra EM, He R, Geng YJ, Willerson JT, Perin EC. Mesenchymal stem cells differentiate into an endothelial phenotype, enhance vascular density, and improve heart function in a canine chronic ischemia model. *Circulation* 2005; 111(2): 150–156.

59 Quirici N, Soligo D, Caneva L *et al.* Differentiation and expansion of endothelial cells from human bone marrow CD133+ cells. *Br J Haematol* 2001; 115: 186–194.

60 Thomson JA, Itskovitz-Eldor J, Shapiro SS *et al.* Embryonic stem cell lines derived from human blastocysts. *Science* 1998; 282: 1145–1147.

61 Hristov M, Erl W, Weber PC. Endothelial progenitor cells: mobilization, differentiation, and homing. *Arterioscler Thromb Vasc Biol* 2003; 23: 1185–1189.

62 Harraz M, Jiao C, Hanlon HD *et al.* CD34 blood-derived human endothelial cell progenitors. *Stem Cells* 2001; 19: 304–312.

63 Yoon CH, Hur J, Park KW *et al.* Synergistic neovascularization by mixed transplantation of early endothelial progenitor cells and late outgrowth endothelial cells: the role of angiogenic cytokines and matrix metalloproteinases. *Circulation* 2005; 112: 1618–1627.

64 Kucia M, Dawn B, Hunt G *et al.* Cells expressing early cardiac markers reside in the bone marrow and are mobilized into the peripheral blood after myocardial infarction. *Circ Res* 2004; 95: 1191–1199.

65 Deb A, Wang S, Skelding KA *et al.* Bone marrow-derived cardiomyocytes are present in adult human heart: A study of gender-mismatched bone marrow transplantation patients. *Circulation* 2003; 107: 1247–1249.

66 Dowell JD, Rubart M, Pasumarthi *et al.* Myocyte and myogenic stem cell transplantation in the heart. *Cardiovasc Res* 2003; 58: 336–350.

67 Menasche P. Cellular transplantation: hurdles remaining before widespread clinical use. *Curr Opin Cardiol* 2004; 19: 154–161.

68 Leor J, Patterson M, Quinones MJ *et al.* Transplantation of fetal myocardial tissue into the infarcted myocardium of rat. A potential method for repair of infarcted myocardium? *Circulation* 1996; 94: II332–II336.

69 Tambara K, Sakakibara Y, Sakaguchi F *et al.* Transplanted skeletal myoblasts can fully replace the infarcted myocardium when they survive in the host in large numbers. *Circulation* 2003; 108: II259–II263.

70 Williams RL, Hilton DJ, Pease S *et al.* Myeloid leukaemia inhibitory factor maintains the developmental potential of embryonic stem cells. *Nature* 1988; 336: 684–687.

71 Cowan CA, Klimanskaya I, McMahon J *et al.* Derivation of embryonic stem-cell lines from human blastocysts. *N Engl J Med* 2004; 350: 1353–1356.

72 Itskovitz-Eldor J, Schuldiner M, Karsenti D *et al.* Differentiation of human embryonic stem cells into embryoid bodies compromising the three embryonic germ layers. *Mol Med* 2000; 6: 88–95.

73 Sachinidis A, Fleischmann BK, Kolossov E *et al.* Cardiac specific differentiation of mouse embryonic stem cells. *Cardiovasc Res* 2003; 58: 278–291.

74 Lovell MJ, Mathur A. The role of stem cells for treatment of cardiovascular disease. *Cell Prolif* 2004; 37: 67–87.

75 Anversa P, Sussman MA, Bolli R. Molecular genetic advances in cardiovascular medicine: focus on the myocyte. *Circulation* 2004; 109: 2832–2838.

76 Messina E, De Angelis L, Frati G *et al.* Isolation and expansion of adult cardiac stem cells from human and murine heart. *Circ Res* 2004; 95: 911–921.

77 Dragoo JL, Choi JY, Lieberman JR *et al.* Bone induction by BMP-2 transduced stem cells derived from human fat. *J Orthop Res* 2003; 21: 622–629.

78 Safford KM, Hicok KC, Safford SD *et al.* Neurogenic differentiation of murine and human adipose-derived stromal cells. *Biochem Biophys Res Commun* 2002; 294: 371–379.

79 Zuk PA, Zhu M, Ashjian P *et al.* Human adipose tissue is a source of multipotent stem cells. *Mol Biol Cell* 2002; 13: 4279–4295.

80 Rehman J, Traktuev D, Li J *et al.* Secretion of angiogenic and antiapoptotic factors by human adipose stromal cells. *Circulation* 2004; 109: 1292–1298.

81 Gronthos S, Franklin DM, Leddy HA *et al.* Surface protein characterization of human adipose tissue-derived stromal cells. *J Cell Physiol* 2001; 189: 54–63.

82 Wulf GG, Viereck V, Hemmerlein B *et al.* Mesengenic progenitor cells derived from human placenta. *Tissue Eng* 2004; 10: 1136–1147.

83 Erices A, Conget P, Minguell JJ. Mesenchymal progenitor cells in human umbilical cord blood. *Br J Haematol* 2000; 109: 235–242.

84 Jaiswal RK, Jaiswal N, Bruder SP *et al.* Adult human mesenchymal stem cell differentiation to the osteogenic or adipogenic lineage is regulated by mitogen-activated protein kinase. *J Biol Chem* 2000; 275: 9645–9652.

85 Sims DE. Diversity within pericytes. *Clin Exp Pharmacol Physiol* 2000; 27: 842–846.

86 Zuk PA, Zhu M, Mizuno H *et al.* Multilineage cells from human adipose tissue: implications for cell-based therapies. *Tissue Eng,* 2001; 7: 211–228.

87 Lew WY. Mobilizing cells to the injured myocardium: a novel rescue strategy or an unwelcome intrusion? *J Am Coll Cardiol* 2004; 44: 1521–1522.

88 Maekawa Y, Anzai T, Yoshikawa T *et al.* Effect of granulocyte-macrophage colony-stimulating factor inducer on left ventricular remodeling after acute myocardial infarction. *J Am Coll Cardiol* 2004; 44: 1510–1520.

89 Toma C, Pittenger MF, Cahill KS *et al.* Human mesenchymal stem cells differentiate to a cardiomyocyte phenotype in the adult murine heart. *Circulation* 2002; 105: 93–98.

90 Aicher A, Brenner W, Zuhayra M *et al.* Assessment of the tissue distribution of transplanted human endothelial progenitor cells by radioactive labeling. *Circulation* 2003; 107: 2134–2139.

91 Barbash IM, Chouraqui P, Baron J *et al.* Systemic delivery of bone marrow-derived mesenchymal stem cells to the infarcted myocardium: feasibility, cell migration, and body distribution. *Circulation* 2003; 108: 863–868.

92 Boekstegers P, von Degenfeld G, Giehrl W *et al.* Myocardial gene transfer by selective pressure-regulated retroinfusion of coronary veins. *Gene Ther* 2000; 7: 232–240.

93 Murad-Netto S, Moura R, Romeo LJ *et al.* Stem cell therapy with retrograde coronary perfusion in acute myocardial infarction. A new technique. *Arq Bras Cardiol* 2004; 83: 352–354; 349–351.

94 von Degenfeld G, Raake P, Kupatt C *et al.* Selective pressure-regulated retroinfusion of fibroblast growth factor-2 into the coronary vein enhances regional myocardial blood flow and function in pigs with chronic myocardial ischemia. *J Am Coll Cardiol* 2003; 42: 1120–1128.

95 Herity NA, Lo ST, Oei F *et al.* Selective regional myocardial infiltration by the percutaneous coronary venous route: A novel technique for local drug delivery. *Catheter Cardiovasc Interv* 2000; 51: 358–363.

96 Hou D, Maclaughlin F, Thiesse M *et al.* Widespread regional myocardial transfection by plasmid encoding Del-1 following retrograde coronary venous delivery. *Catheter Cardiovasc Interv* 2003; 58: 207–211.

97 Raake P, von Degenfeld G, Hinkel R *et al.* Myocardial gene transfer by selective pressure-regulated retroinfusion of coronary veins: comparison with surgical and percutaneous intramyocardial gene delivery. *J Am Coll Cardiol* 2004; 44: 1124–1129.

98 Assmus B, Schachinger V, Teupe C *et al.* Transplantation of progenitor cells and regeneration enhancement in acute myocardial infarction (TOPCARE-AMI). *Circulation* 2002; 106: 3009–3017.

99 Britten MB, Abolmaali ND, Assmus B *et al.* Infarct remodeling after intracoronary progenitor cell treatment in patients with acute myocardial infarction (TOPCARE-AMI): mechanistic insights from serial contrast-enhanced magnetic resonance imaging. *Circulation* 2003; 108: 2212–2218.

100 Chen SL, Fang WW, Ye F *et al.* Effect on left ventricular function of intracoronary transplantation of autologous bone marrow mesenchymal stem cell in patients with acute myocardial infarction. *Am J Cardiol* 2004; 94: 92–95.

101 Fernandez-Aviles F, San Roman JA, Garcia-Frade J *et al.* Experimental and clinical regenerative capability of human bone marrow cells after myocardial infarction. *Circ Res* 2004; 95: 742–748.

102 Schachinger V, Assmus B, Britten MB *et al.* Transplantation of progenitor cells and regeneration enhancement in acute myocardial infarction: final one-year results of the TOPCARE-AMI Trial. *J Am Coll Cardiol* 2004; 44: 1690–1699.

103 Strauer BE, Brehm M, Zeus T *et al.* Repair of infarcted myocardium by autologous intracoronary mononuclear bone marrow cell transplantation in humans. *Circulation* 2002; 106: 1913–1918.

104 Wollert KC, Meyer GP, Lotz J *et al.* Intracoronary autologous bone-marrow cell transfer after myocardial infarction: the BOOST randomised controlled clinical trial. *Lancet* 2004; 364: 141–148.

105 Hamano K, Nishida M, Hirata K *et al.* Local implantation of autologous bone marrow cells for therapeutic angiogenesis in patients with ischemic heart disease: clinical trial and preliminary results. *Jpn Circ J* 2001; 65: 845–847.

106 Herreros J, Prosper F, Perez A *et al.* Autologous intramyocardial injection of cultured skeletal muscle derived stem cells in patients with non-acute myocardial infarction. *Eur Heart J* 2003; 24: 2012–2120.

107 Menasche P, Hagege AA, Vilquin JT *et al.* Autologous skeletal myoblast transplantation for severe postinfarction left ventricular dysfunction. *J Am Coll Cardiol* 2003; 41: 1078–1083.

108 Siminiak T, Kalawski R, Fiszer D *et al.* Autologous skeletal myoblast transplantation for the treatment of postinfarction myocardial injury: phase I clinical study with 12 months of follow-up. *Am Heart J* 2004; 148: 531–537.

109 Stamm C, Kleine HD, Westphal B *et al.* CABG and bone marrow stem cell transplantation after myocardial infarction. *Thorac Cardiovasc Surg* 2004; 52: 152–158.

110 Stamm C, Westphal B, Kleine HD *et al.* Autologous bone-marrow stem-cell transplantation for myocardial regeneration. Lancet 2003; 361: 45–46.

111 Hill JM, Dick AJ, Raman VK *et al.* Serial cardiac magnetic resonance imaging of injected mesenchymal stem cells. *Circulation* 2003; 108: 1009–1014.

112 Thompson CA, Nasseri BA, Makower J *et al.* Percutaneous transvenous cellular cardiomyoplasty. A novel nonsurgical approach for myocardial cell transplantation. *J Am Coll Cardiol* 2003; 41: 1964–1971.

113 Siminiak T, Fiszer D, Jerzykowska O *et al.* Percutaneous trans-coronary-venous transplantation of autologous skeletal myoblasts in the treatment of post-infarction myocardial contractility impairment: the POZNAN trial. *Eur Heart J* 2005; 26: 1188–1195.

114 Sarmento-Leite R, Silva GV, Dohman HF *et al.* Comparison of left ventricular electromechanical mapping and left ventricular angiography: defining practical standards for analysis of NOGA maps. *Tex Heart Inst J* 2003; 30: 19–26.

115 Perin EC, Silva GV, Sarmento-Liete R *et al.* Assessing myocardial viability and infarct transmurality with left ventricular electromechanical mapping in patients with stable coronary artery disease: validation by delayed-enhancement magnetic resonance imaging. *Circulation* 2002; 106: 957–961.

116 Fuchs S, Baffour R, Zhou YF *et al.* Transendocardial delivery of autologous bone marrow enhances collateral perfusion and regional function in pigs with chronic experimental myocardial ischemia. *J Am Coll Cardiol* 2001; 37: 1726–1732.

117 Fuchs S, Satler LF, Kornowski R *et al.* Catheter-based autologous bone marrow myocardial injection in no-option patients with advanced coronary artery disease: a feasibility study. *J Am Coll Cardiol* 2003; 41: 1721–1724.

118 Kamihata H, Matsubara H, Nishiue T *et al.* Improvement of collateral perfusion and regional function by implantation of peripheral blood mononuclear cells into ischemic hibernating myocardium. *Arterioscler Thromb Vasc Biol* 2002; 22: 1804–1810.

119 Kawamoto A, Tkebuchava T, Yamaguchi J *et al.* Intramyocardial transplantation of autologous endothelial progenitor cells for therapeutic neovascularization of myocardial ischemia. *Circulation* 2003; 107: 461–468.

120 Kornowski R, Fuchs S, Tio FO *et al.* Evaluation of the acute and chronic safety of the biosense injection catheter system in porcine hearts. *Catheter Cardiovasc Interv* 1999; 48: 447–453; discussion 454–455.

121 Kornowski R, Leon MB, Fuchs S *et al.* Electromagnetic guidance for catheter-based transendocardial injection: a platform for intramyocardial angiogenesis therapy. Results in normal and ischemic porcine models. *J Am Coll Cardiol* 2000; 35: 1031–1039.

122 Perin EC, Dohmann HF, Borojevic R *et al.* Improved exercise capacity and ischemia 6 and 12 months after transendocardial injection of autologous bone marrow mononuclear cells for ischemic cardiomyopathy. *Circulation* 2004; 110: II213–II218.

123 Tse HF, Kwong YL, Chan JK *et al.* Angiogenesis in ischaemic myocardium by intramyocardial autologous bone marrow mononuclear cell implantation. *Lancet* 2003; 361: 47–49.

124 Kraitchman DL, Tatsumi M, Gilson WD *et al.* Dynamic imaging of allogeneic mesenchymal stem cells trafficking to myocardial infarction. *Circulation* 2005; 112: 1451–1461.

125 Hou D, Youssef EA, Brinton TJ *et al.* Radiolabeled cell distribution after intramyocardial, intracoronary, and interstitial retrograde coronary venous delivery: implications for current clinical trials. *Circulation* 2005; 112: I150–I156.

126 Kinnaird T, Stabile E, Burnett MS *et al.* Local delivery of marrow-derived stromal cells augments collateral perfusion through paracrine mechanisms. *Circulation* 2004; 109: 1543–1549.

127 Bartunek J, Vanderheyden M, Vandekerckhove B *et al.* Intracoronary injection of CD133-positive enriched bone marrow progenitor cells promotes cardiac recovery after recent myocardial infarction: feasibility and safety. *Circulation* 2005; 112: I178–I183.

128 Kang HJ, Kim HS, Zhang SY *et al.* Effects of intracoronary infusion of peripheral blood stem-cells mobilised with granulocyte-colony stimulating factor on left ventricular systolic function and restenosis after coronary stenting in myocardial infarction: the MAGIC cell randomised clinical trial. *Lancet* 2004; 363: 751–756.

129 Kinnaird T, Stabile E, Burnett MS *et al.* Marrow-derived stromal cells express genes encoding a broad spectrum of arteriogenic cytokines and promote *in vitro* and *in vivo* arteriogenesis through paracrine mechanisms. *Circ Res* 2004; 94: 678–685.

130 Goodell MA. Stem-cell "plasticity": befuddled by the muddle. *Curr Opin Hematol* 2003; 10: 208–213.

131 Hocht-Zeisberg E, Kahnert H, Guan K *et al.* Cellular repopulation of myocardial infarction in patients with sex-mismatched heart transplantation. *Eur Heart J* 2004; 25: 749–758.

132 Murry CE, Soonpaa MH, Reinecke H *et al.* Haematopoietic stem cells do not transdifferentiate into cardiac myocytes in myocardial infarcts. *Nature* 2004; 428: 664–668.

133 Losordo DW, Dimmeler S. Therapeutic angiogenesis and vasculogenesis for ischemic disease: part II: cell-based therapies. *Circulation* 2004; 109: 2692–2697.

134 Yeh ET, Zhang S, Wu HD *et al.* Transdifferentiation of human peripheral blood CD34$^+$-enriched cell population into cardiomyocytes, endothelial cells, and smooth muscle cells in vivo. *Circulation* 2003; 108: 2070–2073.

135 Zhang S, Wang D, Estrov Z *et al.* Both cell fusion and transdifferentiation account for the transformation of human peripheral blood CD34-positive cells into cardiomyocytes *in vivo. Circulation* 2004; 110: 3803–3807.

136 Forrester JS, Price MJ, Makkar RR. Stem cell repair of infarcted myocardium: an overview for clinicians. *Circulation* 2003; 108: 1139–1145.

137 Duan HF, Wu CT, Wu DL *et al.* Treatment of myocardial ischemia with bone marrow-derived mesenchymal stem cells overexpressing hepatocyte growth factor. *Mol Ther* 2003; 8: 467–474.

138 Kudo M, Wang Y, Wani MA *et al.* Implantation of bone marrow stem cells reduces the infarction and fibrosis in ischemic mouse heart. *J Mol Cell Cardiol* 2003; 35: 1113–1119.

139 Tang YL, Zhao Q, Zhang YC *et al.* Autologous mesenchymal stem cell transplantation induce VEGF and neovascularization in ischemic myocardium. *Regul Pept* 2004; 117: 3–10.

140 Dohmann HF, Perin EC, Takiya CM *et al.* Trans-endocardial autologous bone marrow mononuclear cell injection in ischemic heart failure: postmortem anatomicopathologic and immunohistochemical findings. *Circulation* 2005; 112: 521–526.

141 Yoon YS, Park JS, Tkebuchava T *et al.* Unexpected severe calcification after transplantation of bone marrow cells in acute myocardial infarction. *Circulation* 2004; 109: 3154–3157.

142 Chachques JC, Herreros J, Trainini J, Juffe A, Rendal E, Prosper F, Genovese J. Autologous human serum for cell culture avoids the implantation of cardioverter-defibrillators in cellular cardiomyoplasty. *Int J Cardiol.* 2004; 95(Suppl I): 29–33.

143 Smits PC, van Geuns RJ, Poldermans D, Bountioukos M, Onderwater EE, Lee CH, Maat AP, Serruys PW. Catheter-based intramyocardial injection of autologous skeletal myoblasts as a primary treatment of ischemic heart failure: clinical experience with six-month follow-up. *J Am Coll Cardiol.* 2003; 42: 2063-2069.

144 Assmus B, Honold J, Lehmaun R, Pistorius K, Hoffman WK, Martin H, Schachinger V, Zeiher AM. Transcoronary transplantation of progenitor cells and recovery of left ventricular function in patients with chronic ischemic heart disease: results of a randomized, controlled trial. *Circulation.* 2004; 110 (Suppl III): 238.

145 Wollert KC, Drexler H. Clinical applications of stem cells for the heart. *Circ Res.* 2005 Feb 4; 96 (2): 151–163. (with permission)

146 Perin EC, Dohmann HF, Borojevic R, Silva SA, Sousa AL, Silva GV, Mesquita CT, Belem L, Vaughn WK, Rangel FO, Assad JA, Carvalho AC, Branco RV, Rossi MI, Dohmann HJ, Willerson JT. Improved exercise capacity and ischemia 6 and 12 months after transendocardial injection of autologous bone marrow mononuclear cells for ischemic cardiomyopathy. *Circulation.* 2004 Sep 14; 110(11 Suppl 1): II213–8. (with permission)

CHAPTER 11

Pharmacogenetics and personalized medicine

Julie A. Johnson, PharmD, FCCP, BCPS, *& Issam Zineh,* PharmD

Introduction

Pharmacogenetics and pharmacogenomics aim to provide an understanding of the genetic basis for interpatient variability in drug response. *Pharmacogenetics* is an older term that has typically been attributed to studies focused on genetic associations between drug response and a single gene (the majority of studies to date). Pharmacogenetics and pharmacogenomics are increasingly being used synonymously, although some will argue that *pharmacogenomics* describes approaches that investigate a plethora of genes or the entire genome, as it relates to drug response. Because most of the studies in the literature to date focus on one or a few genes, we will use the term pharmacogenetics.

Knowledge gained through pharmacogenetics and pharmacogenomics research has numerous potential benefits. For example, pharmacogenetics may lead to a better understanding of the mechanisms by which drugs provide their benefit in various disease states. While the drug pharmacology provides some of this understanding, there are numerous examples of drugs (e.g., angiotensin-converting enzyme [ACE] inhibitors, statins) where the benefits seem to extend beyond the purported primary pharmacologic mechanism. When genetic polymorphisms are associated with variable response to a given drug, this suggests that the gene's encoded protein has a role in the drug's mechanism. An example of this is the α-adducin protein, whose genetic variability has been associated with diuretic response in several studies (discussed below). α-Adducin's physiologic function is consistent with involvement in the action of diur-

etics, but it is not a protein that has historically been linked to the mechanism of action of diuretics. As genome-wide approaches are more commonly used in pharmacogenomics investigations, there will be increasing numbers of such examples.

In similar ways, pharmacogenetics may help provide insights into the genetic basis for disease. For example, genes that are found to be associated with drug response in a given disease state could then be very strong candidates as disease genes, assuming that the drug works by providing more than just symptomatic benefit.

Pharmacogenomics may also be used in drug discovery and/or drug development. In the case of drug discovery, one approach is to use genomic approaches in individuals with (cases) and without (controls) the disease of interest. For those genes with different signals between cases and controls, their encoded proteins may represent rational drug targets for that disease. It is estimated that there are approximately 10,000 potential drug targets in the body, but all of the currently available drugs use only 500 of those targets. Thus, there is potential for many additional drugs with unique mechanisms of action, possibly identified through a pharmacogenomics approach. There are currently several cancer drugs that have been discovered in this way, but no such examples yet among the cardiovascular drugs.

Pharmacogenetics and pharmacogenomics may also lead to improvements in drug development. Within cardiovascular diseases, this is perhaps easiest to envision in heart failure. There are numerous heart failure drugs that have failed in late Phase III clinical trials (e.g., vasopeptidase inhibitors, endothelin blockers, tumor necrosis factor α [TNF-α] blockers), and so use of pharmacogenetics in the drug development process might be beneficial in such situations. Specifically, it is believed that in a broad population of heart failure patients, additional neurohormonal blockade cannot achieve sufficient benefit across the population. However, there is likely to be a smaller subpopulation that might benefit from such therapy. The potential approach is that the genetic polymorphisms associated with response would be identified in Phase II studies, and then the Phase III studies would be enriched with patients with the target genotypes. Although enrollment was not based on genotype per se, this is in essence what led to the successful clinical trial with isosorbide dinitrate (ISDN) hydralazine (and the subsequent Food and Drug Administration [FDA] approval of the combination product) [1]. Whether companies that were developing these drugs will go back and attempt to resurrect them using pharmacogenetics remains to be seen.

To the clinician, the greatest promise of pharmacogenetics is the potential to optimize drug therapy for a specific patient based on their genetic information. In some cases this will mean using genetic information to optimize efficacy, in others to minimize adverse drug effects. Currently, drug therapy of cardiovascular diseases is essentially either empirical (e.g., trial and error approach to hypertension, angina, dyslipidemia) or protocol-driven (e.g., acute coronary syndromes, heart failure, stroke prevention). In either approach, there will be patients who will not derive the desired benefit or will experience adverse effects from a specific drug. In the case of diseases treated by trial and error, use of genetic information might streamline this process, such that more rapid identification of the optimal therapy for a given patient might be accomplished. In the case of diseases managed per protocol, use of genetic information might facilitate the decision not to treat a certain patient with a given drug if it is predicted to provide little benefit, so as to allow for other therapies that might be beneficial. This is easiest to envision in heart failure, where the list of standard or recommended drugs continues to grow, and there are increasing concerns about how these regimens might be simplified.

Thus, the promise of pharmacogenetics is highlighted in Plate 11.1. In this paradigm, those likely to have an efficacious response, no response or toxicity can be predicted based on genetic information prior to initiation of therapy. This is in contrast to the current approach of prescribing the drug, and then once the patient is on therapy, determining to which of these three response groups they belong.

Herein, we review the advances in pharmacogenetics in the various therapeutic areas of cardiovascular medicine, highlighting studies in the literature and future clinical potential of pharmacogenetics for the given area. Examples of genes that have been studied with some significant associations between genotype and drug response are shown in Table 11.1.

Pharmacogenetics of dyslipidemia

Dyslipidemia has been long recognized as an important risk factor for cardiovascular disease. Consequently, treatment of elevated low density lipoprotein (LDL) cholesterol and mixed dyslipidemia (elevated LDL, low high density lipoprotein [HDL] and elevated triglycerides) is a cornerstone of both primary and secondary prevention strategies [2,3]. Despite an armamentarium of cholesterol-modifying drugs that includes HMG-CoA reductase inhibitors (statins), fibric acid derivatives (e.g., gemfibrozil, fenofibrate), bile acid sequestrants (e.g., colestyramine, colestipol, colesevelam), niacin, intestinal cholesterol absorption inhibitors (ezetimibe) and others, achieving consensus guideline-recommended lipoprotein levels in patients remains challenging. The inability to routinely achieve cholesterol goals may be related to multiple biologic and nonbiologic factors, with variability in pharmacokinetic and pharmacodynamic genes perhaps contributing.

Pharmacogenetics of HMG-CoA reductase inhibitors

Statin drugs are the most commonly used cholesterol-modifying agents in clinical practice. Furthermore, studies evaluating contribution of genetic variability to differential cholesterol-lowering response to drugs are most abundant for the statin drug class [4]. To date, there have been approximately

Table 11.1 Sample candidate pharmacokinetic and pharmacodynamic genes for statin pharmacogenetic studies.

Gene symbol	Common name of encoded protein	OMIM No.*	Drugs
Pharmacokinetics			
ABCB1	P-glycoprotein	171050	Statins, digoxin, CCBs
CYP3A4	Cytochrome P450 3A4	124010	Statins, CCBs
CYP3A5	Cytochrome P450 3A5	605325	Statins, CCBs
CYP2C9	Cytochrome P450 2C9	601130	Statins, warfarin, ARBs
CYP2D6	Cytochrome P450 2D6	124030	Statins, beta-blockers, CCBs
SLCO1B1	OATP1B1 aka OATP-C	604843	Statins
Pharmacodynamics			
ABCA1	ATP-binding cassette, subfamily A, member 1	600046	Statins
ABCG5	ATP-binding cassette, subfamily G, member 5	605459	Statins
ABCG8	ATP-binding cassette, subfamily G, member 8	605460	Statins
ACE	Angiotensin I-converting enzyme	106180	Statins, fibrates, ACE-I, spironolactone
APOB	Apolipoprotein B	107730	Statins
APOE	Apolipoprotein E	107741	Statins, fibrates
ADRB1	β_1-adrenergic receptor	109630	Beta-blockers, β-agonists
ADRB2	β_2-adrenergic receptor	109690	Beta-blockers, β-agonists
ADD1	α-adducin	102680	Diuretics
CETP	Cholesteryl ester transfer protein	118470	Statins, fibrates
FCGR2A	Immunoglobulin G Fc receptor II	14670	Heparin
GNB3	G protein β3 subunti	139130	Diuretics
HMGCR	HMG CoA reductase	142910	Statins
ITGB3	Platelet glycoprotein IIIa	173470	Statins, aspirin, GPIIb/IIIa blockers
LDLR	Low density lipoprotein receptor	606945	Statins
LPL	Lipoprotein lipase	238600	Statins, fibrates
NOS3	Endothelial nitric oxide synthase	163729	Hydralazine/nitrate
NPC1L1	Neimann-Pick C1-like 1 protein	608010	Ezetimibe
P2RY12	Platelet ADP receptor	600515	Clopidogrel
PPARA	Peroxisome proliferator-activated receptor α	170998	Statins, fibrates
PTGS1	COX-1	176805	Aspirin
SREBF1	Sterol regulatory element-binding transcription factor 1 (SREBP1)	184756	Statins
VKORC1	Vitamin K epoxide reductase complex, subunit 1	608547	Warfarin

ACE-I, angiotensin converting enzyme inhibitors; ARB, angiotensin receptor blocker; ADP, adenosine diphosphate; CCB, calcium channel blocker; COX, cyclo-oxygenase; GPIIb/IIIa, glycoprotein IIb/IIIa; OATP, organic anion transporter protein; SREBP, sterol regulatory element-binding protein.

* OMIM, Online Mendelian Inheritance of Man (http://www.ncbi.nlm.nih.gov/omim/omimfaq.html).

50 statin pharmacogenetic studies involving polymorphisms in more than 30 genes. The phenotype of interest is overwhelmingly lipoprotein response, although phenotypes also include pharmacokinetics, drug tolerability and safety, compliance/adherence and clinical outcomes.

Candidate gene lists for statin pharmacogenetic studies invariably include genes whose protein products are thought to be important in statin pharmacokinetics/drug disposition (Table 11.2). The variability in response and tolerability to statins may be related to variability in hepatic uptake and metabolism of the drugs. To the extent that the liver is the site of action for LDL-lowering effects of statins, heterogeneity in hepatic uptake of these agents may contribute to heterogeneity in patient responses [5]. Furthermore, as many statins are metabolized by the hepatic cytochrome P450 (CYP) isoenzyme

Table 11.2 Select statin pharmacogenetics studies of involving genes related to pharmacodynamics.

Gene/protein	Drug	Polymorphism	End-point(s)	Genetic association with end-point*	Reference
ACE	Fluvastatin	I/D	Lipids; CAD progression/ regression	Yes (both outcomes)	[170]
	Pravastatin	I/D	Lipids; fatal CAD or non-fatal MI	No	[171]
	Atorvastatin	I/D	Fibrinolytic markers; adhesion markers; C-reactive protein (CRP)	Yes (D-dimer and CRP)	[172]
ABCA1	Fluvastatin	-477C/T; -320G/C	Lipids; CAD progression/ regression	Yes (apoAI)	[173]
ABCG5/ABCG8	Atorvastatin	Q604E; D19H; Y54C; T400K; Δ632V	Lipids	Yes (LDL, D19H of ABCG8; effect enhanced in CYP7A1 -204A/A genotype)	[174,175]
APOB	Fluvastatin	I/D	Lipids	Yes (LDL, women)	[176]
		Xbal; EcoRI	Lipids	No	[176]
	Simvastatin	Xbal	Lipids	Yes	[177]
	Lovastatin	Xbal	Lipids	No	[178]
APOE	Pravastatin	ε2; ε3; ε4	Lipids	No	[179]
	Atorvastatin	ε2; ε3; ε4	Lipids	Yes (gender-specific)	[29]
	Simvastatin	ε2; ε3; ε4	Lipids; mortality	Yes	[31,34]
	Lovastatin	ε2; ε3; ε4	Lipids	Yes	[180]
	Fluvastatin	ε2; ε3; ε4	Lipids; CAD progression	Yes (lipids)	[30]
CETP	Atorvastatin	B1/B2 (TaqIB)	Lipids	Yes (HDL)	[181]
	Pravastatin	B1/B2 (TaqIB)	Lipids; CAD progression; fatal CAD or non-fatal MI	Yes (angiographic changes)	[182]
			CAD incidence	No	[183,184]
	Miscellaneous	B1/B2 (TaqIB)	Lipids, CVD events	Yes (events)	[185]
	Fluvastatin	Various haplotypes	Lipids	Yes	[18]
HMGCR	Pravastatin	SNP 12A/T; 29T/G	Lipids	Yes (TC and LDL)	[27]
IL-6	Fluvastatin	-174G/C	Lipids; CAD progression; events	Yes (Lp(a) concentrations)	[186]
	Pravastatin	-174G/C	Lipids; first fatal or nonfatal CV event	Yes (both outcomes)	[187]
		-572G/C	Lipids; first fatal or nonfatal CV event	No	[187]
SCAP	Pravastatin	Ile796Val	Lipids; coronary reactivity	No	[188]
	Fluvastatin	Ile796Val	Lipids; CAD progression; events	No	[189]
	Simvastatin	Ile796Val	Lipids	Yes (TC and triglycerides)	[190]

Table 11.2 (cont'd)

Gene/protein	Drug	Polymorphism	End-point(s)	Genetic association with end-point*	Reference
SREBF1	Fluvastatin	-36G+/-	Lipids; CAD progression; events	Yes (apoAI concentrations)	[189]
	Simvastatin	-36G+/-	Lipids	No	[190]
TLR4	Pravastatin	Asp299Gly	Lipids; CAD progression; events	Yes (events)	[184]
		Thr399Ile	Lipids; CAD progression; events	No	[184]

ACE, angiotensin-converting enzyme; APOE, apolipoprotein E; CETP, cholesteryl ester transfer protein; HDL, high density lipoprotein; HMGCR, HMG CoA reductase; IL-6, interleukin-6; LDL, low density lipoprotein; MI, myocardial infarction; SCAP, SREBF cleavage activating protein; SNP, single nucleotide polymorphism; SREBF1, sterol regulatory element-binding transcription factor; TLR, toll-like receptor.

* In the case of genetic associations with hard clinical end-points, the association may denote excess cardiovascular risk in a genotype group given placebo is attenuated in the presence of statin therapy.

system, variability in drug metabolism may also be an important factor in both statin efficacy and toxicity [6].

Hepatic uptake of statins is an important first step in the drug response pathway in that this step results in statin transport to both their site of action and elimination. The organic anion transporter (OAT) group of proteins appears to be important in various processes that could affect statin exposure *in vivo* including drug absorption, hepatic uptake and renal elimination [7–9]. OATP-C (encoded by the *SLCO1B1* gene) is a transporter protein found in hepatocytes, and is postulated to influence the uptake of pravastatin, simvastatin, atorvastatin and rosuvastatin.

SLCO1B1 is genetically polymorphic with at least 14 non-synonymous (causing an amino acid change) single nucleotide polymorphisms (SNPs) and many other synonymous variants reported in the literature [10]. Furthermore, many of these polymorphisms have been shown to affect the expression or activity of OATP-C [11]. The impact of *SLCO1B1* polymorphisms on differing statin phenotypes was initially investigated in pharmacokinetic studies of pravastatin [12–14]. *SLCO1B1* genotypes and haplotypes were associated with differences in pravastatin drug exposure (e.g., area under the plasma concentration–time curve, non-renal clearance and total clearance). However, it was

unclear as to whether this variability in drug exposure translates into differences in drug effect. For example, if individuals carry variant *SLCO1B1* alleles that reduce hepatic uptake of statins and increase systemic exposure to these agents, it can be hypothesized that these patients would perhaps be more likely to experience reduced efficacy in terms of cholesterol modulation and increased systemic toxicity (e.g., myopathic syndromes) when compared with their homozygous wild-type counterparts.

In an attempt to answer this question, one study evaluated the role of *SLCO1B1* polymorphisms on statin pharmacodynamics [15]. Forty-one healthy Caucasian individuals were treated with a single dose of 40 mg pravastatin, and surrogate markers of hepatic HMG-CoA reductase inhibition (i.e., plasma cholesterol and lathosterol concentrations) were measured 12 hours later. It was found that pravastatin inhibition of *in vivo* cholesterol synthesis was attenuated among individuals with variant *SLCO1B1* alleles ($n = 3$) when compared with non-variant carriers ($n = 38$). A retrospective analysis of 66 individuals treated with pravastatin, atorvastatin or simvastatin also showed a genetic association with drug response (total cholesterol changes from baseline) [16].

It should be noted that the majority of studies investigating *SLCO1B1* genetic associations with variable statin pharmacokinetics and pharmacodynamics have had small sample sizes and have not generally looked at long-term therapy. Nonetheless, transport protein pharmacogenetics continues to grow in number and sophistication of study design, and these initial studies have undoubtedly formed a foundation for continued investigation.

The impact of polymorphisms in transport protein genes other that *SLCO1B1* has also been investigated [13,17,18]. Because transport proteins are additionally found on the apical surfaces of the intestines and kidneys, it is conceivable that variants in genes encoding such proteins as P-glycoprotein (*ABCB1*), multidrug resistance-associated protein-2 (*ABCC2*), OATs in enterocytes and renal tubules (*SLCO2B1* and *SLC22A8*) and others could influence statin responses in patients.

A major area of focus in statin pharmacogenetics is hepatic drug metabolism. Specifically, five of the six commonly used statins are metabolized predominantly through hepatic (and extrahepatic) cytochrome P450 enzymes, a major pathway of drug metabolism in general. Simvastatin, atorvastatin and lovastatin are largely metabolized by CYP3A4/5, while the predominant enzyme in fluvastatin and rosuvastatin metabolism is CYP2C9. It has also been submitted that various statins are additionally substrates for other isoenzymes such as CYP2D6 and CYP2C8.

Studies have investigated the impact of polymorphisms in drug metabolizing enzymes on both efficacy and tolerability to statins. The earliest study of healthy volunteers suggested potential association between *CYP2D6* polymorphisms and variable lipid response to simvastatin [19]. In particular, it was concluded that *CYP2D6* variant carriers had decreased clearance of active simvastatin metabolites, resulting in greater drug response in these individuals. However, as other studies with larger sample sizes were conducted, it became clear that no consistent relationship between *CYP* polymorphisms and statin response could be discerned [20,21]. In general, consideration of metabolizing enzyme polymorphisms alone does not sufficiently explain variability in statin responses [20–25].

There have been suggestive recent studies of *CYP3A5* that perhaps warrant further investigation. For example, in a retrospective analysis of 69 Caucasians with coronary disease treated with statins known to be metabolized by CYP3A (i.e., simvastatin, atorvastatin or lovastatin), it was shown that individuals with at least one functional *CYP3A5*1* allele had a 13% and 14% diminishment of 1-year total cholesterol and LDL reduction, respectively [24]. The attenuated response is hypothesized to be brought about by enhanced expression of CYP3A5 in *1 carriers, resulting in increased statin metabolism. Interestingly, polymorphisms in *CYP3A5* have also been associated with severity of atorvastatin-induced myopathy [26]. While incidence of atorvastatin-induced myopathy was not related to *CYP3A5* genotypes, the magnitude of creatine kinase elevations among individuals experiencing myalgia with myopathy was higher in those with the *CYP3A5*3* allele. These findings will need to be replicated and the gene-effect further refined prior to use of genotyping for prediction of myopathy severity. However, they are consistent with the expected effects of these genotypes on statin plasma drug concentrations.

One of the major limitations of pharmacogenetic studies is that they generally do not simultaneously investigate the associations between multiple genes and drug response [4]. Two notable studies have attempted to use a multiple candidate gene approach to elucidate genetic contributions to variable statin lipoprotein responses [27,28]. Chasman *et al.* [27] retrospectively evaluated whether any of nearly 150 SNPs in 10 candidate genes were associated with response to pravastatin among over 1500 patients in the Pravastatin Inflammation/CRP Evaluation (PRINCE) study [27]. After establishment of strict statistical criteria for significance, it was found that two highly linked SNPs in the HMG-CoA reductase gene (*HMGCR*) were associated with differences in pravastatin response. Specifically, carriers of at least one variant allele at either SNP locus had an approximately 20% smaller relative reduction in total cholesterol and LDL compared with wild-type homozygotes.

In another retrospective multigene study (43 SNPs in 16 candidate genes) of 2735 individuals receiving statins from the Atorvastatin Comparative Cholesterol Efficacy and Safety Study (ACCESS) database, Thompson *et al.* [28] did not replicate the findings from PRINCE. In fact, while identifying several gene polymorphisms related to variability in statin responses, only one association (LDL response by apolipoprotein E [*APOE*] genotype) was consistent with previous literature reports. These data highlight the importance of conducting multigene studies and considering methodologic differences among these studies in pushing the field of statin pharmacogenetics forward.

Genes involved in absorption, distribution, metabolism and excretion of statins represent one side of the pharmacologic coin with respect to statin pharmacogenetics. There is tremendous diversity in the choice of pharmacodynamic (i.e., drug action) candidate genes that may have a role in disparate therapeutic responses between patients. For example, nondrug metabolism genes investigated to date include those whose protein products are involved in cholesterol biosynthesis and lipoprotein metabolism, nonlipoprotein related enzyme targets and inflammatory processes (e.g., cytokines) [4]. Pharmacogenetic studies of statins investigating select pharmacodynamic genes are presented in Table 11.2.

One of the best studied nonpharmacokinetic genes in statin pharmacogenetics is *APOE*. ApoE is a surface ligand found on lipoproteins that include very low density lipoprotein (VLDL) and HDL. Individuals carry two (parental) copies of any of three commonly occurring *APOE* alleles: ε2, ε3 (wild-type) and ε4. The majority of studies have shown lipid-lowering response to statins is highest among those with the ε2 allele and lowest among those with the ε4 allele [29–33]. However, as a dramatic departure from this observation, it appears that ε4 carriers are at the greatest risk of cardiovascular mortality, and that these individuals have the greatest relative reduction in mortality when treated with a statin [34]. Furthermore, while these individuals in general are more likely to receive statin benefits in the form of hard end-point reduction, they are also more likely to discontinue therapy, possibly as a consequence of perceived attenuated response of the LDL surrogate [35]. These data highlight the importance of consideration of appropriate phenotype in pharmacogenetic studies. The clinical implications of each observed genotype–phenotype relationship should be carefully considered, as there is undoubtedly a complex interaction between genetics, changes in surrogate end-points, variable hard end-point risks, physician prescribing and patient adherence.

Pharmacogenetics of nonstatin cholesterol modulators

Pharmacogenetic studies of cholesterol-modulating agents other than statins have also been conducted. For example, data continue to accumulate on relationships between fibrate response and polymorphisms in such genes as *ACE*, *APOE*, *CETP* (cholesteryl ester transfer protein), *LPL* (lipoprotein lipase), *PPARA* (peroxisome proliferator-activated receptor alpha) and *FABP1* (liver fatty acid binding protein) [36–45]. Because activation of PPARα by fibrates is seen as the main pharmacologic catalyst for downstream lipoprotein modulation, *PPARA* represents the main candidate gene in fibrate pharmacogenetic studies.

The data regarding associations between *PPARA* polymorphisms and variability in fibrate response are conflicting [38,41–43]. In a study of 32 abdominally obese men, carriers of the variant L162→V allele ($n = 6$, 19% of the population) had a dramatically higher increase in the HDL_2 subfraction in response to gemfibrozil compared with L162 homozygotes (50% vs. 5.5%, $P = 0.03$); no differences in triglyceride responses were seen [43]. However, in a larger subset of patients, variability in response by L162V genotypes was seen for both triglycerides and HDL, although the exact contribution of *PPARA* polymorphisms to these phenotypes in this study is complicated by consideration of additional genotypes (e.g., *APOE* and *LPL*), age, sex and anthropometric factors [38].

Other polymorphisms in *PPARA* (e.g., intron 7 G→C) have also been positively associated with variable triglyceride responses to fibrates [41]. However, whether or not *PPARA* polymorphisms influence cardiovascular risk reduction by fibrates is unclear. For example, in an analysis of the Lopid Coronary Angiography Trial (LOCAT), while *PPARA* polymorphisms were associated with differences in progression of atherosclerotic coronary disease, there was no effect of these polymorphisms on treatment outcomes [42]. As with the example of statins and *APOE*, the use of surrogate markers in fibrate pharmacogenetic studies must be corroborated by evidence from outcomes trials.

Pharmacogenetic data are emerging for newer agents used in management of dyslipidemia. For example, data suggest that genetic polymorphisms might have a role in the ezetimibe response pathway. Ezetimibe is a relatively new compound that prevents the intestinal absorption of cholesterol and related dietary phytosterols [46]. It is available alone or as a combination product with simvastatin. Genetic control of intestinal cholesterol absorption is complex, and the exact molecular target of ezetimibe has been elusive [47]. However, it has recently been shown that ezetimibe's target of action is the Niemann–Pick C1-like 1 protein (NPC1L1) found in enterocytes [48]. The exact function of this protein is unknown. Nonetheless, contribution of genetic variability in *NPC1L1* to ezetimibe response has been recently described. In particular, a case series of 52 patients receiving ezetimibe revealed a combination of variant genotypes associated with nonresponse to drug therapy [49]. In a larger study ($n = 101$), *NPC1L1* haplotype analyses revealed that individuals without two copies of the most commonly occurring haplotype exhibited the most significant response to ezetimibe in terms of LDL cholesterol (35% vs. 24% reduction; $P = 0.02$) with trends toward significance for total cholesterol ($P = 0.07$) [50].

Relevance of pharmacogenetics to treatment of dyslipidemias

There are several limitations to pharmacogenetic studies of cholesterol lowering agents. These include use of surrogate rather than hard clinical endpoints, little control for genetic and nongenetic confounders (e.g., population stratification, diet, adherence) and variability in study design [4]. Furthermore, even in scenarios of genotype-associated diminishment of drug response (i.e., LDL lowering), the magnitude of the differences among differing genotype groups may not be clinically meaningful [51]. Nonetheless, pharmacogenetics of lipid modulation is a relatively new field of study. To the extent that dyslipidemia is a major antecedent risk factor for vasculopathic manifestations of disease, and many patients receive cholesterol drugs, it is important to identify the sources of patient-specific variability in drug response. While many lipid modulators confer significant relative risk reduction in adverse cardiovascular events based on data from randomized clinical trials, there is still a significant proportion of individuals that experiences major morbidity and mortality at the end of follow-up. As such, genetic and non-genetic determinants of response to dyslipidemia management (both dietary and drug) are likely to be investigated further.

Pharmacogenetics of hypertension

Hypertension is the most common chronic disease in adults, estimated to affect approximately 65 million Americans. Treatment of hypertension is a trial and error process, with drug therapy typically selected from among the five first line drug classes, which include thiazide diuretics, beta-blockers, ACE inhibitors, angiotensin receptor blockers (ARBs) and calcium-channel blockers (CCBs). There are a number of pharmacogenetics studies for all of the first line drug classes in hypertension, with the exception of CCBs, and this literature has been recently reviewed [52]. Among the studies published to date, there have been several genes that have been extensively studied and in some cases have shown positive associations with response in multiple studies. It is these on which we will focus.

Diuretics in hypertension
The most extensively studied gene relative to the diuretic response is the gene for α-adducin (*ADD1*), which is a ubiquitously expressed cytoskeletal protein involved in ion transport. There is a nonsynonymous polymorphism in *ADD1* (Gly460Trp), and several studies have shown that carriers of the Trp460 allele have greater blood pressure lowering than Gly460 homozygotes [53–55]. Further adding to the interest in this polymorphism was a case–control study that suggested that the reduction in nonfatal myocardial infarction and stroke with diuretics was limited to carriers of the Trp460 allele [56]. In a recent study of patients with cardiovascular disease and hypertension, we found the Trp460 allele to be associated with adverse cardiovascular outcomes, but this was not influenced by the presence or absence of diuretics [57]. Additionally, not all studies that have examined the relationship of Gly460Trp with the blood pressure response have seen positive associations. Nonetheless, *ADD1* appears to be an intersting gene/polymorphism that may well have an important role in the diuretic response.

Beta-blockers in hypertension

For beta-blockers, the most extensively studied gene is that for the β_1-adrenergic receptor (*ADRB1*). There are two common polymorphisms in *ADRB1* that change encoded amino acids (Ser49Gly and Arg389Gly), and both have been shown through *in vitro* studies to have functional consequences, where the Ser49 and Arg389 forms have greater responsiveness to receptor stimulation by agonists (e.g., norepinephrine). Several studies have suggested that Arg389 homozygotes have the greatest blood pressure lowering with beta-blockers [58–60], although not all studies have found this association [61,62]. Additionally, data from our laboratory suggest that consideration of both the codon 49 and codon 389 polymorphisms is more informative than consideration of codon 389 alone. Specifically, our data suggested that consideration of baseline blood pressure and the genotypes at the two sites explained 57% of the antihypertensive response variabilty with metoprolol, with genotype at the two *ADRB1* polymorphic sites explaining 20% of the variability [58]. It is also interesting to note that these polymorphisms may help to explain the well-recognized difference in antihypertensive response between African-Americans and white people. Specifically, the most responsive diplotype (genotype considering both codon 49 and 389) occurs in 45% of white people, but only about 10% of African-Americans. This highlights that pharmacogenetics may shed light on some clinically recognized differences in response, and will allow clinicians to move away from using race to help make clinical decisions. Thus, while the data on the *ADRB1* SNPs and antihypertensive response to beta-blockers are not to the point of being useful in the clinical setting, they highlight that it might be possible in the future to predict antihypertensive response through use of genetic and nongenetic information in a predictive model.

ACE inhibitors and angiotensin receptor blockers in hypertension

The most extensively studied gene for the ACE inhibitors and ARBs is the *ACE* gene, and its intron 11 insertion/deletion (I/D) polymorphism, with studies numbering into the twenties. Several early,

smaller studies of this polymorphism showed positive associations, although were inconsistent in that some showed the D allele to be associated with the greatest response [63,64] while others showed the I allele to have the greatest response [65,66]. However, many other studies, including more recent large studies, have not shown this polymorphism to be associated with the antihypertensive response to ACE inhibitors or ARBs [67–70]. Thus, evaluating the body of literature on the *ACE* I/D polymorphism leads to the conclusion that it does not importantly influence the response to ACE inhibitors or ARBs. This is not to say that *ACE* genetic variability does not influence response to these drugs. It is possible that study of other polymorphisms in the gene, or a haplotype tagging SNP approach (which more comprehensively captures the gene's variability) might be strongly associated with response to these drugs. However, the extensively studied I/D polymorphism is not likely to be useful in the future for predicting responses to ACE inhibitors or ARBs.

There are several other genes that have been studied (some with positive associations) for the antihypertensive drugs, but not with positive findings in more than one study.

Relevance of pharmacogenetics to the treatment of hypertension

The *ADD1* and *ADRB1* genes appear to have future potential for predicting antihypertensive response to diuretics and beta-blockers, respectively. Additionally, there are other genes that appear to be strong candidates, for which there are interesting data that require validation. Based on the hypertension pharmacogenetics studies to date, it seems clear that like hypertension, the antihypertensive response is a complex phenotype, governed by a large number of genes. The future challenge is to identify the constellation of genes that contribute to variability in response to the major antihypertensive drug classes. This might be accomplished through either study of comprehensive lists of candidate genes, or through a genome-wide approach. Studies utilizing both of these approaches are underway, and are likely to yield important data for hypertension pharmacogenetics in the future.

Pharmacogenetics of heart failure

There is a mounting body of literature regarding the influence of genetic polymorphisms on the cornerstones of therapy for heart failure, namely beta-blockers, ACE inhibitors, diuretics, digoxin, spironolactone and the ISDN–hydralazine combination. The majority of these papers have been published since 2003, highlighting the rapid pace of accumulation of knowledge in this area. There are also numerous studies that have addressed genetic associations with risk of heart failure, or prognosis in heart failure. As these are addressed in Chapters 3, 4 and 7, we will focus only on the pharmacogenetic studies in heart failure. There are few pharmacogenetics studies in the setting of acute myocardial infarction, which is perhaps not surprising given the limited clinical potential for genetically guided therapy in the acute setting. However, there are some pharmacogenetic data on post-myocardial infarction therapies, and because one of their major benefits is prevention of heart failure, we discuss those here as well.

Beta-blockers in heart failure

Among the heart failure drugs, there are the most pharmacogenetic data on beta-blockers. It is well recognized that beta-blockers are beneficial in heart failure through their ability to directly block the detrimental effects of sympathetic nervous system (SNS) activation, and the majority of the studies center on the β-adrenergic receptor genes.

A number of different studies suggest that the genes for the β_1-adrenergic receptor (*ADRB1*) and the β_2-adrenergic receptor (*ADRB2*) influence the response to beta-blockers in heart failure. These studies have focused on genetic associations with beta-blocker induced change in left ventricular ejection fraction (LVEF), clinical outcomes and early tolerability of beta-blockers (Table 11.3).

At least two studies have shown that the improvement in LVEF with beta-blockade was associated with *ADRB1* genotype. Specifically, both studies found Arg389 homozygotes had the greatest improvement in LVEF [71,72]. These findings are consistent with the *in vitro* data, as it was expected that Arg389 homozygotes would have the greatest harm from SNS activation and so might derive the greatest benefit from beta-blockers. Another study considered only *ADRB2* polymorphisms and found Gln27 homozygotes had much smaller improvements in LVEF than Glu27 carriers (heterozygotes and Glu27 homozygotes), as assessed by percentage of those who had a 10-unit increase in LVEF [73]. A study similar in design to these others considered both *ADRB1* and *ADRB2* SNPs, but did not show such an association between any of the SNPs studied and the LVEF response [74]. One potential explanation is that the patients in this latter study were generally healthier than those in the positive studies. For example, in the three studies with positive findings, baseline LVEFs were 26%, 23% and 21%, whereas in the negative study it was 30%. Whether this difference in heart failure severity explains the negative findings in the latter study can only be resolved through further well-powered studies.

The study from our laboratory that assessed the relationship between genotype and reverse remodeling also tested the influence of *ADRB1* and *ADRB2* genotypes on initial tolerability of beta-blockers, during the titration period [75]. This study found that there were significant differences by *ADRB1* codon 49 and 389 genotype in the need for an increase in diuretic dose during the titration period. For example, in one diplotype group (combination of both polymorphisms), 52% of patients required an increase in diuretic dose, whereas in another, none of the individuals required this intervention. If these findings could be replicated, they would suggest that a clinician might be able to determine the aggressiveness of dose titration or need for close follow-up based on genotype.

Focus of studies on LVEF is based on data suggesting that change in LVEF is a strong surrogate for survival benefits of drug therapy in heart failure, and several additional studies have looked at clinical outcomes with beta-blockers, relative to genotype. Two separate reports from Magnusson *et al.* [76] and Borjesson *et al.* [77] provide evidence that in the absence of beta-blocker therapy, the *ADRB1* Ser49Ser genotype is associated with worse outcomes (death and transplant) than Gly49 carriers. Additionally, beta-blockers provide a greater improvement in event-free survival for Ser49Ser than Gly49 carriers. This does not mean that Gly49

Table 11.3 Pharmacogenetic studies of beta-blockers in heart failure.

Gene	Drug	Polymorphism studied	End-point/findings	Reference
ADRB1	Carvedilol	Arg389Gly	Arg389Arg had greatest improvement in LVEF	[72]
	Metoprolol CR/XL	Arg389Gly Ser49Gly	Arg389Arg had greatest improvement in LVEF. No association between Ser49Gly genotype and LVEF change	[71]
	Carvidolol or bisoprolol	Arg389Gly Ser49Gly	LVEF change – NA LVEF change – NA	[74]
	Metoprolol CR/XL	Arg389Gly Ser49Gly	Arg389Arg had better initial tolerability; Codon 49 and 389 haploptypes also Associated with different initial tolerability	[75]
	Various	Arg389Gly Ser49Gly	Death – NA Death – NA	[80]
	Metoprolol CR/XL	Arg389Gly	Death – NA	[79]
	Various	Arg389Gly Ser49Gly	Transplant-free survival. NA for Arg389 Gly; Ser49Ser genotype on low dose beta-blocker associated with similar outcomes as no beta-blocker; those on high dose beta-blocker derived significant benefit	[76]
ADRB2	Carvedilol	Gln27Glu Arg16Gly	Gln27Gln less likely to have ≥10 unit increase in LVEF. No association between LVEF change and Arg16Gly	[73]
	Carvedilol	Gln27Glu Arg16Gly	Change in LVEF – NA Change in LVEF – NA	[74]
	Various	Gln27Glu Arg16Gly	Gln27Gln and Arg16Arg treated w/ beta-blockers post-MI had lowest 2 year survival	[80]
CYP2D6	Carvedilol	EM/PM	PMs had significantly higher carvedilol plasma concentrations; no assessment of clinical response	[81]
	Metoprolol	EM/PM	PMs had significantly higher plasma metoprolol concentrations; no difference by genotype in adverse effects of initial tolerability	[75]

CR/XL, controlled release/extended release (Toprol XL®); EM, extensive metabolizer; LVEF, left ventricular ejection fraction; MI, myocardial infarction; NA, no association; PM, poor metabolizer.

carriers fair worse, but that beta-blockers narrow the disparity between the two genotype groups in terms of clinical adverse events. Their more recent report also shows a dose effect by genotype with the beta-blockers. Specifically, for Ser49 homozygotes, low dose beta-blocker was associated with similar outcomes as no beta-blocker, while high dose beta-blocker in this group had significant improvements in their event-free survival (Fig. 11.1) [76]. In con-

trast, Gly49 carriers derived similar benefit from low and high dose beta-blockade. If these data are replicated, they could have important clinical implications for the use of beta-blockers in heart failure, where genotype may help guide the need for aggressive beta-blocker dosing.

Although not yet published in full, data have been presented from the BEST trial [78], showing that in *ADRB1* Arg389 homozygotes, there was a

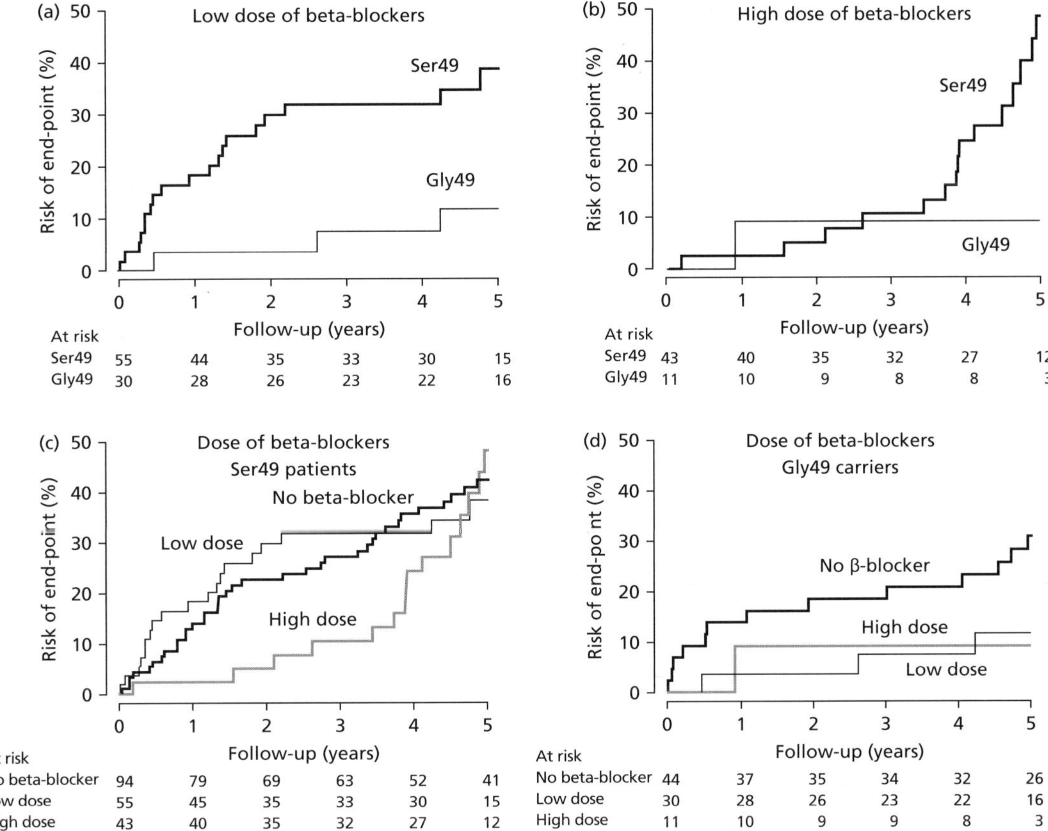

Figure 11.1 Influence of β_1AR genotype on outcome in relation to doses of beta-blockers. (a) Low dose of beta-blocker (≤50% of target dose) by codon 49 genotype. (b) High dose of beta-blocker (>50% of target dose) by codon 49 genotype. (c) Influence of beta-blocker dose on outcomes among Ser49 homozygotes. (d) Influence of beta-blocker dose on outcomes among Gly49 carriers. Reprinted from [76] with permission from American Society for Clinical Pharmacology and Therapeutics.

significant benefit associated with bucindolol therapy, but there was no such benefit in Gly389 carriers. These data are consistent with the studies showing Arg389 homozygotes had the greatest improvements in ejection fraction. However, a 600 subject substudy of MERIT-HF tested whether the *ADRB1* Arg389Gly polymorphism was associated with different outcomes [79]. They did not see any differences in outcomes by genotype in either the metoprolol treated, or the placebo-treated groups, but did not conduct an analysis that would have allowed them to detect a genotype–drug (i.e., gene–environment) interaction.

Finally, in a recent study a post-myocardial infarction patient population was prospectively enrolled during myocardial infarction hospitalization, then followed for all-cause 3-year mortality [80]. It was found that *ADRB2* Arg16 homozygotes, Gln27 homozygotes and the Arg16/Gln27 haplotype were associated with a significantly increased risk of death in beta-blocker treated patients. These findings are consistent with the previous study that found Gln27 homozygous heart failure patients had smaller improvements in LVEF with beta-blocker therapy. Lanfear *et al.* [80] found no associations between the *ADRB1* polymorphisms and adverse outcomes.

Both carvedilol and metoprolol are extensively metabolized by the drug metabolizing enzyme CYP2D6, for which 7% of Causasians lack functional protein due to genetic polymorphisms in the gene. Studies have clearly documented that plasma drug concentrations of metoprolol and carvedilol are significantly influenced by CYP2D6 genotype

[75,81], with poor metabolizers having drug concentrations that are 3–5 times higher than extensive metabolizers. However, a study from our laboratory suggests that these differences in plasma drug concentration do not translate into important clinical differences in tolerability of the beta-blocker [75]. Data similarly suggest that adverse effects of metoprolol in hypertension are unrelated to *CYP2D6* genotype [82].

Relevance of beta-blocker pharmacogenetics to practice

There are a number of studies that suggest the β_1- and β_2-adrenergic receptor gene polymorphisms may influence response to beta-blockers, including initial tolerability, improvements in left ventricular remodeling and clinical outcomes. While these findings have not been replicated in all studies where they have been tested, when there are positive findings, they are highly consistent across the studies (e.g., *ADRB1* Arg389Arg and Ser49Ser and *ADRB2* Glu27Glu genotypes having the greatest benefits from beta-blocker therapy). These data are not to the point of being utilized clinically, but several of the studies highlight the future clinical potential, either in assessing those who will need close follow-up during therapy initiation, those who need to be treated aggressively or those in whom beta-blockers might offer minimal benefit and so alternative therapy could be considered. To get to this point will require large clinical trials, which is most efficiently accomplished by addressing pharmacogenetic hypotheses as substudies to large clinical trials. It is not anticipated that there will be numerous additional beta-blocker trials. Thus, unless the recent beta-blocker trials (such as COMET and COPERNICUS) included collection of genetic samples, it may be difficult to achieve the level of evidence that will be necessary to translate this information into clinical practice.

ACE inhibitors in heart failure

While there are numerous pharmacogenetic studies of ACE inhibitors in hypertension, they are more limited in the heart failure population. The bulk of the data come from the laboratory of McNamara *et al.* [83] who have established a heart failure genetics study population that they are following prospectively. In this population (for whom treatment at baseline included ACE inhibitor in about 85%, ARB in about 10% and beta-blocker in about 40%), they noted that the *ACE* I/D polymorphism was associated with event-free survival. Specifically, those with the (D/D) genotype had the worst event-free survival. This was most evident in the subgroups on low-dose ACE inhibitor therapy, or those not on beta-blocker therapy. Interestingly, the presence of beta-blocker therapy, irrespective of the presence or dosage of the ACE inhibitor therapy, negated any of the negative effects of the D/D genotype on survival. This highlights that in the presence of contemporary pharmacotherapy, the genes/genotypes that may adversely affect prognosis in heart failure may be more difficult to discern, as the drug therapy may overcome the risk associated with the genotype.

Pharmacogentics of other heart failure drugs

Digoxin

Digoxin has been shown to be a substrate of the P-glycoprotein (P-gp) drug efflux pump. Because P-gp is responsible for ejection of xenobiotics from cells in the intestine, liver, kidney, brain and other highly sequestered tissues, it has been hypothesized that polymorphisms in the gene that encodes P-gp (*ABCB1* aka *MDR1*) might affect digoxin absorption and distribution [84,85].

The vast majority of *ABCB1*-digoxin pharmacogenetic studies have investigated the impact of the C3435T polymorphism (synonymous SNP in exon 26) on digoxin pharmacokinetics [86–93]. While several studies have demonstrated a significant effect of the C3435T polymorphism on serum digoxin concentrations and other measures of drug exposure, data are conflicting. In fact, a meta-analysis of eight studies investigating the impact of C3435T genotypes on oral digoxin pharmacokinetics revealed no significant associations between genotype and either P-gp expression, digoxin AUC_{0-4h}, or AUC_{0-24h}. Overall, borderline significance was demonstrated between genotypes and peak digoxin concentrations (Cmax) [94]. While polymorphic *ABCB1* represents a biologically plausible mediator of digoxin response variability, studies have not robustly established this gene as clinically significant. Because of the predictable, linear nature of digoxin pharmacokinetics, considering *ABCB1* genotype to guide use of digoxin is unlikely to occur clinically.

Diuretics

There is a broad literature on the pharmacogenetics of thiazide diuretics, particularly in hypertension, as discussed above. The literature is far more limited for loop diuretics, with little information on pharmacogenetics of loop diuretics in heart failure. One group studied the association of the BP response to furosemide, relative to the *ADD1* polymorphism. Consistent with the findings on this gene with thiazides, the Trp460 carriers had a greater BP response than the Gly460 homozygotes [53]. The clinical use of diuretics in heart failure is relatively straightforward, and so it seems less likely that there would be clinical value to guiding therapy with genetic information in this setting. This is in contrast with the potential benefit of genetic information to guide the use of diuretics in hypertension. Thus, this body of literature may remain limited. However, it seems likely that the genetic association studies of thiazides in hypertension may still inform, to some degree, the genes that contribute importantly to the loop diuretic efficacy in heart failure (e.g., *ADD1*).

Spironolactone

Spironolactone is an old drug that has seen a recent rebirth as an aldosterone antagonist, useful for preventing the progressive remodeling of the left ventricle (LV) and associated adverse outcomes in heart failure. To our knowledge there is a single study on spironolactone pharmacogenetics, which investigated the relationship between *ACE* I/D genotype and the LV remodeling with spironolactone [95]. In this study, the investigators found that those with the *ACE* non-D/D genotype (i.e., I/D or I/I) were the ones that showed significant improvements in LV reverse remodeling with spironolactone therapy. These data need to be replicated, but are similar to some of the beta-blocker studies in that they suggest that certain genotypes may derive greater reverse remodeling effects than others.

Isosorbide dinitrate–hydralazine

The combination of ISDN and hydralazine represented the first vasodilators to be documented to

have survival benefits in heart failure (in V-HeFT), although a few years later it was shown clearly that ACE inhibitors were superior to the ISDN–hydralazine combination [96]. Thus, the hydralazine–nitrate combination has seen limited use in heart failure. However, the original V-HeFT investigators noted on *post hoc* analysis that the noted benefits of this combination therapy seemed to be largely confined to the African-American subjects. Thus, they recently tested ISDN–hydralazine against placebo in a population of African-American Class III heart failure patients receiving standard pharmacotherapy [1]. This study showed significant clinical benefits of the therapy in this population, including an improvement in survival. This study was fairly controversial for its exclusive focus in an African-American population, and the fixed dose combination product has since received FDA approval, with the labeled indication being only in African-Americans.

The proposed hypothesis surrounding this therapy is that in some patients (particularly African-Americans), oxidative stress and reduced availability of nitric oxide may have an important role in heart failure pathophysiology. Recently presented pharmacogenetic data may provide some support for that hypothesis (see summary at http://www.theheart.org/viewArticle.do?primaryKey=570543&from=/searchLayout.do). Specifically, genetic polymorphisms in the gene for nitric oxide synthase (*NOS3*), which is the enzyme responsible for nitric oxide production, were studied. It was shown that Glu298 homozygotes derived benefits from ISDN–hydralazine that nearly achieved statistical significance, while the Asp298 carriers had no benefit from this therapy. Consistent with the hypothesis that this therapy is more beneficial in African-Americans, it was noted that about 70–80% of African-Americans are Glu298 homozygotes, whereas only about 40% of Caucasians carry this genotype. Thus, these data are compelling for several reasons. First, they provide support for the nitric oxide hypothesis surrounding ISDN–hydralazine therapy. Secondly, they highlight that with the increasing numbers of drugs available for treatment of heart failure patients, it might be possible to optimize therapies in the future through use of genetic information. Finally, they highlight (like the beta-blocker in hypertension data) that use of genotype data at specific genes relevant to drug pharmacology may be a much more appropriate method of guiding therapy than making decisions based on a patient's skin color. For example, if these data are correct, they suggest that 20–30% of African-Americans might not benefit from ISDN–hydralazine, whereas 40% of Caucasians might benefit from such therapy. Whether this therapy will be guided in the future with genetic information remains to be seen, but these data provide evidence that such an approach may be in store.

Relevance of pharmacogenetics to management of heart failure

As highlighted above, there are some interesting pharmacogenetic data surrounding heart failure therapies. However, to move these data to the point of clinical utility will require much additional research. This would be most effectively accomplished by having pharmacogenetic substudies of large heart failure clinical trials. However, a concern in this regard is that large studies of our standard therapies are likely to be very limited in the future, given their well-documented benefits. Thus, achieving the potential for pharmacogenomics in heart failure may present a particular challenge. However, because of the routine practice of polypharmacy in heart failure, this is a disease state whose management could likely benefit greatly from use of genetic information to help guide therapy. Over the last two decades, management of heart failure has become increasingly complex, such that nearly all patients are on a minimum of four heart failure drugs, and some are on as many as seven. However, we know with some certainty that not all patients benefit from all of those drugs, and so the challenge is to have a way to identify those patients most likely to derive benefit from a particular therapy. Pharmacogenomics represents an appealing manner of achieving this goal.

Pharmacogenetics of thrombosis

Dysregulation of hemostatic equilibrium can manifest in myriad pathologic conditions including deep vein thrombosis of the extremities, pulmonary embolism, cerebrovascular events and acute coronary syndromes. Because these conditions are variations in essentially the same prothrombotic processes, common pharmacologic agents are used in prevention and treatment. Drug therapy for thrombosis-related conditions include antiplatelet agents (e.g., aspirin, clopidogrel and glycoprotein IIb/IIIa inhibitors), anthrombin agents (e.g., unfractionated and low molecular weight heparin) and anticoagulants such as warfarin. Several studies of antithrombotic agents are available in the pharmacogenetics literature.

Antiplatelet pharmacogenetics

Aspirin

Aspirin remains a cornerstone of therapy for prevention and treatment of myriad thromboembolic conditions. While beyond the scope of the current discussion, there is a fair amount of data regarding associations between various genetic polymorphisms and aspirin intolerance or hypersensitivity in noncardiac, predominantly asthmatic populations. When focusing on cardiovascular pharmacogenetic studies of aspirin, data largely relate to a handful of genes encoding pharmacodynamic proteins and their influence on surrogate markers of aspirin antiplatelet activity.

Interest in genetic contribution to variable aspirin response began to emerge in the 1990s. Increasing interest has been paid to the phenomenon of "aspirin resistance"; that is, the observation that aspirin does not reduce ischemic end-points in many individuals, and that aspirin fails to inhibit platelet aggregation *ex vivo* in up to 25% of individuals [97]. As such, in the last decade increasingly rigorous attempts have been made to elucidate the impact of polymorphisms in genes that encode such proteins as cyclo-oxygenase-1 (COX-1), subunits of the platelet glycoprotein IIb/IIIa (GP IIb/IIIa) receptor, and the platetelet adenosine diphosphate (ADP) receptor on aspirin resistance [98–101].

COX-1 is an important upstream enzyme in the eicosanoid production pathway, and is a pharmacodynamic target of aspirin therapy. A limited number of studies have investigated COX-1 gene (*PTGS1*) haplotypes on aspirin action (e.g., inhibition of platelet aggregation or eicosanoid production) [102,103]. In general, there appear to be at least certain "at-risk" genotype combinations in which carriers of these haplotypes are differentially responsive to aspirin.

A more often studied target in antithrombotic pharmacogenetics is the platelet GP IIb/IIIa receptor. This receptor on the activated platelet's surface allows for fibrinogen binding with subsequent platelet cross-linking and aggregation. A C→T SNP at nucleotide 1565 of the IIIa subunit results in a change from Leu→Pro at amino acid 33. The wild-type allele is designated PlA1 while the variant is denoted PlA2. It has been suggested that carriers of the A2 allele are at higher risk for cardiovascular events as well as diminished response to aspirin. Several small investigations have found aspirin's ability to inhibit thrombin generation and prolong bleeding times is attenuated in those with the A2 allele [104–106]. However, results regarding the A1/A2 polymorphism are not consistent, vary by phenotype studied, and may be confounded by differential responses of A2-positive platelets to agonists *ex vivo* [107–111]. Further, the use of *ex vivo* platelet aggregation studies to determine the contribution of this polymorphism to aspirin resistance *in vivo* remains under debate [112]. Despite these limitations, the biologic plausibility of the GP IIb/IIIa receptor as a candidate gene, the associations of the PlA1/A2 polymorphism with variable cardiovascular risk and aspirin effects in some studies, and relatively high A2 variant allele frequency (12–15% among Caucasians) make this polymorphism a likely candidate for continued study into the mechanisms of aspirin resistance.

GP IIb/IIIa inhibitors

Predictably, the IIIa subunit of GP IIb/IIIa is a major candidate gene for pharmacogenetic studies of GP IIb/IIIa inhibitors such as abciximab, eptifibatide and tirofiban. In 87 individuals undergoing coronary revascularization, Wheeler *et al.* [113] investigated whether or not platelets from P1A2 carriers had diminished response to abciximab compared with those from A1 homozygotes. Indeed, A2 carriers exhibited a partially responsive phenotype as defined by several *ex vivo* assays. The diminished response may be explained by fewer fibrinogen receptors seen in A2-positive platelets.

Similar findings of an attenuated drug response have been demonstrated in A2-positive platelets in the presence of eptifibatide [114]. However, it is important to note that genotype–drug interactions with regard to variability in platelet response (particularly with eptifibatide and tirofiban) have recently been shown to be dependent on the type of anticoagulant used, complicating interpretation of association studies using platelet activity as a phenotypic measure [115]. Furthermore, consideration of surrogate markers for GP IIb/IIIa inhibitor efficacy by genotype other than platelet activity (e.g., infarct size, mortality) has revealed no significant impact of PlA genotype on drug effect, and studies of drug toxicity (i.e., bleeding) are limited [116,117].

Clopidogrel

Clopidogrel, an ADP receptor antagonist, has emerged as a standard antiplatelet treatment modality in acute coronary syndromes and coronary revascularization procedures. Because ADP acts as a platelet activator by binding to $P2Y_{12}$ receptors on the platelet's surface, inhibition of $P2Y_{12}$ receptors by clopidogrel blocks an initial step in platelet degranulation, upregulation of GP IIb/IIIa expression and platelet aggregation.

As with aspirin, resistance to clopidogrel effects have been described with an upper prevalence estimate of 30% [118]. It has been hypothesized that polymorphisms in *P2RY12* encoding for the pharmacodynamic target of clopidogrel action may be an important determinant of drug response. Several fairly large studies have investigated *P2RY12* polymorphisms and clopidogrel responses. Ziegler *et al.* [119] studied whether either of two exonic SNPs in *P2RY12* (C34→T or G52→T) were associated with variability in clopidogrel response for secondary prevention of cerebrovascular events in patients with peripheral arterial disease. After adjustment for known risk factors for cerebrovascular events and statin use, it was found that clopidogrel users who were 34T variant carriers had a fourfold higher risk of the primary end-point when compared with 34C homozygotes (95% CI, 1.08–14.92; $P - 0.04$). The G52→T SNP was not associated with variable response. Neither SNP was associated with variable responses among aspirin users. Of note, these SNPs have not been associated with different responses to clopidogrel loading doses in the setting of revascularization (platelet aggregation phenotype) [120].

A study by Angiolillo *et al.* [121] looked at the effect of a different *P2RY12* polymorphism (T744→C) on *ex vivo* platelet aggregation and activation in two groups of patients with coronary disease: group 1 consisted of 36 individuals undergoing revascularization with stent placement who received a 300-mg clopidogrel loading dose, while group 2 consisted of 83 individuals on long-term clopidgrel therapy (75 mg/day). There was no differential response to clopidogrel by genotype for any phenotypic measure, and the authors conclude that this SNP alone is not likely to significantly contribute to clopidgrel resistance.

While *P2RY12* has been the most consistently studied gene in clopidogrel pharmacogenetics, single studies exist for other candidate genes such as those encoding protease-activated receptor-1 (PAR-1), GP IIb/IIIa (PlA polymorphism) and GP Ia, all of which suggest a role of genotype in modulating antiplatelet effects of clopidogrel [122–124]. Furthermore, as *CYP3A4* is responsible for conversion of clopidogrel to its pharmacologically active metabolite, polymorphisms in this gene are likely to be studied in the future.

Anticoagulants

Anticoagulation with warfarin is standard of care for preventing thromboembolic events in most patients with chronic atrial fibrillation, and is used in numerous other populations for either short- or long-term thromboembolism prophylaxis. However, warfarin has a narrow therapeutic window in which efficacy can be maximized and risk for excessive bleeding minimized. Furthermore, appropriate warfarin dosing to achieve target international normalized ratio (INR) is problematic because of the myriad factors that contribute to variability in drug exposure (e.g., age, drug interactions, vitamin K intake).

Warfarin pharmacogenetic studies have largely focused on two candidate genes: *CYP2C9*, responsible for warfarin metabolism, and *VKORC1*, which encodes vitamin K epoxide reductase, the site of warfarin action. As such, studies in aggregate have attempted to elucidate the association between variants in these pharmacokinetic and pharmacodynamic genes and such phenotypes as warfarin dosage requirements and risk for bleeding.

Major *CYP2C9* alleles include the *1 wild-type, and *2 and *3 variants. While the *1 allele confers full metabolic activity, the *2 and *3 alleles are associated with decreased *CYP2C9* activity and decreased clearance of S-warfarin (the more pharmacologically active enantiomer) [125–127]. Consequently, it has been shown that individuals with these variant alleles require lower maintenance doses of warfarin [128–130]. In addition, it takes variant carriers longer to achieve therapeutic INRs, and variant carriers are at increased risk of over-anticoagulation upon initiation of therapy; in particular, *2 or *3 carriers are more likely to have supratherapeutic INRs (INR >4) and increased risk for bleeding [131].

In an attempt to push pharmacogenetics a step closer to clinical practice, Voora *et al.* [132] prospectively tested a warfarin dosing algorithm containing *CYP2C9* genotype along with non-genetic variables to see whether 48 surgical patients could achieve therapeutic INR without significant delay. The equation to predict warfarin dosage included inputs for age, race, sex, body surface area, target INR, genotype and presence of amiodarone and/or a statin. Use of genotype-guided therapy resulted in equivalent effectiveness between *CYP2C9* wild-type homozygotes and variant carriers (i.e., time to stable therapeutic INR). However, there was still a greater incidence of overanticoagulation (INR >4) among variant carriers, suggesting use of *CYP2C9* genotype alone to guide warfarin regimens does not currently eliminate the need for traditional monitoring parameters such as INR. Further prospective studies using *CYP2C9* genotype to dose warfarin are likely [133].

A limitation of most studies investigating genetic associations with warafrin dosage requirements is that they fail to account for variability in pharmacodynamic genes such as *VKORC1*. Two seminal papers published in 2004 identified *VKORC1* as the target of warfarin therapy and suggested *VKORC1* polymorphisms are important factors in warfarin resistance [134,135]. Since then, several studies have correlated polymorphisms in this gene to warfarin dosage requirements [136–142]. Studies have demonstrated that *VKORC1* and *CYP2C9* genotypes, along with known nongenetic predictors of warfarin dosage, account for over 50% of the patient–patient variability in weekly warfarin dosage requirements [139,141]. It appears that *VKORC1* polymorphisms explain more of this variability than *CYP2C9* variants [142,143]. Clotting factor polymorphisms have also been considered by some groups, and while some seem to exhibit differences in warfarin sensitivity by genotype, they contribute minimally to the variability, particularly in relation to *CYP2C9* and *VKORC1*. Regimen algorithms utilizing both *CYP2C9*, *VKORC1* and nongenetic variables are under development. It seems likely that genotype-guided warfarin therapy will be the first example among cardiovascular drugs to enter clinical practice within the next couple of years.

Antithrombin pharmacogenetics

Despite the well-studied relationship between variants in genes such as factor V and prothrombin and hypercoagulable states, pharmacogenetics studies of antithrombin agents such as unfractionated and low molecular weight heparin are exceedingly scarce. This may perhaps be because these agents are usually used short term in acute settings. The majority of studies are small investigations assessing whether a polymorphism in the Fcγ receptor IIa (*FCGR2A*), which cross-links platelet factor 4, heparin and IgG, contributes to heparin-induced thrombocytopenia (HIT) [144–149]. In particular, an Arg131→His polymorphism has been contradictorily implicated in various HIT phenotypes, whereby both the 131His and 131Arg alleles have been associated with HIT; furthermore, neutral studies also exist.

In a particularly interesting study of HIT with thromboembolic complications, Carlsson *et al.* [150] employed a multigene approach to determine whether variant alleles in GP IIb/IIIa, GP Ia/IIa, GPIb/IX/V, factor V, prothrombin and methyltetrahydrofolate reductase genes were overrepresented in HIT patients with thromboembolic complications ($n = 79$) compared with HIT patients without complications ($n = 63$). There were no significant genetic associations with risk for thromboembolic complications in patients with HIT. While sample size and power are always of concern when conducting pharmacogenetic analyses of multiple polymorphisms, the study represents an important undertaking in trying to elucidate the contribution

of multiple genes to a potentially life-threatening complication of heparin therapy.

Pharmacogenetic studies of antithrombin efficacy are even scarcer. In a biomarker study of acute coronary syndromes, Ray *et al.* [151] determined whether or not a variable number of tandem repeat (VNTR) polymorphism in the interleukin-1 receptor antagonist gene (*IL1RN*) was associated with responses to unfractionated heparin (UFH) or low molecular weight heparin (LMWH). The marker of response was change in circulating von Willibrand factor (vWF) concentrations 24 and 48 hours after treatment. It was found that LMWH was superior over UFH in attenuating increased vWF concentrations overall. However, it appeared that the differential benefit of LMWH was largely limited to carriers if the *IL1RN**2 variant allele. Differences in response between UFH and LMWH were not seen among the majority non-*2 carriers (i.e., *1/*1 wild-type homozygotes). These data offer interesting insights as to how genotyping or molecular profiling might be used in the future. For example, understanding that treatment effects between UFH and LMWH were only seen for *2 carriers (roughly 43% of the population) may help guide treatment policy for agents such as LMWH that tend to be more expensive than UFH. Further studies are required.

Pharmacogenetics of arrhythmia and proarrhythmia

Treatment strategies for the management of arrhythmias have dramatically evolved since the early 1990s, such that class I and III antiarrhythmic drugs are now rarely used as first line therapy [152]. Thus, while pharmacologic rhythm control still remains a therapeutic option, the risks often outweigh the benefits in chronic management and prevention of ventricular dysrhythmia.

It has been long appreciated that polymorphic Phase I and II drug metabolism contribute to variable drug exposure and effects for antiarrhythmic drugs such as procainamide, propafenone, flecainide, encainide and others [153,154]. Furthermore, genetically mediated polymorphic drug metabolism may be responsible for enhancing pharmacodynamic activity (and presumably likelihood for drug-induced arrhythmias) when multiple antiarrhythmics are used simultaneously [155–159]. As such, serum drug concentrations are used to guide therapy for certain antiarrhythmics.

Potentially significant clinical implications lie in the pharmacogenetics of drug-induced QT prolongation and torsades de pointes (TdP). The genetics of acquired (i.e., drug-induced) long QT syndrome have been well described [160,161]. The ability of both cardiovascular and noncardiovascular drugs (e.g., terfenadine, astemizole, fluoroquinolone antibiotics and antispychotics) to prolong the QT interval has been cited as a major contributor to arrhythmogenicity of these agents.

Because of the phenotypical similarities between acquired and congenital long QT syndromes (e.g., a predominance in women over men and near identical ECG changes), polymorphisms in genes associated with congenital long QT syndrome have served as candidate genes for the study of drug-induced QT prolongation and TdP. For example, in an analysis of a fairly large data set, genetic variants in protein subunits of the cardiac K^+ channel encoded by *KCNQ1* (or *KvLQT1*), *KCNH2* (or *HERG*) and *SCN5A*, have been identified to occur in roughly 5–10% of individuals experiencing drug-induced prolongation of the QT interval [162]. These and other findings suggest that polymorphisms in pore-forming subunits of electrocurrent-related proteins, as well as protein modifiers subunit function (e.g., *KCNE2* or *MiRP1*) may be important in *a priori* risk assessment for arrhythmia in a patient or in the drug development process [163–166].

Because intrinsic regulatory control of cardiac rhythm is so complex, it has been promulgated that no one set of genetic or nongenetic factors can predict the likelihood of an agent to cause QT prolongation or TdP. Rather, variants in drug metabolism genes (*CYP2D6, CYP2C9, CYP2C19, CYP3A4, CYP3A5*) should be considered in conjunction with genes for proteins involved in potassium-related action potentials (i.e., I_{TO}, I_{Kr} and I_{Ks}) and sodium channels, along with nongenetic factors (e.g., sex, renal function, serum potassium and interacting drugs) [167,168].

The usefulness of arrhythmia pharmacogenetics does not lie solely in understanding the molecular basis of drug-induced arrhythmias. Associations between single polymorphisms, haplotypes and sets of multiple polymorphisms with dysrhythmia phenotypes and *in vitro* analyses could potentially lead to:

1 Improved diagnosis of individuals with both congenital and acquired long QT syndrome;

2 More rational use of existing medications (both cardiac and non-cardiac); and

3 Enhancement of the drug development process [169].

By genomically profiling novel compounds against compounds with known QT prolonging effects, comparative toxicogenomics can perhaps be used to triage novel compounds with likely QT prolonging effects in the preclinical phases of drug discovery and development.

The potential clinical relevance of the work on drug-induced QT prolongation is that it may improve drug development for drugs with QT prolongation properties. Additionally, it might be possible to identify those patients at highest risk for drug-induced arrhythmia, although reaching this goal will be a significant challenge.

Future directions in cardiovascular pharmacogenetics

The examples discussed above provide clear evidence that genetics contributes to the variable efficacy and toxicity that is encountered across a population of individuals given a certain drug. The hope is that this information might someday be used to optimize the pharmacotherapy management of patients, and of the examples provided, warfarin is the first in this group for which the promise might be realized. However, in most cases it is likely to be another 10–15 years before there are a large number of drugs for which their use can be guided by genetic information, and even longer for this to gain widespread use in the practice of medicine.

To achieve the goals of pharmacogenetics requires several steps, as highlighted in Fig. 11.2. For many of the examples discussed in this chapter, our knowledge is at the base of the pyramid – there has been documentation that genetics contributes to variable drug response. However, this knowledge must be built upon to move pharmacogenetics to the point of use in clinical practice. The next step on the pyramid is to determine the group

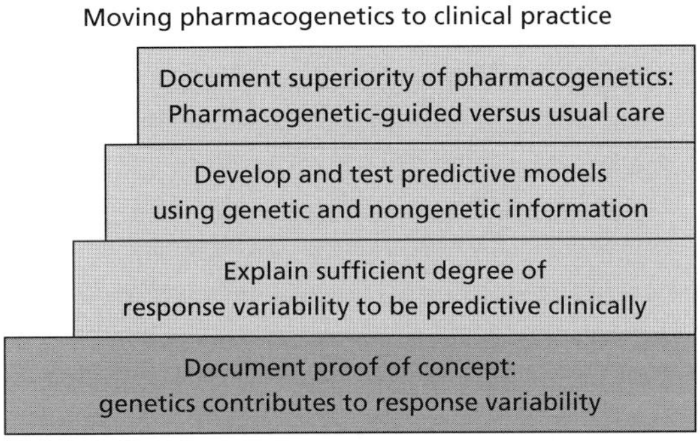

Moving pharmacogenetics to clinical practice

Document superiority of pharmacogenetics:
Pharmacogenetic-guided versus usual care

Develop and test predictive models
using genetic and nongenetic information

Explain sufficient degree of
response variability to be predictive clinically

Document proof of concept:
genetics contributes to response variability

Figure 11.2 Moving pharmacogenetics to clinical practice. The pyramid of knowledge that will likely need to be built for use of pharmacogenetic data to be translated into practice. See text for details. Reprinted from [192] with permission from Future Medicine Ltd.

of genes that will allow for a sufficient degree of the response variability to be explained or predicted. While we have probably reached this threshold for warfarin, this is not yet the case for our other examples. Once the various genes that contribute to variable drug response are known, then mathematical models that incorporate genetic and nongenetic information to predict response must be developed and tested. This is currently underway for warfarin. Finally, to achieve the evidence base that will be needed for wide adoption into practice, there will need to be studies documenting that a pharmacogenetic-guided approach is superior to the usual approach of treating diseases with drug therapy. For those drugs where the level of evidence can be accumulated to reach the top of this knowledge pyramid, it seems likely that use of pharmacogenetics will be embraced in clinical practice.

References

1 Taylor AL, Ziesche S, Yancy C *et al.* the African-American Heart Failure Trial Investigators. Combination of isosorbide dinitrate and hydralazine in blacks with heart failure. *N Engl J Med* 2004; 351: 2049–2057.

2 Third Report of the National Cholesterol Education Program (NCEP) Expert Panel on Detection, Evaluation, and Treatment of High Blood Cholesterol in Adults (Adult Treatment Panel III) final report. *Circulation* 2002; 106: 3143–3421.

3 Grundy SM, Cleeman JI, Merz CN *et al.* Implications of recent clinical trials for the National Cholesterol Education Program Adult Treatment Panel III guidelines. *Circulation* 2004; 110: 227–239.

4 Zineh I. HMG-CoA reductase inhibitor pharmacogenomics: overview and implications for practice. *Clin Cardiol* 2005; 1: 191–206.

5 Kim RB. 3-Hydroxy-3-methylglutaryl-coenzyme A reductase inhibitors (statins) and genetic variability (single nucleotide polymorphisms) in a hepatic drug uptake transporter: what's it all about? *Clin Pharmacol Ther* 2004; 75: 381–385.

6 Corsini A, Bellosta S, Baetta R, Fumagalli R, Paoletti R, Bernini F. New insights into the pharmacodynamic and pharmacokinetic properties of statins. *Pharmacol Ther* 1999; 84: 413–428.

7 Hsiang B, Zhu Y, Wang Z *et al.* A novel human hepatic organic anion transporting polypeptide (OATP2). Identification of a liver-specific human organic anion transporting polypeptide and identification of rat and human hydroxymethylglutaryl-CoA reductase inhibitor transporters. *J Biol Chem* 1999; 274: 37161–37168.

8 Kobayashi D, Nozawa T, Imai K, Nezu J, Tsuji A, Tamai I. Involvement of human organic anion transporting polypeptide OATP-B (SLC21A9) in pH-dependent transport across intestinal apical membrane. *J Pharmacol Exp Ther* 2003; 306: 703–708.

9 Hasegawa M, Kusuhara H, Sugiyama D *et al.* Functional involvement of rat organic anion transporter 3 (rOat3; Slc22a8) in the renal uptake of organic anions. *J Pharmacol Exp Ther* 2002; 300: 746–753.

10 Tirona RG, Leake BF, Merino G, Kim RB. Polymorphisms in OATP-C: identification of multiple allelic variants associated with altered transport activity among European- and African-Americans. *J Biol Chem* 2001; 276: 35669–35675.

11 Kameyama Y, Yamashita K, Kobayashi K, Hosokawa M, Chiba K. Functional characterization of SLCO1B1 (OATP-C) variants, SLCO1B1*5, SLCO1B1*15 and SLCO1B1*15+C1007G, by using transient expression systems of HeLa and HEK293 cells. *Pharmacogenet Genomics* 2005; 15: 513–522.

12 Mwinyi J, Johne A, Bauer S, Roots I, Gerloff T. Evidence for inverse effects of OATP-C (SLC21A6) 5 and 1b haplotypes on pravastatin kinetics. *Clin Pharmacol Ther* 2004; 75: 415–421.

13 Nishizato Y, Ieiri I, Suzuki H *et al.* Polymorphisms of OATP-C (SLC21A6) and OAT3 (SLC22A8) genes: consequences for pravastatin pharmacokinetics. *Clin Pharmacol Ther* 2003; 73: 554–565.

14 Niemi M, Schaeffeler E, Lang T *et al.* High plasma pravastatin concentrations are associated with single nucleotide polymorphisms and haplotypes of organic anion transporting polypeptide-C (OATP-C, SLCO1B1). *Pharmacogenetics* 2004; 14: 429–440.

15 Niemi M, Neuvonen PJ, Hofmann U *et al.* Acute effects of pravastatin on cholesterol synthesis are associated with SLCO1B1 (encoding OATP1B1) haplotype *17. *Pharmacogenet Genomics* 2005; 15: 303–309.

16 Tachibana-Iimori R, Tabara Y, Kusuhara H *et al.* Effect of genetic polymorphism of OATP-C (SLCO1B1) on lipid-lowering response to HMG-CoA reductase inhibitors. *Drug Metab Pharmacokinet* 2004; 19: 375–380.

17 Kajinami K, Brousseau ME, Ordovas JM, Schaefer EJ. Polymorphisms in the multidrug resistance-1 (MDR1) gene influence the response to atorvastatin treatment in a gender-specific manner. *Am J Cardiol* 2004; 93: 1046–1050.

18 Bercovich D, Friedlander Y, Korem S *et al.* The association of common SNPs and haplotypes in the CETP and MDR1 genes with lipids response to fluvastatin in familial hypercholesterolemia. *Atherosclerosis* 2006; 185: 97–107.

19 Nordin C, Dahl ML, Eriksson M, Sjoberg S. Is the cholesterol-lowering effect of simvastatin influenced by CYP2D6 polymorphism? *Lancet* 1997; 350: 29–30.

20 Mulder AB, van Lijf HJ, Bon MA, *et al.* Association of polymorphism in the cytochrome CYP2D6 and the efficacy and tolerability of simvastatin. *Clin Pharmacol Ther* 2001; 70: 546–551.

21 Geisel J, Kivisto KT, Griese EU, Eichelbaum M. The efficacy of simvastatin is not influenced by CYP2D6 polymorphism. *Clin Pharmacol Ther* 2002; 72: 595–596.

22 Kajinami K, Brousseau ME, Ordovas JM, Schaefer EJ. CYP3A4 genotypes and plasma lipoprotein levels before and after treatment with atorvastatin in primary hypercholesterolemia. *Am J Cardiol* 2004; 93: 104–107.

23 Wang A, Yu BN, Luo CH *et al.* Ile118Val genetic polymorphism of CYP3A4 and its effects on lipid-lowering efficacy of simvastatin in Chinese hyperlipidemic patients. *Eur J Clin Pharmacol* 2005; 60: 843–848.

24 Kivisto KT, Niemi M, Schaeffeler E *et al.* Lipid-lowering response to statins is affected by CYP3A5 polymorphism. *Pharmacogenetics* 2004; 14: 523–525.

25 Kirchheiner J, Kudlicz D, Meisel C *et al.* Influence of CYP2C9 polymorphisms on the pharmacokinetics and cholesterol-lowering activity of (−)-3S,5R-fluvastatin and (+)-3R,5S-fluvastatin in healthy volunteers. *Clin Pharmacol Ther* 2003; 74: 186–194.

26 Wilke RA, Moore JH, Burmester JK. Relative impact of CYP3A genotype and concomitant medication on the severity of atorvastatin-induced muscle damage. *Pharmacogenet Genomics* 2005; 15: 415–421.

27 Chasman DI, Posada D, Subrahmanyan L, Cook NR, Stanton VP Jr, Ridker PM. Pharmacogenetic study of statin therapy and cholesterol reduction. *JAMA* 2004; 291: 2821–2827.

28 Thompson JF, Man M, Johnson KJ *et al.* An association study of 43 SNPs in 16 candidate genes with atorvastatin response. *Pharmacogenomics J* 2005; 5: 352–358.

29 Pedro-Botet J, Schaefer EJ, Bakker-Arkema RG *et al.* Apolipoprotein E genotype affects plasma lipid response to atorvastatin in a gender specific manner. *Atherosclerosis* 2001; 158: 183–193.

30 Ballantyne CM, Herd JA, Stein EA *et al.* Apolipoprotein E genotypes and response of plasma lipids and progression-regression of coronary atherosclerosis to lipid-lowering drug therapy. *J Am Coll Cardiol* 2000; 36: 1572–1578.

31 Nestel P, Simons L, Barter P *et al.* A comparative study of the efficacy of simvastatin and gemfibrozil in combined hyperlipoproteinemia: prediction of response by baseline lipids, apo E genotype, lipoprotein(a) and insulin. *Atherosclerosis* 1997; 129: 231–239.

32 Ordovas JM, Mooser V. The APOE locus and the pharmacogenetics of lipid response. *Curr Opin Lipidol* 2002; 13: 113–117.

33 O'Neill FH, Patel DD, Knight BL *et al.* Determinants of variable response to statin treatment in patients with refractory familial hypercholesterolemia. *Arterioscler Thromb Vasc Biol* 2001; 21: 832–837.

34 Gerdes LU, Gerdes C, Kervinen K *et al.* The apolipoprotein ε4 allele determines prognosis and the effect on prognosis of simvastatin in survivors of myocardial infarction: a substudy of the Scandinavian simvastatin survival study. *Circulation* 2000; 101: 1366–1371.

35 Maitland-van der Zee AH, Stricker BH, Klungel OH *et al.* Adherence to and dosing of beta-hydroxy-beta-methylglutaryl coenzyme A reductase inhibitors in the general population differs according to apolipoprotein E-genotypes. *Pharmacogenetics* 2003; 13: 219–223.

36 Sanllehy C, Casals E, Rodriguez-Villar C *et al.* Lack of interaction of apolipoprotein E phenotype with the lipoprotein response to lovastatin or gemfibrozil in patients with primary hypercholesterolemia. *Metabolism* 1998; 47: 560–565.

37 Nemeth A, Szakmary K, Kramer J *et al.* Apolipoprotein E and complement C3 polymorphism and their role in the response to gemfibrozil and low fat low cholesterol therapy. *Eur J Clin Chem Clin Biochem* 1995; 33: 799–804.

38 Brisson D, Ledoux K, Bosse Y *et al.* Effect of apolipoprotein E, peroxisome proliferator-activated receptor alpha and lipoprotein lipase gene mutations on the ability of fenofibrate to improve lipid profiles and reach clinical guideline targets among hypertriglyceridemic patients. *Pharmacogenetics* 2002; 12: 313–320.

39 Brousseau ME, O'Connor JJ Jr, Ordovas JM *et al.* Cholesteryl ester transfer protein TaqI B2B2 genotype is associated with higher HDL cholesterol levels and lower risk of coronary heart disease end points in men with HDL deficiency: Veterans Affairs HDL Cholesterol Intervention Trial. *Arterioscler Thromb Vasc Biol* 2002; 22: 1148–1154.

40 Brousseau ME, Goldkamp AL, Collins D *et al.* Polymorphisms in the gene encoding lipoprotein lipase in

men with low HDL-C and coronary heart disease: the Veterans Affairs HDL Intervention Trial. *J Lipid Res* 2004; 45: 1885–1891.

41 Foucher C, Rattier S, Flavell DM *et al.* Response to micronized fenofibrate treatment is associated with the peroxisome-proliferator-activated receptors alpha G/C intron7 polymorphism in subjects with type 2 diabetes. *Pharmacogenetics* 2004; 14: 823–829.

42 Flavell DM, Jamshidi Y, Hawe E *et al.* Peroxisome proliferator-activated receptor alpha gene variants influence progression of coronary atherosclerosis and risk of coronary artery disease. *Circulation* 2002; 105: 1440–1445.

43 Bosse Y, Pascot A, Dumont M *et al.* Influences of the PPAR alpha-L162V polymorphism on plasma HDL(2)-cholesterol response of abdominally obese men treated with gemfibrozil. *Genet Med* 2002; 4: 311–315.

44 Brouillette C, Bosse Y, Perusse L, Gaudet D, Vohl MC. Effect of liver fatty acid binding protein (FABP) T94A missense mutation on plasma lipoprotein responsiveness to treatment with fenofibrate. *J Hum Genet* 2004; 49: 424–432.

45 Bosse Y, Vohl MC, Dumont M *et al.* Influence of the angiotensin-converting enzyme gene insertion/deletion polymorphism on lipoprotein/lipid response to gemfibrozil. *Clin Genet* 2002; 62: 45–52.

46 Smart EJ, De Rose RA, Farber SA. Annexin 2-caveolin 1 complex is a target of ezetimibe and regulates intestinal cholesterol transport. *Proc Natl Acad Sci USA* 2004; 101: 3450–3455.

47 Lammert F, Wang DQ. New insights into the genetic regulation of intestinal cholesterol absorption. *Gastroenterology* 2005; 129: 718–734.

48 Garcia-Calvo M, Lisnock J, Bull HG *et al.* The target of ezetimibe is Niemann–Pick C1-Like 1 (NPC1L1). *Proc Natl Acad Sci USA* 2005; 102: 8132–8137.

49 Wang J, Williams CM, Hegele RA. Compound heterozygosity for two non-synonymous polymorphisms in NPC1L1 in a non-responder to ezetimibe. *Clin Genet* 2005; 67: 175–177.

50 Hegele RA, Guy J, Ban MR, Wang J. NPC1L1 haplotype is associated with inter-individual variation in plasma low-density lipoprotein response to ezetimibe. *Lipids Health Dis* 2005; 4: 16.

51 Zineh I. Genetic polymorphisms and statin therapy. *JAMA* 2004; 292: 1302–1303.

52 Johnson JA, Turner ST. Hypertension pharmacogenomics: current status and future directions. *Curr Opin Mol Ther* 2005; 7: 218–225.

53 Cusi D, Barlassina C, Azzani T *et al.* Polymorphisms of alpha-adducin and salt sensitivity in patients with essential hypertension. *Lancet* 1997; 349: 1353–1357.

54 Glorioso N, Manunta P, Filigheddu F *et al.* The role of alpha-adducin polymorphism in blood pressure and sodium handling regulation may not be excluded by a negative association study. *Hypertension* 1999; 34: 649–654.

55 Sciarrone MT, Stella P, Barlassina C *et al.* ACE and alpha-adducin polymorphism as markers of individual response to diuretic therapy. *Hypertension* 2003; 41: 398–403.

56 Psaty BM, Smith NL, Heckbert SR *et al.* Diuretic therapy, the alpha-adducin gene variant, and the risk of myocardial infarction or stroke in persons with treated hypertension. *JAMA* 2002; 287: 1680–1689.

57 Gerhard T, Gong Y, Beitelshees AL *et al.* Cardiovascular outcomes, diuretic therapy and the alpha-adducin polymorphism: Results for the International Verapamil SR-Trandolapril Study Genetic Substudy (INVEST-GENES) (Abstract). *Circulation* 2005; 112: II608.

58 Johnson JA, Zineh I, Puckett BJ, McGorray SP, Yarandi HN, Pauly DF. β_1-Adrenergic receptor polymorphisms and antihypertensive response to metoprolol. *Clin Pharmacol Ther* 2003; 74: 44–52.

59 Liu J, Liu ZQ, Tan ZR *et al.* Gly389Arg polymorphism of β_1-adrenergic receptor is associated with the cardiovascular response to metoprolol. *Clin Pharmacol Ther* 2003; 74: 372–379.

60 Sofowora GG, Dishy V, Muszkat M *et al.* A common β_1-adrenergic receptor polymorphism (Arg389Gly) affects blood pressure response to beta-blockade. *Clin Pharmacol Ther* 2003; 73: 366–371.

61 O'Shaughnessy KM, Fu B, Dickerson C, Thurston D, Brown MJ. The gain-of-function G389R variant of the β_1-adrenoceptor does not influence blood pressure or heart rate response to beta-blockade in hypertensive subjects. *Clin Sci (Colch)* 2000; 99: 233–238.

62 Karlsson J, Lind L, Hallberg P *et al.* β_1-Adrenergic receptor gene polymorphisms and response to β_1-adrenergic receptor blockade in patients with essential hypertension. *Clin Cardiol* 2004; 27: 347–350.

63 Ohmichi N, Iwai N, Uchida Y, Shichiri G, Nakamura Y, Kinoshita M. Relationship between the response to the angiotensin converting enzyme inhibitor imidapril and the angiotensin converting enzyme genotype. *Am J Hypertens* 1997; 10: 951–955.

64 Stavroulakis GA, Makris TK, Krespi PG *et al.* Predicting response to chronic antihypertensive treatment with fosinopril: the role of angiotensin-converting enzyme gene polymorphism. *Cardiovasc Drugs Ther* 2000; 14: 427–432.

65 Li X, Du Y, Huang X. Correlation of angiotensin converting enzyme gene polymorphism with effect of antihypertensive therapy by angiotensin-converting enzyme inhibitor. *J Cardiovasc Pharmacol Ther* 2003; 8: 25–30.

66 Kurland L, Melhus H, Karlsson J *et al.* Aldosterone synthase (CYP11B2) -344 C/T polymorphism is related to antihypertensive response: result from the Swedish

Irbesartan Left Ventricular Hypertrophy Investigation versus Atenolol (SILVHIA) trial. *Am J Hypertens* 2002; 15: 389–393.

67 Yu H, Zhang Y, Liu G. Relationship between polymorphism of the angiotensin-converting enzyme gene and the response to angiotensin-converting enzyme inhibition in hypertensive patients. *Hypertens Res* 2003; 26: 881–886.

68 Harrap SB, Tzourio C, Cambien F *et al.* The ACE gene I/D polymorphism is not associated with the blood pressure and cardiovascular benefits of ACE inhibition. *Hypertension* 2003; 42: 297–303.

69 Ortlepp JR, Hanrath P, Mevissen V, Kiel G, Borggrefe M, Hoffmann R. Variants of the CYP11B2 gene predict response to therapy with candesartan. *Eur J Pharmacol* 2002; 445: 151–152.

70 Arnett DK, Davis BR, Ford CE *et al.* Pharmacogenetic association of the angiotensin-converting enzyme insertion/ deletion polymorphism on blood pressure and cardiovascular risk in relation to antihypertensive treatment: the Genetics of Hypertension-Associated Treatment (GenHAT) study. *Circulation* 2005; 111: 3374–3383.

71 Terra SG, Hamilton KK, Pauly DF *et al.* Beta$_1$-adrenergic receptor polymorphisms and left ventricular remodeling changes in response to beta-blocker therapy. *Pharmacogenet Genomics* 2005; 15: 227–234.

72 Mialet Perez J, Rathz DA, Petrashevskaya NN *et al.* β_1-Adrenergic receptor polymorphisms confer differential function and predisposition to heart failure. *Nat Med* 2003; 9: 1300–1305.

73 Kaye DM, Smirk B, Williams C, Jennings G, Esler M, Holst D. β_2-Adrenoceptor genotype influences the responses to carvedilol in patients with congestive heart failure. *Pharmacogenetics* 2003; 13.

74 de Groote P, Helbecque N, Lamblin N *et al.* Association between β_1 and β_1-adrenergic receptor gene polymorphisms and the response to beta-blockade in patients with stable congestive heart failure. *Pharmacogenet Genomics* 2005; 15: 137–142.

75 Terra SG, Pauly DF, Lee CR *et al.* β-Adrenergic receptor polymorphisms and responses during titration of metoprolol controlled release/extended release in heart failure. *Clin Pharmacol Ther* 2005; 77: 127–137.

76 Magnusson Y, Levin MC, Eggertsen R *et al.* Ser49Gly of β_1-adrenergic receptor is associated with effective beta-blocker dose in dilated cardiomyopathy. *Clin Pharmacol Ther* 2005; 78: 221–231.

77 Borjesson M, Magnusson Y, Hjalmarson A, Andersson B. A novel polymorphism in the gene coding for the β_1-adrenergic receptor associated with survival in patients with heart failure. *Eur Heart J* 2000; 21: 1853–1858.

78 Beta-Blocker Evaluation of Survival Trial Investigators. A trial of the beta-blocker bucindolol in patients with advanced chronic heart failure. *N Engl J Med* 2001; 344: 1659–1667.

79 White HL, de Boer RA, Maqbool A *et al.* An evaluation of the β_1-adrenergic receptor Arg389Gly polymorphism in individuals with heart failure: a MERIT-HF substudy. *Eur J Heart Fail* 2003; 5: 463–468.

80 Lanfear DE, Jones PG, Marsh S, Cresci S, McLeod HL, Spertus JA. β_2-Adrenergic receptor genotype and survival among patients receiving beta-blocker therapy after an acute coronary syndrome. *JAMA* 2005; 294: 1526–1533.

81 Giessmann T, Modess C, Hecker U *et al.* CYP2D6 genotype and induction of intestinal drug transporters by rifampin predict presystemic clearance of carvedilol in healthy subjects. *Clin Pharmacol Ther* 2004; 75: 213–222.

82 Zineh I, Beitelshees AL, Gaedigk A *et al.* Pharmacokinetics and CYP2D6 genotypes do not predict metoprolol adverse events or efficacy in hypertension. *Clin Pharmacol Ther* 2004; 76: 536–544.

83 McNamara DM, Holubkov R, Postava L *et al.* Pharmacogenetic interactions between angiotensin-converting enzyme inhibitor therapy and the angiotensin-converting enzyme deletion polymorphism in patients with congestive heart failure. *J Am Coll Cardiol* 2004; 44: 2019–2026.

84 Ambudkar SV, Dey S, Hrycyna CA, Ramachandra M, Pastan I, Gottesman MM. Biochemical, cellular, and pharmacological aspects of the multidrug transporter. *Annu Rev Pharmacol Toxicol* 1999; 39: 361–398.

85 Hoffmeyer S, Burk O, von Richter O *et al.* Functional polymorphisms of the human multidrug-resistance gene: multiple sequence variations and correlation of one allele with P-glycoprotein expression and activity *in vivo*. *Proc Natl Acad Sci USA* 2000; 97: 3473–3478.

86 Becquemont L, Verstuyft C, Kerb R *et al.* Effect of grapefruit juice on digoxin pharmacokinetics in humans. *Clin Pharmacol Ther* 2001; 70: 311–316.

87 Sakaeda T, Nakamura T, Horinouchi M *et al.* MDR1 genotype-related pharmacokinetics of digoxin after single oral administration in healthy Japanese subjects. *Pharm Res* 2001; 18: 1400–1404.

88 Gerloff T, Schaefer M, Johne A *et al.* MDR1 genotypes do not influence the absorption of a single oral dose of 1 mg digoxin in healthy white males. *Br J Clin Pharmacol* 2002; 54: 610–616.

89 Johne A, Kopke K, Gerloff T *et al.* Modulation of steady-state kinetics of digoxin by haplotypes of the P-glycoprotein MDR1 gene. *Clin Pharmacol Ther* 2002; 72: 584–594.

90 Kurata Y, Ieiri I, Kimura M *et al.* Role of human MDR1 gene polymorphism in bioavailability and interaction of digoxin, a substrate of P-glycoprotein. *Clin Pharmacol Ther* 2002; 72: 209–219.

91 Morita Y, Sakaeda T, Horinouchi M *et al.* MDR1 genotype-related duodenal absorption rate of digoxin in healthy Japanese subjects. *Pharm Res* 2003; 20: 552–556.

92 Verstuyft C, Schwab M, Schaeffeler E *et al.* Digoxin pharmacokinetics and MDR1 genetic polymorphisms. *Eur J Clin Pharmacol* 2003; 58: 809–812.

93 Verstuyft C, Strabach S, El-Morabet H *et al.* Dipyridamole enhances digoxin bioavailability via P-glycoprotein inhibition. *Clin Pharmacol Ther* 2003; 73: 51–60.

94 Chowbay B, Li H, David M, Bun Cheung Y, Lee EJ. Meta-analysis of the influence of MDR1 C3435T polymorphism on digoxin pharmacokinetics and MDR1 gene expression. *Br J Clin Pharmacol* 2005; 60: 159–171.

95 Cicoira M, Rossi A, Bonapace S *et al.* Effects of ACE gene insertion/deletion polymorphism on response to spironolactone in patients with chronic heart failure. *Am J Med* 2004; 116: 657–661.

96 Cohn JN, Johnson G, Ziesche S *et al.* A comparison of enalapril with hydralazine-isosorbide dinitrate in the treatment of chronic congestive heart failure. *N Engl J Med* 1991; 325: 303–310.

97 Szczeklik A, Musial J, Undas A, Sanak M. Aspirin resistance. *J Thromb Haemost* 2005; 3: 1655–1662.

98 Cambria-Kiely JA, Gandhi PJ. Aspirin resistance and genetic polymorphisms. *J Thromb Thrombolysis* 2002; 14: 51–58.

99 Schafer AI. Genetic and acquired determinants of individual variability of response to antiplatelet drugs. *Circulation* 2003; 108: 910–911.

100 Rozalski M, Boncler M, Luzak B, Watala C. Genetic factors underlying differential blood platelet sensitivity to inhibitors. *Pharmacol Rep* 2005; 57: 1–13.

101 Jefferson BK, Foster JH, McCarthy JJ *et al.* Aspirin resistance and a single gene. *Am J Cardiol* 2005; 95: 805–808.

102 Halushka MK, Walker LP, Halushka PV. Genetic variation in cyclooxygenase 1: effects on response to aspirin. *Clin Pharmacol Ther* 2003; 73: 122–130.

103 Maree AO, Curtin RJ, Chubb A *et al.* Cyclooxygenase-1 haplotype modulates platelet response to aspirin. *J Thromb Haemost* 2005; 3: 2340–2345.

104 Undas A, Sanak M, Musial J, Szczeklik A. Platelet glycoprotein IIIa polymorphism, aspirin, and thrombin generation. *Lancet* 1999; 353: 982–983.

105 Undas A, Brummel K, Musial J, Mann KG, Szczeklik A. Pl(A2) polymorphism of beta(3) integrins is associated with enhanced thrombin generation and impaired antithrombotic action of aspirin at the site of microvascular injury. *Circulation* 2001; 104: 2666–2672.

106 Szczeklik A, Undas A, Sanak M, Frolow M, Wegrzyn W. Relationship between bleeding time, aspirin and the PlA1/A2 polymorphism of platelet glycoprotein IIIa. *Br J Haematol* 2000; 110: 965–967.

107 Cooke GE, Bray PF, Hamlington JD, Pham DM, Goldschmidt-Clermont PJ. PlA2 polymorphism and efficacy of aspirin. *Lancet* 1998; 351: 1253.

108 Andrioli G, Minuz P, Solero P *et al.* Defective platelet response to arachidonic acid and thromboxane A(2) in subjects with Pl(A2) polymorphism of beta(3) subunit (glycoprotein IIIa). *Br J Haematol* 2000; 110: 911–918.

109 Michelson AD, Furman MI, Goldschmidt-Clermont P *et al.* Platelet GP IIIa Pl(A) polymorphisms display different sensitivities to agonists. *Circulation* 2000; 101: 1013–1018.

110 Morawski W, Sanak M, Cisowski M *et al.* Prediction of the excessive perioperative bleeding in patients undergoing coronary artery bypass grafting: Role of aspirin and platelet glycoprotein IIIa polymorphism. *J Thorac Cardiovasc Surg* 2005; 130: 791–796.

111 Macchi L, Christiaens L, Brabant S *et al.* Resistance *in vitro* to low-dose aspirin is associated with platelet PlA1 (GP IIIa) polymorphism but not with C807T(GP Ia/IIa) and C-5T Kozak (GP Ibalpha) polymorphisms. *J Am Coll Cardiol* 2003; 42: 1115–1119.

112 Szczeklik A, Musial J, Undas A. Reasons for resistance to aspirin in cardiovascular disease. *Circulation* 2002; 106: e181–182; author reply e-2.

113 Wheeler GL, Braden GA, Bray PF, Marciniak SJ, Mascelli MA, Sane DC. Reduced inhibition by abciximab in platelets with the PlA2 polymorphism. *Am Heart J* 2002; 143: 76–82.

114 Rozalski M, Watala C. Antagonists of platelet fibrinogen receptor are less effective in carriers of Pl(A2) polymorphism of beta(3) integrin. *Eur J Pharmacol* 2002; 454: 1–8.

115 Aalto-Setala K, Karhunen PJ, Mikkelsson J, Niemela K. The effect of glycoprotein IIIa PIA1/A2 polymorphism on the PFA-100 response to GP IIbIIIa receptor inhibitors: the importance of anticoagulants used. *J Thromb Thrombolysis* 2005; 20: 57–63.

116 Gorchakova O, Koch W, Mehilli J *et al.* PlA polymorphism of the glycoprotein IIIa and efficacy of reperfusion therapy in patients with acute myocardial infarction. *Thromb Haemost* 2004; 91: 141–145.

117 O'Connor FF, Shields DC, Fitzgerald A, Cannon CP, Braunwald E, Fitzgerald DJ. Genetic variation in glycoprotein IIb/IIIa (GPIIb/IIIa) as a determinant of the responses to an oral GPIIb/IIIa antagonist in patients with unstable coronary syndromes. *Blood* 2001; 98: 3256–3260.

118 Nguyen TA, Diodati JG, Pharand C. Resistance to clopidogrel: a review of the evidence. *J Am Coll Cardiol* 2005; 45: 1157–1164.

119 Ziegler S, Schillinger M, Funk M *et al.* Association of a functional polymorphism in the clopidogrel target receptor gene, P2Y12, and the risk for ischemic

cerebrovascular events in patients with peripheral artery disease. *Stroke* 2005; 36: 1394–1399.

120 von Beckerath N, von Beckerath O, Koch W, Eichinger M, Schomig A, Kastrati A. P2Y12 gene H2 haplotype is not associated with increased adenosine diphosphate-induced platelet aggregation after initiation of clopidogrel therapy with a high loading dose. *Blood Coagul Fibrinolysis* 2005; 16: 199–204.

121 Angiolillo DJ, Fernandez-Ortiz A, Bernardo E *et al.* Lack of association between the P2Y(12) receptor gene polymorphism and platelet response to clopidogrel in patients with coronary artery disease. *Thromb Res* 2005; 116: 491–497.

122 Smith SM, Judge HM, Peters G *et al.* PAR-1 genotype influences platelet aggregation and procoagulant responses in patients with coronary artery disease prior to and during clopidogrel therapy. *Platelets* 2005; 16: 340–345.

123 Angiolillo DJ, Fernandez-Ortiz A, Bernardo E *et al.* PlA polymorphism and platelet reactivity following clopidogrel loading dose in patients undergoing coronary stent implantation. *Blood Coagul Fibrinolysis* 2004; 15: 89–93.

124 Angiolillo DJ, Fernandez-Ortiz A, Bernardo E *et al.* 807 C/T Polymorphism of the glycoprotein Ia gene and pharmacogenetic modulation of platelet response to dual antiplatelet treatment. *Blood Coagul Fibrinolysis* 2004; 15: 427–433.

125 Lee CR, Goldstein JA, Pieper JA. Cytochrome P450 2C9 polymorphisms: a comprehensive review of the in-vitro and human data. *Pharmacogenetics* 2002; 12: 251–263.

126 Takahashi H, Kashima T, Nomizo Y *et al.* Metabolism of warfarin enantiomers in Japanese patients with heart disease having different CYP2C9 and CYP2C19 genotypes. *Clin Pharmacol Ther* 1998; 63: 519–528.

127 Takahashi H, Echizen H. Pharmacogenetics of warfarin elimination and its clinical implications. *Clin Pharmacokinet* 2001; 40: 587–603.

128 Furuya H, Fernandez-Salguero P, Gregory W *et al.* Genetic polymorphism of CYP2C9 and its effect on warfarin maintenance dose requirement in patients undergoing anticoagulation therapy. *Pharmacogenetics* 1995; 5: 389–392.

129 Daly AK, Day CP, Aithal GP. CYP2C9 polymorphism and warfarin dose requirements. *Br J Clin Pharmacol* 2002; 53: 408–409.

130 Daly AK, Aithal GP. Genetic regulation of warfarin metabolism and response. *Semin Vasc Med* 2003; 3: 231–238.

131 Aithal GP, Day CP, Kesteven PJ, Daly AK. Association of polymorphisms in the cytochrome P450 CYP2C9 with warfarin dose requirement and risk of bleeding complications. *Lancet* 1999; 353: 717–719.

132 Voora D, Eby C, Linder MW *et al.* Prospective dosing of warfarin based on cytochrome P-450 2C9 genotype. *Thromb Haemost* 2005; 93: 700–705.

133 Hillman MA, Wilke RA, Yale SH *et al.* A prospective, randomized pilot trial of model-based warfarin dose initiation using CYP2C9 genotype and clinical data. *Clin Med Res* 2005; 3: 137–145.

134 Li T, Chang CY, Jin DY, Lin PJ, Khvorova A, Stafford DW. Identification of the gene for vitamin K epoxide reductase. *Nature* 2004; 427: 541–544.

135 Rost S, Fregin A, Ivaskevicius V *et al.* Mutations in VKORC1 cause warfarin resistance and multiple coagulation factor deficiency type 2. *Nature* 2004; 427: 537–541.

136 D'Andrea G, D'Ambrosio RL, Di Perna P *et al.* A polymorphism in the VKORC1 gene is associated with an interindividual variability in the dose-anticoagulant effect of warfarin. *Blood* 2005; 105: 645–649.

137 Harrington DJ, Underwood S, Morse C, Shearer MJ, Tuddenham EG, Mumford AD. Pharmacodynamic resistance to warfarin associated with a Val66Met substitution in vitamin K epoxide reductase complex subunit 1. *Thromb Haemost* 2005; 93: 23–26.

138 Pelz HJ, Rost S, Hunerberg M *et al.* The genetic basis of resistance to anticoagulants in rodents. *Genetics* 2005; 170: 1839–1847.

139 Wadelius M, Chen LY, Downes K *et al.* Common VKORC1 and GGCX polymorphisms associated with warfarin dose. *Pharmacogenomics J* 2005; 5: 262–270.

140 Yuan HY, Chen JJ, Lee MT *et al.* A novel functional VKORC1 promoter polymorphism is associated with inter-individual and inter-ethnic differences in warfarin sensitivity. *Hum Mol Genet* 2005; 14: 1745–1751.

141 Sconce EA, Khan TI, Wynne HA *et al.* The impact of CYP2C9 and VKORC1 genetic polymorphism and patient characteristics upon warfarin dose requirements: proposal for a new dosing regimen. *Blood* 2005; 106: 2329–2333.

142 Rieder MJ, Reiner AP, Gage BF *et al.* Effect of VKORC1 haplotypes on transcriptional regulation and warfarin dose. *N Engl J Med* 2005; 352: 2285–2293.

143 Veenstra DL, You JH, Rieder MJ *et al.* Association of vitamin K epoxide reductase complex 1 (VKORC1) variants with warfarin dose in a Hong Kong Chinese patient population. *Pharmacogenet Genomics* 2005; 15: 687–691.

144 Burgess JK, Lindeman R, Chesterman CN, Chong BH. Single amino acid mutation of Fc gamma receptor is associated with the development of heparin-induced thrombocytopenia. *Br J Haematol* 1995; 91: 761–766.

145 Brandt JT, Isenhart CE, Osborne JM, Ahmed A, Anderson CL. On the role of platelet Fc gamma RIIa phenotype in heparin-induced thrombocytopenia. *Thromb Haemost* 1995; 74: 1564–1572.

146 Denomme GA, Warkentin TE, Horsewood P, Sheppard JA, Warner MN, Kelton JG. Activation of platelets by sera containing IgG1 heparin-dependent antibodies: an explanation for the predominance of the Fc gammaRIIa "low responder" (*HIS131*) gene in patients with heparin-induced thrombocytopenia. *J Lab Clin Med* 1997; 130: 278–284.

147 Arepally G, McKenzie SE, Jiang XM, Poncz M, Cines DB. Fc gamma RIIA H/R 131 polymorphism, subclass-specific IgG anti-heparin/platelet factor 4 antibodies and clinical course in patients with heparin-induced thrombocytopenia and thrombosis. *Blood* 1997; 89: 370–305.

148 Bachelot-Loza C, Saffroy R, Lasne D, Chatellier G, Aiach M, Rendu F. Importance of the FcgammaRIIa-Arg/His-131 polymorphism in heparin-induced thrombocytopenia diagnosis. *Thromb Haemost* 1998; 79: 523–528.

149 Carlsson LE, Santoso S, Baurichter G *et al.* Heparin-induced thrombocytopenia: new insights into the impact of the FcgammaRIIa-R-H131 polymorphism. *Blood* 1998; 92: 1526–1531.

150 Carlsson LE, Lubenow N, Blumentritt C *et al.* Platelet receptor and clotting factor polymorphisms as genetic risk factors for thromboembolic complications in heparin-induced thrombocytopenia. *Pharmacogenetics* 2003; 13: 253–258.

151 Ray KK, Francis S, Crossman DC. A potential pharmacogenomic strategy for anticoagulant treatment in non-ST elevation acute coronary syndromes: the role of interleukin-1 receptor antagonist genotype. *J Thromb Haemost* 2005; 3: 287–291.

152 Bauman JL, Schoen MD. Arrhythmias. In: Dipiro JT, Talbert RL, Yee GC *et al.* eds. *Pharmacotherapy: A Pathophysiological Approach.* McGraw Hill, New York, 2002: 273–303.

153 Wang T, Roden DM, Wolfenden HT, Woosley RL, Wood AJ, Wilkinson GR. Influence of genetic polymorphism on the metabolism and disposition of encainide in man. *J Pharmacol Exp Ther* 1984; 228: 605–611.

154 Lee JT, Kroemer HK, Silberstein DJ *et al.* The role of genetically determined polymorphic drug metabolism in the beta-blockade produced by propafenone. *N Engl J Med* 1990; 322: 1764–1768.

155 Funck-Brentano C, Kroemer HK, Pavlou H, Woosley RL, Roden DM. Genetically determined interaction between propafenone and low dose quinidine: role of active metabolites in modulating net drug effect. *Br J Clin Pharmacol* 1989; 27: 435–444.

156 Funck-Brentano C, Turgeon J, Woosley RL, Roden DM. Effect of low dose quinidine on encainide pharmacokinetics and pharmacodynamics. Influence of genetic polymorphism. *J Pharmacol Exp Ther* 1989; 249: 134–142.

157 Turgeon J, Pavlou HN, Wong W, Funck-Brentano C, Roden DM. Genetically determined steady-state interaction between encainide and quinidine in patients with arrhythmias. *J Pharmacol Exp Ther* 1990; 255: 642–649.

158 Birgersdotter UM, Wong W, Turgeon J, Roden DM. Stereoselective genetically determined interaction between chronic flecainide and quinidine in patients with arrhythmias. *Br J Clin Pharmacol* 1992; 33: 275–280.

159 Morike KE, Roden DM. Quinidine-enhanced beta-blockade during treatment with propafenone in extensive metabolizer human subjects. *Clin Pharmacol Ther* 1994; 55: 28–34.

160 Chiang CE, Roden DM. The long QT syndromes: genetic basis and clinical implications. *J Am Coll Cardiol* 2000; 36: 1–12.

161 Roden DM, Viswanathan PC. Genetics of acquired long QT syndrome. *J Clin Invest* 2005; 115: 2025–2032.

162 Yang P, Kanki H, Drolet B *et al.* Allelic variants in long-QT disease genes in patients with drug-associated torsades de pointes. *Circulation* 2002; 105: 1943–1948.

163 Roden DM. Pharmacogenetics and drug-induced arrhythmias. *Cardiovasc Res* 2001; 50: 224–231.

164 Donger C, Denjoy I, Berthet M *et al.* KVLQT1 C-terminal missense mutation causes a forme fruste long-QT syndrome. *Circulation* 1997; 96: 2778–2781.

165 Napolitano C, Schwartz PJ, Brown AM *et al.* Evidence for a cardiac ion channel mutation underlying drug-induced QT prolongation and life-threatening arrhythmias. *J Cardiovasc Electrophysiol* 2000; 11: 691–696.

166 Sesti F, Abbott GW, Wei J *et al.* A common polymorphism associated with antibiotic-induced cardiac arrhythmia. *Proc Natl Acad Sci USA* 2000; 97: 10613–10618.

167 Roden DM. Genetic polymorphisms, drugs, and proarrhythmia. *J Interv Card Electrophysiol* 2003; 9: 131–135.

168 Roden DM. Proarrhythmia as a pharmacogenomic entity: a critical review and formulation of a unifying hypothesis. *Cardiovasc Res* 2005; 67: 419–425.

169 Roden DM. Human genomics and its impact on arrhythmias. *Trends Cardiovasc Med* 2004; 14: 112–116.

170 Marian AJ, Safavi F, Ferlic L, Dunn JK, Gotto AM, Ballantyne CM. Interactions between angiotensin-I converting enzyme insertion/deletion polymorphism and response of plasma lipids and coronary atherosclerosis to treatment with fluvastatin: the lipoprotein and coronary atherosclerosis study. *J Am Coll Cardiol* 2000; 35: 89–95.

171 Bray PF, Cannon CP, Goldschmidt-Clermont P *et al.* The platelet Pl(A2) and angiotensin-converting enzyme (ACE) D allele polymorphisms and the risk of recurrent events after acute myocardial infarction. *Am J Cardiol* 2001; 88: 347–352.

172 Potaczek DP, Undas A, Iwaniec T, Szczeklik A. The angiotensin-converting enzyme gene insertion/deletion polymorphism and effects of quinapril and atorvastatin on haemostatic parameters in patients with coronary artery disease. *Thromb Haemost* 2005; 94: 224–225.

173 Lutucuta S, Ballantyne CM, Elghannam H, Gotto AM Jr, Marian AJ. Novel polymorphisms in promoter region of ATP binding cassette transporter gene and plasma lipids, severity, progression, and regression of coronary atherosclerosis and response to therapy. *Circ Res* 2001; 88: 969–973.

174 Kajinami K, Brousseau ME, Ordovas JM, Schaefer EJ. Interactions between common genetic polymorphisms in ABCG5/G8 and CYP7A1 on LDL cholesterol-lowering response to atorvastatin. *Atherosclerosis* 2004; 175: 287–293.

175 Kajinami K, Brousseau ME, Nartsupha C, Ordovas JM, Schaefer EJ. ATP binding cassette transporter G5 and G8 genotypes and plasma lipoprotein levels before and after treatment with atorvastatin. *J Lipid Res* 2004; 45: 653–656.

176 Ojala JP, Helve E, Ehnholm C, Aalto-Setala K, Kontula KK, Tikkanen MJ. Effect of apolipoprotein E polymorphism and XbaI polymorphism of apolipoprotein B on response to lovastatin treatment in familial and non-familial hypercholesterolaemia. *J Intern Med* 1991; 230: 397–405.

177 Guzman EC, Hirata MH, Quintao EC, Hirata RD. Association of the apolipoprotein B gene polymorphisms with cholesterol levels and response to fluvastatin in Brazilian individuals with high risk for coronary heart disease. *Clin Chem Lab Med* 2000; 38: 731–736.

178 Ye P, Shang Y, Ding X. The influence of apolipoprotein B and E gene polymorphisms on the response to simvastatin therapy in patients with hyperlipidemia. *Chin Med Sci J* 2003; 18: 9–13.

179 Pena R, Lahoz C, Mostaza JM *et al.* Effect of apoE genotype on the hypolipidaemic response to pravastatin in an outpatient setting. *J Intern Med* 2002; 251: 518–525.

180 Carmena R, Roederer G, Mailloux H, Lussier-Cacan S, Davignon J. The response to lovastatin treatment in patients with heterozygous familial hypercholesterolemia is modulated by apolipoprotein E polymorphism. *Metabolism* 1993; 42: 895–901.

181 van Venrooij FV, Stolk RP, Banga JD *et al.* Common cholesteryl ester transfer protein gene polymorphisms and the effect of atorvastatin therapy in type 2 diabetes. *Diabetes Care* 2003; 26: 1216–1223.

182 Kuivenhoven JA, Jukema JW, Zwinderman AH *et al.* The role of a common variant of the cholesteryl ester transfer protein gene in the progression of coronary atherosclerosis. The Regression Growth Evaluation Statin Study Group. *N Engl J Med* 1998; 338: 86–93.

183 de Grooth GJ, Zerba KE, Huang SP *et al.* The cholesteryl ester transfer protein (CETP) TaqIB polymorphism in the cholesterol and recurrent events study: no interaction with the response to pravastatin therapy and no effects on cardiovascular outcome: a prospective analysis of the CETP TaqIB polymorphism on cardiovascular outcome and interaction with cholesterol-lowering therapy. *J Am Coll Cardiol* 2004; 43: 854–857.

184 Boekholdt SM, Sacks FM, Jukema JW *et al.* Cholesteryl ester transfer protein TaqIB variant, high-density lipoprotein cholesterol levels, cardiovascular risk, and efficacy of pravastatin treatment: individual patient meta-analysis of 13,677 subjects. *Circulation* 2005; 111: 278–287.

185 Mohrschladt MF, van der Sman-de Beer F, Hofman MK, van der Krabben M, Westendorp RG, Smelt AH. TaqIB polymorphism in CETP gene: the influence on incidence of cardiovascular disease in statin-treated patients with familial hypercholesterolemia. *Eur J Hum Genet* 2005; 13: 877–882.

186 Elghannam H, Tavackoli S, Ferlic L, Gotto AM Jr, Ballantyne CM, Marian AJ. A prospective study of genetic markers of susceptibility to infection and inflammation, and the severity, progression, and regression of coronary atherosclerosis and its response to therapy. *J Mol Med* 2000; 78: 562–568.

187 Basso F, Lowe GD, Rumley A, McMahon AD, Humphries SE. Interleukin-6 -174G→C polymorphism and risk of coronary heart disease in West of Scotland coronary prevention study (WOSCOPS). *Arterioscler Thromb Vasc Biol* 2002; 22: 599–604.

188 Fan YM, Laaksonen R, Janatuinen T *et al.* Effects of pravastatin therapy on serum lipids and coronary reactivity are not associated with SREBP cleavage-activating protein polymorphism in healthy young men. *Clin Genet* 2001; 60: 319–321.

189 Salek L, Lutucuta S, Ballantyne CM, Gotto AM Jr, Marian AJ. Effects of SREBF-1a and SCAP polymorphisms on plasma levels of lipids, severity, progression and regression of coronary atherosclerosis and response to therapy with fluvastatin. *J Mol Med* 2002; 80: 737–744.

190 Fiegenbaum M, Silveira FR, Van der Sand CR *et al.* Determinants of variable response to simvastatin treatment: the role of common variants of SCAP, SREBF-1a and SREBF-2 genes. *Pharmacogenomics J* 2005; 5: 359–364.

191 Johnson JA. Use of pharmacogenetics in clinical medicine: hype or hope? *Personalized Med* 2005; 2: 279–282.

CHAPTER 12

The potential of blood-based gene profiling for disease assessment

Steve Mohr, PhD, *& Choong-Chin Liew,* PhD

Introduction

A major thrust in current biomedical research involves the search for genes that can be used to evaluate the onset, progression and severity of human disease. Such genes are not necessarily the cause of the condition, but markers that will help in diagnosis and in risk assessment [1]. Discovering disease genes, however, can be challenging for several reasons. First, many disease phenotypes are difficult to ascertain, disease may be heterogeneous, and many disorders can be influenced by environmental and behavioral factors. Second, some diseases lack a known anatomic lesion or have lesions that are difficult to sample. Finally, collecting human biologic samples may impose an unacceptable burden on the patient.

Human genetics has been extremely successful in disease gene identification during the past 15 years. Much of this success can be attributed in earlier studies to the availability of genetic tools, such as positional cloning via linkage analysis, and in later research to the "candidate gene" approach in association studies [2–3]. Although mapping strategies have led to the isolation of a number of genes associated with cardiovascular diseases (CVD) (e.g., APO A-I, APO C-III and Leptin genes) such approaches have proven most successful in exploring the monogenic diseases; that is, those caused by a single-gene mutation with high-risk variant alleles segregating in rare families. In cardiovascular medicine such inheritance can be observed in familial forms of cardiomyopathies and hyperlipi-

demias [4]. However, in the more common cardiovascular disorders (e.g., atherosclerosis) and in CVD related risk factors (e.g., hypertension, diabetes and obesity), causation involves a dynamic interplay among multiple genetic and epigenetic factors that make it difficult for researchers to narrow the focus to a single gene [5]. While some progress has been made over the past decade in the genetics of the polygenic and multifactorial cardiovascular disorders, gaps clearly remain.

It is likely that more information about these complex disorders can be gathered from the continued analysis and characterization of changes at the genomic scale. The recent completion of the Human Genome Project [6–8] and innovative research coupling fundamental sciences with microtechnologies [9,10] has provided an unexpected opportunity for researchers to identify disease genes and biomarkers through genome-wide analyses: the use of technologies that integrate the entire genome [11]. Microarrays in particular have revolutionized transcriptomics (transcriptional or messenger RNA [mRNA] profiling) by allowing the simultaneous analysis of the expression of tens of thousands of genes. The output is a molecular "portrait" (or profile) describing which transcripts are "turned-on" and which "turned-off" in the system under study [12]. Microarray data can then be mined systematically to prioritize candidate genes to be tested individually or collectively for their discriminative and predictive power in human diseases; analyzing genome-wide data sets may also provide insight into disease mechanisms.

Microarray studies are often limited by difficulty in obtaining human tissues and by the lack of models that effectively capture clinically relevant disease features. Blood is easily accessible and has long been used for noninvasive biomedical investigations in laboratory medicine. Thus, it was recently hypothesized in the *Sentinel Principle*™ [13] that circulating blood could be regarded as "surrogate" tissue from which informative molecular signatures can be obtained safely. According to this hypothesis, environmental, physiological or disease perturbations anywhere in the body may leave a molecular signature detectable by applying genomic tools, especially microarrays, to profile blood-derived RNA. Thus, transcriptional profiling from circulating blood cells provides an alternative to tissue biopsy in the search for new disease genes and biomarkers (Plate 12.1).

In this chapter we present recent progress in utilizing and integrating blood genomics data to aid in the search for human disease genes and biomarkers. We begin with an overview of the basics of using blood cells. Then we outline the methodology used to develop blood cell expression profiles. Next we review the data available and how they are being used to identify disease genes (focusing mainly on discoveries in CVD). Finally, we discuss validation strategies and the potential of blood as a molecular or genomic diagnostic tool.

The blood option: Concepts and basics

The *Sentinel Principle*™

The human body is nourished by a dynamic circulatory system [14]. Blood is a circulating "connective" tissue composed of a fluid matrix, the plasma (55%), and formed elements (45%): erythrocytes or red blood cells, leukocytes or white blood cells, and platelets or thrombocytes. The main function of blood is to supply nutrients and constitutional elements to tissues and to remove waste products such as carbon dioxide and lactic acid. Blood also enables circulating cells, such as leukocytes, and circulating substances, such as amino acids, lipids, cytokines and hormones, to be transported among tissues and organs. As blood moves through the body, it interacts and communicates with every cell, tissue and organ, providing physiologic connectivity between systems. Thus, circulating blood has a number of critical roles in homeostasis, response to injury and hormonal communication.

The dynamic and interactive properties of blood give rise to the possibility that subtle changes occurring within the body, such as changes in association with a disease process or in response to an injury, may leave "marks" in the blood. The cascade of events induced by temporal and individual factors may alter the set of genes expressed in circulating blood cells, and consequently may be potentially detectable by RNA profiling techniques. We thus hypothesize that circulating blood can act as a biosensor that reflects the health or disease of the body, a concept referred to as the *Sentinel Principle*™ [13]. We propose to capitalize on this property of blood in the hunt for new disease genes and biomarkers, and ultimately for the diagnosis and prognosis of human diseases.

Practical and biologic rationales

The *Sentinel Principle*™ reasons that the blood cells that circulate throughout the body present an ideal "surrogate" tissue for gene expression studies in humans and for subsequent disease evaluation for the following reasons.

1 Blood provides scientists with a relatively safe and easily obtained source of samples to evaluate human health and disease. For many years, peripheral blood measurements, such as cell types and counts, and levels of cholesterol, glucose or hormones have been used in laboratory medicine to guide clinical decision-making and to quantify clinical outcomes [15]. In addition, drawing blood is a procedure that does not depend on specific preparations and causes little discomfort to patients. Therefore blood collection is much more acceptable to the public than procedures such as colonoscopy, which are perceived as unpleasant.

2 Blood sampling allows for the collection of larger sample sizes with better patient matching (avoiding some problems of statistical interpretation and confounding factors that weaken interpretation), and offers more opportunities to standardize the technical procedures.

3 Blood cells express a substantial part of the human genome, an important criterion for blood as a "surrogate" tissue. In a recent study, we found that peripheral blood cells share more than 80% of the transcriptome with each of the nine tissues we tested: brain, colon, heart, kidney, liver, lung, prostate, spleen and stomach [13]. Interestingly, this analysis showed that blood cells also express genes of specific systems such as genes encoding β-myosin heavy chain and insulin [13]. Other studies have documented the expression in blood of genes relevant to cardiomyocyte excitability and contractibility [16] and genes involved in neuroendocrine pathways [17].

4 Even more importantly, blood cells express genes that are responsive to various signals or stimuli. For example, we found that insulin mRNA levels were significantly different between fasting and nonfasting subjects [13], a responsiveness to cellular changes that is another important criterion for a "surrogate" tissue. Other examples of such responsiveness are outlined later in this chapter.

5 Diseases such as chronic fatigue syndrome or schizophrenia remain enigmas because of the lack of a known or accessible anatomic lesion. In these cases, blood would serve as a representative sample of the systemic state allowing for evaluation and profiling of multiple pathologic and physiologic pathways.

The continuous interaction between blood and the body's tissues, together with the capacity of peripheral blood cells to specifically respond to various stimuli or signals, and the accessibility of blood for obtaining samples suggests that the "blood option" represents a convenient, rigorous and high-throughput method of profiling human disease and discovering new disease genes and biomarkers (Plate 12.1) [18].

Blood and cardiovascular pathogenesis

Besides the technical and methodologic advantages of using blood over traditional biopsy samples, blood components have been implicated in the pathogenesis of a variety of disorders: immune disorders such as asthma, rheumatoid arthritis, systemic lupus erythematosus [19], hematologic malignancies [20] in which peripheral blood cells have a central role, and in many other diseases associated with inflammatory processes, such as cancer [21].

Circulating blood cells are in direct contact with the cardiovascular system and therefore may play a pathogenic role in many cardiovascular disorders.

For example, peripheral blood cells have been implicated in vascular disorders such as ischemia/reperfusion, atherosclerosis and vasculitis, and dysfunction of blood cells has been associated with many of the risk factors for CVD such as hypertension, diabetes, obesity and hyperhomocysteinemia [22–26]. An extensive discussion of these complex phenomena is beyond the scope of this chapter; rather, we provide epidemiologic and clinical evidence that attests to the association between blood components and cardiovascular diseases and risk factors.

Atherosclerosis is the major underlying cause of myocardial infarction, stroke and other CVD. Current evidence clearly indicates that atherosclerosis is not simply an inevitable consequence of aging, but rather a chronic inflammatory fibroproliferative disease of the vessel wall [24,27]. The paradigm emerging with this hypothesis incorporates biochemical cross-talk among blood cells, their descendents such as tissue macrophages, and the endothelial cells and vascular smooth muscle cells [28,29]. Inflammatory cells are recruited to the injured vessel wall initially as a reparative mechanism; however, the same inflammatory process are also pivotal in the development of lesions [30].

Studies in both humans and animals show that innate immunity – including phagocytic leukocytes, complement and proinflammatory cytokines – is essential for atherogenesis, whereas adaptive immunity – involving T and B cells, antibody and immunoregulatory cytokine – is an important modulator of disease activity and progression [31]. One of the earliest and most crucial steps in atherogenesis is the attachment of circulating monocytes and T lymphocytes to the injured endothelium, followed by their migration into the intima. This recruitment of blood-derived cells is a constant feature found in atherosclerotic lesions, with substantial numbers of cells in all stages [32].

Recent advances in our understanding of the vascular biology of atherosclerosis and its clinical manifestations make more attractive the hypothesis that white blood cells are major contributors to microvascular injury and atherogenesis. According to Hoffman *et al.* [26] leukocytes may act by plugging microvessels, impairing vascular flow (altered rheologic properties of leukocytes) and injuring endothelial cells. Leukocytes serve as the primary inflammatory cells, and we now know that other blood elements also have an important, although often under-recognized, role in CVD. For example, new data from mouse models have indicated that platelets not only contribute to acute thrombotic vascular occlusion, but also participate in the inflammatory and matrix degrading processes of coronary atherosclerosis [33]. The pivotal role of platelets in the pathogenesis of CVD is further emphasized by the fact that platelet hyperaggregability is associated with risk factors for coronary artery disease (CAD), such as smoking, hypertension and hypercholesterolemia [34]. Both thrombosis and inflammation are intrinsically linked processes in which blood components are key mediators [35].

Apart from local processes in the vessel wall, systemic signs of inflammation are also associated with the development of cardiovascular lesions. Plasma levels of several inflammatory proteins, including proinflammatory cytokines and acute-phase reactants, and indicators of cellular response to inflammation, such as white blood cell counts, have been positively correlated to the risk of future cardiovascular events both in patients with CVD and in healthy individuals [36,37]. It has been shown that a chronic leukocytosis reflects ischemic risk in a direct manner and that patients with elevated white blood cell counts are at higher risk of developing acute myocardial infarction and acute coronary and vascular events [26,38]. Circulating markers may consist of cytokines directly released from inflammatory cells present in the plaques and tissues exposed to recurrent ischemia (IL-1, TNF-α, IL-6, IL-8 and MCP1), as well as other reactants produced in response to those cytokines such as adhesion molecules (ICAM-1, VCAM-1, L, P and S selectin) and acute phase proteins (CRP, SAA and fibrinogen) [39]. Measurement of these molecules in serum can provide information about an individual's inflammatory status. Furthermore, adhesion molecules released in soluble form into the peripheral blood stream can serve as markers of vascular inflammation. In fact, it is possible to measure blood-based biomarkers at different levels of atherosclerotic inflammatory reactions [39].

More generally, evidence is increasing that a state of mild, chronic and systemic inflammation, with abnormal leukocyte counts and inflammatory

biomarker levels, is present not only in atherothrombotic CVD but also in various cardiovascular risk situations, including hypertension, diabetes and obesity. For example, C-reactive protein (CRP) level increases in the presence of a growing number of CVD risk factors constitutive of metabolic syndrome abnormalities [40]. Interestingly, the inflammatory state is not only present, but in fact precedes and may predict the development of cardiovascular risk factors such as type 2 diabetes [41]. Similarly, inflammatory mechanisms may underlie the pathogenesis of many cardiovascular risk conditions and associated lesion formation and organ dysfunction [42]. A variety of additional blood-based markers that reflect either lipoprotein metabolism (i.e., lipoprotein [a]), or endothelial dysfunction (i.e., homocysteine) have been linked to an excess risk of cardiovascular disease [43,44], suggesting that these circulating molecules could also contribute to the pathogenic processes underlying abnormal metabolic and vascular profiles of risk situations and their associated complications.

Consonant with the well-established inflammatory and immune nature of cardiovascular diseases and risk factors, blood components are now viewed as major contributors to the biochemical and clinical features of atherothrombotic CVD. Peripheral blood cell gene expression profile studies in CVD will be extremely valuable. Not only will such investigations directly assist in the discovery of cardiovascular disease genes and new targets for clinical purpose, but they will also generate useful data to enhance our understanding of the biology of the disease process and to shed light on etiologic pathways. Therefore, the potential utility of peripheral blood cell expression profiling as a new way to probe cardiovascular disease and risk factors is even more attractive.

Blood gene expression profiling: Methodology

Gene expression profiling

The potential involvement of a gene in a disease can be inferred from quantitative or functional analyses of human genome data. One of the simplest clues that a gene may be a "disease gene" is that its expression is altered in disease samples as compared with healthy controls. Changes in gene expression have traditionally been monitored by a "candidate gene" approach, in which molecules of interest have been analyzed one or a few at a time using techniques such as reverse-transcription polymerized chain reaction (RT-PCR) or Northern blot.

The early twenty-first century has seen the rise of genomic technologies that examine the entire genome of an organism. The goal of each of these approaches is to measure gene expression profiles of normal samples and of samples affected by a disease [45]. If many genes are monitored at once, their combined expression pattern can be viewed as a "molecular portrait" of the sample [12]. A comparison of these "portraits" in the two states (normal vs. abnormal) can identify a "gene expression signature" that identifies and characterizes a specific physiological state [18,46].

So far, gene expression profiling has relied to a large extent on technologies for analyzing the composition of complex mRNA samples (the transcriptome) and, less frequently, on technologies for analyzing protein samples (the proteome). Investigation of the human transcriptome has become particularly important in cardiovascular medicine. Cardiovascular genomics began in the late 1980s and early 1990s with the sequencing of clones obtained from human heart cDNA libraries. Named "expressed sequence tags" (EST) these clones, which represent the coding part of the genome, were an attempt to obtain directly "biologically relevant" sequences and were an alternative to sequencing DNA, which contains introns [47]. In 1997, Liew *et al.* [48] analyzed about 76,000 ESTs from different cDNA libraries of the cardiovascular system. This study formed the basis for a comprehensive, annotated inventory of the genes expressed in the human cardiovascular system [49], illustrating the potential of transcriptomic studies to detect genes and markers of human CVD. Subsequent research established that up to 27,000 distinct genes are expressed in this system [50] and that cardiovascular-related genes cluster at specific chromosomal locations [51].

Other methods currently in use for the analysis of RNA-based gene expression include differential display, cDNA and oligonucleotide microarrays, serial analysis of gene expression (SAGE), massively parallel signature sequencing (MPSS) and total gene expression analysis (TOGA). An excellent overview of these methods can be found in Scheel *et al.* [52].

Microarray technologies

Since the mid-1990s, studies in genomics have mainly involved microarray technologies. While simple in theory and design – a microarray is essentially a high-throughput Southern blot on a very small scale – this technology has revolutionized the study of global gene expression. With the ability to represent up to 60,000 distinct transcripts on a single chip, microarray is a robust platform for global expression profiling from genes all across the genome.

Early microarrays used nylon membranes as a solid support, mimicking traditional blotting strategies [53,54]. Today, the most common microarray format uses defined sequences (cDNA or oligonucleotide) spotted or directly synthesized in a grid-like fashion on a solid support such as a glass or silicone slide, and hybridized with a solution phase containing labeled nucleic acids (cDNA or aRNA) that represent the mRNA expressed in the biologic sample tested. By monitoring the amount of label (signal intensity) associated with each location, it is possible to infer the abundance of each mRNA species (Plate 12.2). (For further information about microarray techniques, the reader is referred to the *The Chipping Forecast* series of special issues published by *Nature Genetics* and highlighting microarray theory, concepts, manufacture, applications and perspectives [55–57].)

Using microarrays, researchers may choose to study the entire genome or to scan only a specific subset of the transcriptome. Those transcripts that are particularly relevant to a disease (e.g., a specific pathway or system) can be explored using customized support containing sequences that match only these transcripts. For example, the "LymphoChip" [58] was designed to assess gene expression in normal and malignant lymphocytes. More recently, Barrens *et al.* [59] constructed a cardiac-specific cDNA microarray called the "CardioChip" and containing more than 10,000 distinct transcripts derived from cardiac cDNA libraries. Using this customized system, gene expression profiles were compiled from human fetal and adult heart to draw specific "portraits" of gene expression alteration in dilated and hypertrophic cardiomyopathy [60,61]. Since then, specific or more general microarrays have been used in several other studies to characterize the human heart transcriptome and investigate the cardiovascular system [18,46,47].

Microarray technologies are a powerful method whereby complex gene expression patterns can be distilled to identify specific genes and pathways involved in a given disease. A hierarchy of importance can be determined to prioritize disease genes for subsequent biologic or clinical validation. In general, microarray data can be used for three different purposes. First, gene expression profiles can help to identify new "disease genes" and to characterize the basic molecular pathways regulated by etiological disease processes. Second, gene expression profiling can provide clues about the mechanisms underlying the effects of an intervention. Third, gene expression profiling can help to identify key genes that are altered in pre-disease states or "at risk" phenotypes. Such early stage alterations might therefore act as molecular biomarkers for early disease detection. Microarray data quickly yielded quite impressive results, mostly in cancer research, with applications that range from molecular nosology to the identification of differentially expressed genes, prospective markers and intervention targets [62,63].

From blood collection to disease gene

The protocol outlined here, typical of the protocol used in our laboratory, illustrates step by step the method for determining gene expression profile in blood using microarrays. While other methods exist and may prove beneficial to the individual user, this protocol has offered consistent results. We use the Affymetrix GeneChip systems, currently the popular platform of choice for the commercial oligonucleotide-based arrays consumer. With the most recent human array, the HG-U133Plus 2.0 (Affymetrix; Santa Clara, CA), the expression level of more than 47,000 human transcripts and variants, including 38,500 well-characterized human genes, can be assessed in a single experiment.

Step 1: Collect blood

Approximately 10 mL peripheral whole blood is collected by standardized venipuncture in EDTA Vacutainer™ tubes (Becton Dickinson, Franklin Lakes, NJ), and immediately stored at 4°C until processing (within 6 hours) for RNA isolation.

Step 2: Isolate total RNA

Upon centrifugation, the plasma is removed and a hypotonic buffer (1.6 mmol/L EDTA, 10 mmol $KHCO_3$, 153 mmol/L NH_4Cl, pH 7.4) is added at a 3 : 1 volume ratio to lyse the red blood cells. The sample is spun at 1,400 rpm for 10 min at 4°C. The resulting cell pellet is washed with the hemolysis buffer several times and then resuspended into 1.0 mL TRIzol® Reagent (Invitrogen Corp., Carlsbad, CA) and 0.2 mL chloroform to isolate total RNA according to the manufacture's instructions. The quality of the total RNA (i.e., purity and integrity) is assessed by microcapillary electrophoresis on an Agilent 2100 Bioanalyzer using the RNA 6000 Nano Chip (Agilent Technologies, Palo Alto, CA) according to the manufacturer's instructions. Total RNA quantity is determined by absorbance at 260 nm in a spectrophotometer.

Step 3: Make labeled RNA and apply to the chip

Five micrograms of each purified total RNA is labeled and hybridized onto an Affymetrix HG-U133Plus 2.0 GeneChip array (Affymetrix; Santa Clara, CA) following the manufacturer's instructions (see overview in Plate 12.2). Briefly, double-stranded cDNA is synthesized from 5 µg blood total RNA using SuperScript RTII (Invitrogen) and the T7-Oligo(dT) primer (GeneChip T7-Oligo(dT) Promoter Primer kit, Affymetrix) as described in the manual. *In vitro* transcription is performed with the BioArray™ HighYield™ RNA Transcript Labeling Kit (Enzo Life Sciences, Inc. for distribution by Affymetrix), followed by cRNA fragmentation, and hybridization overnight.

Step 4: Scan the chip and measure expression

The next day, the array is washed, stained with streptavidin-phycoerythrin and biotinylated anti-streptavidin antibody, and scanned using the Affymetrix GeneChip Scanner 3000. Hybridization signals are scaled in the Affymetrix GCOS software, using a scaling factor determined by adjusting the global trimmed mean signal intensity value to 500 for each array, and imported into GeneSpring software (Silicon Genetics, Redwood City, CA). For normalization, signal intensities are then centered to the 50th percentile of each chip and, for each individual gene, to the median intensity of each specific subset first to minimize the possible technical bias, then to the whole sample set. Only genes called "present" or "marginal" by the GCOS software in all samples are used for further analysis.

Step 5: Analyze the expression data

Based upon specific clinical information gathered from the patients, correlations between a patient's clinical parameters (disease activity, stage, etc.) and changes in gene expression are performed using dedicated algorithms and statistical tools. From the normalized microarray data the relative expression level of each transcript is calculated as the ratio of the mean signal in the two groups of samples (normal vs. abnormal), and the genes differentially expressed are identified using the Wilcoxon–Mann–Whitney nonparametric test (P <0.05). Hierarchical cluster analysis is performed to assess correlations among samples for each significant gene set [64]. The output graphic representation shows a colored matrix (typically, red and green) in which each column represents an individual sample and its measured gene expression levels, and each row represents a gene and its expression levels in all samples. The color and the intensity represent the direction (up or down) and the magnitude of fold-change relative to the control group (see Plate 12.2, for an example). On the left side is the generated clustering tree (or gene dendrogram) for the significant gene set used. A similar clustering tree is generated for the samples (or sample dendrogram). Genes or samples in a tree are joined by very short branches if they are very similar to each other, and by increasingly longer branches as their similarity decreases. The colored matrix itself is arranged according to the result of the hierarchical clustering. Starting with an original set of more than 35,000 genes, this number is typically reduced to a few hundred using signal filtering and statistical analysis. These gene, are ranked based on fold-change (higher is better) and P value (lower is better), and the top 50 genes are selected as initial candidates to be evaluated by quantitative RT-PCR. Other criteria such as known biologic function can also be used at this step to prioritize genes for further analyses.

Step 6: Verify candidate gene expression using quantitative real-time RT-PCR

The expression patterns of the selected candidate genes are verified using quantitative real-time RT-PCR (qRT-PCR). Forward and reverse primers are designed for the selected genes, and an internal control or housekeeping gene (e.g., beta-actin). Total blood RNA (2 µg) is reversed-transcribed into single-stranded complementary DNA (cDNA) using High-Capacity cDNA Archive Kit (Applied Biosystems, Foster City, CA) according to the manufacturer's protocol. Then 5 ng cDNA is amplified and quantified by the Quantitect SYBR Green® PCR kit (Qiagen, Valencia, CA) using an ABI 7500 Real-time PCR System (Applied Biosystems, Foster City, CA). The PCR reaction is performed in a 96-well format with the following cycling parameters: 2 min at 50°C; 15 min at 95°C; 40 cycles of 15 s at 94°C, 35 s at 55°C and 30 s at 72°C; and measuring melting curve at 60–95°C, at 0.2°C intervals.

An automatically calculated melting point dissociation curve is examined to ensure the specific PCR amplification and the lack of primer-dimer formation in each well. The amount of amplified product is given as a calculated threshold cycle (Ct) for each gene tested (see Plate 12.3). The Ct reflects the cycle number at which the fluorescence generated within a reaction crosses the threshold. The Ct value assigned to a particular well thus indicates the point during the reaction at which a sufficient number of amplicons have accumulated, in that well, to be at a statistically significant point above the baseline. Differences in gene expression are then estimated using the "comparative Ct method" of relative quantitation [65], normalizing the Ct values relative to the housekeeping gene (Plate 12.3), beta-actin. To ensure reproducibility of the results, all genes are tested in triplicate and the averaged Ct value is used for relative expression quantification.

For the "comparative Ct method" to be valid, the efficiency of the target (candidate gene) amplification and the efficiency of the reference (housekeeping gene) amplification must be relatively similar. Before the quantitation, a validation experiment is performed to determine the amplification efficiency and specificity of the primer pairs using serial dilution of a reference cDNA generated from a normal blood RNA pool with confirmation by agarose gel electrophoresis to ensure that the values were within linear range and the amplification efficiency was approximately equal for each of the target gene tested (for details of the strategies used to analyze the results of qRT-PCR, please refer to the section of this chapter entitled "Discovery validation").

Critical factors and issues

Microarrays are powerful tools for gene expression profiling in blood, but researchers must be aware that there are specific limitations and critical issues when using leukocytes as tissue source and microarray as the analytical technology. Some major concerns are discussed below.

Microarray technology

First of all, microarrays have certain limitations, crucially:

1 The presence of a large number of false positive and false negative discovery rates, which can be minimized – but not completely eliminated – through careful experimental design to reduce background noise generated by technologic and biologic factors and by the use of appropriate statistical methods to estimate and control multiple testing error rates [66,67].

2 The low reliability of conclusions drawn from only a few microarray analyses, which leads to the need to replicate measurements [68,69] and to validate the primary screening by alternative methods such as real-time quantitative PCR [70].

3 The need for clear standards and consensus, in particular to ensure reliable and consistent use of results across different laboratories and platforms [71–73].

Blood variability and confounding factors

Blood is one of the most variable tissues in the body and displays both intra- and intersubject variations [74]. Variations such as the number and relative proportion of blood cell types at different times because of hormonal or diurnal changes, and variations between donors, may give rise to differences in the blood transcriptome that reflect differences in cellular composition rather than differences underlying disease processes [75]. These issues can be minimized by optimizing the study design (e.g., by increasing the sample size, harmonizing collec-

tion time points and requesting that patients fast before sample collection), and by developing approaches to deal with the blood cell count in parallel with gene expression analysis [76]. Another factor that can make blood gene expression profiles difficult to interpret is the presence of confounding variables such as age, sex or environmental exposures [77,78]. For example, Whitney *et al.* [79] reported significant gender bias in a number of genes. In addition, Wang *et al.* [80] showed that smoking status influences gene expression signatures from whole blood and is an important confounder in the toxicogenomic studies of particulate exposure. As in most human studies, linking outcomes with genomic markers can be biased if the populations are not well characterized or the study is not well designed. This wide range of variability also reinforces the need for a sufficient number of both normal donor as well as disease samples to generate a representative gene expression profile [81].

Blood sample handling and processing

Peripheral blood cell preparation requires a time delay before RNA stabilization. Such delay exposes RNA to numerous pre-analytical factors that potentially induce *ex vivo* changes in gene expression profiles – changes not relevant to the disease status or the clinical response under study [80, 82–84]. Critical aspects of blood transcriptomics concern blood collection devices, cell and RNA isolation procedures, time and temperature [76,82, 85–88]. Currently, several blood sampling methods are available for gene expression profiling, including PAXgene™ tube (PreanalytiX GmbH), TEMPUS™ tube (ABI) for whole blood, CPT™ tube (BD) and Ficoll-Hypaque density gradient for peripheral blood mononuclear cells. The benefits and drawbacks of each of these methods have been reviewed elsewhere [75,82,89,90]. Although some methods might be more convenient for day-to-day clinical use, others are more adapted to the study of specific types of blood cells. Overall, no one of these methods outperformed the others and therefore users should choose the most suitable procedure for their research goals. Requesting that subjects fast 3 hours before venipuncture, storing blood at 4°C and isolating peripheral blood cells within 2 hours following venipuncture may help to minimize possible effects on gene expression caused by sample handling and processing [91]. RNA preparation and microarray hybridization are substantial sources of additional variability in gene expression [92]. The great diversity of instruments, reagents and protocols available today should be taken into account for the reproducibility, validity and generalizability of the results.

RNA quality

The key to successful microarray experiments is to start with high-quality RNA. As variations in RNA quality can alter the assumed relative expression levels of many genes, the RNA must be rigorously tested. Various methods have traditionally been used to analyze RNA quality, including spectrophotometers to ensure an optical density (OD) ratio at 260–280 nm of at least 1.8, and formaldehyde gel electrophoresis, in which clear 28S, 18S and 5S bands should be visible with no genomic DNA contamination. The Agilent 2100 Bioanalyzer provides more sensitive qualitative analysis from less RNA than do traditional methods. This system uses a fluorescent assay involving electrophoretic separation to evaluate RNA samples qualitatively and creates a graph called an electropherogram, which diagrams fluorescence over time. High-quality RNA electropherograms show clear 28S and 18S peaks, low noise between the peaks and minimal low molecular weight contamination. Additional quality control criteria can be used to confirm the integrity of the mRNA hybridized on the chip. For example, the Affymetrix GeneChip system includes different probe sets recognizing the 5′ and 3′ regions of such housekeeping genes as glyceraldehyde dehydrogenase (GAPD), and if the mRNA is degraded the ratio of the 5′ : 3′ signal drops dramatically.

To obtain the best results using blood as source for gene expression analysis, experiments should be carefully designed, the quality of the biologic materials must be strictly controlled and factors that may cause variability in gene expression not related to what is being analyzed should be assessed. In fact, all steps from blood collection to RNA analysis require standardization.

Biomedical applications

A demonstration of the utility of blood for gene expression profiling and biomarker discovery would

have important medical and health implications. Tests of the *Sentinel Principle*™ in a number of recent studies provide clear evidence that unique gene expression patterns in peripheral blood reflect both static (inherited) and dynamic (acquired) changes that occur within the cells or tissues of the body. Below we summarize the key findings, with emphasis on CVD-related studies.

Blood reflects individual factors

Expression profiling of peripheral blood cells in healthy subjects was carried out in two studies, both of which reported that stable patterns of gene expression differentiate among individuals. Whitney *et al.* [79] found variations associated with the composition of the peripheral blood cell population and constitutive factors such as gender and age. This group found a set of 340 genes that could describe individual samples in 13 of 16 people, sampled more than once over a 1-month period. In a longitudinal study, Radich *et al.* [93] reported stable variations of gene expression in peripheral blood leukocytes, reflecting biologic differences between individuals such as responsiveness in immunologic and/or inflammatory pathways. Such individual-specific genes will be helpful in identifying or predicting a person's status (drug response, disease risk or environmental exposure).

Blood reflects environmental and behavioral factors

Several studies have shown that behavioral and environmental factors, such as drugs, diet, exercise, alcohol, tobacco and industrial pollutants, alter gene expression in peripheral leukocytes. For example, Lampe *et al.* [94] demonstrated in an observational study that on the basis of mRNA expression profiling in peripheral blood leukocytes it was possible to distinguish between two groups of 85 people exposed and unexposed to tobacco smoke. Furthermore, they observed a gene expression signature indicative of cigarette smoking, a promising finding with regard to the need to monitor CVD risk status and detect early biologic effects of cigarette smoke exposure. Van Leeuwen *et al.* [95] used cigarette smoke and some of its constituents to exemplify the general applicability of peripheral blood mononuclear cell expression profiling as a toxicologic model to detect biomarkers for carcinogen exposure. Overall, the studies of external factors suggest that expression profiling in peripheral blood cells can be extended to estimate environmental exposures or to evaluate the host response to different forms of environmental exposures or behaviors (e.g., tobacco smoke, alcohol intake, dietary, exercise and medication). Thus, it is reasonable to think that by exploring peripheral blood cell transcriptomes in human observational studies, it may be possible to identify and classify individuals according to their "environmental" status. Such differences may in turn help to determine the likelihood of late effects of an exposure or a behavior, including disease susceptibility and outcomes.

Blood reflects disease factors

The most important evidence in support of using peripheral blood cell profiles as surrogate markers in human disease and disease risk derives from studies of diseases. The most impressive examples include microarray applications to hematologic malignancies, such as large B-cell lymphoma, chronic lymphocytic leukemia, acute leukocyte leukemia and myeloma [20], in which peripheral blood cells are the cells actually affected by the disease. These studies showed that peripheral blood cells can display disease-specific gene expression signatures that are accurate enough to identify relevant patient subgroups [96]. Gene expression assays have also been used in peripheral blood to diagnose nonhematologic tumors based on specific transcripts derived from either circulating tumor cells [97–99] or circulating cancer-related RNA molecules [100,101].

Our laboratory and other groups are now profiling gene expression of peripheral blood cells in a wide range of nonhematologic disorders. The general objective of these studies is to determine whether peripheral blood cell gene expression can distinguish between patients and healthy controls. As a prerequisite for the use of peripheral blood cell-based expression profile for disease detection, Whitney *et al.* [79] found that temporal and individual variations in healthy subjects were different from variations found in patients with cancer or with bacterial infection. The latter finding was subsequently confirmed by Cobb *et al.* [102], who reported that traumatic injuries induced changes in blood gene expression of a magnitude sufficiently great to be distinuishable from analytical noise or interindividual variations.

We have now demonstrated that monitoring gene expression in blood results in distinct transcriptional signatures for more than 35 different conditions in humans [13], including various types of cancer (i.e., bladder [103], colorectal [104] and prostate [105]), schizophrenia [106,107], osteoarthritis [108] and cardiovascular diseases (as below). Other laboratories have also shown, independently, that blood gene expression can reveal unique gene expression profiles in a range of diseases, disorders and injuries: for example, renal carcinoma [109] and breast cancer [110]; chronic fatigue syndrome [111]; acute ischemic stroke [112] and other neurologic injuries [113–116]; asthma [19,117]; severe lupus erythematosus [118,119]; kidney disease [74,120]; and Crohn's disease [121]. Rather than discussing each of these studies in detail, we present some studies investigating the usefulness of peripheral blood cell gene expression in cardiovascular disease and risk factors.

Our laboratory has taken advantage of our EST resources and our custom-made cDNA microarrays, such as "CardioChip" [59], to design an in-house "blood chip" from an EST database of peripheral blood cells [122]. This specific 10 K cDNA microarray was used to screen peripheral blood samples from CAD patients, and we showed that profound changes occur in CAD blood samples compared with healthy controls. We found that 108 genes were differentially expressed in peripheral blood cells of patients with CAD, including 43 downregulated and 65 upregulated genes. Some of these changes could be interpreted in terms of earlier observations and in terms of the potential contribution of peripheral blood cells to the pathogenesis of CVD. For example, we identified three CAD-upregulated genes (*PBP*, *F13A* and *PF4*), whose encoded protein levels had previously been found to be elevated in CAD plasma [122]. This preliminary study (four CAD patients and three normal controls) has since been confirmed in 22 CAD patients vs. seven patients diagnosed with Chagas' heart disease and 33 normal controls (Liew CC., unpublished data).

More recently, using SAGE technology on blood-derived monocytes, Patino *et al.* [123] compared patients with atherosclerosis with normal controls and identified a disease expression signature compatible with stress response and inflammation. This study paid particular attention to the *FOS* gene, the expression of which was strongly increased in monocytes of atherosclerotic patients, and was significantly associated with severe forms of atherosclerosis.

CVD risk factors were investigated by Chon *et al.* [124], who studied changes in mRNA levels in pooled samples of leukocytes from untreated or treated hypertensive patients. Their findings indicate that microarray gene expression profiling of peripheral leukocytes can distinguish patients with essential hypertension from age- and sex-matched controls. Interestingly, the hypertension-specific disturbances were missing in the expression profile from hypertensive patients who had become normotensive with treatment. Microarray analysis of peripheral blood cells is thus also a potentially promising tool to assess drug efficacy and for individualizing therapy with maximal effects and minimal side effects (pharmacogenomics) [89].

In a recent study, our laboratory investigated another well-known CVD risk factor: plasma lipids. When we examined the correlative relationships between plasma levels of high-density lipoprotein (HDL), low-density lipoprotein (LDL) and Triglycerides (TG) and circulating leukocyte gene expression, we found that peripheral blood cells respond to changing plasma lipid levels by regulating a network of genes, including genes involved in immune responses and inflammation, and in lipid and fatty acid metabolism [125]. We found for example that total cholesterol, TG and LDL levels have a positive correlation with genes involved in inflammatory responses (*NFE2L1* and *MGLL*), while HDL level showed a negative correlation (*NFATC3* and *MGLL*), suggesting systemic inflammation as a potential mechanism of CVD. Similarly, genes involved in lipid metabolism were positively correlated with plasma levels of TG and LDL, while the opposite pattern was seen with plasma levels of HDL. This study provides further data on the interrelationships between plasma lipid levels – atherogenic risk factors – and leukocytes – inflammatory executors – in the atherogenic process. As this study shows, peripheral blood gene expression profiling is not only useful for identifying biomarkers, but also promises to shed light on etiologic pathways.

Finally, the applicability of blood gene expression signatures in human disease is well illustrated in a recent study on cardiac transplant rejection. Deng *et al.* [126] showed that an 11-blood gene classifier accurately discriminates allograft recipients with moderate to severe rejection from a quiescent state. This predictive panel was subsequently optimized into a kit format [127] bringing peripheral blood expression profiling even closer to clinical application.

Overall, the findings presented here demonstrate that distinct subsets of genes are specifically induced or inhibited in peripheral blood cells in relation to specific physiologic, pathologic, behavioral and environmental factors. The studies also confirm the hypothesis that peripheral blood cells carry a gene expression signature that identifies the presence of a disease, and therefore is a valuable way to identify new disease genes and markers.

Validation strategies

Despite the many issues that remain to be resolved, microarray expression profiling approaches have identified a wealth of candidate genes and signaling pathways correlated with biologic dysregulation and disease. However, in most cases microarray studies represent only a first line of screening and require suitable validation strategies to select potential disease candidate genes and biomarkers for further in-depth analysis and for intervention efforts.

Overall the validation process requires:
1 "discovery" validation: that the candidate gene expression level is effectively associated with the disease; and
2 "functional" validation: perhaps most important, that the alteration of this gene affects the relevant disease phenotype.
For genes selected as disease markers, validation efforts focus on showing that the changes observed are sufficiently specific enough and sensitive enough to provide a test of predictive value better than those currently implemented in clinical medicine.

The next section provides clues in some validation strategies used to delineate the reliability and the biologic significance of disease genes and markers found in blood. In the case of blood, we focus on strategies used to validate disease genes in the sense of biomarkers.

"Discovery" validation

Whole-genome mRNA profiling approaches used in the discovery phase typically yield lists of tens to hundreds of candidate genes requiring confirmation [128]. Confirmation is of critical importance given the high rate of false discovery using such techniques, especially microarrays. Therefore, the first step in follow-up analysis involves validating differential expression and distribution of the potential disease gene by alternative, often single-gene techniques. Because microarrays are currently not a mature enough technology for widespread use as clinical tools, the validation step may also serve to translate expression biomarkers to a platform for future clinical application.

Validation methods used vary depending on the scientific question; one commonly used technique for gene expression change validation is qRT-PCR [129].

Quantitative RT-PCR basics

RT-PCR involves PCR amplification of segments of mRNA that have been turned into cDNA by reverse transcription. RT-PCR is the most sensitive and the most flexible of the quantification methods. Besides conventional RT-PCR, the application of fluorescence techniques has led to the development of automated real time RT-PCR methodologies, which combine the process of amplification and detection to permit the monitoring of the reaction in real-time during the PCR [129]. Advantages include ease and reliability of the quantification, a broad dynamic range of detection (up to 5 logs), the small amount of RNA required (5–500 ng total RNA) and the number of samples (96 or 384 well plate format) that can be processed in parallel. Real time RT-PCR measures product amount after each cycle of amplification, based on the association of the fluorescence signal with the amount of amplicon accumulated during the PCR.

Real time RT-PCR formats

There are available three real-time chemistries that detect amplified product with about the same sensitivity [130]. The simplest method uses a fluorescent dye (SYBR Green®) that binds non-

specifically to double-stranded DNA. During the elongation phase of the PCR increased amounts of dye bind to the nascent DNA, and fluorescence increases proportionally (Plate 12.3). This assay can suffer from lack of specificity, as the presence of any double-stranded DNA (specific amplicon, as well as primer dimers) generates fluorescence and therefore will be detected and entered in the quantification.

The two other methods rely on the hybridization of target-specific fluorescent probes, thereby obviating the need for post-PCR Southern analysis or sequencing to confirm the identity of the amplicon. In the Taqman® system, a fluorescence resonance energy transfer (FRET) oligonucleotide probe complementary to the target sequence is used as the reporter system. The fluorescence of the reporter molecule at the 5′ end of the oligonucleotide is interfered with by a quencher molecule attached at 3′ end. This probe hybridizes to the specific target amplicon after the denaturation step, and when strand synthesis occurs the 5′-nuclease activity of Taq DNA polymerase degrades the FRET probe and releases the reporter from the quencher, producing fluorescence. Alternatively, the reporter and the quencher dyes can be carried by two different probes to maximize the specificity [130].

In the Molecular Beacon method, the 3′ quencher and 5′ reporter of FRET probes initially exhibit no fluorescence because the oligonucleotide forms a hairpin loop that brings these two factors into close proximity. At the annealing step of PCR, the probe forms a hybrid with the target sequence that separates the two fluorochromes, allowing the reporter to fluoresce. The main drawback with the molecular beacon method is associated with the design of the hybridization probes. Straightforward quantification is accomplished by any of the methods described above.

The validation process: step by step

"Discovery" validation provides independent verification of the primary data and typically begins with the same samples studied in the initial investigation. The investigators extend their findings to additional samples in order to demonstrate that the gene is a "universal" feature of the disease under study.

Differential expression and statistical significance do not necessarily mean that the disease genes identified in blood will translate to the clinic. In defining the value of any novel disease gene, or multiple disease genes, the evidence of its utility must be critically assessed against criteria that define potential clinical applicability (i.e., discriminatory power and predictive performance). Such criteria will be evaluated during the validation process.

Because the number of potential genes or combination of genes that have the power to discriminate between "disease" and "control" (a gene expression classifier) is extremely large, a major concern in the validation analyses is overfitting; that is, if enough classification rules are investigated then, by chance, one of them is likely to perform well. To minimize overfitting the data, the split-sample approach is a common statistical practice that uses a training sample set to formulate the classification rules and a test sample set for evaluating the accuracy and the reproducibility of the classifier. This "internal validation" can also be accomplished using a cross-validation (e.g., leave-one-out method; for review see [131]).

Step 1. Validation of differentially expressed genes via real-time qRT-PCR: The Comparative Ct method

For each sample, the expression level of a target gene is quantified by its threshold cycle (Ct), which is the PCR cycle number at which the increase in fluorescent signal associated with the exponential growth of PCR products becomes distinguishable over the background. To analyze the results of qRT-PCR, the "comparative Ct method" is used (Plate 12.3); thus, no standard curves are required to run in each assay. This involves comparing the Ct values of the disease samples with controls (normal samples). The Ct values of both the disease and the normal samples are first normalized to an appropriate endogenous housekeeping gene (e.g., beta-actin gene). This is done by calculating a ΔCt_{sample} = Ct (target gene) − Ct (housekeeping gene). Then the relative fold changes (disease vs. controls) is represented as $2^{-\Delta\Delta Ct}$ where $\Delta\Delta Ct$ = mean ΔCt (disease samples) − mean ΔCt (control samples) (see Livak and Schmittgen [65] for a review; and the ABI-7700 User Bulletin 2, Applied Biosystems [132] for further details of quantitation methods). A nonparametric Mann–Whitney test is then used to evaluate the statistical significance of the differences in mRNA levels between controls and patients with disease.

Disease genes are validated on the basis of correlation with microarray results – in terms of differential expression (under- or overexpressed) and statistical significance (P <0.05) – and selected to evaluate their diagnostic discriminative power.

Step 2. Discriminatory power of blood biomarker panel: Logistic regression and ROC curve

To test the discriminatory power of the validated genes, a logistic regression analysis is performed for the ΔCt values of the validated disease genes (Plate 12.4). The goal of logistic regression is to find the best fitting model to describe the relationship between the outcome variable, "disease" or "control," and a set of independent variables, the ΔCt values (the predictor variables). Logistic regression generates the coefficients "bk" (and its standard errors and significance levels) of a formula to predict a *logit transformation* of the probability of presence of the characteristic of interest: Logit [p ("disease")/p ("control")] = $b_0 + b_1 \cdot \Delta Ct_1 + b_2 \cdot \Delta Ct2 + \ldots + b_k \cdot \Delta Ct_k$ [133]. In other words, logistic regression analysis generates from the complete panel of validated disease genes, different gene combinations or classifiers with the power to discriminate between "disease" and "control" samples.

To visualize the efficacy of each of these gene combinations, the logistic regression data are summarized in a receiver operating characteristic (ROC) curve [134]. This curve plots the sensitivity (true positives, or cases with disease classified as disease/all disease cases) on the y axis against "1 − the specificity" (false positives, or "1 − control cases classified as controls/all control cases") on the x axis, considering each value as a possible cutoff value. The area under curves (AUC) is calculated as a single measure for the discriminate efficacy of the selected gene combination. When a gene set has no discriminative value, the ROC curve will lie close to the diagonal and the AUC will be close to 0.5. By contrast, when a gene set has strong discriminative value, the ROC curve will move up to the upper left-hand corner and the AUC will be close to 1.0 (Plate 12.4).

An optimal ROC curve (highest AUC) is identified based on the sample from the training set, and a "Logit cutoff value" is selected in order to get a consensus on sensitivity and specificity closest to the targeted discriminatory power. For example, if biomarkers are selected for early detection of a disease, when a low false positive rate is required, the "Logit cutoff value" would be set to target a high sensitivity. However, in other situations, such as detecting biomarkers to be used in conjunction with other tests, a different part of the ROC curve would be selected and the "Logit cutoff value" would be set accordingly (e.g., high specificity; true negative fraction).

Step 3. Predictive performance of the disease gene panel: The blind test
To test the predictive performance of the optimal disease gene panel, the corresponding Logit equation and cutoff value are evaluated in the test samples (Plate 12.4). Briefly, the mRNA level for these genes is quantified by real-time qRT-PCR, using the independent samples of the test set. The equation Logit [p ("disease")/p ("control")] generated from the training set is used for the sample prediction. If the Logit value is less than the selected cutoff, then the sample is predicted as "control," and if the Logit value is more than the selected cutoff, then the sample is predicted as "disease." In addition to estimating the overall predictive accuracy, the blind test creates a definitive ROC curve, which allows the researcher to validate other important operating characteristics of the test, such as sensitivity, specificity, and positive and negative predictive value.

Ultimately, if the predictive performance of the disease gene panel fits the intended use, the expression biomarker panel requires further validation in a setting that simulates broad clinical use: "external validation," ideally, in prospective, well-controlled clinical studies of independent samples across multiple sites with well-established standards for all steps in the testing process [135].

"Functional" validation

"Functional" validation involves identifying the function of individual disease genes or gene products, deducing their causal relationship to the disease under study, and defining the biochemical mechanisms and pathways they could disrupt or through which the genes could exert their influence and participate in the disease process.

The "functional validation" of disease genes selected as biomarkers raises the question: Does an expression signature and its components need to be understood mechanistically before it can be concluded that the genes represent valid disease biomarkers? RNA expression is not a biologic function; candidate genes are statistically associated with the event in question but nothing is implied thereby about potential mechanistic involvement. Observed changes may be causative, coincidental or may simply reflect cellular response to a perturbation. In fact, functional criteria are not usually the basis of biomarker development, and the biologic plausibility of the candidate biomarker is considered only in retrospect [136]. In addition, in the case of blood, little is know about the mechanisms by which nonblood disorders such as schizophrenia leave their "marks" on blood cells such that the disease state can readily be detected by peripheral blood cell expression profiling. However, to more thoroughly understand the mechanistic relationship and biologic meaning of the blood expression signature, functional validation is encouraged.

The typical analysis of microarray expression data is performed by clustering the expression profiles (Plate 12.2). This approach allows for the identification of coordinately expressed genes with biologic or clinical associations. The challenge that arises is how to weigh biologic relevance and to measure the strength of the candidate gene or gene panel issuing from microarray experiments. To make microarray data powerful, it may be worthwhile to link microarray data to existing biomedical data cataloged over past decades by previous experimentation (e.g., cloning and discovery of gene new functions, protein expression analysis, RT-PCR, and so forth), and retrieved from the most complete databases [137,138].

In the following section we present several complementary ways to functionally annotate microarray-derived disease genes, and thereby to expand the dimensionality of the array data to encompass functional and biologic validation of the blood-based disease genes and biomarkers.

Chromosome analysis

Disease genes identified can be functionally evaluated in terms of their chromosomal location [51] and possible overlap with regions of suggestive linkage or association highlighted by genetic approaches. In our recent work on the relationship between plasma lipid levels and circulating leukocytes gene expression, we discovered that a number of genes correlating with plasma lipid levels were located in the chromosomal regions of known quantitative trait loci (QTLs, often referred to as "susceptibility" genes) associated with hyperlipemia [125]. These genes are thus excellent candidates for eQTLs (i.e., the mapping of gene expression levels as quantitative trait loci) of hyperlipemia [139]. Indeed, it is highly conceivable that new candidate genes will be defined initially on the basis of their differential expression in a disease state, and subsequently will be determined to be new genetic (susceptibility) markers. This approach is particularly suited to the study of complex polygenic diseases, because it allows researchers to take a genome-wide "snapshot" at different progressive stages of a disease and then nail down the analysis to specific region of the chromosomes.

Pathway analysis

Pathway analysis involves looking for changes in gene expression by incorporating either pathway or functional annotations. Different software packages can be used to annotate disease genes by cross-referencing to public biomedical databases.

One of these tools, DAVID (Database for Annotation, Visualization, and Integrated Discovery) [140], allows users to access a relational database of functional annotation. Functional annotations are derived primarily from LocusLink at the *National Center for Biotechnology Information* (NCBI). DAVID then uses LocusLink accession numbers to link gene accessioning systems like Genbank, Unigene and Affymetrix identifiers to biologic annotations including gene names and aliases, functional summaries, Gene Ontologies (controlled vocabularies that describe gene products in terms of their associated biologic processes, cellular components and molecular functions), protein domains and biochemical and signal transduction pathways. Annotation pedigrees and pathways maps are provided via direct links to the primary sources of annotation, which also provide additional gene-specific information.

In our case, the Affymetrix identifiers of the blood-derived disease genes can directly be downloaded in the DAVID software to observe the complex changes that occur in known metabolic and regulatory pathways as tracked by the changing expression levels of those candidate genes. This may lead to identify potential mechanisms that underlie diseases and new avenues for investigation.

Literature analysis

The published literature is a major source of information about genes, and interpretation of gene expression signatures relies largely on manual literature searches. To allow better surveying of the literature, several authors have initiated literature-searching tools to automatically retrieve, organize and analyze the tremendous wealth of knowledge stored in the scientific literature that enabled, for example, automatic protein annotations or association of keywords related to diseases [141].

Such applications can be of use for researchers to compare microarray data for differentially expressed genes with literature data available in Medline© (available through [142]) [143]. Jenssen *et al.* [144] described a tool for automatic literature extraction using co-citation index. Their hypothesis was that a "biological meaningful relationship" between two genes exists if they were co-cited in the MeSH descriptors of Medline©. The automatic analysis of co-citations allowed them to establish a virtual gene network, called Pubgene [145], and to rank genes according to biologic processes. This gene network can be used to find groups of genes that had co-occurred in the literature together with the disease genes identified by blood profiling, thus providing another way to structure gene expression data and extract potential biologically meaningful relationships.

In addition, several databases containing inventories of genes expressed in different systems have been made available publicly, such as Gene Expression Omnibus [146], Gene Expression Atlas [147], SAGEmap [148], ONCOMINE [149] and ArrayExpress [150]. Mining such repositories may also help to construct informative and structured networks among human disease genes and to validate candidate genes for specific conditions.

Animal and cell models

The above mentioned analyses clearly show that by linking microarray data to existing scientific literature and to biomedical databases it is possible to evaluate the disease gene expression signature from a functional point of view. Successes using these approaches usually rely on raising functional hypotheses and confirming them by direct experimentation in model systems. These assays involve manipulation of the candidate gene in cell or animal systems, with the aim of producing a modified phenotype or behavior, which is then examined using functional tests for detecting changes in the disease-relevant phenotype. If the intervention results in a disease phenotype or behavior, the involvement of the target gene is further confirmed. Several complementary strategies for manipulating gene expression and activity have been developed. Loss-of-function strategies focus on decreasing or eliminated gene expression and activity, while gain-of-function strategies focus on increasing gene expression and activity. Below, we provide a brief overview of the systems currently in use.

Cell cultures are extensively used to design *in vitro* validation assays through modifying the expression or activity of a candidate gene in a cell type of interest. Cell-based strategies to disrupt gene expression mostly use RNA interference (RNAi) (for review see Milhavet *et al.* [151]). Primary cell-based approaches for gain-of-function (gene overexpression) use a cDNA expression vector to overexpress a gene in a cell type of interest. Alternatively, overexpression of endogenous gene can be achieved by using insertional mutagenesis. In these assays, a specialized plasmid or retroviral vector containing a promoter is integrated in the genome (by transfection or viral infection) resulting in the transcriptional activation of the endogenous gene of interest downstream of the insertion site [152].

Animal models, especially murine models, continue to be the option of choice in the functional validation of human disease genes. Alteration of a target gene can be achieved with different levels of sophistication. Basic systems use knockout or transgenic animals, either in isolation or in conjunction with disease models. The goal is to reproduce as precisely as possible the human gene alteration in mice and to assess its phenotypic consequences. "Transgenic" refers to the introduction (random integration) of a human gene, the transgene, into the genetic material of an animal, a technique used to create mice that express more than normal amounts of the candidate gene product or to introduce a different form of the gene in question. Transgenic technologies have made possible great advances in numerous fields including CVD research [153]. Knockout and knockin mice are created by gene-targeting techniques (targeted mutagenesis) that produce animals in which a specific gene has been deleted (knocked out) or mutated (knocked in) [154].

In addition to the time required to produce knockout or transgenic animals, the most significant problems this approach faces are embryonic lethality and the induction of compensatory mechanisms during development. These problems can be substantially overcome through the construction of conditional expression systems, which allow spatial and temporal control over the expression of the introduced genotypic alteration. Such inducible systems better mimic the type of gene expression changes occurring in late-onset human pathologies. With transgenic animals, several systems are available to regulate the expression of the transgene using external inducers (e.g., tetracycline and ecdysone), and the addition of the corresponding inducible promoter in front of the transgene [155]. The Cre-lox system from bacteriophage P1 is the most popular current system to produce conditional knockout mice. There are numerous variations on this technique, and the Cre-lox system has been widely used to uncover gene function [156,157].

It is important to note that these functional approaches are not only useful in validation studies, but also open up opportunities in the discovery phase [158]. To this end, various approaches to undertake genome-wide functional screens in mouse models have been initiated in areas such as cardiology, central nervous system and neurology, metabolism and obesity, osteoporosis, reproductive biology and oncology [159]. Moreover, these approaches can be used to identify new disease models [66]. In this way, genome-wide model screens specifically designed to identify disease phenotypes are highly complementary to expression profiling and genetic studies.

Computational modeling

Deciphering biologic mechanisms and disease pathways requires looking beyond the single-gene biologic context and the unidimensional view of gene expression. Moreover, the intersection of multiple datasets of diverse biologic and clinical information provides higher dimensional views of gene function and the clinical significance of gene function. To achieve this aim, biomedical sciences, including cardiology, have been moving towards the integration of many disciplines formalized by the emergence of systems biology [160]. Greater benefits for mechanistic discoveries are expected collecting and integrating complete datasets from genetics (DNA level), transcriptomics (RNA level), proteomics (protein level) and even metabolomics (metabolite level) measurements, and by interpreting these data in the context of underlying biologic systems and in conjunction with patient information [161]. Using this integrated systems biology approach, changes in transcript abundance and/or protein expression can be related to modification in cellular function and to tissue pathology. For example, Schadt *et al.* [162] combined microsatellite genotyping, 23,000 gene expression profiling and detailed phenotypic data to identify the genetic loci controlling the mRNA levels and phenotypic traits associated with common multigenic diseases such as obesity. This method promises to be an order of magnitude more efficient than conventional genetic (linkage) analysis for finding alleles with a causal role in disease.

This combinatorial approach to disease gene discovery and validation raises significant challenges in term of compiling, warehousing and synergizing the vast amount of data produced. Bioinformatics and biostatistics will play a significant part in addressing the abundance of data through developing new computerized methods (*in silico* analyses) that maximize the information made available to the researchers (data basing) and that integrate diverse types of data at a deeper level in order to better model cells, organs and ultimately the entire body (data mining). The common theme here is the integration of biochemical, anatomic, profiling and physiologic information together with current biomedical research in order to develop detailed dynamic models and decision schemes. These models in turn will be used:

1 to refine our understanding of the role of specific genes in disease pathways;

2 to establish predictive rules based on key genes and physiological parameters; and

3 to select the best point of intervention [163].

Because CVD involve a combination of bioactive factors acting upon complex biologic systems, this group of diseases is ideally positioned to profit from developments in systems biology. Integrative approaches to model the human heart and cardiovascular function have been initiated in the Physiome Project [164,165]. As applied to blood as a "surrogate" model for investigation into the biology of the entire body, systems biology has the potential to add to our current understanding of biologic pathways in disease and in health.

Conclusions and perspectives

Microarray-based gene expression profiling is, unquestionably, now established in the study of human disease and merits consideration as a technique to assist in disease gene discovery. However, few if any of the candidate genes identified using microarray technology have yet been integrated into clinical practice. In fact, it is much easier to develop a disease gene or a multigene panel than it is to translate such a disease signature into a robust clinical tool that benefits medical practice [135].

In this chapter we have outlined an alternative strategy that may have significant long-term implications for genomic research, clinical diagnostics and disease management. The approach is based on gene expression profiling in peripheral blood cells to detect disease genes and biomarkers. First, we use microarray platform and appropriate software packages to profile mRNA expression in peripheral blood cells and compile a comprehensive list of candidate disease genes. Second, we explore those candidates using an alternative technique to measure RNA expression, real-time qRT-PCR, and we apply logistic regression analysis and the ROC curve representation to score and rank the resulting gene combinations and define a predictive expression signature. Third, we apply (blind) the "locked-down" panel of disease genes to a new set of independent samples to evaluate the performance of the gene expression classifier. Finally, we attempt to annotate the functions of each blood-based disease gene according to information gleaned from existing biologic databases and biomedical bibliography.

This cycle of screening, molecular signature analysis, expression classifier generation and biologic and clinical validation has become the standard operating approach to disease gene discovery from transcriptome assessment. The novelty described in this chapter is in the use of peripheral blood cell-derived RNA, as opposed to tissue biopsy as the source of samples to detect disease-driven changes. While we focus on CVD and associated risk factors as our major example in this chapter, our group has confirmed the feasibility of this strategy and identified disease genes and biomarkers from more than 35 conditions in humans [13]. Our results have affirmed our initial hypothesis, the *Sentinel Principle*™, which holds that changes in circulating leukocytes, a readily accessible tissue source, reflect disease changes occurring in human body tissues [166], and that such changes may have a potential role in the disease process. Blood is thus an excellent model for both the characterization and the monitoring of human disease.

One of the main goals of these types of studies is to optimize the disease gene panels generated from blood-based RNA profile into an accurate expression assays and to implement them as molecular diagnostic tools to determine clinical outcomes based on specific blood gene expression signatures (Plate 12.1). The challenge will lie in the development of appropriate detection and monitoring systems including quality control and standard operating procedure for sample collection and processing, and data generation and analysis. Nonetheless, we anticipate that in the very near future gene expression profiling in blood will became as diagnostically routine as histologic examination of tissue, and will be used in a wide range of diseases, delivering for the first time the promise of genomics in human healthcare.

Blood genomics is in a position not only to make important contributions in how disease can be better diagnosed, but also to provide important new insights into how disease develops. As we embark on this exciting new era, other strategies emerging in the field focus on circulating tumor cells and endothelial cells [98,167], circulating DNA and RNA [100], serum or plasma proteomics (protein profiling) [168–170] to cover different levels of blood dynamics. Such strategies may prove complementary to identify subtle changes that can be translated to multimolecular measurements in blood to comprehensively evaluate disease. Hence we propose to introduce the field of "bloodomics," corresponding to the "omics" investigations (i.e., genomic, transcriptomic, proteomic or metabolomic studies) of blood components to search for molecular signatures and biomarkers that will define patients' clinical features and guide decision-making.

Acknowledgments

The authors wish to thank Isolde Prince for her efforts in editing this manuscript. Blood profiling work reported in this paper was supported by GeneNews Corp. (Toronto, Ontario, Canada).

References

1 Vasan RS. Biomarkers of cardiovascular disease: Molecular basis and practical considerations. *Circulation* 2006; 113: 2335–2362.

2 Giallourakis C, Henson C, Reich M *et al.* Disease gene discovery through integrative genomics. *Ann Rev Genomics Hum Genet* 2005; 6: 381–406.

3 Hirschhorn JN, Daly MJ. Genome-wide association studies for common diseases and complex traits. *Nat Rev Genet* 2005; 6: 95–108.

4 Gibbons GH, Liew CC, Goodarzi MO *et al.* Genetic markers: progress and potential for cardiovascular disease. *Circulation* 2004; 109 (Supplement IV): 47–58.

5 Stephens JW, Humphries SE. The molecular genetics of cardiovascular disease: clinical implications. *J Intern Med* 2003; 253: 120–127.

6 International Human Genome Sequencing Consortium. Finishing the euchromatic sequence of the human genome. *Nature* 2004; 431: 931–945.

7 Lander ES, Linton LM, Birren B *et al.* Initial sequencing and analysis of the human genome. *Nature* 2001; 409: 860–921.

8 Venter JC, Adams MD, Myers EW *et al.* The sequence of the human genome. *Science* 2001; 291: 1304–1351.

9 Kopp MU, Crabtree HJ, Manz A. Developments in technology and applications of microsystems. *Curr Opin Chem Biol* 1997; 1: 410–419.

10 Wang J. From DNA biosensors to gene chips. *Nucleic Acids Res* 2000; 28: 3011–3016.

11 Sauer S, Lange BM, Gobom J *et al.* Miniaturization in functional genomics and proteomics. *Nat Rev Genet* 2005; 6: 465–476.

12 Chung CH, Bernard PS, Perou CM. Molecular portraits and the family tree of cancer. *Nat Genet* 2002; 32 (Supplement): 533–540.

13 Liew CC, Ma J, Tang HC *et al.* The peripheral blood transcriptome dynamically reflects system wide biology: a potential diagnostic tool. *J Lab Clin Med* 2006; 147: 126–132.

14 Ogawa M. Differentiation and proliferation of hematopoietic stem cells. *Blood* 1993; 81: 2844–2853.

15 Frank R, Hargreaves R. Clinical biomarkers in drug discovery and development. *Nat Rev Drug Discov* 2003; 2: 566–580.

16 Seiler PU, Stypmann J, Breithardt G *et al.* Real-time RT-PCR for gene expression profiling in blood of heart failure patients: a pilot study: gene expression in blood of heart failure patients. *Basic Res Cardiol* 2004; 99: 230–238.

17 Nicholson AC, Unger ER, Mangalathu R *et al.* Exploration of neuroendocrine and immune gene expression in peripheral blood mononuclear cells. *Brain Res Mol Brain Res* 2004; 129: 193–197.

18 Liew CC. Expressed genome molecular signatures of heart failure. *Clin Chem Lab Med* 2005; 43: 462–469.

19 Gladkevich A, Nelemans SA, Kauffman HF *et al.* Microarray profiling of lymphocytes in internal diseases with an altered immune response: potential and methodology. *Mediators Inflamm* 2005; 2005: 317–330.

20 Margalit O, Somech R, Amariglio N *et al.* Microarraybased gene expression profiling of hematologic malignancies: basic concepts and clinical applications. *Blood Rev* 2005; 19: 223–234.

21 Coussens LM, Werb Z. Inflammation and cancer. *Nature* 2002; 420: 860–867.

22 Chon H, Verhaar MC, Koomans HA *et al.* Role of circulating karyocytes in the initiation and progression of atherosclerosis. *Hypertension* 2006; 47: 803–810.

23 Afshar-Kharghan V, Thiagarajan P. Leukocyte adhesion and thrombosis. *Curr Opin Hematol* 2006; 13: 34–39.

24 Hansson GK. Inflammation, atherosclerosis, and coronary artery disease. *N Engl J Med* 2005; 352: 1685–1695.

25 Toker S, Rogowski O, Melamed S *et al.* Association of components of the metabolic syndrome with the appearance of aggregated red blood cells in the peripheral blood. An unfavorable hemorheological finding. *Diabetes Metab Res Rev* 2005; 21: 197–202.

26 Hoffman M, Blum A, Baruch R *et al.* Leukocytes and coronary heart disease. *Atherosclerosis* 2004; 172: 1–6.

27 Ross R. Atherosclerosis; an inflammatory disease. *N Engl J Med* 1999; 340: 115–126.

28 Kher N, Marsh JD. Pathobiology of atherosclerosis: a brief review. *Semin Thromb Hemost* 2004; 30: 665–672.

29 Osterud B, Bjorklid E. Role of monocytes in atherogenesis. *Physiol Rev* 2003; 83: 1069–1112.

30 Davis C, Fischer J, Ley K *et al.* The role of inflammation in vascular injury and repair. *J Thromb Haemost* 2003; 1: 1699–1709.

31 Hansson GK, Libby P. The immune response in atherosclerosis: a double-edged sword. *Nat Rev Immunol* 2006; 6: 508–519.

32 de Boer OJ, Becker AE, van der Wal AC. T lymphocytes in atherogenesis-functional aspects and antigenic repertoire. *Cardiovasc Res* 2003; 60: 78–86.

33 Massberg S, Schulz C, Gawaz M. Role of platelets in the pathophysiology of acute coronary syndrome. *Semin Vasc Med* 2003; 3: 147–162.

34 Willoughby S, Holmes A, Loscalzo J. Platelets and cardiovascular disease. *Eur J Cardiovasc Nurs* 2002; 1: 273–288.

35 Wagner DD. New links between inflammation and thrombosis. *Arterioscler Thromb Vasc Biol* 2005; 25: 1321–1324.

36 Tanigawa T, Iso H, Yamagishi K *et al.* Association of lymphocyte sub-populations with clustered features of metabolic syndrome in middle-aged Japanese men. *Atherosclerosis* 2004; 173: 295–300.

37 Ohshita K, Yamane K, Hanafusa M *et al.* Elevated white blood cell count in subjects with impaired glucose tolerance. *Diabetes Care* 2004; 27: 491–496.

38 Madjid M, Awan I, Willerson JT *et al.* Leukocyte count and coronary heart disease: implications for risk assessment. *J Am Coll Cardiol* 2004; 44: 1945–1956.

39 Saadeddin SM, Habbab MA, Ferns GA. Markers of inflammation and coronary artery disease. *Med Sci Monit* 2002; 8: RA5–RA12.

40 Festa A, D'Agostino R Jr, Howard G *et al.* Chronic subclinical inflammation as part of the insulin resistance syndrome: the Insulin Resistance Atherosclerosis Study (IRAS). *Circulation* 2000; 102: 42–47.

41 Hu FB, Meigs JB, Li TY *et al.* Inflammatory markers and risk of developing type 2 diabetes in women. *Diabetes* 2004; 53: 693–700.

42 Schmidt MI, Duncan BB. Diabesity: an inflammatory metabolic condition. *Clin Chem Lab Med* 2003; 41: 1120–1130.

43 Tsimikas S, Willerson JT, Ridker PM. C-reactive protein and other emerging blood biomarkers to optimize risk stratification of vulnerable patients. *J Am Coll Cardiol* 2006; 47 (Supplement C): 19–31.

44 Fruchart JC, Nierman MC, Stroes ES *et al.* New risk factors for atherosclerosis and patient risk assessment. *Circulation* 2004; 109 (Supplement 3): 15–19.

45 Liotta L, Petricoin E. Molecular profiling of human cancer. *Nat Rev Genet* 2000; 1: 48–56.

46 Kittleson MM, Hare JM. Molecular signature analysis: using the myocardial transcriptome as a biomarker in cardiovascular disease. *Trends Cardiovasc Med* 2005; 15: 130–138.

47 Barrans DB, Liew CC. "Chip"ing away at Heart Failure. In: Kearns-Jonker M, ed. *Methods in Molecular Medicine,*

Vol. 126: *Congenital Heart Disease: Molecular Diagnostics*, Humana Press Inc., Totowa NJ, 2006: 157–169.

48 Liew CC, Hwang DM, Wang RS *et al.* Construction of a human heart cDNA library and identification of cardiovascular-based genes (CvBest). *Mol Cell Biochem* 1997; 172: 81–87.

49 Hwang DM, Dempsey AA, Wang RX *et al.* A genome-based resource for molecular cardiovascular medicine: Towards a compendium of cardiovascular genes. *Circulation* 1997; 96: 4146–4203.

50 Dempsey AA, Dzau VJ, Liew CC. Cardiovascular genomics: Estimating the total number of genes expressed in the human cardiovascular system. *J Mol Cell Cardiol* 2001; 33: 1879–1886.

51 Barrans JD, Ip J, Lam CW *et al.* Chromosomal distribution of the human cardiovascular transcriptome. *Genomics* 2003; 81: 520–525.

52 Scheel J, Von Brevern MC, Horlein A *et al.* Yellow pages to the transcriptome. *Pharmacogenomics* 2002; 3: 791–807.

53 Bertucci F, Bernard K, Loriod B *et al.* Sensitivity issues in DNA array-based expression measurements and performance of nylon microarrays for small samples. *Human Mol Genet* 1999; 8: 1715–1722.

54 Granjeaud S, Nguyen C, Rocha D *et al.* From hybridization image to numerical values: a practical, high throughput quantification system for high density filter hybridizations. *Genet Anal* 1996; 12: 151–162.

55 The Chipping Forecast III. *Nat Genet* 2005; 37 (Supplement): 1–45.

56 The Chipping Forecast II. *Nat Genet* 2002; 32 (Supplement): 465–552.

57 The Chipping Forecast. *Nat Genet* 1999; 21 (Supplement): 1–60.

58 Alizadeh A, Eisen M, Davis RE *et al.* The lymphochip: a specialized cDNA microarray for the genomic-scale analysis of gene expression in normal and malignant lymphocytes. *Cold Spring Harbor Symp Quant Biol* 1999; 64: 71–78.

59 Barrans JD, Stamatiou D, Liew C. Construction of a human cardiovascular cDNA microarray: portrait of the failing heart. *Biochem Biophys Res Commun* 2001; 280: 964–969.

60 Barrans JD, Sc MH, Allen PD *et al.* Global gene expression of end-stage dilated cardiomyopathy using a human cardiovascular-based cDNA microarray. *Am J Pathol* 2002; 160: 2035–2043.

61 Hwang JJ, Allen PD, Tseng GC *et al.* Microarray gene expression profiles in dilated and hypertrophic cardiomyopathic end-stage heart failure. *Physiol Genomics* 2002; 10: 31–44.

62 Quackenbush J. Microarray analysis and tumor classification. *N Engl J Med* 2006; 354: 2463–2472.

63 Mohr S, Leikauf GD, Keith G *et al.* Microarrays as cancer keys: an array of possibilities. *J Clin Oncol* 2002; 20: 3165–3175.

64 Tsai TH, Milhorn DM, Huang SK. Microarray and gene-clustering analysis. *Methods Mol Biol* 2006; 315: 165–174.

65 Livak KJ, Schmittgen TD. Analysis of relative gene expression data using real-time quantitative PCR and the 2(-Delta Delta C(T)) Method. *Methods* 2001; 25: 402–408.

66 Lindsay MA. Target discovery. *Nat Rev Drug Discov* 2003; 2: 831–838.

67 Pounds SB. Estimation and control of multiple testing error rates for microarray studies. *Brief Bioinform* 2006; 7: 25–36.

68 Le Meur N, Lamirault G, Bihouee A *et al.* A dynamic, web-accessible resource to process raw microarray scan data into consolidated gene expression values: importance of replication. *Nucleic Acids Res* 2004; 32: 5349–5358.

69 Lee ML, Kuo FC, Whitmore GA *et al.* Importance of replication in microarray gene expression studies: statistical methods and evidence from repetitive cDNA hybridizations. *Proc Natl Acad Sci USA.* 2000; 97: 9834-9839.

70 Dallas PB, Gottardo NG, Firth MJ *et al.* Gene expression levels assessed by oligonucleotide microarray analysis and quantitative real-time RT-PCR: how well do they correlate? *BMC Genomics* 2005; 6: 59.

71 Tumor Analysis Best Practices Working Group. Expression profiling: best practices for data generation and interpretation in clinical trials. *Nat Rev Genet* 2004; 5: 229–237.

72 Allison DB, Cui X, Page GP *et al.* Microarray data analysis: from disarray to consolidation and consensus. *Nat Rev Genet* 2006; 7: 55–65.

73 Bammler T, Beyer RP, Bhattacharya S *et al.* Standardizing global gene expression analysis between laboratories and across platforms. *Nat Methods* 2005; 2: 351–356.

74 Alcorta D, Preston G, Munger W *et al.* Microarray studies of gene expression in circulating leukocytes in kidney diseases. *Exp Nephrol* 2002; 10: 139–149.

75 Fan H, Hegde PS. The transcriptome in blood: challenges and solutions for robust expression profiling. *Curr Mol Med* 2005; 5: 3–10.

76 Baechler EC, Batliwalla FM, Karypis G *et al.* Expression levels for many genes in human peripheral blood cells are highly sensitive to *ex vivo* incubation. *Genes Immun* 2004; 5: 347–353.

77 Bakay M, Chen YW, Borup R *et al.* Sources of variability and effect of experimental approach on expression profiling data interpretation. *BMC Bioinformatics* 2002; 3: 4.

78 Anderson D. Factors that contribute to biomarker responses in humans including a study in individuals taking Vitamin C supplementation. *Mutat Res* 2001; 480–481: 337–347.

79 Whitney AR, Diehn M, Popper SJ *et al.* Individuality and variation in gene expression patterns in human blood. *Proc Natl Acad Sci USA* 2003; 100: 1896–1901.

80 Wang Z, Neuburg D, Li C *et al.* Global gene expression profiling in whole-blood samples from individuals exposed to metal fumes. *Environ Health Perspect* 2005; 113: 233–241.

81 Dobbin KK, Simon RM. Sample size planning for developing classifiers using high dimensional DNA microarray data. *Biostatistics* 2006; Apr 13 [Epub ahead of print].

82 Debey S, Schoenbeck U, Hellmich M *et al.* Comparison of different isolation techniques prior gene expression profiling of blood derived cells: impact on physiological responses, on overall expression and the role of different cell types. *Pharmacogenomics J* 2004; 4: 193–207.

83 Pahl A, Brune K. Gene expression changes in blood after phlebotomy: implications for gene expression profiling. *Blood* 2002; 100: 1094–1095.

84 Pahl A, Brune K. Stabilization of gene expression profiles in blood after phlebotomy. *Clin Chem* 2002; 48: 2251–2253.

85 Marteau JB, Mohr S, Pfister M *et al.* Collection and storage of human blood cells for mRNA expression profiling: a 15-month stability study. *Clin Chem* 2005; 51: 1250–1252.

86 Feezor RJ, Baker HV, Mindrinos M *et al.* Whole blood and leukocyte RNA isolation for gene expression analyses. *Physiol Genomics* 2004; 19: 247–254.

87 Tanner MA, Berk LS, Felten DL *et al.* Substantial changes in gene expression level due to the storage temperature and storage duration of human whole blood. *Clin Lab Haematol* 2002; 24: 337–341.

88 Hartel C, Bein G, Muller-Steinhardt M *et al.* *Ex vivo* induction of cytokine mRNA expression in human blood samples. *J Immunol Methods* 2001; 249: 63–71.

89 Burczynski ME, Dorner AJ. Transcriptional profiling of peripheral blood cells in clinical pharmacogenomic studies. *Pharmacogenomics* 2006; 7: 187–202.

90 Rainen L, Oelmueller U, Jurgensen S *et al.* Stabilization of mRNA expression in whole blood samples. *Clin Chem* 2002; 48: 1883–1890.

91 Campbell C, Vernon SD, Karem KL *et al.* Assessment of normal variability in peripheral blood gene expression. *Dis Markers* 2002; 18: 201–206.

92 Pahl A. Gene expression profiling using RNA extracted from whole blood: technologies and clinical applications. *Expert Rev Mol Diagn* 2005; 5: 43–52.

93 Radich JP, Mao M, Stepaniants S *et al.* Individual-specific variation of gene expression in peripheral blood leukocytes. *Genomics* 2004; 83: 980–988.

94 Lampe JW, Stepaniants SB, Mao M *et al.* Signatures of environmental exposures using peripheral leukocyte gene expression: tobacco smoke. *Cancer Epidemiol Biomarkers Prev* 2004; 13: 445–453.

95 van Leeuwen DM, Gottschalk RW, van Herwijnen MH *et al.* Differential gene expression in human peripheral blood mononuclear cells induced by cigarette smoke and its constituents. *Toxicol Sci* 2005; 86: 200–210.

96 Ebert BL, Golub TR. Genomic approaches to hematologic malignancies. *Blood* 2004; 104: 923–932.

97 Nebozhyn M, Loboda A, Kari L *et al.* Quantitative PCR on 5 genes reliably identifies CTCL patients with 5% to 99% circulating tumor cells with 90% accuracy. *Blood* 2006; 107: 3189–3196.

98 Smirnov DA, Zweitzig DR, Foulk BW *et al.* Global gene expression profiling of circulating tumor cells. *Cancer Res* 2005; 65: 4993–4997.

99 Bosma AJ, Weigelt B, Lambrechts AC *et al.* Detection of circulating breast tumor cells by differential expression of marker genes. *Clin Cancer Res* 2002; 8: 1871–1877.

100 Li Y, Elashoff D, Oh M *et al.* Serum circulating human mRNA profiling and its utility for oral cancer detection. *J Clin Oncol* 2006; 24: 1754–1760.

101 Wong SC, Lo SF, Cheung MT *et al.* Quantification of plasma beta-catenin mRNA in colorectal cancer and adenoma patients. *Clin Cancer Res* 2004; 10: 1613–1617.

102 Cobb JP, Mindrinos MN, Miller-Graziano C *et al.* Application of genome-wide expression analysis to human health and disease. *Proc Natl Acad Sci USA* 2005; 102: 4801–4806.

103 Osman I, Bajorin D, Sun TT *et al.* Novel blood biomarkers of human urinary bladder cancer. *Clin Cancer Res* 2006; 12: 3374–3380.

104 Han M, Liew CT, Zhang HW *et al.* Novel blood biomarker panel detects human colorectal cancer (Poster Abstract). ASCO Meeting, Atlanta, Georgia, June 2006.

105 Nam RK, Marshall KW, Zheng R *et al.* Blood-based biomarkers for detecting aggressive prostate cancer at time of biopsy (Poster Abstract). ASCO Meeting, Atlanta, Georgia, June 2006.

106 Tsuang MT, Nossova N, Yager T *et al.* Assessing the validity of blood-based gene expression profiles for the classification of schizophrenia and bipolar disorder: a preliminary report. *Am J Med Genet B Neuropsychiatr Genet* 2005; 133: 1–5.

107 Glatt SJ, Everall IP, Kremen WS *et al.* Comparative gene expression analysis of blood and brain provides concurrent validation of SELENBP1 up-regulation in schizophrenia. *Proc Natl Acad Sci USA* 2005; 102: 15533–15538.

108 Marshall KW, Zhang H, Yager TD *et al.* Blood-based biomarkers for detecting mild osteoarthritis in the human knee. *Osteoarthritis Cartilage* 2005; 13: 861–871.

109 Twine NC, Stover JA, Marshall B *et al.* Disease-associated expression profiles in peripheral blood mononuclear cells from patients with advanced renal cell carcinoma. *Cancer Res* 2003; 63: 6069–6075.

110 Sharma P, Sahni NS, Tibshirani R *et al.* Early detection of breast cancer based on gene-expression patterns in peripheral blood cells. *Breast Cancer Res* 2005; 7: R634–R644.

111 Vernon SD, Unger ER, Dimulescu IM *et al.* Utility of the blood for gene expression profiling and biomarker discovery in chronic fatigue syndrome. *Dis Markers* 2002; 18: 193–199.

112 Moore DF, Li H, Jeffries N *et al.* Using peripheral blood mononuclear cells to determine a gene expression profile of acute ischemic stroke: a pilot investigation. *Circulation* 2005; 111: 212–221.

113 Du X, Tang Y, Xu H *et al.* Genomic profiles for human peripheral blood T cells, B cells, natural killer cells, monocytes, and polymorphonuclear cells: comparisons to ischemic stroke, migraine, and Tourette syndrome. *Genomics* 2006; 87: 693–703.

114 Tang Y, Gilbert DL, Glauser TA *et al.* Blood gene expression profiling of neurologic diseases: a pilot microarray study. *Arch Neurol* 2005; 62: 210–215.

115 Tang Y, Nee AC, Lu A *et al.* Blood genomic expression profile for neuronal injury. *J Cereb Blood Flow Metab* 2003; 23: 310–319.

116 Tang Y, Lu A, Aronow BJ *et al.* Blood genomic responses differ after stroke, seizures, hypoglycemia, and hypoxia: blood genomic fingerprints of disease. *Ann Neurol* 2001; 50: 699–707.

117 Hansel NN, Hilmer SC, Georas SN *et al.* Oligonucleotide-microarray analysis of peripheral-blood lymphocytes in severe asthma. *J Lab Clin Med* 2005; 145: 263–274.

118 Rus V, Chen H, Zernetkina V *et al.* Gene expression profiling in peripheral blood mononuclear cells from lupus patients with active and inactive disease. *Clin Immunol* 2004; 112: 231–234.

119 Baechler EC, Batliwalla FM, Karypis G *et al.* Interferon-inducible gene expression signature in peripheral blood cells of patients with severe lupus. *Proc Natl Acad Sci USA* 2003; 100: 2610–2615.

120 Preston GA, Waga I, Alcorta DA *et al.* Gene expression profiles of circulating leukocytes correlate with renal disease activity in IgA nephropathy. *Kidney Int* 2004; 65: 420–430.

121 Burczynski ME, Peterson RL, Twine NC *et al.* Molecular classification of Crohn's disease and ulcerative colitis patients using transcriptional profiles in peripheral blood mononuclear cells. *J Mol Diagn* 2006; 8: 51–61.

122 Ma J, Liew CC. Gene profiling identifies secreted protein transcripts from peripheral blood cells in coronary artery disease. *J Mol Cell Cardiol* 2003; 35: 993–998.

123 Patino WD, Mian OY, Kang JG *et al.* Circulating transcriptome reveals markers of atherosclerosis. *Proc Natl Acad Sci USA* 2005; 102: 3423–3428.

124 Chon H, Gaillard CA, van der Meijden BB *et al.* Broadly altered gene expression in blood leukocytes in essential hypertension is absent during treatment. *Hypertension* 2004; 43: 947–951.

125 Ma J, Dempsey AA, Stamatiou D, *et al.* Identifying leukocyte gene expression patterns associated with plasma lipid levels in human subjects. *Atherosclerosis* 2006; Jun 26; [Epub ahead of print].

126 Deng MC, Eisen HJ, Mehra MR *et al.* Noninvasive discrimination of rejection in cardiac allograft recipients using gene expression profiling. *Am J Transplant* 2006; 6: 150–160.

127 http://www.allomap.com/

128 Chuaqui RF, Bonner RF, Best CJ *et al.* Post-analysis follow-up and validation of microarray experiments. *Nat Genet* 2002; 32 (Supplement): 509–514.

129 Bustin SA, Mueller R. Real-time reverse transcription PCR (qRT-PCR) and its potential use in clinical diagnosis. *Clin Sci (Lond)* 2005; 109: 365–379.

130 Bustin SA. Absolute quantification of mRNA using real-time reverse transcription polymerase chain reaction assays. *J Mol Endocrinol* 2000; 25: 169–193.

131 Simon R. Roadmap for developing and validating therapeutically relevant genomic classifiers. *J Clin Oncol* 2005; 23: 7332–7341.

132 http://docs.appliedbiosystems.com/pebiodocs/ 04303859.pdf

133 Pampel FC. *Logistic Regression: A Primer*. Sage Publications (Quantitative applications in the social sciences; no. 07-132), Thousand Oaks, California, 2000.

134 Baker SG. The central role of receiver operating characteristic (ROC) curves in evaluating tests for the early detection of cancer. *J Natl Cancer Inst* 2003; 95: 511–515.

135 Ludwig JA, Weinstein JN. Biomarkers in cancer staging, prognosis and treatment selection. *Nat Rev Cancer* 2005; 5: 845–856.

136 Liu ET. Mechanism-derived gene expression signatures and predictive biomarkers in clinical oncology. *Proc Natl Acad Sci USA* 2005; 102: 3531–3532.

137 Masys DR. Linking microarray data to the literature. *Nat Genet* 2001; 28: 9–10.

138 Gaasterland T, Bekiranov S. Making the most of microarray data. *Nat Genet* 2000; 24: 204–206.

139 Hubner N, Wallace CA, Zimdahl H *et al.* Integrated transcriptional profiling and linkage analysis for

identification of genes underlying disease. *Nat Genet* 2005; 37: 243–253.

140 Dennis G Jr, Sherman BT, Hosack DA *et al.* DAVID: Database for annotation, visualization, and integrated discovery. *Genome Biol* 2003; 4: P3.

141 Andrade MA, Bork P. Automated extraction of information in molecular biology. *FEBS Lett* 2000; 476: 12–17.

142 Rihn BH, Vidal S, Nemurat C *et al.* From transcriptomics to bibliomics. *Med Sci Monit* 2003; 9: MT89–MT95.

143 http://www.ncbi.nlm.nih.gov/PubMed/

144 Jenssen TK, Laegreid A, Komorowski J *et al.* A literature network of human genes for high-throughput analysis of gene expression. *Nat Genet* 2001; 28: 21–28.

145 http://www.pubgene.uio.no

146 Barrett T, Suzek TO, Troup DB *et al.* NCBI GEO: mining millions of expression profiles: database and tools. *Nucleic Acids Res* 2005; 33 (Database issue): D562–D566.

147 Su AI, Cooke MP, Ching KA *et al.* Large-scale analysis of the human and mouse transcriptomes. *Proc Natl Acad Sci USA* 2002; 99: 4465–4470.

148 Lash AE, Tolstoshev CM, Wagner L *et al.* SAGEmap: a public gene expression resource. *Genome Res* 2000; 10: 1051–1060.

149 Rhodes DR, Yu J, Shanker K *et al.* ONCOMINE: a cancer microarray database and integrated data-mining platform. *Neoplasia* 2004; 6: 1–6.

150 Parkinson H, Sarkans U, Shojatalab M *et al.* ArrayExpress: a public repository for microarray gene expression data at the EBI. *Nucleic Acids Res* 2005; 33 (Database issue): D553–D555.

151 Milhavet O, Gary DS, Mattson MP. RNA interference in biology and medicine. *Pharmacol Rev* 2003; 55: 629–648.

152 Jackson PD, Harrington JJ. High-throughput target discovery using cell-based genetics. *Drug Discov Today* 2005; 10: 53–60.

153 Carmeliet P, Collen D. Transgenic mouse models in angiogenesis and cardiovascular disease. *J Pathol* 2000; 190: 387–405.

154 van der Weyden L, Adams DJ, Bradley A. Tools for targeted manipulation of the mouse genome. *Physiol Genomics* 2002; 11: 133–164.

155 Albanese C, Hulit J, Sakamaki T *et al.* Recent advances in inducible expression in transgenic mice. *Semin Cell Dev Biol* 2002; 13: 129–141.

156 Tornell J, Snaith M. Transgenic systems in drug discovery: from target identification to humanized mice. *Drug Discov Today* 2002; 7: 461–470.

157 Metzger D, Chambon P. Site- and time-specific gene targeting in the mouse. *Methods* 2001; 24: 71–80.

158 Ilyin SE, Belkowski SM, Plata-Salaman CR. Biomarker discovery and validation: technologies and integrative approaches. *Trends Biotechnol* 2004; 22: 411–416.

159 Zambrowicz BP, Sands AT. Knockouts model the 100 best-selling drugs: will they model the next 100? *Nat Rev Drug Discov* 2003; 2: 38–51.

160 Kitano H. Systems biology: a brief overview. *Science* 2002; 295: 1662–1664.

161 Liu ET. Systems biology, integrative biology, predictive biology. *Cell* 2005; 121: 505–506.

162 Schadt EE, Lamb J, Yang X *et al.* An integrative genomics approach to infer causal associations between gene expression and disease. *Nat Genet* 2005; 37: 710–717.

163 Kramer R, Cohen D. Functional genomics to new drug targets. *Nat Rev Drug Discov* 2004; 3: 965–972.

164 Hunter PJ, Borg TK. Integration from proteins to organs: the Physiome Project. *Nat Rev Mol Cell Biol* 2003; 4: 237–243.

165 Noble D. Modeling the heart: from genes to cells to the whole organ. *Science* 2002; 295: 1678–1682.

166 Liew CC, inventor; GeneNews, Inc., assignee. Method for the detection of gene transcripts in blood and uses thereof. United States patent US 20040014059. 2004 Jan 22 (priority date January 1999).

167 Cristofanilli M, Budd GT, Ellis MJ *et al.* Circulating tumor cells, disease progression, and survival in metastatic breast cancer. *N Engl J Med* 2004; 351: 781–791.

168 Wang X, Yu J, Sreekumar A *et al.* Autoantibody signatures in prostate cancer. *N Engl J Med* 2005; 353: 1224–1235.

169 Thadikkaran L, Siegenthaler MA, Crettaz D *et al.* Recent advances in blood-related proteomics. *Proteomics* 2005; 5: 3019–3034.

170 Veenstra TD, Conrads TP, Hood BL *et al.* Biomarkers: mining the biofluid proteome. *Mol Cell Proteomics* 2005; 4: 409–418.

Index